Advances in Experimental Medicine and Biology

Volume 1048

More information about this series at http://www.springer.com/series/5584

Quaiser Saquib • Mohammad Faisal
Abdulaziz A. Al-Khedhairy
Abdulrahman A. Alatar
Editors

Cellular and Molecular Toxicology of Nanoparticles

Springer

Editors
Quaiser Saquib
Zoology Department, College of Sciences
King Saud University
Riyadh, Saudi Arabia

Mohammad Faisal
Department of Botany and Microbiology, College of Sciences
King Saud University
Riyadh, Saudi Arabia

Abdulaziz A. Al-Khedhairy
Zoology Department, College of Sciences
King Saud University
Riyadh, Saudi Arabia

Abdulrahman A. Alatar
Department of Botany and Microbiology, College of Sciences
King Saud University
Riyadh, Saudi Arabia

ISSN 0065-2598 ISSN 2214-8019 (electronic)
Advances in Experimental Medicine and Biology
ISBN 978-3-030-10150-3 ISBN 978-3-319-72041-8 (eBook)
https://doi.org/10.1007/978-3-319-72041-8

Printed on acid-free paper

This Springer imprint is published by the registered company Springer International Publishing AG part of Springer Nature.
The registered company address is: Gewerbestrasse 11, 6330 Cham, Switzerland

Preface

We are pleased to introduce the book entitled *Cellular and Molecular Toxicology of Nanoparticles* in the Advances in Experimental Medicine and Biology series. This body of work is the melding of a number of ideas and perspectives. Hence, we must thank all the contributors who wholeheartedly welcomed our invitation and agreed to contribute chapters to embellish toxicological information on nanoparticles (NPs). We express our special thanks to King Saud University for all the support. NPs research within the last decade has received tremendous attention due to their unique properties in optics, physical, chemical, and biological processes. Larger NPs surface area and more surface atoms endow them with high chemical reactivity and intrinsic toxicity. As a result, NPs have been adopted in medical biology as an antibacterial agent, DNA structure probes, biosensors in drug and delivery formulations, tissue engineering, and cancer therapy to eliminate tumors through hyperthermia. Notwithstanding the benefits, several in vitro and in vivo studies have demonstrated that NPs exposure can provoke inflammatory responses, oxidative stress, myocardial infarction, and thrombosis. They can also alter the permeability of blood-brain barrier and re-translocate from the site of deposition to other parts of the body via circulatory or lymphatic system. A majority of NPs get internalized in cells through phagocytosis, macropinocytosis, receptor-mediated endocytosis, and passive penetration. Undeniably, NPs due to their high biological reactivity promote preferential toxicity toward genetic material. Therefore, the importance of their toxicological impact and of the development of early indicators for detection of possible adverse health effects arising from nanomaterial exposure is strongly realized. For these grounds, nano-genotoxicology and nanotoxicology are expanding as novel fields that are looking for the potential risk and mechanism of toxicity in various assay systems. Despite the above hazards, comprehensive information on NPs interaction with biological macromolecules (DNA and protein) is lacking. Especially, the toxicogenomic responses altering the normal cellular functioning and link with molecular pathways to trigger cell death and carcinogenesis have not been compiled. Therefore, the purpose of this book is to gather up-to-date and state-of-the-art toxicological effects of NPs in different in vivo and in vitro test models. The chapters have been organized to provide a crisp information on the cellular and molecular toxicity of different types of NPs. Special attention has been given to explore the mechanism of NPs toxicity which can lead to cell death. In addition, transcriptomic approach has also been provided to explain a clear picture on NPs effects on global gene expression and its connection to alter the varying pathways, vital for cell survival. The book has been designed for scientists engaged in NPs toxicity research. Nonetheless, it should be of interest to a variety of scientific disciplines including marine biologist, environmentalists, genetics, pharmacology, medicine, drug and food material sciences, and consumer products. Also, the compilations will be of interest to the environmental watchdogs, federal regulators, risk assessors, and the policy makers.

Riyadh, Saudi Arabia

Quaiser Saquib
Mohammad Faisal
Abdulaziz A. Al-Khedhairy
Abdulrahman A. Alatar

Acknowledgments

The authors are grateful to the Deanship of Scientific Research, King Saud University, for funding through Vice Deanship of Scientific Research Chairs.

Contents

1 **Toxicity Assessment in the Nanoparticle Era** 1
Valeria De Matteis and Rosaria Rinaldi

2 **Mechanisms of Uptake and Translocation of Nanomaterials in the Lung** 21
Chloé Puisney, Armelle Baeza-Squiban, and Sonja Boland

3 **Transmucosal Nanoparticles: Toxicological Overview** 37
Swapnil Talkar, Sagar Dhoble, Anuradha Majumdar, and Vandana Patravale

4 **The Toxicity of Nanoparticles to Human Endothelial Cells** . . . 59
Yi Cao

5 **The Role of Autophagy in Nanoparticles-Induced Toxicity and Its Related Cellular and Molecular Mechanisms** 71
Yubin Li and Dianwen Ju

6 **Nanoparticles-Caused Oxidative Imbalance** 85
Mariusz Zuberek and Agnieszka Grzelak

7 **Toxicity of Metal Oxide Nanoparticles** 99
Koyeli Girigoswami

8 **Relevance of Physicochemical Characterization of Nanomaterials for Understanding Nano-cellular Interactions** 123
Henriqueta Louro

9 **Toxicogenomics: A New Paradigm for Nanotoxicity Evaluation** 143
Sourabh Dwivedi, Quaiser Saquib, Bilal Ahmad, Sabiha M. Ansari, Ameer Azam, and Javed Musarrat

10 **Nickel Oxide Nanoparticles Induced Transcriptomic Alterations in HEPG2 Cells** 163
Quaiser Saquib, Maqsood A. Siddiqui, Javed Ahmad, Sabiha M. Ansari, Mohammad Faisal, Rizwan Wahab, Abdulrahman A. Alatar, Abdulaziz A. Al-Khedhairy, and Javed Musarrat

11 Nanoparticle-Protein Interaction: The Significance and Role of Protein Corona 175
Saad Mohammad Ahsan, Chintalagiri Mohan Rao, and Md. Faiz Ahmad

12 Cellular and Molecular Toxicity of Iron Oxide Nanoparticles 199
Blanca Laffon, Natalia Fernández-Bertólez, Carla Costa, Fátima Brandão, João Paulo Teixeira, Eduardo Pásaro, and Vanessa Valdiglesias

13 Detection of DNA Damage Induced by Cerium Dioxide Nanoparticles: From Models to Molecular Mechanism Activated 215
Tiago Alves Jorge de Souza, Thiago Lopes Rocha, and Leonardo Pereira Franchi

14 Mechanisms Underlying Neurotoxicity of Silver Nanoparticles 227
Lidia Strużyńska and Joanna Skalska

15 Toxic and Beneficial Potential of Silver Nanoparticles: The Two Sides of the Same Coin 251
Lilian Rodrigues Rosa Souza, Veronica Santana da Silva, Leonardo Pereira Franchi, and Tiago Alves Jorge de Souza

16 Molecular and Cellular Toxicology of Nanomaterials with Related to Aquatic Organisms 263
Mohd Ashraf Rather, Irfan Ahmad Bhat, Niti Sharma, and Rupam Sharma

17 Cytotoxicity and Physiological Effects of Silver Nanoparticles on Marine Invertebrates 285
Adriano Magesky and Émilien Pelletier

18 A *Drosophila* Model to Decipher the Toxicity of Nanoparticles Taken Through Oral Routes 311
S. Aurosman Pappus and Monalisa Mishra

19 Using of Quantum Dots in Biology and Medicine 323
Svetlana Pleskova, Elza Mikheeva, and Ekaterina Gornostaeva

Index 335

Contributors

Bilal Ahmad Department of Agricultural Microbiology, Faculty of Agricultural Sciences, Aligarh Muslim University, Aligarh, Uttar Pradesh, India

Javed Ahmad A.R. Al-Jeraisy Chair for DNA Research, Zoology Department, College of Sciences, King Saud University, Riyadh, Saudi Arabia

Md. Faiz Ahmad Department of Pharmacology, School of Medicine, Case Western Reserve University, Cleveland, OH, USA

Saad Mohammad Ahsan Centre for Cellular and Molecular Biology (CSIR-CCMB), Hyderabad, Telangana State, India

Abdulrahman A. Alatar Department of Botany and Microbiology, College of Sciences, King Saud University, Riyadh, Saudi Arabia

Abdulaziz A. Al-Khedhairy Zoology Department, College of Sciences, King Saud University, Riyadh, Saudi Arabia

Sabiha M. Ansari Department of Botany and Microbiology, College of Sciences, King Saud University, Riyadh, Saudi Arabia

S. Aurosman Pappus Department of Biological Sciences, IISER Kolkata, Nadia, West Bengal, India

Ameer Azam Department of Applied Physics, Faculty of Engineering and Technology, Aligarh Muslim University, Aligarh, Uttar Pradesh, India

Armelle Baeza-Squiban CNRS UMR 8251, Unit of Functional and Adaptive Biology (BFA), Laboratory of Molecular and Cellular Responses to Xenobiotics, University Paris Diderot (Sorbonne Paris Cité), Paris, France

Irfan Ahmad Bhat Division of Fish Genetics and Biotechnology, Central Institute of Fisheries Education, Mumbai, India

Sonja Boland CNRS UMR 8251, Unit of Functional and Adaptive Biology (BFA), Laboratory of Molecular and Cellular Responses to Xenobiotics, University Paris Diderot (Sorbonne Paris Cité), Paris, France

Fátima Brandão Department of Environmental Health, Portuguese National Institute of Health, Porto, Portugal

ISPUP-EPIUnit, Universidade do Porto, Porto, Portugal

Yi Cao Key Laboratory of Environment-Friendly Chemistry and Applications of Ministry Education, Laboratory of Biochemistry, College of Chemistry, Xiangtan University, Xiangtan, People's Republic of China

Carla Costa Department of Environmental Health, Portuguese National Institute of Health, Porto, Portugal

ISPUP-EPIUnit, Universidade do Porto, Porto, Portugal

Veronica Santana da Silva Department of Biology, ICB-UFMG, Federal University of Minas Gerais – UFMG, Belo Horizonte, MG, Brazil

Valeria De Matteis Dipartimento di Matematica e Fisica "Ennio De Giorgi", Università del Salento, Lecce, Italy

Tiago Alves Jorge de Souza Department of Genetics, Faculty of Medicine of Ribeirão Preto; USP, Ribeirão Preto, São Paulo, Brazil

Sagar Dhoble Department of Pharmaceutical Sciences and Technology, Institute of Chemical Technology, Mumbai, India

Sourabh Dwivedi Department of Applied Physics, Faculty of Engineering and Technology, Aligarh Muslim University, Aligarh, Uttar Pradesh, India

Mohammad Faisal Department of Botany and Microbiology, College of Sciences, King Saud University, Riyadh, Saudi Arabia

Natalia Fernández-Bertólez Department of Psychology, Area of Psychobiology, Universidade da Coruña, DICOMOSA Group, A Coruña, Spain

Department of Cell and Molecular Biology, Universidade da Coruña, A Coruña, Spain

Leonardo Pereira Franchi Department of Chemistry, Faculty of Philosophy, Sciences and Letters of Ribeirão Preto; USP, Ribeirão Preto, São Paulo, Brazil

Koyeli Girigoswami Department of Medical Bionanotechnology, Chettinad Hospital and Research Institute, Chettinad Academy of Research and Education, Rajiv Gandhi Salai, Kelambakkam, Kanchipuram, Tamil Nadu, India

Ekaterina Gornostaeva Lobachevsky Nizhny Novgorod State University, Nizhniy Novgorod, Russia

Agnieszka Grzelak Department of Molecular Biophysics, Faculty of Biology and Environmental Protection, University of Lodz, Lodz, Poland

Dianwen Ju Department of Microbiological and Biochemical Pharmacy & The Key Laboratory of Smart Drug Delivery, Ministry of Education, School of Pharmacy, Fudan University, Shanghai, People's Republic of China

Blanca Laffon Department of Psychology, Area of Psychobiology, Universidade da Coruña, DICOMOSA Group, A Coruña, Spain

Yubin Li Department of Microbiological and Biochemical Pharmacy & The Key Laboratory of Smart Drug Delivery, Ministry of Education, School of Pharmacy, Fudan University, Shanghai, People's Republic of China

Perelman School of Medicine, University of Pennsylvania, Philadelphia, PA, USA

Henriqueta Louro Department of Human Genetics, National Institute of Health Dr. Ricardo Jorge (INSA), Lisbon, Portugal

Toxicogenomics and Human Health (ToxOmics), Nova Medical School/ Faculdade de Ciências Médicas, Universidade Nova de Lisboa, Lisbon, Portugal

Adriano Magesky Institut de sciences de la mer de Rimouski (ISMER), Université du Québec à Rimouski, Rimouski, QC, Canada

Anuradha Majumdar Department of Pharmacology and Toxicology, Bombay College of Pharmacy, Mumbai, India

Elza Mikheeva Lobachevsky Nizhny Novgorod State University, Nizhniy Novgorod, Russia

Monalisa Mishra Neural Developmental Biology Lab, Department of Life Science, National Institute of Technology, Rourkela, Odisha, India

Javed Musarrat Department of Agricultural Microbiology, Faculty of Agricultural Sciences, Aligarh Muslim University, Aligarh, Uttar Pradesh, India

Department of Biosciences and Biotechnology, Baba Ghulam Shah Badshah University, Rajouri, Jammu and Kashmir, India

Eduardo Pásaro Department of Psychology, Area of Psychobiology, Universidade da Coruña, DICOMOSA Group, A Coruña, Spain

Vandana Patravale Department of Pharmaceutical Sciences and Technology, Institute of Chemical Technology, Mumbai, India

Émilien Pelletier Institut de sciences de la mer de Rimouski (ISMER), Université du Québec à Rimouski, Rimouski, QC, Canada

Svetlana Pleskova Lobachevsky Nizhny Novgorod State University, Nizhniy Novgorod, Russia

Chloé Puisney CNRS UMR 8251, Unit of Functional and Adaptive Biology (BFA), Laboratory of Molecular and Cellular Responses to Xenobiotics, University Paris Diderot (Sorbonne Paris Cité), Paris, France

Chintalagiri Mohan Rao Centre for Cellular and Molecular Biology (CSIR-CCMB), Hyderabad, Telangana State, India

Mohd Ashraf Rather Department of Fisheries Biology, College of Fisheries Shirgaon, Rantagiri, Maharashtra, India

Rosaria Rinaldi Dipartimento di Matematica e Fisica "Ennio De Giorgi", Università del Salento, Lecce, Italy

Thiago Lopes Rocha Institute of Tropical Pathology and Public Health, Federal University of Goiás, Goiania, Goiás, Brazil

Quaiser Saquib Zoology Department, College of Sciences, King Saud University, Riyadh, Saudi Arabia

A.R. Al-Jeraisy Chair for DNA Research, Zoology Department, College of Sciences, King Saud University, Riyadh, Saudi Arabia

Niti Sharma Regional Centre, Central Inland Fisheries Research Institute, Guwahati, Assam, India

Rupam Sharma Division of Fish Genetics and Biotechnology, Central Institute of Fisheries Education, Mumbai, India

Maqsood A. Siddiqui A.R. Al-Jeraisy Chair for DNA Research, Zoology Department, College of Sciences, King Saud University, Riyadh, Saudi Arabia

Joanna Skalska Laboratory of Pathoneurochemistry, Department of Neurochemistry, Mossakowski Medical Research Centre, Polish Academy of Sciences, Warsaw, Poland

Lilian Rodrigues Rosa Souza Department of Chemistry, FFCLRP-USP, University of São Paulo – USP, Ribeirão Preto, SP, Brazil

Lidia Strużyńska Laboratory of Pathoneurochemistry, Department of Neurochemistry, Mossakowski Medical Research Centre, Polish Academy of Sciences, Warsaw, Poland

Swapnil Talkar Department of Pharmaceutical Sciences and Technology, Institute of Chemical Technology, Mumbai, India

João Paulo Teixeira Department of Environmental Health, Portuguese National Institute of Health, Porto, Portugal

ISPUP-EPIUnit, Universidade do Porto, Porto, Portugal

Vanessa Valdiglesias Department of Psychology, Area of Psychobiology, Universidade da Coruña, DICOMOSA Group, A Coruña, Spain

ISPUP-EPIUnit, Universidade do Porto, Porto, Portugal

Rizwan Wahab A.R. Al-Jeraisy Chair for DNA Research, Zoology Department, College of Sciences, King Saud University, Riyadh, Saudi Arabia

Mariusz Zuberek Department of Molecular Biophysics, Faculty of Biology and Environmental Protection, University of Lodz, Lodz, Poland

About the Editors

Dr. Quaiser Saquib is currently working as an assistant professor in Zoology Department, College of Sciences, King Saud University, Riyadh, Saudi Arabia. He is also serving as coordinator of Al-Jeraisy Chair for DNA Research in KSU. He earned his PhD from AMU, India. His interest includes molecular toxicity and biophysical research. Dr. Saquib has been conferred the "2016 Scientist of the Year Award" by the National Environmental Science Academy, India, and the "2008 Young Scientist Award" by the Environmental Mutagen Society of India. A total of 49 research publications, 2 books, and 2 book chapters have been added to his credentials. He is the principal and co-investigator in several projects funded by the National Plan for Science and Technology, Saudi Arabia. Dr. Saquib is also engaged in teaching and training of graduate and postgraduate students at KSU.

Dr. Mohammad Faisal is an associate professor at the Department of Botany and Microbiology, King Saud University, Riyadh, Saudi Arabia. He received his PhD degree in botany from AMU, India. He is a well-known plant biotechnologist and an active researcher; he also worked as postdoctoral fellow at the Centro de Investigaciones Biologicas, Madrid, Spain, and young scientist at AMU, Aligarh, India. He availed several national fellowships during his doctoral work. Dr. Faisal has been awarded the 2017 Plant Biotechnologist Award by the SESR, India, and the 2015 Scientist of the Year Award by the NESA. He has published over 64 research articles, and these papers have been cited over 1362 times with Google h-index of 23. He holds life membership of several academic societies and is fellow of the SESR. He is in the editorial board of several peer-reviewed international journals.

Abdulaziz A. Al-Khedhairy is currently serving as director of Twasol Research Excellence Program in King Saud University, Riyadh, Saudi Arabia. He has been appointed at several key academic and administrative positions in KSU, including dean of Community College and assistant vice-president for Graduate Studies and Research. Professor Al-Khedhairy is also supervising Al-Jeraisy Chair for DNA Research. His research interest includes molecular genetics of inherited diseases in the Saudi community and interactions of eukaryotic DNA with genotoxic chemicals and their prolonged effect on human genes. He has published 160 research articles, 9 book chapters, and 2 books. He also has one US patent. He has been conferred with KACST Golden Award for Excellence in Research. As a principal and co-investigator, he has successfully supervised several projects funded by KACST, KSA.

Dr. Abdulrahman A. Alatar is professor at the Department of Botany and Microbiology, King Saud University, Riyadh, Saudi Arabia. He received his PhD degree in plant ecology in the year 2008 from KSU. He has been the principal and co-investigator of various projects funded by different agencies and supervised three PhD students and seven MSc students. He holds membership of several academic societies and participated in several national and international conferences. He published more than 65 scientific articles in different international journals published by Elsevier, Springer, Wiley, Taylor and Francis, etc. His work has been cited in various leading journals. He is the recipient of several awards and honors and has various academic and administrative contributions in the KSU. Currently, he is also serving as the vice dean of Graduate Studies, KSU, Riyadh, Saudi Arabia.

1 Toxicity Assessment in the Nanoparticle Era

Valeria De Matteis and Rosaria Rinaldi

Abstract

The wide use of engineered nanomaterials in many fields, ranging from biomedical, agriculture, environment, cosmetic, urged the scientific community to understand the processes behind their potential toxicity, in order to develop new strategies for human safety. As a matter of fact, there is a big discrepancy between the increased classes of nanoparticles and the consequent applications versus their toxicity assessment. Nanotoxicology is defined as the science that studies the effects of engineered nanodevices and nanostructures in living organisms. This chapter analyzes the physico-chemical properties of the most used nanoparticles, the way they enter the living organism and their cytoxicity mechanisms at cellular exposure level. Moreover, the current state of nanoparticles risk assessment is reported and analyzed.

Keywords

Nanotoxicity · Cytotoxicity · Physico-chemical properties · Biodistribution · In vitro and In vivo studies · Biodistribution

1.1 Introduction

Nanomaterials are defined as objects having a size less than 100 nm; when in the form of nanoparticles, they are different from nanostructures deriving by industrial processes as welding fumes, fire smoke, or carbon black [1]. For this reason, to distinguish the manufactured origin of these products from natural particles with similar size, they were defined "engineered nanoparticles" (ENPs). A "Nanomaterial" has been defined as a "material with any external dimension in the nanoscale or having internal structure or surface structure in the nanoscale" (ISO 2010) [2] and a "nanoparticle" as a "nano-object with all three external dimensions in the nanoscale", where nanoscale was defined as the size ranging approximately from 1 to 100 nm (ISO 2008) [3]. Engineered

V. De Matteis (✉) · R. Rinaldi
Dipartimento di Matematica e Fisica "Ennio De Giorgi", Università del Salento, Lecce, Italy
e-mail: valeria.dematteis@unisalento.it

Q. Saquib et al. (eds.), *Cellular and Molecular Toxicology of Nanoparticles*, Advances in Experimental Medicine and Biology 1048, https://doi.org/10.1007/978-3-319-72041-8_1

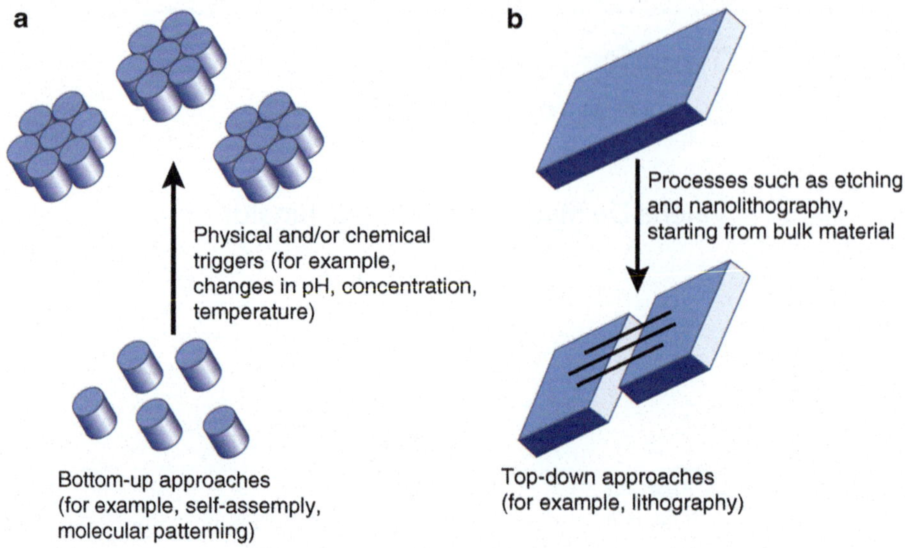

Fig. 1.1 Bottom-up (**a**) and top-down (**b**) approach. (Reprinted by permission from Macmillan Publishers Ltd: Nature Reviews–Neuroscience) [6] copyright 2006

nanoparticles (ENPs) can be manifactured by means of different synthetic routes, starting from chemical elements such as metals, carbon, metal oxides, polymers and biological molecules [4]. They can be produced by '*Top-down*' approach, where bulk materials are broken down to the nano size by milling or etching and by '*Bottom-up*' approach, that makes nano-sized objects by combining atomic scale materials [5] (Fig. 1.1).

ENPs have already been used in lots of commercial products for specific applications (electronics, biomedicine, pharmaceuticals, cosmetics, energy, constructions and catalytic/environmental detection) [7]. To this aim, the synthesized ENPs undergo a surface modification, in order to passivate, stabilize [8] and ehnance their biocompatibility. ENPs have unique properties compared to larger-sized particles of the same material (bulk materials); consequently, the toxicity of a base material is different from that of the same material in form of nanoparticles [9]. On the basis of their composition, the ENPs can be classified in three distinct, large categories:

1. *Carbon based nanoparticles:*
   Fullerenes (C60), Carbon nanofibers, Carbon nanofoams, Graphene, Graphene nanofoils, Carbon Nanotubes (CNT), Single-walled carbon nanotubes (SWCNTs), Multi-walled carbon nanotubes (MWCNTs), Carbon Black (CB)
2. *Organic nanoparticles*
   Organic polymers, Biologically-Inspired Nanoparticles
3. *Inorganic nanoparticles*
   Metal nanoparticles, Metal oxide nanoparticles (MONs), Quantum dots (QDs)

Based on their shape, the most synthetized nanomaterials were classified as nanowires, nanorods, nanoprisms, nanoparticles, nanocubes and nanotubes (Fig. 1.2).

The physico-chemical properties of ENPs, such as size, shape, surface charge, solubility and other ones, influence the biological responses [12]. In fact, greater surface area per unit mass renders ENPs more biologically active than larger-sized particles showing the same chemistry [13]. Smaller particles take up less volume and exhibit greater surface area per unit mass [14]. The engineering and functionalization of NPs, can strongly alter their properties: the distribution of nanoparticles is, therefore, closely connected with surface characteristics [15]. Today, the toxicological effects of ENPs on the human

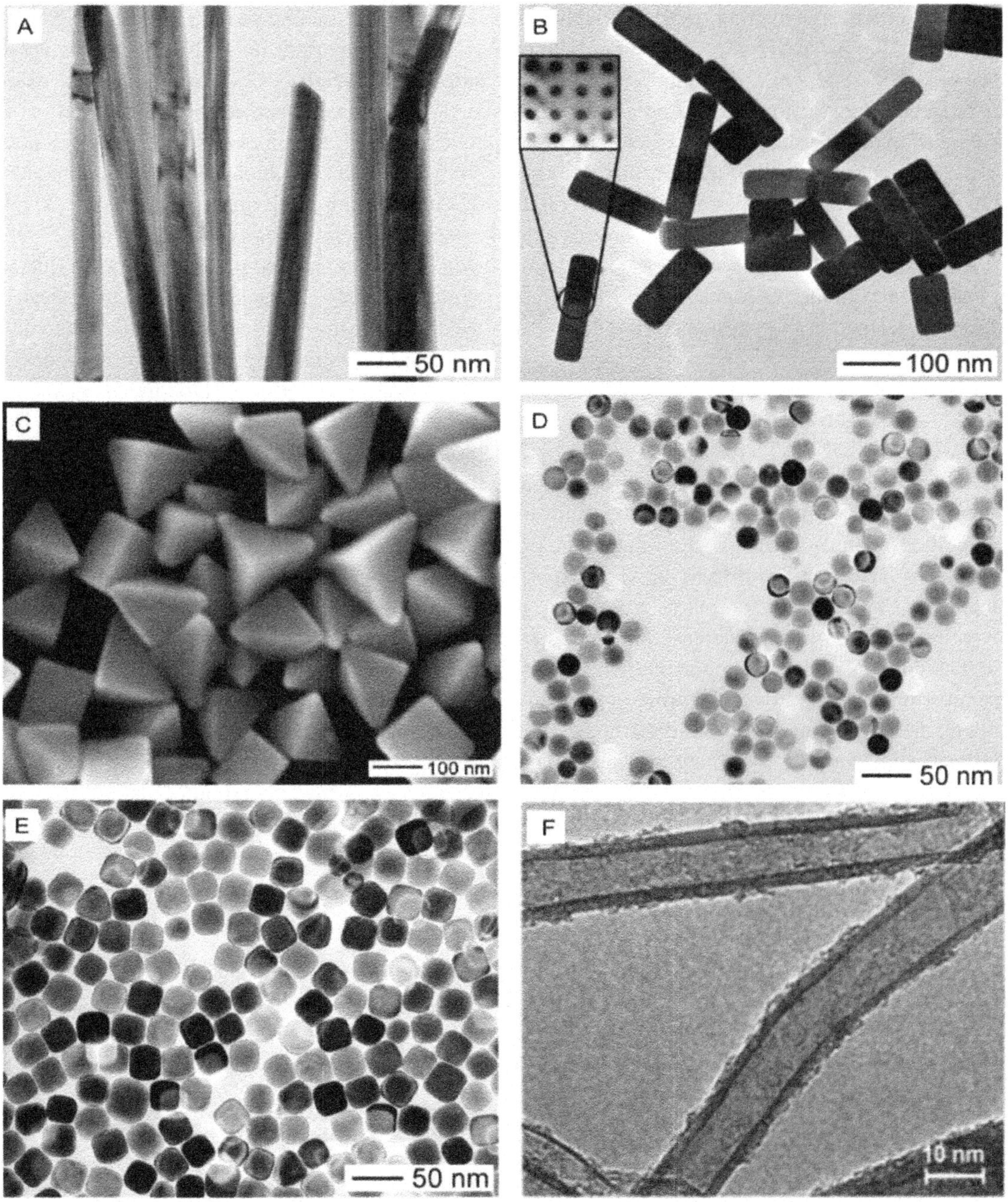

Fig. 1.2 TEM images of different morphological shapes of silver-based nanomaterials: (**a–e**) nanowires, nanorods, nanoprisms, nanoparticles, nanocubes, (**f**) carbon nanotubes (Figures **a–e** reprinted (adapted) with permission from [10]. Copyright 2007 American Chemical Society. Figure **f** reproduced (adapted) from [11] with permission of The Royal Society of Chemistry)

body have not been fully understood, and the mechanism of their toxicity actions is still unclear [16]. The generation of free radicals and the consequent oxidative stress occurs, togheter with cardiovascular diseases, cancer and fibrosis [4]. While the business of products based on ENPs is growing continuously, more studies are oriented to consider the risk assessment as integral part of nanoparticles design in order to protect the health of living organisms and environment [17]. Nanotoxicology is a new discipline with the aim of characterizing the toxicity of ENPs by identificating of exposure route and dose assessment, as well as investigating the relationship between physico-chemical properties and hazard effects [9]. To reach this, research activities are strongly oriented to meet the needs of increased epidemiological and clinical information, and to obtain from in vivo and in vitro systems [18, 19].

1.2 In Vitro and in Vivo Studies

It was estimated that, for toxicology investigations, a large number of animals should be employed, raising ethical questions due to the EC Directive 2010/63/EU [20], on in vivo experimentation. Therefore, also in vitro investigations were relevant in order to increase the epidemiological/clinical studies and contribute to the dissemination of toxicological information [18]. In detail, the development of in vitro methods was a critical point not only to achieve research goals, but also as a screening tool to envisage potential toxicological effects of ENPs that will be synthetized over the coming years. These investigations were also important as useful guidelines about safety and legal issues for companies using ENPs. The in vivo methods were preferred in the field of chronic toxicity investigation as well as to study mechanisms which cannot be simulated by in vitro systems [21]; among these, toxicokinetics and the identification of target tissues [22, 23] (Fig. 1.3). The in vitro methods were conducted to study specific characteristics of nanomaterials used as indicators of toxicity, in order to understand the potential adverse mechanisms induced by ENPs [21]. In general, at the first stage, nanomaterials cross tissue membranes and enter the bloodstream at several sites [24]. In particular, the main route of exposure is unavoidably the respiratory system [25]. Small size ENPs may enter the deep part of the respiratory tract [26]. Moreover, ENPs can enter the body through contaminated food or drink, as well as through the skin, that is the largest organ of the human body [27]. The skin forms a barrier versus the external environment, performing a function of protection and homeostasis. Despite the skin is a large barrier, it represents a potential route of exposure because the dermal absorption of small nanoparticles is possible [28].

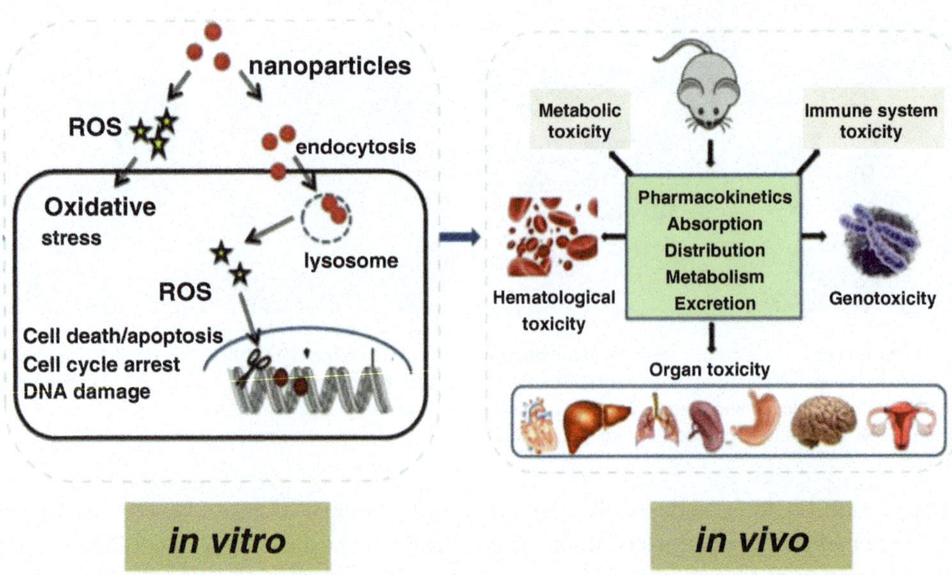

Fig. 1.3 Schematic representation of ENPs toxicity in vitro and in vivo systems (Adapted from reference [29])

1.3 Physico-chemical Properties of Nanostructures

1.3.1 Size

Size is a determining factor for the uptake and translocation of ENPs, as reported in the case of titanium dioxide nanoparticles (TiO_2) particle translocation across the rat alveolar epithelium [30] and nanoferrite targeting tumors in athymic mice [31]; in the interaction between nanoparticels and byological systems, the size and surface area represent critical factors [32]. Compared with larger particles, the smaller ones show a greater surface area per unit volume, making them more chemically reactive [33]. A lot of cellular processes in mammalians, such as the cellular uptake, endocytosis and inflammatory response, are closely connected with the size of nanomaterials [21]. In vitro cytotoxicity studies are conducted at different concentrations of ENPs, exposure times, cell types, while in vivo studies are more difficult, owing to the complexity of living systems [34]. In fact, in many cases, nanomaterials undergo modifications due to biological macromolecules, inside the organisms [35]. The induction of reactive oxigen species (ROS), responsible of free radicals release, resulted stronger when the small nano-objects were endocyted by cells [36]. The free radicals trigger the inflammatory responses inducing DNA damage and lipid oxidations [37]. In addition, the size of ENPs is responsible for their behaviour inside living organisms from the pharmacological point of view [38]. The intravenous injection of 50 nm nanoparticles, induces toxic effects in various tissues, whereas 100–200 nm NPs can be stopped by the action of the reticuloendothelial system, including liver and spleen [39]; however, in these organs, NPs start the oxidative stress phenomenon. In the respiratory system, NPs with hydrodynamic size less than 100 nm induce more respiratory damage compared to particles of the same material, but with a greater size [40, 41]. Nanomaterials with a size range between 10 and 20 nm sediment in the tracheobronchial region whereas particles smaller than 10 nm can deposit in the alveolar region [42]. Indeed, when smaller particles move into the respiratory tract, they reach the pulmonary interstitium with impairment of alveolar macrophages function [43]. In tumor cells, Kettler *et al.* suggested a grater uptake of 40–50 nm ENPs concluding that the size ~50 nm was suitable of uptake by non-phagocytic eukaryotic cells [44]. Baharifar *et al.* investigated the toxicity of Cs/Sk NPs on artifical neurons, showing that smaller nanoparticles were more toxic than the largest [45]. The injection in mice of AgNPs induced a hepatobilary toxicity, in particular due to those having an average size of 10 nm, respect to larger ones (40–100 nm) [46]. In recent studies, zebrafish was used to assess the in vivo toxicity of different Au and AgNPs in the size range of 3, 10, 50, and 100 nm. The experimental results demonstrated the size dependent mortality induced by AgNPs, whereas the toxic effect of AuNPs did not depend on size [47]. Seiffert *et al.*, using Brown-Norway (BN) and Sprague-Dawley (SD) rats, showed that 20 nm AgNPs triggered proinflammatory phenomenon in terms of neutrophil influx, but not in eosinophilic influx [48]. In all these examples, the predominant role of size in cellular uptake mechanism was unequivocal.

1.3.2 Shape

Shape dependent toxicity has been studied for several ENPs, such as carbon nanotubes, silica, allotropies, nickel, gold, and titanium nanomaterials [49–51]. In general, the shape influences the membrane wrapping processes during the endocytosis or phagocytosis pathway [52, 53]. Computational models described the acceleration of ENPs traslocation across cell membranes depending on shape and surface charge. This accelaration was up to 60 orders of magnitude [54]. Differently shaped AgNPs were reported to affect the cells in a different way [55]. When the size and surface area were the same, the shape of ENPs was the principal feature influencing the toxicity. Nanorods of ZnO induced more toxic results than ZnO NPs in A549 cell lines [56]. Instead, AuNPs spherical shape contributed to toxicity than rod-type particles in ephitelial cells

[57]. Generally, endocytosis of spherical ENPs by cells is easier respect to rod shaped ones and relatively less toxic independently if they are homogenous or heterogeneous [58]. Besides, also TiO_2 fibres are more cytotoxic than spherical shaped NPs [59]. Asbestos fibres of about 10 μm caused lung carcinoma whereas 5 μm fibres induced mesothelioma. Smaller fibres, having a size around 2 μm, caused asbestosis [60]. In general, the effect of the shape dependent NPs can be due to the number of edges that induce oxidative stress [61], but additional evidences will be required to support this hypothesis.

1.3.3 Charge

The uptake of positively charged nanoparticles is more evident than negatively charged and neutral ENPs for the opsonization process due to plasma proteins. Unfortunately, they induce hemolysis and platelet aggregation [62, 63], resulting more toxic for living systems. For example, anionic amine-terminated poly(amidoamine) (PAMAM) showed no toxic effects, while positively charged PAMAM dendrimers (G4) triggered time-dependent toxicity in mice embryos and zebrafish [64]. In a recent study, Poly (lactic-co-glycolic acid) (PLGA) NPs with different surface charges, (positive, neutral or negative), were synthesized, and toxicity effects were analyzed in L5178Y mouse lymphoma cells, TK6 human B-lymphoblastoid cells and 16HBE14o- human bronchial epithelial cells. The data demonstrated that positive PLGANPs were more cytotoxic and induced chromosomal aberrations [65]. Therefore, the surface charge influences the uptake of NPs in cell; since the cellular membrane and the intracellular compartment is characterized by negative charge, the positive NPs are more internalized than negative nanoparticles. Wang *et al.* followed the effects of surface charges of gold nanoclusters AuNCs and showed the ability of positively charged nano-objects to influence pharmacokinetics. In particular, they ehanced the renal excretion and accumulation in kidney, liver, spleen, and testis. On the contrary, negative charge AuNCs were kept in renis, increasing tumor uptake [66]. CDs in vitro toxicity was studied on standard mouse fibroblasts (NIH/3T3). The data obtained highlighted the possibility to use neutral CDs-PEG for biological applications. On the contrary, negatively charged CDs-Pri induce a sudden arrest of the cell cycle, in particular the block of G2/M phase and the increase of oxidative stress. Positively charged CDs-PEI are the most cytotoxic, inducing the largest changes in the G0/G1 phase of cell cycle [67]. The blood-brain barrier modification and transmembrane permeability alteration is a peculiarity of surface charge of nanoparticles. 50 nm NPs have showed a high ability to penetrate skin due to their small size. On the other hand, the 500 nm NPs cross the skin barrier thanks to a high charge concentration, exceeding that of the skin barrier itself [68].

1.3.4 Surface Modification

The toxicity of ENPs is also influenced by the surface, which dictates the interactions with biological entities that can alter the pharmacokinetics, distribution, accumulation, and toxicity of nano-objects [69, 70]. In fact, several types of ZnONPs, different in size and surface chemistry, were used to assess the development of zebrafish after exposure to NPs. Zhou *et al.* illustrated that the surface functionalization of ZnONPs was crucial in the induction of toxicity in zebrafish embryos, while the size showed a minimum impact on animal model [71]. PEG and citrate surface functionalization of AgNPs strongly influenced viability of HaCaT cells and mechanisms of cell death. Citrate-coated AgNPs displayed more severe cytotoxic effects than PEG-coated AgNPs [72]. The cytotoxicity of silica was associated with the presence of surface radicals and reactive oxygen species [73]. Indeed, surface coatings were useful to inhibit the toxic heavy metals release in QDs structures in order to make them safer for biomedical applications. Surface coatings concerned also the QD core in order to avoid less water dispersibility and ions release [74]. CHO-K1 cells (Chinese Hamster ovaries derived cell lines) were used to assess the cytotoxicity of Superparamagnetic iron oxide nanoparticles (SPIONs) functionalized with different chemical groups. The coatings using amino groups increased the uptake and cytotoxic effects compared to SPIONs with PEG functionalization. These data emphasized the importance of surface coatings for biomedical applications [75]. In another work, a thin silica shell was used to cover the Fe_3O_4NPs and to test the toxicity on A549 and HeLa cells. The authors showed the critical role of surface engineering: in fact, it was responsible of increased cytotoxic and genotoxic effects, if compared to bare Fe_3O_4NPs [76]. Moreover, it has been demonstrated that spherical gold nanoparticles with various surface coatings have been found to be non toxic to human cells [77].

1.3.5 Crystalline Structure

The crystalline structure of nanoparticles is another important factor that influences the toxicity in cells, also evident in nanoparticles having the same size. For example, titanium dioxide, which has typically three crystalline structures (rutile, anatase and brookite), shows different in vitro different toxicity responses [78]. A recent work, investigating the impact of 50 nm rutile and anatase TiO_2NPs in A549 and MCF7 cell lines, showed that the anatase form is more toxic due to its susceptibility to pH and solar light, promoting severe NPs degradation [79]. Several studies showed a change of crystalline structure after interaction with water or other media. In particular, in presence of water, ZnSNPs underwent a rearrangement of crystalline structure [80]. Silica exibhited different effects inside cells, in fact, while the amorphous silica was used as a food additive, its crystalline form could be human carcinogen [81].

1.3.6 Effects of Media

The toxic effects of ENPs were also dependent on the media conditions, that could induce agglomeration or a specific dispersion rate. Prasad *et al.* analyzed the adverse effects in BEAS-2B cell line due to TiO_2NPs dispersed in three different media: two types of keratinocyte growth media (KGM), with different amount of BSA, and a synthetic medium. DNA damage was found using dispersions of TiO_2NPs in all three media. Moreover, they induced cytokinesis-blocked micronucleus only in one medium type (KF) inducing the lowest amount of agglomeration. This phenomenon triggered a cells accumulation in S phase cell cycle compared to another medium [82]. PBS promoted the increase of size in TiO_2NPs, ZnONPs and carbon black. This effect was not evident in water. In general, different size of nanoparticles showed different size in biological media [83].

1.4 Route of Entry of Nanoparticles

1.4.1 Inhalation

Inhalation was widely recognized as a major route of exposure to nanomaterials. The lung consists of two functional parts, the airways (trachea, bronchi, and bronchioles) and the alveoli (gas exchange areas). In general, the exposure of humans to nanomaterials promotes pleural granuloma formation, characterized by pulmonary fibrosis [84]. The function of mucus layer is to safeguard the epithelium of tracheobronchial region [85], so that when the ENPs deposit in this area, they are expelled from the lung by mucociliary action [86]. Firstly, the mechanism of sedimentation in the respiratory tract is due to inertial impact, followed by gravitational sedimentation and also Brownian diffusion [87]. This last phenomenon is typical of small ENPs because they behave like gas molecules and are able to penetrate in deeper tracts [88]. For the above mentioned reasons, the deposition becomes greater respect to the inhalation of equivalent amounts of coarse or fine particles (Fig. 1.4). ENPs can inhibit the phagocytic capacity of alveolar macrophages, leading to apoptosis [88]. Additionally, results of several studies showed that ENPs deposited in the respiratory tract interact with epithelial cells and interstitial fluid [89]. The interstitial translocation is a phenomenon occurring when the capacity of macrophages to incorporate ENPs is saturated or when smaller ENPs escape from the phagocytosis system [15]. Nevertheless, if on one hand the mechanisms by

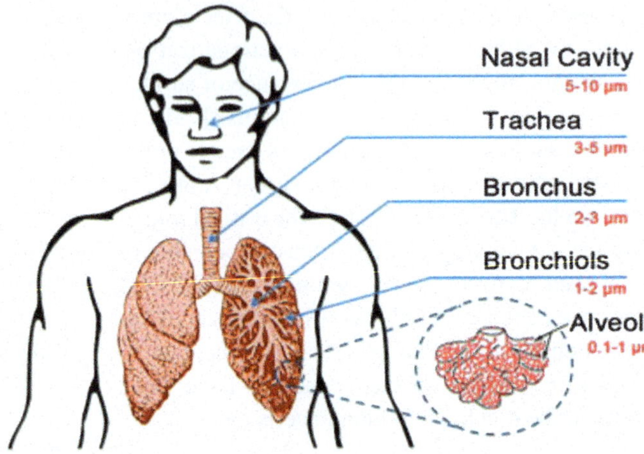

Fig. 1.4 Different size deposition of ENPs within the airway after the inhalation. (Reproduced from [90] with permission of The Royal Society of Chemistry)

which ultrafines penetrate cellular membranes were widely studied, there are only a few investigations about the process of penetration from pulmonary tissue to capillaries [15].

1.4.2 Ingestion

The ingestion of food additives, water and drugs allows ENPs to reach the gastrointestinal (GI) tract [91]. Via the GI tract, they enter lymphatic tissues containing phagocytic cells according to their physicochemical properties and surface ligand [92]. In general, most ENPs enter the GI system and are rapidly expelled [15]. The change in pH can induce agglomeration of nanomaterials inside the intestinal tract [93]. In the case of small agglomerates, they are expelled through faeces with little absorption; whereas, in presence of big aggregates, they could obstruct the GI tract as shown in an in vivo study in which mice developed severe diarrhoea, vomiting and finally death [92]. ENPs that were not eliminated in short time, translocated into blood capillaries in a size dependent manner to move into organs [94].

1.4.3 Skin Penetration

The skin is structurally divided into three layers: the outside layer, or epidermis, the dermis and the inside layer or hypodermis. The stratum corneum (SC), is constituted of cells called corneocytes, that desquamate and form a keratinized layer which is considered a tight barrier against ENPs penetration [95]. A review [96] published in 2010 provided an overview of the most common nanomaterials found in consumer products. The paper evidenced a lot of products (make-up, sunscreens, moisturizers) containing metal nanomaterials and destined to topical applications for consumers. Micron and submicron-sized TiO_2NPs cross the skin and reach hair follicles and dermis [97]. Nano-sized TiO_2 (10–50 nm), with a process depending on nanoparticles shape, confirmed the skin penetration in human volunteers [98]. Some researchers showed the penetration of TiO_2NPs in deeper layers via intercellular channels and hair follicles [99, 100]. Tak *et al.* [101] studied the skin permeation of Ag nano objects with different shape in in vitro and in vivo cell models. They hypotyzed two different possible pathways: the follicular penetration pathway and the intercellular penetration pathway as illustrated in Fig. 1.5.

1.5 Biodistribution

The most common approach to investigate the distribution of ENPs in vivo was to remove the main organs or tissues after animal sacrifice [102]. In general, once in the circulatory system, nanoparticles can be carried into the body and

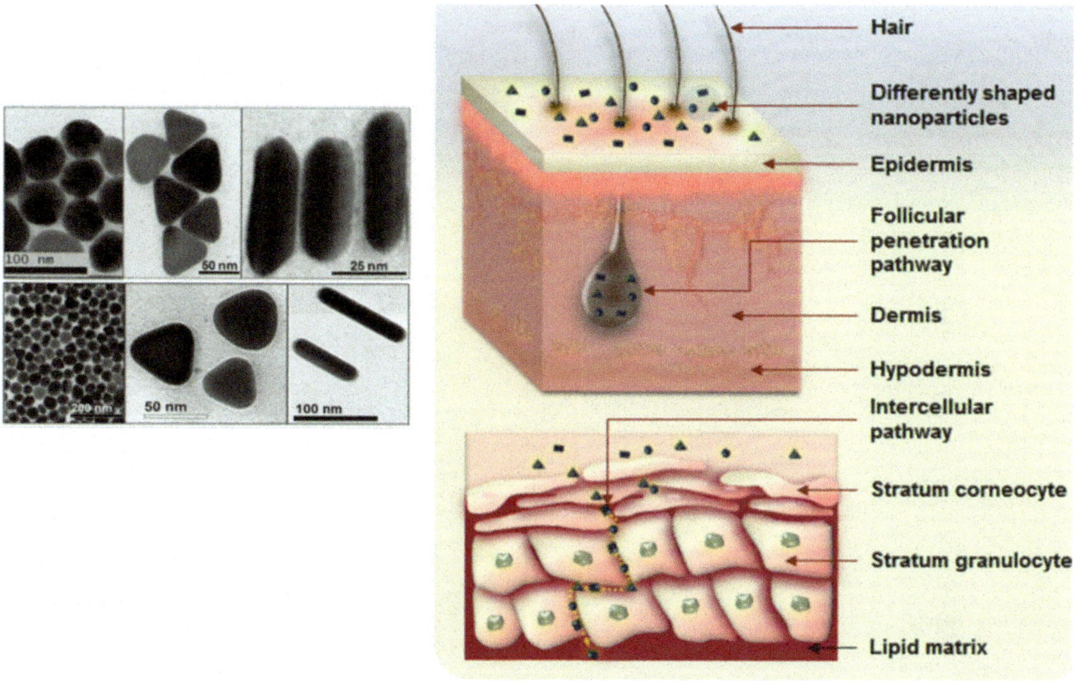

Fig. 1.5 ENPs shape (left) influences different skin penetration. (Adapted from reference [101], copyright 2015, Nature publishing group)

absorbed by organs and tissues, including kidneys, spleen, heart, liver, bone marrow and nervous system [15] (Fig. 1.6). The distribution in these organs is influenced by the physicochemical properties of nanoparticles and the rate of distribution is determined by blood flow and rate of diffusion out of capillaries and into organs and tissues [103]. ENPs reach the blood by phagocytosis process and are then filtered off by the spleen [104, 105]. Larger particles are accumulated in the liver [106] causing necrosis of liver tissues [107]. Recent evidences suggested that the ENPs can cross the blood brain barrier (BBB) and accumulate in the brain [108]. Nanomaterials can be successfully excreted by kidneys, faeces sweat, breast milk and saliva [15]. The capillaries of kidneys show a pore size of 70 nm, so the excretion is dependentent on the size of ENPs [109]. The adsorption phenomena of biomolecules onto the ENPs surface promote an increase of size particles, preventing filtration through urine [109]. In general, ENPs accumulate in many organs, damaging the excretory system. Kidney is a target organ of nanoparticles toxicity that can alter metabolic functions [110].

1.6 Nanoparticles Cellular Uptake

The uptake of nanomaterials by cells in vitro is an important factor to assess nanotoxicity [113]. Plasma membrane is a hydrophobic barrier regulating the traffiking of molecules [114]. The route of molecules and particles (including ENPs and microparticles) uptake is dependent on their characteristics and cell type [115]. Hydrophobic or hydrophilic ENPs enter cells by different mechanisms [116], ranging from active (receptor-mediated) to passive transport across cell membrane [117]. In general, bigger molecules enter the cells by active endocytosis, occurring in two different mechanisms: phagocytosis or pinocytosis [118]. Phagocytosis provides the uptake of large particles and is conducted by phagocytes including macrophages and neutro-

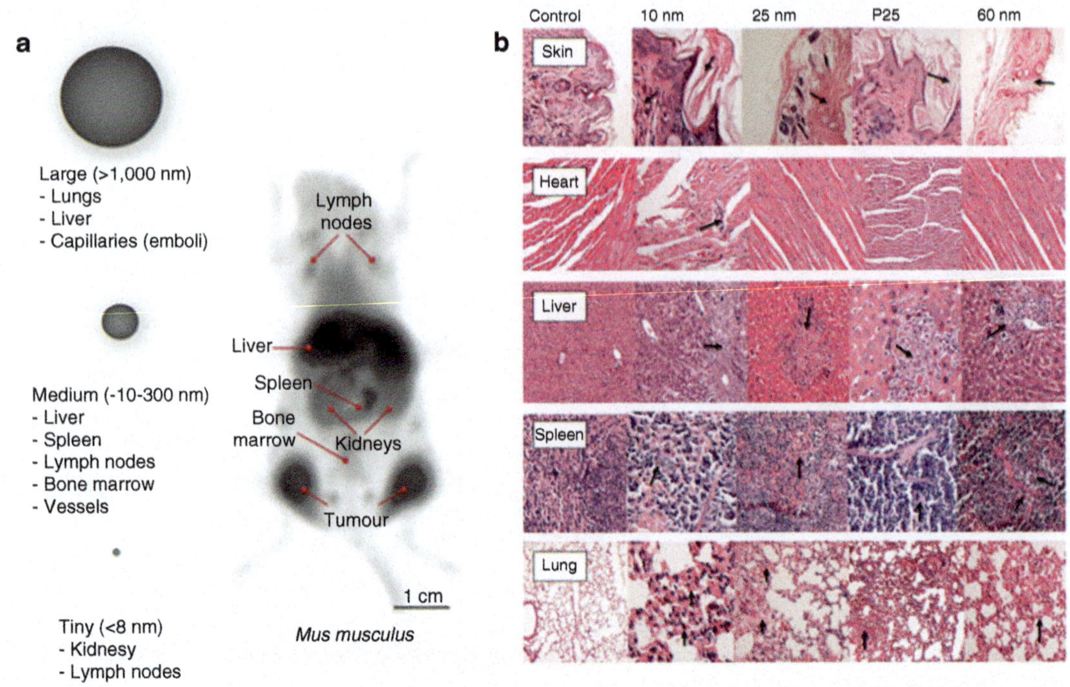

Fig. 1.6 (**a**) Left: different size NPs distribution in organs of *mus musculus*. (Reprinted by permission from Macmillan Publishers Ltd.: Nature Materials) [111] copyright 2014. (**b**) Histopathological evaluation after dermal exposure of TiO_2 NPs at different size for 2 months. Pathological changes were indicated with arrows. The tissues were stained with hematoxilin and eosin and observed at 100X of magnification. (Reprinted from [112] Copyright 2009, with permission from Elsevier)

philes [119]. It has been observed that fluids are inglobated in all cell types through pinocytoses [120]. Clathrin-mediated and non-clathrin-mediated, macropinocytosis, endocytosis mainly via caveolae were included in the pinocytosis process [121]. The endocytosis process is size-dependent: ENPs of around 200 nm were internalized by clathrin-coated phenomenon. When the size increased, the uptake was configurated as caveolae-mediated process. All these processes are characterized by energy consumption, whereas in the passive transport no chemical energy is involved [122]. In a recent work, the effect of protein corona on 40–80 nm AuNP was investigated on human epidermal keratinocytes. The AuNPs were functionalized using lipoic acid, polyethylene glycol (PEG) and branched polyethyleneimine (BPEI) coatings with and without a protein corona, up to 48 h. Most AuNPs were internalized by clathrin and lipid raft-mediated endocytosis, except for 40 nm AuNPs@PEG for which the predominant factor was raft/noncaveolae mediated endocytosis. The uptake efficency was high for BPEI-AuNP, concluding that the protein corona influenced cellular uptake and inflammatory responses of AuNPs [49]. Saikia *et al.* analyzed the influence of size and porosity of SNPs on protein adsorption, cellular uptake and toxicity in RAW 264.7 macrophages, demonstrating that both in presence and absence of proteins, the SNPs entered the cells by vacuole-like structures. In the absence of protein, some particles were found in free cytosol, suggesting phagocytosis [123]. Other studies investigated the uptake mechanism of silica NPs (SiO_2NPs). In one of these works, the authors observed that 50 nm SiO_2NPs were efficiently internalized. They also identified macropinocytosis as the dominant mechanism of uptake and showed the localization of NPs, in the early endosome [124]. Clathrin dependent endocytosis and macropinocytosis played major role in AgNPs uptake [125].

Thurn *et al.* investigated the uptake of TiO_2NPs by their conjugation with a fluorescent dye Alizarin Red S (ARS) [126]. Fluorescence microscopy and flow cytometry were used to characterize the uptake of TiO_2-ARS nanoconjugates in PC-3 M cells, finding a mechanism depending on more factors: temperature, time and concentration. The experiments confirmed a complex nanoparticle-cell interaction involving many endocytic mechanisms: clathrin-mediated endocytosis, caveolin-mediated endocytosis and macropinocytosis. In some cases, the nanoparticles uptake was dependent by pH. Mizuhara *et al.* [127] functionalized AuNPs nanoparticles with acylsulfonamide zwitterionic, demonstrating the ehnancement of cellular uptake at tumor pH (<6.5) in Hela. A representative image of cellular uptake of nanosilver is represented in Fig. 1.7.

1.7 Common Toxicity Tests In vitro

1.7.1 MTT Assay

The most used cytotoxicity test is MTT assay, [129, 130] that provides information about the mitochondria activity. The MTT tetrazolium salt crosses the plasma membrane of metabolically active cells, in which it is reduced to formazan and produces a colored solution [131] spectrophotometrically measured. The intensity of coloration is proportional to live cells, thus the MTT is a standard assay, used to evaluate cell viability. Other variations of the MTT assay are WST-1 [132], WST-8 (CCK-8), MTS (Cell Titer 96) [133, 134] and XTT tetrazolium salts.

1.7.2 Lactate Dehydrogenase (LDH) Leakage Assay

The LDH is a stable cytosolic enzyme which is present in all cells and is released into the extracellular space when the plasma membrane is damaged, measuring the number of lysed cells in the presence of toxic materials [132]. When LDH is free in the extracellar space, it induces the oxidation of lactate to pyruvate and the NAD to NADH. The formazan is producted by reaction between NADH and tetrazolium salt iodonitrotetrazolium (INT).

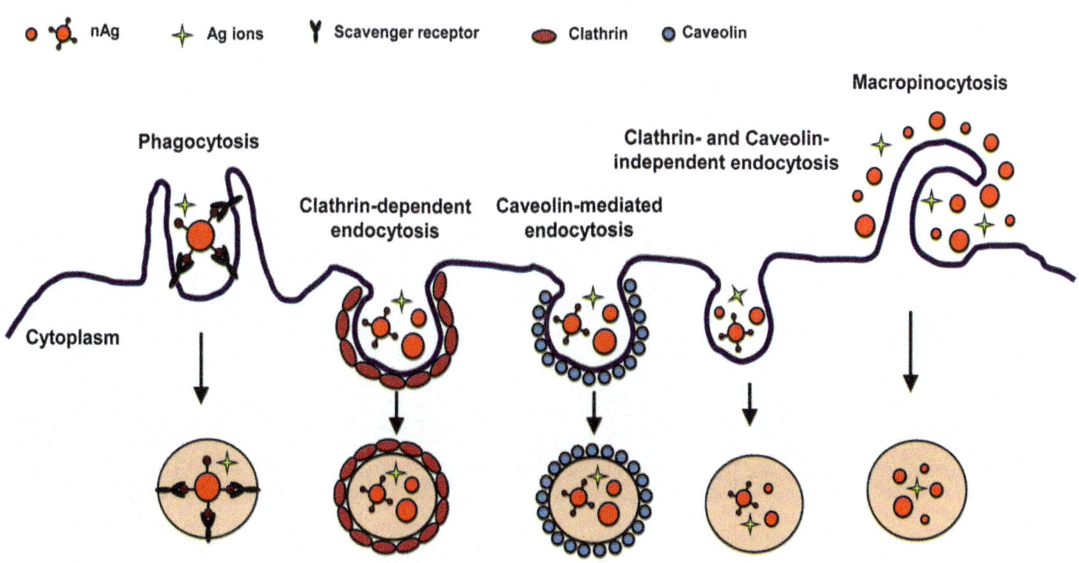

Fig. 1.7 Different mechanisms of NPs cellular uptake. (Reproduced from [128] with permission of The Royal Society of Chemistry)

1.7.3 2′,7′-Dichlorofluorescein Diacetate (DCF-DA) Assay

The DCF assay is used to estimate the presence of oxidative stress after an injury on cells [135]. To this aim, the cells are incubated with nonionic, nonpolar 2,7-dichlorofluorescin diacetate (DCFH-DA). The dye penetrates the cell membranes and becames non fluorescent DCFH due to hydrolyzation of intracellular esterases. The DCFH is oxidized in presence of reactive oxygen species (ROS) to fluorescent dichlorofluorescein (DCF). Then, the intensity of measured fluorescence is proportional to the amount of ROS in cell.

1.7.4 COMET Assay

The comet assay is widely used for in vitro and in vivo genotoxicity testing. Ostling and Johanson in 1984 [136] developed this method to measure DNA damage, later modified, in 1988, by Singh, who used alkaline conditions [137]. This method is a gel electrophoresis–based assay that can be used to measure DNA single-strand break in individual eukaryotic cells. The images obtained look like a "comet". The comet presents a head that corresponds to intact genomic materials and a tail consisting of damaged or broken pieces of DNA (Fig. 1.8).

1.8 Current State of Nano Risk Assessment

The development of nanotechnology evoked increasing societal and public concerns on the safety of ENPs and associated technologies, emphasizing the importance of setting goals on research concerning their safety. This aspect is very critical to minimize the uncertainties about safety and health issues concerning these materials and nanotechnologies [14]. The final goal of current researches, oriented to understand the toxicity of ENPs, is to make users expert to manipulate the nanomaterials in order to develop safer products. Indeed the safety evaluation must be assessed during the entire life cycle of nanoproducts. The scientific community should provide the final users with clear information on the risk associated to nanomaterials together with a risk management [139]. In 2004 the EC, with a Communication, encouraged the interaction between universities and industries in order to stimulate productions and industrial processes based on nanomaterials [140]. Afterwards, in 2005, [141] an action plan regarding the implementation of studies about the toxicity evaluation of nanomaterials was edited, followed by an evaluation of European Parliament (EP) in 2006 [142]. The EP suggested the importance to identify and recognize ENPs in commercial products. Two years later, the EC, on the basis of EP

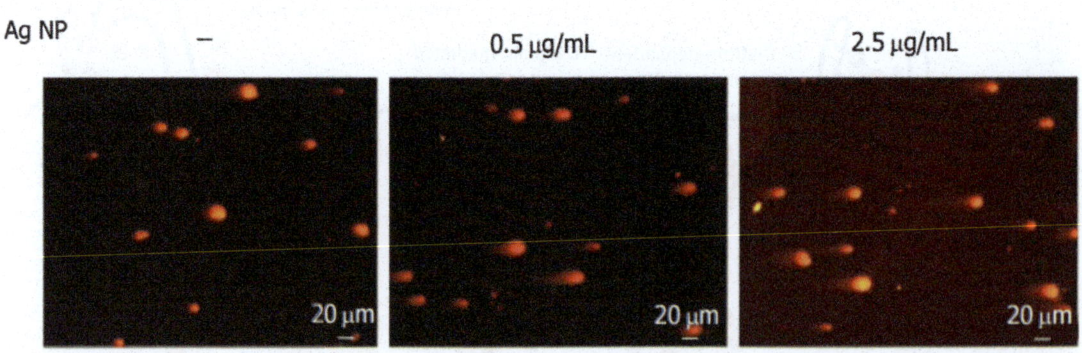

Fig. 1.8 Comet assay performed on HMEC cell lines doped with AgNPs for 24 h. The images cleary show the classic "comet" morphology due to genotoxicity effects of NPs. The slides were analyzed with a fluorescence microscope. (Adapted from reference [138] with permission of Baishideng Publishing Group Inc.)

Resolution, formulated another Communication [143], that put in evidence the effectiveness of laws regarding the risk associated to nanomaterials. The double face of nanomaterials that include both advantages and disadvantages, caused by their toxicity, was misleading for the absence of clear information on products based on "nanomaterials". The European legislative framework was implemented after the EP Resolution; in fact, the introduction of several characteristics for the use of nanomaterials in the cosmetics field was actuated by Regulation (EC) No 1223/2009, while the same attention was conducted for medical products by European Medicine Agency (EMA) [144]. In 2011, the concept of "nanomaterial" was defined in EEA when the EC published Recommendation 2011/696/EU [145], so that in 2012 a draft for the medical device regulation was formulated by EC, then revised in 2016 [146]. The International Organization of Standardization (ISO) [3] defined standards that were implemented by the Scientific Committee on Emerging and Newly Identified Health Risks (SCENIHR) [147]. In addition, the implementation was also actuated by the European Commission Joint Research Centre (JRC) [148] in order to adapt the ISO standard to the European regulatory framework. A Nanomaterial was defined as "A natural, incidental or manufactured material containing particles, in an unbound state or as an aggregate or as an agglomerate and where, for 50% or more of the particles in the number size distribution, one or more external dimensions is in the size range 1–100 nm. In specific cases and where warranted by concerns for the environment, health, safety or competitiveness the number size distribution threshold of 50% may be replaced by a threshold between 1% and 50%". By derogation from the above, fullerenes, graphene flakes and single wall carbon nanotubes with one or more external dimensions below 1 nm should be considered as nanomaterials. The EC indicated size, distribution, and physical state as univocal parameters to identify a nanomaterial besides a specific surface area by volume greater than 60 m^2/cm^3 [145]. However, the Recommendation indicated fullerenes, single-wall carbon nanotubes and graphene flakes toxicity associated to <1 nm size. Another critical point is represented by size measurement methods, that in several of cases were incomparable [149]. The most appropriate measurement methods for specific nanomaterials were recommended by JRC Reference Report in 2012, followed, two years later, by an admission of relevant gaps among the analytic methods that could be overcome by means of a fruitful interaction between scientific Academia and European Community [150]. Outside Europe, the National Nanotechnology Initiative's Environmental, Health, and Safety (EHS) Research Strategy and the recent report by US National Research Council [151] underlined the relevance of understanding the processes used to develop ENPs.

1.9 In Silico Toxicology

Computational methods are considered very important tools to assess the toxicity of materials in order to exceed the in vivo studies that would require long experimental times, ethical limitations as well as resource-intensive nanotoxicity tests [152]. In silico toxicology is used to predict and analyze toxicity in order to strengthen the standard tests and to guarantee an optimal drug design [153]. The quantitative structure activity (QSAR) model was originally a computational chemistry method; in nanotoxicology nano-QSAR (nano quantitative structure activity relationship) or quantitative nanostructure activity relationship (QNAR) were developed, using nanomaterials physico-chemical properties as suitable descriptors [154]. Several researches carefully described the procedure in QNAR modeling [155, 156]. Fourches *et al.* (2010) developed two QNARs using as many datasets from in vitro experiments [154]. These two models were statistically validated. With this in mind, structural descriptors were important to improve the QNAR models, to bypass the lack of nanomaterials' experimental data. The descriptors included 0D-descriptors, 1D-descriptors, 2D-descriptors, 3D-descriptors, and 4D-descriptors [157] but it was different for nanomaterials because they were not classificated as simple chemical compounds

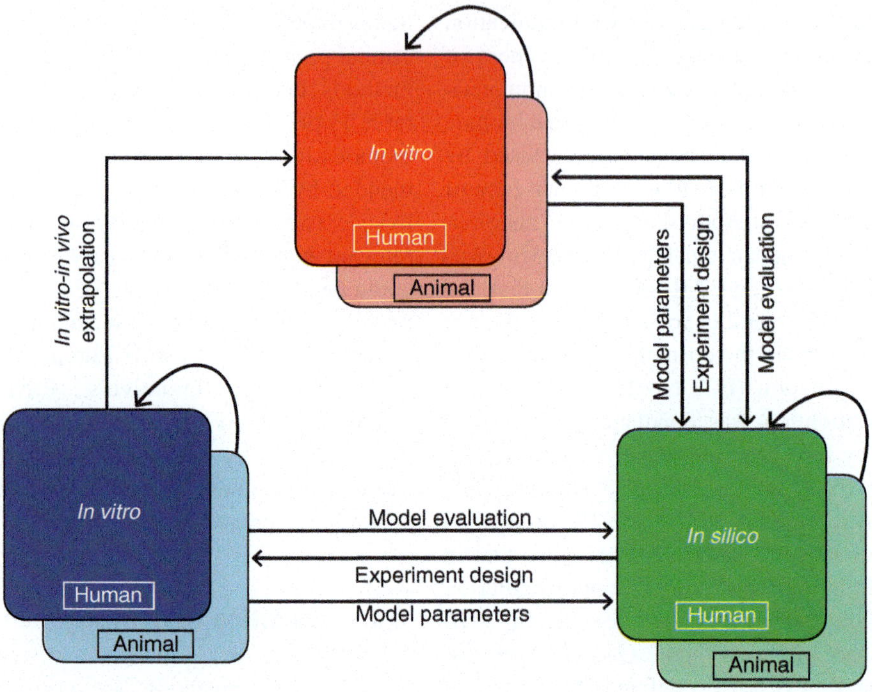

Fig. 1.9 Correlation among in vivo, in vitro and in silico investigations. (Reproduced from reference [158] Hindawi Publishing Corporation 2014)

[158]. The goal of the future toxicology research will be to connect the "big data" obtained by in vitro and in vivo studies with computational modeling in order to generate more and more precise human risk estimates (Fig. 1.9).

1.10 Conclusions

The use of nanostructured materials in many fields of human life has led to investigate the toxic effects of materials getting in touch with living organism by different routes of entry (ingestion, respiration and skin penetration). Hovewer, in several cases toxicological data are conflicting, so that a lot of studies are still required to evaluate the critical factors that contribute to NPs toxicity, in vivo and in vitro. This challenge is fundamental in order to standardize the analysis methods and to create a database containing the risks associated to NPs that will be accessible by scientists, manufacturings and consumers as well.

References

1. Hoyt VW, Mason E (2008) Nanotechnology. Emerging health issues. J Chem Health Saf 15:10–15
2. ISO, 2010. International Organization for Standardization. Nanotechnologies vocabulary part 1: Core Terms. ISO/TS 80004-1:2010
3. ISO (2008) International Organization for Standardization. Technical specification: nanotechnologies terminology and definitions for nanoobjects nanoparticle, nanofibre and nanoplate. ISO/TS 80004-2:2008
4. Oberdörster G, Stone V, Donaldson K (2007) Toxicology of nanoparticles: a historical perspective. Nanotoxicology 1:2–25
5. Wang Y, Xia Y (2004) Bottom-up and top-down approaches to the synthesis of monodispersed spherical colloids of low melting-point metals. Nano Lett 4(10):2047–2050
6. Silva GA (2006) Neuroscience nanotechnology: progress, opportunities and challenges. Nat Rev Neurosci 7:65–74
7. Ray PC, Fu PP (2010) Toxicity and environmental risks of nanomaterials: challenges and future needs. J Env Sci Heal C Env Carcinog Ecotoxicol Rev 27:1–35

8. Kango S, Kalia S, Celli A et al (2013) Surface modification of inorganic nanoparticles for development of organic-inorganic nanocomposites – a review. Prog Polym Sci 38:1232–1261
9. Oberdörster G, Oberdörster E, Oberdörster J (2005) Nanotoxicology: an emerging discipline evolving from studies of ultrafine particles. Environ Health Perspect 113:823–839
10. Wiley B, Sun Y, Xia Y (2007) Synthesis of silver nanostructures with controlled shapes and properties. Acc Chem Res 40:1067–1076
11. Prasek J, Drbohlavova J, Chomoucka J et al (2011) Methods for carbon nanotubes synthesis—review. J Mater Chem 21:15872
12. Mu Q, Jiang G, Chen L et al (2014) Chemical basis of interactions between engineered nanoparticles and biological systems. Chem Rev 114(15):7740–7781
13. Evans SJ, Clift MJD, Singh N et al (2017) Critical review of the current and future challenges associated with advanced in vitro systems towards the study of nanoparticle (secondary) genotoxicity. Mutagenesis 32(1):233–241
14. Oberdörster G (2010) Safety assessment for nanotechnology and nanomedicine: concepts of nanotoxicology. J Intern Med 267:89–105
15. Oberdörster G, Maynard A, Donaldson K et al (2005) Principles for characterizing the potential human health effects from exposure to nanomaterials: elements of a screening strategy. Part Fibre Toxicol 2:8
16. Ma DD, Yang WX (2015) Engineered nanoparticles induce cell apoptosis: potential for cancer therapy. Oncotarget. 28 7(26):40882–40903
17. Hund-Rinke K, Herrchen M, Schlich K et al (2015) Test strategy for assessing the risks of nanomaterials in the environment considering general regulatory procedures. Environ Sci Eur 27:1–12
18. Stone V, Johnston H, Schins RPF (2009) Development of in vitro systems for nanotoxicology: methodological considerations. Crit Rev Toxicol 39:613–626
19. Fröhlich E, Salar-Behzadi S (2014) Toxicological assessment of inhaled nanoparticles: role of in vivo, ex vivo, in vitro, and in silico studies. Int J Mol Sci 15:4795–4822
20. (2010) Directive 2010/63/EU. http://ec.europa.eu/environment/chemicals/lab_animals/index_en.htm
21. Aillon KL, Xie Y, El-Gendy N et al (2009) Effects of nanomaterial physicochemical properties on in vivo toxicity. Adv Drug Deliv Rev 61:457–466
22. Dechsakulthorn F, Hayes A, Bakand S et al (2007) In vitro cytotoxicity assessment of selected nanoparticles using human skin fibroblasts. Proceeding 6th World Congr Altern Anim Use Life Sci 397–400
23. Jones CF, Grainger DW (2009) In vitro assessments of nanomaterial toxicity. Adv Drug Deliv Rev 61(6):438–5636. 490–499
24. Barua S, Mitragotri S (2014) Challenges associated with penetration of nanoparticles across cell and tissue barriers: a review of current status and future prospects Sutapa. Nano Today 9:223–243
25. Yah CS, Simate GS, Iyuke SE (2012) Nanoparticles toxicity and their routes of exposures. Pak J Pharm Sci 25(2):477–491
26. Qiao H, Liu W, Gu H et al (2015) The transport and deposition of nanoparticles in respiratory system by inhalation. J Nanomater 2015:394507. 8
27. Bakshi S, He ZL, Harris WG (2014) Natural nanoparticles: implications for environment and human health. Crit Rev Environ Sci Technol 45:861–904
28. Schneider M, Stracke F, Hansen S, Schaefer UF (2009) Nanoparticles and their interactions with the dermal barrier. Dermatoendocrinol 1:197–206
29. Li J, Chang X, Chen X et al (2014) Toxicity of inorganic nanomaterials in biomedical imaging. Biotechnol Adv 32:727–743
30. Oberdörster G, Sharp Z, Atudorei V et al (2004) Translocation of inhaled ultrafine particles to the brain. Inhal Toxicol 16:437–445
31. Natarajan A, Gruettner C, Ivkov R et al (2008) Nanoferrite particle based radioimmunonanoparticles: binding affinity and in vivo pharmacokinetics. Biophys Chem 19:1211–1218
32. Powers KW, Palazuelos M, Moudgil BM, Roberts SM (2007) Characterization of the size, shape, and state of dispersion of nanoparticles for toxicological studies. Nanotoxicology 1:42–51
33. Jiang J, Oberdörster G, Elder A et al (2008) Does nanoparticle activity depend upon size and crystal phase? Nanotoxicology 2(1):33–42
34. Souza TAJ, Franchi LP, Rosa LR et al (2016) Cytotoxicity and genotoxicity of silver nanoparticles of different sizes in CHO-K1 and CHO-XRS5 cell lines. Mutat Res Genet Toxicol Environ Mutagen 795:70–83
35. Yallapu MM, Ebeling MC, Jaggi M, Chauhan SC (2013) Plasma proteins interaction with curcumin nanoparticles: implications in cancer therapeutics. Curr Drug Metab 14:504–515
36. Gatoo MA, Naseem S, Arfat MY et al (2014) Physicochemical properties of nanomaterials: implication in associated toxic manifestations. Biomed Res Int. 498420, 8
37. Nemmar A, Yuvaraju P, Beegam S et al (2016) Oxidative stress, inflammation, and DNA damage in multiple organs of mice acutely exposed to amorphous silica nanoparticles. Int J Nanomedicine 11:919–928
38. Sonavane G, Tomoda K, Makino K (2008) Biodistribution of colloidal gold nanoparticles after intravenous administration: effect of particle size. Colloids Surf B: Biointerfaces 66:274–280
39. De Jong WH, Hagens WI, Krystek P et al (2008) Particle size-dependent organ distribution of gold nanoparticles after intravenous administration. Biomaterials 29:1912–1919
40. Braakhuis HM, Park MVDZ, Gosens I et al (2014) Physicochemical characteristics of nanomateri-

als that affect pulmonary inflammation. Part Fibre Toxicol 11:18
41. Ahamed M, Akhtar MJ, Alhadlaq HA, Alrokayan SA (2015) Assessment of the lung toxicity of copper oxide nanoparticles: current status. Nanomedicine(Lond) 10(15):2365–2377
42. Asgharian B, Price OT (2007) Deposition of ultrafine (nano) particles in the human lung. Inhal Toxicol 19:1045–1054
43. Von Garnier C (2013) Nanoparticles in the respiratory tract: modulation of antigen-presenting cell function. J Enviromental Immunol Toxicol 1:140
44. Kettler K, Veltman K, van de Meent D et al (2014) Cellular uptake of nanoparticles as determined by particle properties, experimental conditions, and cell type. Environ Toxicol Chem 33:481–492
45. Baharifar H, Amani A (2016) Cytotoxicity of chitosan/streptokinase nanoparticles as a function of size: An artificial neural networks study. Nanomedicine 12(1):171–180
46. Recordati C, De Maglie M, Bianchessi S et al (2016) Tissue distribution and acute toxicity of silver after single intravenous administration in mice: nano-specific and size-dependent effects. Part Fibre Toxicol 13:12
47. Chen Z, Meng H, Xing G et al (2006) Acute toxicological effects of copper nanoparticles in vivo. Toxicol Lett 163:109–120
48. Seiffert J, Hussain F, Wiegman C et al (2015) Pulmonary toxicity of instilled silver nanoparticles: Influence of size, coating and rat strain. PLoS One 10:1–17
49. Li Y, Monteiro-Riviere NA (2016) Mechanisms of cell uptake, inflammatory potential and protein corona effects with gold nanoparticles. Nanomedicine 11:3185–3203
50. Allegri M, Bianchi MG, Chiu M et al (2016) Shape-related toxicity of titanium dioxide nanofibres. PLoS One 11:1–21
51. Yokel RA (2016) Physicochemical properties of engineered nanomaterials that influence their nervous system distribution and effects. Nanomedicine Nanotechnol Biol Med 12:2081–2093
52. Verma A, Stellacci F (2010) Effect of surface properties on nanoparticle-cell interactions. Small 6:12–21
53. Oh N, Park JH (2014) Endocytosis and exocytosis of nanoparticles in mammalian cells. Int J Nanomedicine 9:51–63
54. Nangia S, Sureshkumar R (2012) Efects of nanoparticle charge and shape anisotropy on traslocation through cell membranes. Langmuir 28:1766
55. Zhang X-F, Liu Z-G, Shen W, Gurunathan S (2016) Silver nanoparticles: synthesis, characterization properties, applications, and therapeutic approaches. Int J Mol Sci 17:1534
56. Hsiao I-L, Huang Y-J (2011) Effects of various physicochemical characteristics on the toxicities of ZnO and TiO_2 nanoparticles toward human lung epithelial cells. Sci Total Environ 409:1219–1228
57. Tarantola M, Pietuch A, Schneider D et al (2011) Toxicity of gold-nanoparticles: Synergistic effects of shape and surface functionalization on micromotility of epithelial cells. Nanotoxicology 5:254–268
58. Lee M-K, Lim S-J, Kim C-K (2007) Preparation, characterization and in vitro cytotoxicity of paclitaxel-loaded sterically stabilized solid lipid nanoparticles. Biomaterials 28:2137–2146
59. Chellappa M, Anjaneyulu U, Manivasagam G, Vijayalakshmi U (2015) Preparation and evaluation of the cytotoxic nature of TiO2 nanoparticles by direct contact method. Int J Nanomedicine 10:31–41
60. Lippmann M (1990) Effects of fiber characteristics on lung deposition, retention, and disease. Environ Health Perspect 88:311–317
61. Yamamoto A, Honma R, Sumita M, Hanawa T (2004) Cytotoxicity evaluation of ceramic particles of different sizes and shapes. J Biomed Mater Res A 68:244–256
62. Goodman CM, McCusker CD, Yilmaz T, Rotello VM (2004) Toxicity of gold nanoparticles functionalized with cationic and anionic side chains. Bioconjug Chem 15:897–900
63. Huang H, Lai W, Cui M et al (2016) An evaluation of blood compatibility of silver nanoparticles. Sci Rep 6:25518
64. King Heiden TC, Dengler E, Kao WJ et al (2007) Developmental toxicity of low generation PAMAM dendrimers in zebrafish. Toxicol Appl Pharmacol 225:70–79
65. Platel A, Carpentier R, Becart E et al (2016) Influence of the surface charge of PLGA nanoparticles on their in vitro genotoxicity, cytotoxicity, ROS production and endocytosis. J Appl Toxicol 36:434–444
66. Wang JY, Chen J, Yang J et al (2016) Effects of surface charges of gold nanoclusters on long-term in vivo biodistribution, toxicity, and cancer radiation therapy. Int J Nanomedicine 11:3475–3485
67. Havrdova M, Hola K, Skopalik J et al (2016) Toxicity of carbon dots-effect of surface functionalization on the cell viability, reactive oxygen species generation and cell cycle. Carbon NY 99:238–248
68. Kohli AK, Alpar HO (2004) Potential use of nanoparticles for transcutaneous vaccine delivery: Effect of particle size and charge. Int J Pharm 275:13–17
69. Bartczak D, Baradez M-O, Merson S et al (2013) Surface ligand dependent toxicity of zinc oxide nanoparticles in HepG2 cell model. J Phys Conf Ser 429:12015
70. Yin H, Too HP, Chow GM (2005) The effects of particle size and surface coating on the cytotoxicity of nickel ferrite. Biomaterials 26:5818–5826
71. Zhou Z, Son J, Harper B et al (2015) Influence of surface chemical properties on the toxicity of engineered zinc oxide nanoparticles to embryonic zebrafish. Beilstein J Nanotechnol 6:1568–1579
72. Bastos V, Ferreira de Oliveira JMP, Brown D et al (2016) The influence of Citrate or PEG coating on

silver nanoparticle toxicity to a human keratinocyte cell line. Toxicol Lett 249:29–41
73. Fubini B, Hubbard A (2003) Reactive oxygen species (ROS) and reactive nitrogen species (RNS) generation by silica in inflammation and fibrosis. Free Radic Biol Med 34:1507–1516
74. Chen H, Wang T, Li K et al (2017) Effects of surface modification of quantum dots on viability and migration of triple-negative breast cancer cells. J Colloid Interface Sci 485:51–58
75. Hanot CC, Choi YS, Anani TB et al (2016) Effects of iron-oxide nanoparticle surface chemistry on uptake kinetics and cytotoxicity in CHO-K1 cells. Int J Mol Sci 17(1):54
76. Malvindi MA, De Matteis V, Galeone A et al (2014) Toxicity assessment of silica coated iron oxide nanoparticles and biocompatibility improvement by surface engineering. PLoS One 9:1–11
77. Connor EE, Mwamuka J, Gole A et al (2005) Gold nanoparticles are taken up by human cells but do not cause acute cytotoxicity. Small 1:325–327
78. Uboldi C, Urbán P, Gilliland D et al (2016) Role of the crystalline form of titanium dioxide nanoparticles: rutile, and not anatase, induces toxic effects in Balb/3T3 mouse fibroblasts. Toxicol in Vitro 31:137–145
79. De Matteis V, Cascione M, Brunetti V et al (2016) Toxicity assessment of anatase and rutile titanium dioxide nanoparticles: the role of degradation in different pH conditions and light exposure. Toxicol In Vitro 37:201–210
80. Zhang H, Gilbert B, Huang F, Banfield JF (2003) Water driven structure transformation in nanoparticles at room temperature. Nature 424:1025–1029
81. Johnston CJ, Driscoll KE, Finkelstein JN et al (2000) Pulmonary chemokine and mutagenic responses in rats after subchronic inhalation of amorphous and crystalline silica. Toxicol Sci 56:405–413
82. Prasad RY, Wallace K, Daniel KM et al (2013) Effect of treatment media on the agglomeration of titanium dioxide nanoparticles: impact on genotoxicity, cellular interaction, and cell cycle. ACS Nano 7:1929–1942
83. Jiang J, Oberdörster G, Biswas P (2009) Characterization of size, surface charge, and agglomeration state of nanoparticle dispersions for toxicological studies. J Nanopart Res 11:77–89
84. Song Y, Li X, Du X (2009) Exposure to nanoparticles is related to pleural effusion, pulmonary fibrosis and granuloma. Eur Respir J 34:559–567
85. Whitsett JA, Alenghat T (2015) Respiratory epithelial cells orchestrate pulmonary innate immunity. Nat Immunol 16:27–35
86. Yang W, Peters JI, Williams RO (2008) Inhaled nanoparticles—a current review. Int J Pharm 356:239–247
87. Darquenne C, Paiva M, Prisk GK (2000) Effect of gravity on aerosol dispersion and deposition in the human lung after periods of breath holding. J Appl Physiol 89:1787–1792
88. Tsuda A, Henry FS, Butler JP (2013) Particle transport and deposition: basic physics of particle kinetics. Compr Physiol 3:1437–1471
89. Geiser M, Jeannet N, Fierz M, Burtscher H (2017) Evaluating adverse effects of inhaled nanoparticles by realistic in vitro technology. Nanomaterials 7:49
90. Kumar A, Chen F, Mozhi A et al (2013) Innovative pharmaceutical development based on unique properties of nanoscale delivery formulation. Nanoscale 5:8307–8325
91. Bergin IL, Witzmann FA (2013) Nanoparticle toxicity by the gastrointestinal route: evidence and knowledge gaps. Int J Biomed Nanosci Nanotechnol 3(1–3). doi:https://doi.org/10.1504/IJBNN.2013.054515
92. JAni P, Halbert GW, Langridge J, Florence AT (1990) Nanoparticle uptake by the rat gastrointestinal mucosa: quantitation and particle size dependency. J Pharm Pharmacol 42:821–826
93. Fröhlich EE, Fröhlich E (2016) Cytotoxicity of nanoparticles contained in food on intestinal cells and the gut microbiota. Int J Mol Sci 17(4):509
94. Heringa MB, Geraets L, Van Eijkeren JCH et al (2016) Risk assessment of titanium dioxide nanoparticles via oral exposure, including toxicokinetic considerations. Nanotoxicology 10:1515–1525
95. Randall Wickett R, Visscher MO (2006) Structure and function of the epidermal barrier. Am J Infect Control 34:s98–s110
96. Robertson TA, Sanchez WY, Roberts MS (2010) Are commercially available nanoparticles safe when applied to the skin? J Biomed Nanotechnol 6:452–468
97. Lademann J, Weigmann HJ, Rickmeyer C et al (1999) Penetration of titanium dioxide microparticles in a sunscreen formulation into the horny layer and the follicular orifice. Skin Pharmacol Appl Skin Physiol 12:247–256
98. Tan MH, Commens CA, Burnett L, Snitch PJ (1996) A pilot study on the percutaneous absorption of microfine titanium dioxide from sunscreens. Australas J Dermatol 37:185–187
99. Crosera M, Prodi A, Mauro M et al (2015) Titanium dioxide nanoparticle penetration into the skin and effects on HaCaT cells. Int J Environ Res Public Health 12:9282–9297
100. Larese F, Mauro M, Adami G, Bovenzi MCM (2015) Nanoparticles skin absorption: new aspects for a safety profile evaluation. Regul Toxicol Pharmacol 72:310–322
101. Tak YK, Pal S, Naoghare PK, Rangasamy S (2015) Shape-dependent skin penetration of silver nanoparticles: does it really matter? Sci Rep 5:16908
102. Balogh L, Nigavekar SS, Nair BM et al (2007) Significant effect of size on the in vivo biodistribution of gold composite nanodevices in mouse

tumor models. Nanomedicine Nanotechnol Biol Med 3:281–296
103. Yang Y, Qin Z, Zeng W et al (2017) Toxicity assessment of nanoparticles in various systems and organs. Nanotechnol Rev 6(3):279–289
104. Longmire M, Choyke PL, Kobayashi H (2008) Clearance properties of nano-sized particles and molecules as imaging agents: considerations and caveats. Nanomedicine (Lond) 3(5):703–717
105. Blanco E, Shen H, Ferrari M (2015) Principles of nanoparticle design for overcoming biological barriers to drug delivery. Nat Biotechnol 33(9):941–951
106. Doudi M, Setorki M (2014) The acute liver injury in rat caused by gold nanoparticles. Nanomedicine J 1:248–257
107. Gaiser BK, Hirn S, Kermanizadeh A et al (2012) Effects of silver nanoparticles on the liver and hepatocytes *in vitro*. Toxicol Sci 131:537–547
108. Tang J, Xiong L, Zhou G et al (2010) Silver nanoparticles crossing through and distribution in the blood-brain barrier In vitro. J Nanosci Nanotechnol 10(10):6313–6317
109. Choi HS, Liu W, Misra P et al (2007) Renal clearance of nanoparticles. Nat Biotechnol 25:1165–1170
110. Lei R, Wu C, Yang B et al (2008) Integrated metabolomic analysis of the nano-sized copper particle-induced hepatotoxicity and nephrotoxicity in rats: a rapid in vivo screening method for nanotoxicity. Toxicol Appl Pharmacol 232:292–301
111. Weissleder R, Nahrendorf M, Pittet MJ (2014) Imaging macrophages with nanoparticles. Nat Mater 13:125–138
112. Wu J, Liu W, Xue C et al (2009) Toxicity and penetration of TiO_2 nanoparticles in hairless mice and porcine skin after subchronic dermal exposure. Toxicol Lett 191:1–8
113. Zhao Y, Xing G, Chai Z (2008) Nanotoxicology: are carbon nanotubes safe? Nat Nanotechnol 3:191–192
114. Cooper GM (2000) The cell: molecular approach. ASM Press, Washington, DC
115. Kafshgari MH, Harding FJ, Voelcker NH (2015) Insights into cellular uptake of nanoparticles. Curr Drug Deliv 12(1):63–77
116. Zhao F, Zhao Y, Liu Y et al (2011) Cellular uptake, intracellular trafficking, and cytotoxicity of nanomaterials. Small 7:1322–1337
117. Shang L, Nienhaus K, Nienhaus GU (2014) Engineered nanoparticles interacting with cells: size matters. J Nanobiotechnol 12:5
118. Johannes L, Lamaze C (2002) Clathrin-dependent or not: is it still the question? Traffic 3(7):443–451
119. Aderem A, Underhill DM (1999) Mechanisms of phagocytosis in macrophages. Annu Rev Immunol 17:593–562
120. Conner SD, Schmid SL (2003) Regulated portals of entry into the cell. Nature 422(6927):37–44
121. Pelkmans L, Helenius A (2002) Endocytosis via caveolae. Traffic 3(5):311–320
122. Rejman J, Oberle V, Zuhorn IS et al (2004) Size-dependent internalization of particles via the pathways of clathrinand caveolae-mediated endocytosis. Biochem J 1:159–169
123. Saikia J, Yazdimamaghani M, Pouya S et al (2016) Differential protein adsorption and cellular uptake of silica nanoparticles based on size and porosity. ACS Appl Mater Interfaces 8(50):34820–34832
124. Pritz CO, Bitsche M, Salvenmoser W et al (2013) Endocytic trafficking of silica nanoparticles in a cell line derived from the organ of Corti. Nanomedicine 8:239–252
125. Greulicha C, Diendorfb J, Simonc T et al (2011) Uptake and intracellular distribution of silver nanoparticles in human mesenchymal stem cells. Acta Biomater 7:347–354
126. Thurn KT, Arora H, Paunesku T et al (2011) Endocytosis of titanium dioxide nanoparticles in prostate cancer PC-3M cells. Nanomedicine Nanotechnol Biol Med 7:123–130
127. Mizuhara T, Saha K, Moyano DF et al (2015) Acylsulfonamide-Functionalized zwitterionic gold nanoparticles for enhanced cellular uptake at tumor pH. Angew Chemie-Int Ed 54:6567–6570
128. Wang Z, Xia T, Liu S (2015) Mechanisms of nanosilver-induced toxicological effects: more attention should be paid to its sublethal effects. R Soc Chem 7:7470–7481
129. Verissimo TV, Santos NT, Silva JR et al (2016) In vitro cytotoxicity and phototoxicity of surface-modified gold nanoparticles associated with neutral red as a potential drug delivery system in phototherapy. Mater Sci Eng C 65:199–204
130. Hashemi E, Akhavan O, Shamsara M et al (2016) Synthesis and cyto-genotoxicity evaluation of graphene on mice spermatogonial stem cells. Colloids Surf B: Biointerfaces 146:770–776
131. Mosmann T (1983) Rapid colorimetric assay for cellular growth and survival: application to proliferation and cytotoxicity assays. J Immunol Methods 65:55–63
132. Olbrich C, Bakowsky U, Lehr CM et al (2001) Cationic solid-lipid nanoparticles can efficiently bind and transfect plasmid DNA. J Control Release 77:345–355
133. Wong Shi Kam N, Dai H (2005) Carbon nanotubes as intracellular protein transporters: generality and biological functionality. J Am Chem Soc 127:6021–6026
134. Shenoy D, Fu W, Li J (2005) Surface functionalization of gold nanoparticles using hetero-bifunctional poly(ethylene glycol) spacer for intracellular tracking and delivery. Int J Nanomedicine 1(1):51–57
135. Wang H, Joseph JA (1999) Quantifying cellular oxidative stress by dichlorofluorescein assay using microplate reade. Free Radic Biol Med 27:612–616
136. Ostling JK (1984) Microelectrophoretic study of radiation-induced DNA damages in individual

mammalian cells. Biochem Biophys Res Commun 30:291–298
137. Singh NP, McCoy MT, Tice RR et al (1988) A simple technique for quantitation of low levels of DNA damage in individual cells. Exp Cell Res 175:184–191
138. Castiglioni S, Caspani C, Cazzaniga A et al (2014) Short- and long-term effects of silver nanoparticles on human microvascular endothelial cells. World J Biol Chem 26:457–464
139. Boverhof DR, Bramante CM, Butala JH et al (2015) Comparative assessment of nanomaterial definitions and safety evaluation considerations. Regul Toxicol Pharmacol 73:137–150
140. Musazzi UM, Marini V, Casiraghi A et al (2017) Is the European regulatory framework sufficient to assure the safety of citizens using health products containing nanomaterials? Drug Discov Today 22(6):870–882
141. Commission of the European Communities (2005) Communication from the Commission to the Council, the European Parliament and the Economic and Social Committee. Nanosciences and nanotechnologies: an action plan for Europe 2005–2009
142. European Parliament (2006) Resolution on Nanosciences and nanotechnologies: an action plan for Europe 2005–2009
143. Commission of the European Communities (2008) Communication from the Commission to the European Parliament, the Council and the European Economic and Social Committee. Regulatory aspects of nanomaterial
144. Regulation (EC) No 1223/2009 of the European Parliament and of the Council of 30 November 2009 on cosmetic products. Off. J. EU L.342, 59–209. J. EU L.342, 59–209
145. Commission E (2011) Commission Recommendation of 18 October 2011 on the definition of nanomaterial (2011/696/EU). EU, ff. J. L. 275, 38–40.
146. Council of the European Union (2016 Proposal for a regulation of the European parliament and of the council on medical devices, and amending directive 2011/83/EC, Regulation (EC) No 178/2002 and Regulation (EC) No 1223/2009. Council of the European Union
147. SCENIHR (2010) Scientific basis for the definition of the term "Nanomaterial"
148. Lövestam G et al. (2010) JRC reference report: considerations on a definition of nanomaterial for regulatory purposes (EUR 24403 EN), European Union
149. Domingos RF, Baalousha MA, Ju-Nam Y et al (2009) Characterizing manufactured nanoparticles in the environment: multimethod determination of particle sizes. Environ Sci Technol 43:7277–7284
150. Roebben G et al. (2014) JRC science and policy report: towards a review of the EC recommendation for a definition of the term "Nanomaterial". Part 2: assessment of collected information concerning the experience with the definition. Eur. Union
151. (2011) https://www.nano.gov/sites/default/files/pub_resource/nni_2011_ehs_research_strategy.pdf
152. Raies AB, Bajic VB (2016) In silico toxicology: computational methods for the prediction of chemical toxicity. Wiley Interdiscip Rev Comput Mol Sci 6:147–172
153. Ying J, Zhang T, Tang M (2015) Metal oxide nanomaterial QNAR models: available structural descriptors and understanding of toxicity mechanisms. Nanomaterials 5:1620–1637
154. Fourches D, Pu D, Tropsha A (2011) Exploring quantitative nanostructure—activity relationships (QNAR) modeling as a tool for predicting biological effects of manufactured nanoparticles. Comb Chem High Throughput Screen 1;14(3):217–225
155. Damoiseaux R, George S, Li M et al (2011) No time to lose—high throughput screening to assess nanomaterial safety. Nanoscale 3:1345–1360
156. Clark KA, White RH, Silbergeld EK (2011) Predictive models for nanotoxicology: current challenges and future opportunities. Regul Toxicol Pharmacol 59:361–363
157. Todeschini RCV (2000) Handbook of molecular descriptors. Wiley-VCH, Weinheim, pp 927–933
158. Mukherjee D, Royce SG, Sarkar S et al (2014) Modeling in vitro cellular responses to silver nanoparticles. J Toxicol 2014:852890. 13

2 Mechanisms of Uptake and Translocation of Nanomaterials in the Lung

Chloé Puisney, Armelle Baeza-Squiban, and Sonja Boland

Abstract

Nanomaterials are invading our environment due to their increasing use in a very broad range of sectors making human exposure foreseeable during the life cycle of these materials. Inhalation is one of the most frequent routes of exposure in case of unintentional exposure and the small size of nanomaterials allows them to reach the deep lung. Understanding the fate and effects of nanomaterials is a great challenge for scientists as they exhibit a huge physico-chemical diversity that drives their biological reactivity. It is critical to determine the fate of nanomaterials at their route of entry in the organism as this will determine their local and/or systemic effects. In this review we will describe the epithelial barriers and the clearance processes of the respiratory tract. The mechanisms involved in the internalization of nanomaterials by respiratory cells and their ability to cross the epithelial barrier will be presented, highlighting methodologies and the role of the nanomaterial physico-chemical properties.

Keywords

Nanomaterials · Nanotoxicity · Lung toxicity · Nanoparticles uptake

2.1 Introduction

The development of nanotechnologies puts on the market a wide diversity of nanomaterials (NM). NMs can be of very various compositions (metals, metal oxides, carbons, polymers) and for each composition they can exhibit different sizes, shapes, charges, crystallinity, coating, and functionalization that are designed to display unique properties for specific applications. Until now, risk assessment of NMs is far to be stated. Due to

C. Puisney · A. Baeza-Squiban (✉) · S. Boland
CNRS UMR 8251, Unit of Functional and Adaptive Biology (BFA), Laboratory of Molecular and Cellular Responses to Xenobiotics, University Paris Diderot (Sorbonne Paris Cité), Paris, France
e-mail: baeza@univ-paris-diderot.fr

Q. Saquib et al. (eds.), *Cellular and Molecular Toxicology of Nanoparticles*, Advances in Experimental Medicine and Biology 1048, https://doi.org/10.1007/978-3-319-72041-8_2

the recent use of NMs, no epidemiological studies are available to demonstrate an association between NM exposure and health effects, making toxicological studies mandatory to establish hazards related to NM exposure. The diversity of NMs to be evaluated and the difficulties encountered for a proper hazard characterization due to their inherent properties make experimental studies more complicated than for chemicals and have for now limited risk assessment to few NMs.

Throughout their life cycle, NMs can be in powder, liquid suspension, aerosol, or embedded inside or at the surface of a material making the probability of exposure quite different. The frequency of exposure will also be very different between occupational settings where NMs are either produced, included in manufactured goods or released during waste recycling. Consumers, aware or not, may also be exposed to NMs included in cosmetics, food, clothes…. Ingestion, inhalation and skin applications are the main routes of exposure to NMs but the respiratory route is likely the most frequent in case of unintentional exposure. Lungs can be the target organ of NMs and/or just a route of entry of NMs into the organism where they could be further distributed by circulation and reach other organs distant from the lungs where they will elicit adverse effects. Nanotoxicology has get benefit from lessons learned in the toxicology of particles originated from air pollution [1]. Fine and ultrafine (nanosize range) particles are regulated in ambient air as it was clearly demonstrated that they contribute to mortality and morbidity that are not limited to respiratory causes. Their ability to induce inflammation as well as to cross the air-blood barrier is supposed to be involved in their systemic effects. In nanotoxicology it is thus crucial to characterize the ability of NMs to enter and cross the epithelial barrier by identifying the physico-chemical properties of NMs favouring or preventing this process as well as by determining the cellular and molecular mechanisms involved.

This chapter will first describe the anatomic and tissular specificities of the respiratory tract as well as the defense mechanisms involved in particle clearance. We will then present the state of the art on NM internalization, translocation and biodistribution based on in vitro and in vivo studies.

2.2 The Respiratory Tract: Target Organ of Inhaled Nanomaterials

Respiration is an autonomic function and even being aware of the presence of a pollutant in the air we cannot stay without breathing for a long time. The only way to escape from exposure is to be able to get away from the polluted area. Air being frequently contaminated especially in specific environments, inhalation is a frequent route of exposure to pollutants among which NMs are emergent particulate air pollutants.

2.2.1 Structure of The Respiratory Tract

The respiratory tract is a pipe system with airways allowing air conduction from the nose and the mouth to the lung and pulmonary alveoli (Fig. 2.1a). Alveoli are tiny air sacs at the ending of the smallest airways and constitute the respiratory zone where gas exchanges between air and blood take place. There are many alveoli, 300 millions in adult humans, representing a large exchange surface (around 140 m^2) and they are covered by very thin capillaries. This unit constitutes the air-blood barrier characterized by a very short distance (0.3–0.5 μm) between air and blood allowing oxygen entry and CO_2 release.

Conducting airways from trachea to bronchioli are delimited by a mucociliary epithelium (Fig. 2.1b). It is a pseudostratified epithelium composed of ciliated cells bearing 200–300 cilia at their apical surface, goblet cells secreting mucus and basal cells involved in epithelium regeneration. The integrity of the epithelial barrier is provided by different intercellular junctions: tight junctions at the apical zone, adherens junctions and desmosomes at the lateral membranes. The epithelium is separated from air by a

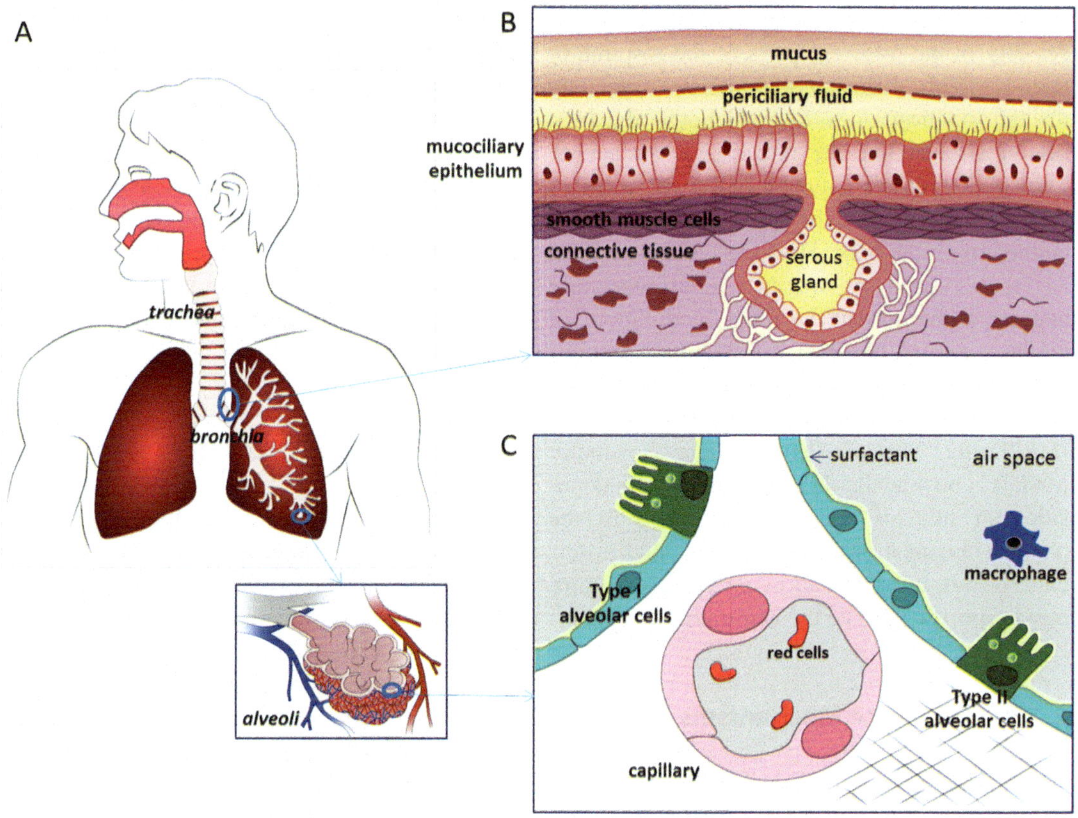

Fig. 2.1 The respiratory tract. (**a**) Scheme of the respiratory tract. (**b**) Details of the mucociliary epithelium lining the airways. The mucus layer produced by the goblet cells entraps particles and goes up towards the pharynx due to ciliary beating. (**c**) Details of the alveolar epithelium at the air-blood barrier. The alveolar epithelium is composed of Type II alveolar cells involved in surfactant synthesis and Type I alveolar cells that cover 90–95% of the alveolar surface. The air – blood distance is around 0.2–0.3 μm. Macrophages, that are free in the alveolar lumen, are involved in particle phagocytosis

mucus layer that is continually swept from the distal to the proximal airways by the synchronous beating of cilia and finally expelled out of the lungs by expectoration or swallowed, removing entrapped particles, microorganisms and dissolved chemicals. The mucus layer is a mobile, essential barrier that protects the lungs when ciliary activity and mucus properties are maintained. At the level of the bronchiole, an additional cell types appears: club cells (previously called Clara cells) involved in xenobiotic metabolism.

Alveoli (Fig. 2.1c) are delimited by the alveolar epithelium lying on a basal membrane. Type I and type II alveolar cells that are joined by tight junctions, form this epithelium. Type I alveolar cells are large (about 400 μm^2/cell) and flat (0.1/0.2 μm thickness) cells covering 90–95% of the surface of the alveolus. There is an equivalent number of type II alveolar cells, that fill the 10–5% remaining surface and which are specialized in the production of the surfactant, a tension-active substance covering the alveolar epithelium.

2.2.2 Particle Deposition

According to their physico-chemical characteristics, airborne particles may enter different regions of the respiratory tract. Three main mechanisms can occur in particle deposition: impaction, sedimentation, and diffusion which depend on particle size, directional changes of air flow, and air velocity in the different parts of the respiratory tract. For NMs, diffusion predominates especially for the smallest ones for which the dynamic behavior looks like the one of gas molecules. This is the reason why an additional route of entry is suspected at the level of the olfactory epithelium in the nose. However, NMs can form aggregates and agglomerates modifying their deposition throughout the lung. Modulations of air flow due to physical efforts or respiratory diseases can profoundly change the levels and amount of deposited NMs. A computational model (MPPD for Multiple-Path Particle Dosimetry Model) is now available to predict particle deposition in humans and rats according to the physico-chemical properties of NMs and different exposure scenarios, allowing a precise dosimetry in specific parts of the respiratory tract [2].

2.2.3 Clearance Mechanisms

The clearance of insoluble particles occurs through two distinct processes: the mucociliary clearance and the alveolar clearance.

The mucociliary clearance results from the joined activity of goblet and ciliated cells allowing the removal of particles trapped in the mucus layer. Within 24 h, particles are eliminated from the human airways by this efficient mechanism.

In alveoli, macrophages remove particles by phagocytosis and migrate towards the mucociliary escalator to be eliminated, this whole process lasting several days. However, macrophages are less efficient in the phagocytosis of NMs compared to microparticles, making NMs dwelling longer in the alveolar space increasing their probability to interact and enter the alveolar epithelium and/or to cross the air-blood barrier.

2.3 Nanomaterial Internalisation

Excepting pathological situations or pollutant exposure that can affect epithelium integrity allowing paracellular passage, NM translocation implies a transcellular passage. Internalization of NMs will be the first step for this NM translocation and systemic distribution, especially for insoluble NMs. This internalization will also lead to their accumulation in the epithelium where they could induce adverse effects. A better knowledge of the determinants and molecular mechanisms involved in NM uptake may provide clues to create safer NMs. Determining the subcellular localization of NMs is also particularly important to better understand potential mechanisms of toxicity.

2.3.1 Mechanism of Internalization

The plasma membrane delineating cells is a lipid bilayer including proteins that regulate and coordinate the entry and exit of molecules to maintain the specificity of the intracellular milieu relatively to the extracellular environment. Nutrients and other physiological relevant molecules as well as pathogens or pollutants are captured from the extracellular medium by endocytosis leading to their internalization into vesicles originated from the plasma membrane. Endocytosis can be divided into phagocytosis which concerns the uptake of large particles, molecular complexes or bacteria, and pinocytosis dedicated to the capture of fluids and solutes.

Phagocytosis is classically used by the so called "professional phagocytes" (i.e. macrophages, monocytes, polynuclear neutrophiles) to protect the organism against invading pathogens or to take up cell debris and large particles. Their binding to specific receptors at the cell surface initiates the assembly of actin filaments leading to the engulfment of the particle like a zip. This will lead to the formation of large intracellular vesicles of 0.5–10 μm called phagosomes.

Concerning the process of pinocytosis, which is used by all cell types for the fluid phase uptake, different mechanisms are distinguished depending on the molecular mechanism involved: macropinocytosis, clathrin-mediated endocytosis, caveolae-mediated endocytosis or clathrin and caveolae independent mechanisms. Macropinocytosis involves the formation of cell protrusions through the assembly of actin filaments but, in contrast to phagocytosis, the plasma membrane is not in close contact with the particle. This allows the uptake of big volumes of extracellular fluids into so called macropinosomes of 0.5–5 μm. In the contrary, the clathrin-mediated endocytosis is used for the specific uptake of molecules via receptor-ligand interactions. This process takes place at specific plasma membrane microdomains containing receptors linked to a submembrane network of proteins called clathrins. The formation of clathrin coated pits allows the invagination of the plasma membrane and finally the formation of intracellular vesicles of 100–150 nm. Caveolae-mediated endocytosis also involves specific receptors present in specialized lipid microdomains. These rafts are enriched in glycosphingolipids and cholesterol and contain the protein caveolin that allows the formation of small vesicles or tubules of about 80 nm in diameter. Beside these three main pinocytosis processes exist some other receptor mediated endocytosis mechanisms, involving proteins such as flotillin, IL2Rbeta or Arf-6 that allow the detachment of vesicles from the plasma membrane [3]. All these internalization mechanisms are active processes leading to the formation of intracellular vesicles. Some molecules could however also enter the cells via other processes such as through specific channels or passive diffusion through the membrane allowing their direct interaction with the cytoplasm.

2.3.2 Endocytic Pathways Used by Nanomaterials

Several studies have investigated the mechanisms involved in the uptake of NMs and the role of their physico-chemical characteristics by using different techniques and experimental approaches. Qualitative approaches, used to visualize the interactions of NMs with cells and their intracellular localization, are often combined to quantitative approaches to enable the relative or absolute quantification of NM uptake. Electron microscopy is particularly interesting as their high optical resolution is compatible with the study of nanosized objects. In case of non-carbonaceous NMs, energy dispersive X-ray analysis could even be used to distinguish the NM from cellular components. Scanning electron microscopy (SEM) and atomic force microscopy (AFM) are other high resolution techniques commonly used to study the interactions of NMs with membranes [4]. However, as these high resolution techniques are very time-consuming, confocal fluorescence microcopy is often used to study the uptake mechanisms of NMs. For quantitative analysis, fluorescence plate readers or flow cytometry have often been used but these techniques did not distinguish between surface bound and actually internalized NMs. Fluorescence quenching molecules which did not enter viable cells such as Trypan blue have been used to selectively quench the fluorescence of NMs which stick to the cell surface and thus to only quantify the intracellular signal [5]. Imaging flow cytometry also allows eliminating the signal of the adsorbed NM by using image analyzing tools for high-throughput analysis of the images of each individual cell analyzed by the flow cytometer [6]. Fluorescence based techniques have however the disadvantage to express results in arbitrary units and could thus only quantify the relative uptake of NMs. This pitfall could be overcome by using confocal Raman microscopy, laser positioning secondary neutral mass spectrometry (laser SNMS), time-of-flight secondary ion mass spectrometry (TOF-SIMS), laser ablation inductively coupled plasma mass spectrometry (ICP-MS), or synchrotron beamlines which allow the elemental analysis of the sample in combination with high-resolution 3D imaging [7, 8]. These techniques can thus not only be used for an absolute quantitative analysis of NM uptake but also for their surface or even subcellular distribution.

There are however several nano-specific challenges which have to be taken into account for a reliable interpretation of the results such as interferences due to optical properties of NMs, necessity of high resolution to distinguish single NMs, changes in their aggregation state due to experimental conditions, dissolution of NMs, or adsorption of biomolecules. In case of labeled NMs it has to be checked that the detected signal is not due to leakage or dissolution of the fluorochrome and that the labeling has not changed the surface properties of the NM. Core-shell labeling techniques have thus been used to insure that the determined uptake mechanisms are representative of the unlabeled NM.

To determine the specific uptake mechanism involved, pharmaceutical inhibitors of the different endocytic pathways have been used. However, these are often highly toxic and their specificity and efficiency have been questioned [9, 10]. Co-localization with specific proteins such as caveolin or clathrin or the use of siRNA of these proteins are also useful tools to identify the endocytosis pathway. Combining inhibition of specific pathways with co-localization studies is crucial as different uptake mechanisms may be used to substitute the inhibited endocytic pathway. Using such a combined approach Vranic et al. [6] have shown that SiO_2 NPs of 100 nm are taken up by bronchial epithelial cells through macropinocytosis. This endocytic pathway seems indeed to be the most predominant uptake mechanism in epithelial cells. On the other hand, the uptake by alveolar macrophages could also involve phagocytosis of the material via receptor interactions. Some studies have also demonstrated the involvement of caveolin or clathrin dependent endocytosis [11–13] or flotillin dependent endocytosis [14] for NM uptake. Some NMs may even be able to enter cells by non active mechanisms as NMs have been found in red blood cells or could cross artificial membranes [4, 15, 16]. Studies using metabolic inhibitors have however shown that the uptake of NMs is most of the time due to active transport through the plasma membrane.

The uptake mechanism is particularly important for the fate of NMs inside the cell. Indeed, macropinosomes or clathrin coated vesicles will fuse with lysosomes which are acidic and contain hydrolases and other enzymes to degrade molecules or invading microorganisms. In contrast, caveolae can either fuse with the acidic lysosomes or with caveosomes which are endosomes with neutral pH which could traffic to the Golgi apparatus or the endoplasmic reticulum. The dissolution of NMs is fostered under acidic conditions and the uptake via caveolae may thus protect NMs from degradation. Some NMs have been shown to destabilize lysosomes such as TiO_2 NPs in pulmonary cells [17] and are thus frequently found free in the cytoplasm. Some NMs have even been observed within the nucleus, for instance silica NPs in macrophages [18].

2.3.3 Physico-chemical Characteristics Influencing NM Uptake

The uptake mechanism will depend upon the physico-chemical properties of the NM. The size of the individual particle or the agglomerates/aggregates is one of the main features influencing the endocytic pathway (Fig. 2.2). Indeed, nanosized objects may be taken up by clathrin dependent endocytosis as this pathway leads to the formation of vesicles of 100–150 nm or could enter caveolae which are less than 80 nm in diameter. Big particles (or agglomerates) will not be able to enter these small vesicles and will only be able to enter the cell by phagocytosis or macropinocytosis. Indeed, phagosomes and macropinosomes measure between 0.5 and 10 μm enabling the uptake of micron-sized structures. It is thus of crucial importance to ensure that the experimental conditions will not artificially modify the agglomerate size. Several pathways may be used simultaneously, especially if NMs or agglomerates of different sizes will coexist. The preferential particle size for uptake will depend upon the cell type. Professional phagocytes such as macrophages are more efficient to take up micrometer particles but non phagocytic cells internalize better NPs in comparison to macrophages [19]. The size of the NM will thus be a very important feature for the efficiency of the alveolar clearance but the size does not seem to play a major role in mucociliary clearance [20].

The shape of the NM will also influence their uptake (Fig. 2.2) such as curvature [21], roughness [4], and the aspect ratio of the particle (relationship between width and height). For instance, rod shaped silica NM with an aspect ratio of 2.1–2.5 were taken up by alveolar epithelial cells

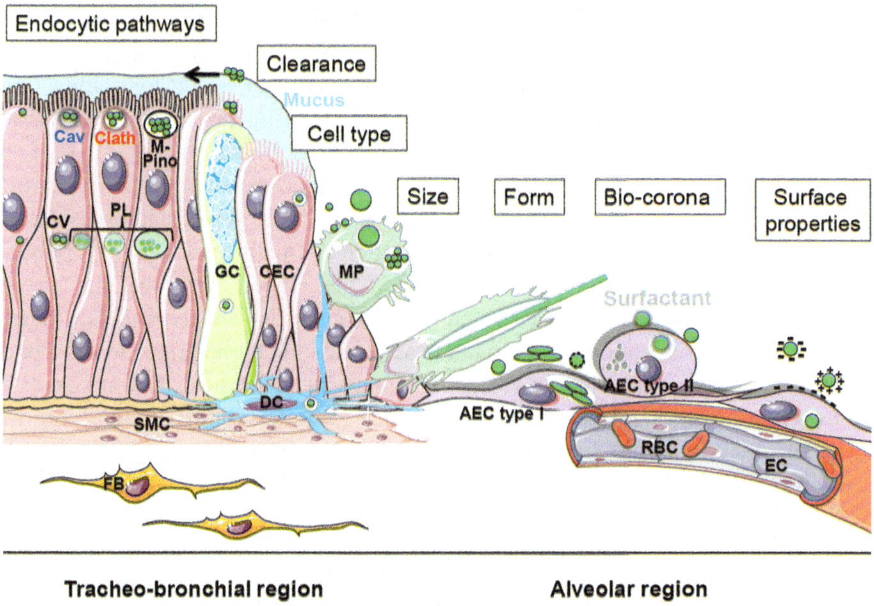

Fig. 2.2 Parameters influencing nanomaterial uptake by pulmonary cells. The fate of NMs depends upon cellular determinants such as endocytic pathways (leading to acidic phagolysosomes or neutral caveosomes), the efficiency of clearance mechanisms (alveolar or mucociliary clearance), and the cell type. The NM determinants of size (of individual NMs or their agglomerates), form (aspect ratio, curvature or roughness), interaction with biomolecules (lipids or proteins), and surface properties (such as charge and hydropobicity) will also influence the uptake of NMs. *AEC* alveolar epithelial cell, *Cav* caveolae, *CEC* ciliated epithelial cell, *Clath* chlathrin coated pit, *CV* Caveosome, *DC* dendritic cell, *EC* endothelial cell, *FB* fibroblast, *GC* goblet cell, *M -Pino*, Macropinosome, *MP* macrophage, *PL* phagolysosome, *RBC* red blood cell, *SMC* smoth muscle cell

in larger quantities compared to shorter or longer length rods [22]. Furthermore, alveolar macrophages could be unable to completely engulf high aspect ratio carbon nanotubes (termed frustrated phagocytosis) leading to an inflammatory reaction and pleural pathology including mesothelioma [23] just like the toxicity mechanism of inhaled asbestos fibers. Once again, the influence of the form on internalization has to be distinguished from its role in the adsorption of the NM to the cell surface. It has for instance been shown that prickly nanodiamonds adhere better to the cell surface but are less internalized than round nanodiamonds [24]. Beside these physical factors of size and shape, the surface properties of the NM will also be important factors for their internalization. Especially their hydrophobicity and surface charge will determine the interaction of NMs with cells. Indeed, the cell surface is negatively charged favoring the interaction with positively charged NMs. Imaging flow cytometry has shown that a negative surface charge reduces the adsorption of TiO_2 NPs to bronchial epithelial cells and their subsequent internalization [6]. Even though most studies are in favor of uptake facilitation by positive surface charges, some contradictory results have been reported which may be explained by concomitant changes in the hydrophobicity of the synthesized particles or preferential interactions with specific proteins [19]. The surface charge or hydrophobicity could also influence the mucus penetration and thus the mucociliary clearance as shown for negative hydrophilic PLGA NPs which diffused across the mucus layer in contrast to positive or neutral hydrophobic NPs [25].

The uptake of NMs will also depend upon the interaction with biomolecules they will encounter before reaching the cells (Fig. 2.2). Lipids or proteins of the lung lining fluids will indeed interact with the NM to form a corona around the particle. The physico-chemical properties of the resulting hybrid bio-nanoparticle will be different from the bare NM. Their so called "biological identity" will thus be determining for the cellular interaction. The biocorona will depend upon the NM characteristics as well as the physico-chemical properties of the proteins and the relative abundance of the proteins in the biological fluids. The composition of the corona can thus be modified when the NM moves from one environment to another. Many studies have investigated the consequences of NM incubation in serum but few studies have focused on the effect of surfactant or mucus. Contradictory results have been observed as surfactant inhibited the uptake of silica NPs in bronchial epithelial cells and macrophages [5]. Besides, incubation of polystyrene NPs with surfactant leads to the adsorption of surfactant protein A and D and enhanced NP uptake in alveolar epithelial cells [12]. Surfactant proteins have also been shown to facilitate the uptake of magnetite [26], polystyrene [27], diesel exhaust [28], and carbon black [28] NPs in macrophages but inhibited the internalization of positively charged (amino-modified) polystyrene NPs [27]. Due to the role of protein corona in NM uptake, the use of serum as a dispersant agent or culture media additive might be a problem for toxicity testing as this may not reflect the actual exposure scenario, especially for inhalation. It has been shown that dispersion of TiO_2 NPs in serum will increase the uptake by bronchial epithelial cells while reducing their in vitro and in vivo toxicity due to a concomitant reduction of their surface reactivity by the protein coating [29].

2.4 Nanomaterial Translocation

Adverse systemic effects induced by inhalation of NPs are suggested to result from lung inflammation but also from the passage of particles through the respiratory barrier [30]. The mechanism responsible for this crossing is called translocation. At the level of alveoli it enables NMs to cross the air-blood barrier to reach systemic circulation and secondary organs. At the level of bronchioles where the distance to capillaries is larger, they may accumulate in the interstitium where they can reside for a long time as it has been shown in the lung parenchyma of women living in polluted cities [31]. Full evidence of NM translocation and further biodistribution was provided by in vivo studies using radiolabeled particles. In vitro models of epithelial barriers were developed with the aim to identify the mechanisms involved in the translocation process as well as to characterize the absorption phase in toxicokinetic studies. It is of paramount importance to understand the underlying mechanisms involved in the ability of NM to translocate epithelial barriers. Knowing the physico-chemical characteristics preventing or favoring such passage would help in designing safer NMs either by reducing their crossing ability or by increasing it in the case of NMs used in nanomedicine.

2.4.1 In Vitro Studies

To determine how NMs can cross epithelial barriers, in vitro models of these barriers are the best way to grasp the direct interaction between the cellular sheet and particles. There are two possibilities for NMs to cross the epithelial barrier: the transcellular route also called transcytosis, or the paracellular route (Fig. 2.3).

Transcytosis consists in the transfer from the apical side of the epithelium to the basal side through cells. This implies that NMs are internalized by cells at the apical side of the epithelium where NM exposure takes place and then be transferred through the cells before a final exocytosis at the basal side. The paracellular route consists in the transfer between cells through the different junctions which are established between each epithelial cell. Among them, tight junctions are located at the apical side of cells and are pivotal to ensure the sealing of the epithelial sheet and its impermeability. In tight connection with the cytoskeleton they also contribute in maintaining the structure of the epithelium. However different insults (pollutants, chronic inflammation, or infection) can disrupt these junctions allowing

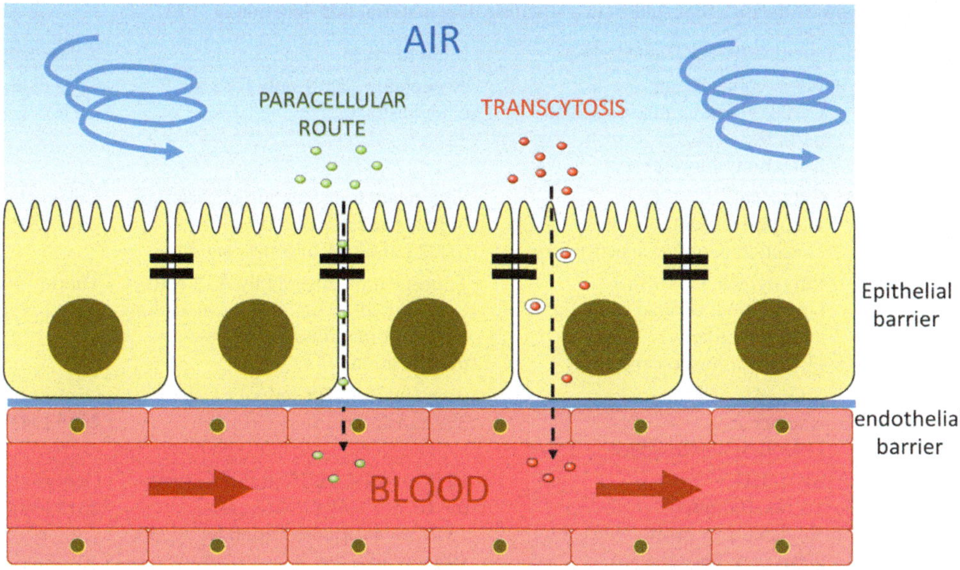

Fig. 2.3 Translocation processes through the epithelial barrier. Epithelial cells form a tight barrier due to the presence of tight junctions. The translocation of particles from the air compartment to the blood can occur either through cells, a process called transcytosis which requires the uptake of the particle by cells and their traffic from the apical part to the basal part of the cells for exocytosis, or by the paracellular route in case of opening of the junctions either by nanomaterials or other insults

NMs to enter the intercellular space and to reach the basal side, as the junctions below the tight junction area are weaker. NMs themselves could disturb these junctions as it was shown for adherens junctions of endothelial cells [32]. TiO_2 as well as SiO_2 and Ag NPs were shown to physically interact with VE-cadherin triggering its phosphorylation and loss of contact with p120 and β-catenin. This induces signaling pathways that lead to actin remodeling and VE-cadherin degradation leading to endothelial cell leakiness [33]. In order to study NM translocation in vitro, cells have to be seeded on porous membranes of an insert device allowing the separation of the apical compartment corresponding to the face exposed to the air, from the basal compartment mimicking the blood. Particles are applied to the apical side and those having crossed the epithelial barrier are recovered in the basal one.

For a better relevance, models established using human cells are favored. Many human airway/alveolar cell lines or primary cells are available but for translocation studies, they have to meet some criteria:

- To be cultivable on insert devices (such as Transwells®) to separate apical and basal sides and to obtain polarized epithelium.
- To grow on insert membranes which exhibit pores large enough to allow NM transfer to the basal compartment.
- To form tight junctions to obtain an efficient epithelial barrier. The strength of the tight junctions can be checked by monitoring the Trans-Epithelial Electric Resistance (TEER) or by measuring the paracellular crossing of marker molecules. For instance, Lucifer Yellow is a fluorescent dye which when applied for a short time, can cross an epithelium only by the paracellular route.

Alveolar epithelial cells could be used to establish an air-blood barrier in vitro. However, the most commonly used human alveolar type II

Table 2.1 Overview of the different studies investigating nanomaterial translocation in vitro

Cellular model(s)	Nanoparticles characteristics	Main findings	References
Calu-3 A549 Murine pneumocytes	Polystyrene 46 nm Carboxyl- and amine-modified particles Detection: fluorescence	Acceptable TEER only for Calu-3 6% translocation	Geys et al. 2006 [37]
Calu-3	CeO_2, ZnO 28–120 nm uncoated and SiO_2-coated Detection: Gamma spectroscopy	Significant translocation of each type of material 0,0025–0,025% of translocation	Cohen et al. 2014 [39]
ATI and ATII cells	Polystyrene 50, 100 nm Unmodified, carboxyl-modified, amine-modified Detection: fluorescence	Translocation ensured by AT1 cells Unmodified-50 nm: 3% translocation Carboxyl-modified-50 nm: 8% translocation	Thorley et al. 2014 [12]
Calu-3 A 549 NCI-H292	SiO_2-FITC 50 nm Uncoated Detection: fluorescence	Acceptable TEER only for Calu-3 3% translocation	George et al. 2015 [34]
Calu-3	SiO_2, TiO_2 16, 50, 100 nm Negatively or positively charged Detection: fluorescence	17.9%/14.4%/12.1% translocation for 16, 50, 100 nm SiO_2 (−) 13%/5%/4% translocation for 16, 50, 100 nm SiO_2 (+)	George et al. 2015 [38]
A549 MLE-12 (murine ATII) 16HBE14o-	Au 2, 7, 18, 46, 80 nm Uncoated Detection: ICP-MS	Translocation rate inversely correlated with size	Bachler et al. 2015 [35]

cell line A549 do not establish strong tight junctions even when grown on inserts for several days [33, 34]. The immunolabeling of the protein ZO-1 (zonula occludens-1) associated to tight junctions did not show consistent staining and the TEER detected for this model was low (around 40 ohm/cm^2) after more than 10 days in culture [34, 35]. An alveolar type I-like cell phenotype was recently obtained by immortalization of primary human alveolar type II cells but they are also characterized by a low TEER (57 ohm/cm^2) [12, 36].

Several studies have compared different human respiratory cell lines to investigate which one could produce a tight epithelium in vitro (Table 2.1) to be suitable for translocation studies [33, 34, 37]. It became apparent that until now, the Calu-3 cell line offers the best compromise albeit it is not an alveolar cell. When grown on inserts, this human bronchial epithelial cell line derived from a lung adenocarcinoma, produced a polarized epithelium with a dense web of microvilli at the apical surface, synthesized and released mucus, and exhibited a high TEER (>500 Ω/cm^2) [34]. With this model, different studies have been performed to characterize the translocation of NMs as summarized in Table 2.1. Geys et al. using polystyrene NPs of 46 nm in diameter and being either carboxyl- or amine modified, found a 6% translocation whatever the modification [37]. George et al., using SiO_2 NPs of different sizes and charges, demonstrated that translocation was increased for the smallest and negatively charged NPs reaching up to 17% [38]. Cohen et al., using ZnO NPs and CeO_2 NPs of two different sizes with or without a SiO_2 coating, observed a translocation for all NMs but at very low amounts (<0.01%) [39]. The common feature of these studies is that the translocation was not associated with an alteration of the integrity of the epithelial monolayer, suggesting transcytosis of internalized NMs. However, altogether these data revealed that the amount of translocation could be quite different according to the physico-chemical properties of the NM. For now, not enough studies have been performed to clearly identify which particle determinants are the most relevant in particle translocation. Characteristics such as size, aggregation, composition, charges and coating are among the factors that could contribute to this process as underlined by the few examples already described above and summarized in Table 2.1. In addition to these intrinsic NM properties, it has also to be taken into account the influence of the cellular environment that can favor the creation of a biocorona on the NMs modulating their fate [38].

2.4.2 In Vivo Studies

For a long time it was suspected that the extrapulmonary effects induced by exposure to ambient particulate matter was in part related to the translocation of their ultrafine fraction without being able to demonstrate it in absence of a sensitive detection method [40]. In 2002 Nemmar et al. claimed that (99 m)Technecium-labeled ultrafine carbon particles pass into the systemic circulation shortly after inhalation by five healthy volunteers [41]. However, two other studies did not confirm significant translocation of these NMs and demonstrated that the gamma camera imaging technique used to detect the NM was in fact detecting unbound (99 m)Tc-pertechnetate [42, 43]. Thanks to stable radiolabeled engineered NPs, it became evident from animal studies that this translocation was possible allowing their biodistribution to secondary organs. The liver is frequently the one accumulating the most important part of NMs. Translocation has now been demonstrated for a large panel of NMs (as summarized in Table 2.2) such as organic ones including polymeric and carbonaceous NPs [41–43, 45, 46], metal oxides [47–49], metals such as gold and iridium [35, 45–47, 49–55]. According to the studies, the translocation is not always fully quantified but limited to the detection of NMs in some secondary organs. Studies from Kreyling's group have performed mass balance analysis showing that the translocation after unique exposure was very low. As an example, maximum *in vivo* translocation of up to 5% was documented for polystyrene latex NPs after pulmonary instillation in rats [56]. Translocation rates are mostly under 1% and can be influenced by the administration route: <0,02% and <0,06% for iridium NPs after respectively inhalation and endotracheal intubation in rats [50, 51].

Size seems to be a critical parameter driving the ability of NMs to cross the air-blood barrier. It was demonstrated by Choi et al. using a large panel of organic, inorganic and hybrid NMs which fate after instillation in rats was detected on line owing to their near-infrared fluorescence that they accumulate in lymph nodes [46]. They concluded that a rapid translocation in the lung required NPs with a hydrodynamic diameter under 38 nm and that below this size the amount of translocation was dependent on charges [46]. The role of the size was also strengthened by studies of Kreyling's group using radiolabeled NPs such as iridium and gold NPs [41, 55]. They showed that the translocated fraction of the 80 nm Ir NPs inhaled by rats for one hour was about an order of magnitude less than that of 15-nm Ir NPs [50]. For Au NPs intratracheally instilled in rats, the highest translocation at 24 h was observed for 1.4 nm Au NPs and the lowest for 80 nm Au NPs but 200 nm Au NPs exhibited a higher translocation rate than 80 nm Au NPs [55]. In addition, they observed that negatively charged NPs had a better translocation rate than positively charged NPs supporting Choi's conclusions. These data underline that size and charge are important determinants of NM translocation but other studies also revealed the role of the coating [46].

In one study it was investigated whether the translocation rate could be modified in compromised animals. A murine model of cystic fibrosis was compared to wild type mice after a one hour inhalation of 30 nm iridium NPs. The biokinetic was not modified but an increased uptake of nanoparticles by alveolar cells was observed in compromised animals [53].

Translocation from the respiratory tract is not limited to the deep lung but can also occur in its upper part at the nasal level where the NMs may reach the brain through the olfactory bulb. This pathway was first suggested by Elder et al. having observed an accumulation of manganese oxide

Table 2.2 *In vivo* studies investigating translocation of nanomaterials from respiratory route

Animal model(s) / Exposure mode / Time of detection	Nanoparticles characteristics	Main findings	Reference
Hamsters	Albumin 100–200 nm	Radioactivity distribution shows low biodistribution (<1%) in secondary organs (liver, heart, spleen, kidneys, brain)	Nemmar et al. 2001 [44]
Endotracheal intubation	Uncoated		
60 min	Detection: radioactivity		
Rats	Iridium 15–80 nm	Translocation in very low amount to liver, spleen, kidney, heart and brain (<0,02%)	Kreyling et al. 2002 [50]
Inhalation	Uncoated		
7 days	Detection: radioactivity		
Human	Carbon 5–10 nm	Detectable translocation to liver (stable with time) and bladder (decreasing with time)	Nemmar et al. 2002 [41]
Inhalation	Uncoated		
60 min	Detection: radioactivity		
Rats	Carbon 35 nm	Small but significant translocation to several part of brain (<0,5 µg ^{13}C/g of tissue)	Oberdörster et al. 2004 [58]
Whole body exposure	Uncoated		
7 days	Detection: radioactivity		
Rats	Iridium 15–20 nm	Rapid translocation to secondary organs (liver, spleen, brain, kidney) very low (<0,06%) decreasing with time (complete excretion after 20 days)	Semmler et al. 2004 [51]
Endotracheal intubation	Uncoated		
180 days	Detection: Radioactivity		
Rats	Manganese oxide 30 nm	Significant increase of Mn in several part of brain increasing with time and number of exposures	Elder et al. 2006 [47]
Whole body + intranasal instillation	Uncoated		
12 days	Detection by ICP-MS	Highest rate of translocation in olfactory bulb	
Human	Carbon 5–25 nm	Translocation into blood, excreted with urine	Mills et al. 2006 [42]
Inhalation	Uncoated	No accumulation of radioactivity detected over liver or spleen	
6 h	Detection: radioactivity		
Human	Carbon 35 nm	Translocation into blood, excreted with urine	Wiebert et al. 2006 [43]
Inhalation	Uncoated		
24 h	Detection: radioactivity	No accumulation of radioactivity detected over liver or spleen	
Rats	Iridium and carbon	24 h retention of NPs shows biodistribution to various secondary organs (liver, spleen, kidneys, heart, brain, blood)	Kreyling et al. 2009 [45]
Inhalation	Primary size: 2–10 nm		
24 h	Uncoated		
	Detection: radioactivity		
Rats	Polystyrene latex 20–1000 nm	Significant translocation to secondary organs (up to 5%)	Sarlo et al. 2009 [56]
Oropharyngeal aspiration	Carboxylated surface		
120 days	Detection: fluorescence	Translocation rate depending organ (retention of several size particles in different organs up to 90 days)	
Rats	Organic/inorganic/hybrid	Translocation to blood and secondary organs	Choi et al. 2010 [46]
Pulmonary instillation	5 à 300 nm		
60 min	ZnCds, Zns, PEG coated		
	Detection: near-infrared fluorescence		

(continued)

Table 2.2 (continued)

Animal model(s) / Exposure mode / Time of detection	Nanoparticles characteristics	Main findings	Reference
Rats	Gold 5–30 nm	No detectable translocation	Lipka et al. 2010 [52]
Intratracheal instillation	Au-Phos, Au-PEG750, Au-PEG10k		
24 h	Detection: radioactivity		
Mice	Europium oxide 80–100 nm	Large fraction remaining in the lung but significant translocation to secondary organs detectable 24 h after instillation (kidney, heart, liver, spleen)	Abid et al. 2013 [48]
Oropharyngeal aspiration	Gadolinium coated		
60 min	Detection by ICP-MS		
Mice (+/− cystic fibrosis)	Iridium 30 nm	Translocation in very low amount detectable 24 h after exposure in both +/− cystic fibrosis mice in several organs (trachea, heart, brain, skin, liver, skin, kidney)	Geiser et al. 2014 [53]
Inhalation	Uncoated		
24 h	Gamma spectroscopy		
Mice	Polymere-lipid 155–450 nm	Translocation to liver and kidney for both 150 and 450 nm	Garbuzenko et al. 2014 [59]
Inhalation	Uncoated		
4 weeks	Detection: fluorescence		
Rats	Barium sulfate 150–350 nm	Translocation to several part of the body	Konduru et al. 2014 [60]
Intratracheal instillation	Uncoated		
7 days	Detection: radioactivity	Long term translocation increasing in bone and bone marrow	
Rats	Gold	Translocation rate is inversely correlated with nanoparticles sizes	Kreyling et al. 2014 [55]
Intratracheal instillation	(1.4, 2.8, 5, 18, 80, 200 nm) Thioglycolic acid, cysteamine coated		
24 h	Detection: Radioactivity	Charge do not influence retention in lung but negatively charges particles are preferentially translocated	
Rats	Titanium Dioxide 20–1680 nm	Translocation rate is correlated with administered dose	Shinohara et al. 2015 [49]
Intratracheal instillation	Uncoated vs $Al(OH)_3$ coated		
90 days	Detection by ICP-SFMS		
Rats and mice	Gold 2–80 nm	Translocation rate is inversely correlated with nanoparticles sizes	Bachler et al. 2015 [35]
Inhalation and intratracheal instillation	Uncoated		
96 h	In *silico* modelisation		
Rats	^{192}Iridium 20 nm	Higher accumulation in brain and more circulating NP in blood after nose-only inhalation	Kreyling et al. 2016 [57]
Nose-only inhalation and intratracheal inhalation	Detection: radioactivity		
1 h		Translocation to the brain is favored by deposition of particle in the upper respiratory tract via theneuronal pathways	

NPs in the olfactory bulb [47]. These findings were recently confirmed by Kreyling et al. who performed different types of exposure (nose-only and ventilation/intubation) to determine the contribution of the olfactory bulb in NP accumulation in the brain. The amount of iridium NPs accumulated in the brain was ten times higher in case of nose-only exposure [57].

References

1. Stone V, Miller MR, Clift MJ et al (2016) Nanomaterials vs ambient ultrafine particles: an opportunity to exchange toxicology knowledge. Environ Health Perspect. https://doi.org/10.1289/EHP424
2. MPPD, Multiple-Path Particle Dosimetry Model (n.d) (MPPD v 3.04) https://www.ara.com/products/multiple-path-particle-dosimetry-model-mppd-v-304
3. Doherty GJ, McMahon HT (2009) Mechanisms of endocytosis. Annu Rev Biochem 78:857–902
4. Roiter Y, Ornatska M, Rammohan AR et al (2009) Interaction of lipid membrane with nanostructured surfaces. Langmuir 25(11):6287–6299
5. Vranic S, Garcia-Verdugo I, Darnis C et al (2013) Internalization of SiO_2 nanoparticles by alveolar macrophages and lung epithelial cells and its modulation by the lung surfactant substitute Curosurf. Environ Sci Pollut Res Int 20(5):2761–2770
6. Vranic S, Boggetto N, Contremoulins V et al (2013) Deciphering the mechanisms of cellular uptake of engineered nanoparticles by accurate evaluation of internalization using imaging flow cytometry. Part Fibre Toxicol 10:2
7. Astolfo A, Arfelli F, Schültke E et al (2013) A detailed study of gold-nanoparticle loaded cells using X-ray based techniques for cell-tracking applications with single-cell sensitivity. Nanoscale 5(8):3337–3345
8. Drescher D, Kneipp J (2012) Nanomaterials in complex biological systems: insights from Raman spectroscopy. Chem Soc Rev 41(17):5780–5799
9. Vercauteren D, Vandenbroucke RE, Jones AT et al (2010) The use of inhibitors to study endocytic pathways of gene carriers: optimization and pitfalls. Mol Ther 18(3):561–569
10. Ivanov AI (2008) Pharmacological inhibition of endocytic pathways: is it specific enough to be useful? Methods Mol Biol 440:15–33
11. Brandenberger C, Mühlfeld C, Ali Z et al (2010) Quantitative evaluation of cellular uptake and trafficking of plain and polyethylene glycol-coated gold nanoparticles. Small 6(15):1669–1678
12. Thorley AJ, Ruenraroengsak P, Potter TE et al (2014) Critical determinants of uptake and translocation of nanoparticles by the human pulmonary alveolar epithelium. ACS Nano 8(11):11778–11789
13. Bohmer N, Jordan A (2015) Caveolin-1 and CDC42 mediated endocytosis of silica-coated iron oxide nanoparticles in HeLa cells. Beilstein J Nanotechnol 6:167–176
14. Kasper J, Hermanns MI, Bantz C et al (2013) Flotillin-involved uptake of silica nanoparticles and responses of an alveolar-capillary barrier in vitro. Eur J Pharm Biopharm 84(2):275–287
15. Rothen-Rutishauser BM, Schürch S, Haenni B et al (2006) Interaction of fine particles and nanoparticles with red blood cells visualized with advanced microscopic techniques. Environ Sci Technol 40(14):4353–4359
16. Zhao Y, Sun X, Zhang G et al (2011) Interaction of mesoporous silica nanoparticles with human red blood cell membranes: size and surface effects. ACS Nano 5(2):1366–1375
17. Hussain S, Thomassen LC, Ferecatu I et al (2010) Carbon black and titanium dioxide nanoparticles elicit distinct apoptotic pathways in bronchial epithelial cells. Part Fibre Toxicol 7:10
18. Nabeshi H, Yoshikawa T, Arimori A et al (2011) Effect of surface properties of silica nanoparticles on their cytotoxicity and cellular distribution in murine macrophages. Nanoscale Res Lett 6(1):93
19. Fröhlich E (2012) The role of surface charge in cellular uptake and cytotoxicity of medical nanoparticles. Int J Nanomedicine 7:5577–5591
20. Kirch J, Guenther M, Doshi N et al (2012) Mucociliary clearance of micro- and nanoparticles is independent of size, shape and charge – an ex vivo and in silico approach. J Control Release 159(1):128–134
21. Florez L, Herrmann C, Cramer JM et al (2012) How shape influences uptake: interactions of anisotropic polymer nanoparticles and human mesenchymal stem cells. Small 8(14):2222–2230
22. Meng H, Yang S, Li Z et al (2011) Aspect ratio determines the quantity of mesoporous silica nanoparticle uptake by a small GTPase-dependent macropinocytosis mechanism. ACS Nano 5(6):4434–4447
23. Donaldson K, Murphy FA, Duffin R et al (2010) Asbestos, carbon nanotubes and the pleural mesothelium: a review of the hypothesis regarding the role of long fibre retention in the parietal pleura, inflammation and mesothelioma. Part Fibre Toxicol 7:5
24. Zhang B, Feng X, Yin H et al (2017) Anchored but not internalized: shape dependent endocytosis of nanodiamond. Sci Rep 7:46462
25. Mura S, Hillaireau H, Nicolas J et al (2011) Biodegradable nanoparticles meet the bronchial airway barrier: how surface properties affect their interaction with mucus and epithelial cells. Biomacromolecules 12(11):4136–4143

26. Ruge CA, Schaefer UF, Herrmann J et al (2012) The interplay of lung surfactant proteins and lipids assimilates the macrophage clearance of nanoparticles. PLoS One 7(7):e40775
27. McKenzie Z, Kendall M, Mackay RM et al (2015) Surfactant protein A (SP-A) inhibits agglomeration and macrophage uptake of toxic amine modified nanoparticles. Nanotoxicology 9(8):952–962
28. Shaw CA, Mortimer GM, Deng ZJ et al (2016) Protein corona formation in bronchoalveolar fluid enhances diesel exhaust nanoparticle uptake and pro-inflammatory responses in macrophages. Nanotoxicology 10(7):981–991
29. Vranic S, Gosens I, Jacobsen NR et al (2017) Impact of serum as a dispersion agent for in vitro and in vivo toxicological assessments of TiO2 nanoparticles. Arch Toxicol 91(1):353–363
30. Braakhuis HM, Park MV, Gosens I et al (2014) Physicochemical characteristics of nanomaterials that affect pulmonary inflammation. Part Fibre Toxicol 11:18
31. Churg A, Brauer M, del Carmen A-CM et al (2003) Chronic exposure to high levels of particulate air pollution and small airway remodeling. Environ Health Perspect 111(5):714–718
32. Setyawati MI, Tay CY, Chia SL et al (2013) TiO2 nanomaterials cause endothelial cells leakiness by disrupting the homophilic interaction of VE-cadherin. Nat Commun 4:1673
33. Geys J, Nemery B, Hoet PH (2007) Optimisation of culture conditions to develop an in vitro pulmonary permeability model. Toxicol In Vitro 21(7):1215–1219
34. George I, Vranic S, Boland S et al (2015) Development of an in vitro model of human bronchial epithelial barrier to study nanoparticle translocation. Toxicol In Vitro 29(1):51–58
35. Bachler G, Losert S, Umehara Y et al (2015) Translocation of gold nanoparticles across the lung epithelial tissue barrier: Combining in vitro and in silico methods to substitute in vivo experiments. Part Fibre Toxicol 12:18
36. Kemp SJ, Thorley AJ, Gorelik J et al (2008) Immortalization of human alveolar epithelial cells to investigate nanoparticle uptake. Am J Respir Cell Mol Biol 39(5):591–597
37. Geys J, Coenegrachts L, Vercammen J et al (2006) In vitro study of the pulmonary translocation of nanoparticles: a preliminary study. Toxicol Lett 160(3):218–226
38. George I, Naudin G, Boland S et al (2015) Metallic oxide nanoparticle translocation across the human bronchial epithelial barrier. Nanoscale 7(10):4529–4544
39. Cohen JM, Derk R, Wang L et al (2014) Tracking translocation of industrially relevant engineered nanomaterials (ENMs) across alveolar epithelial monolayers in vitro. Nanotoxicology 1:216–225
40. Kreyling WG, Semmler-Behnke M, Takenaka S et al (2013) Differences in the biokinetics of inhaled nano- versus micrometer-sized particles. Acc Chem Res 46(3):714–722
41. Nemmar A, Hoet PH, Vanquickenborne B et al (2002) Passage of inhaled particles into the blood circulation in humans. Circulation 105(4):411–414
42. Mills NL, Amin N, Robinson SD et al (2006) Do inhaled carbon nanoparticles translocate directly into the circulation in humans? Am J Respir Crit Care Med 173(4):426–431
43. Wiebert P, Sanchez-Crespo A, Falk R et al (2006) No significant translocation of inhaled 35-nm carbon particles to the circulation in humans. Inhal Toxicol 18(10):741–747
44. Nemmar A, Vanbilloen H, Hoylaerts MF et al (2001) Passage of intratracheally instilled ultrafine particles from the lung into the systemic circulation in hamster. A J Respir Crit Care Med 164:1666–1668
45. Kreyling WG, Semmler-Behnke M, Seitz J et al (2009) Size dependence of the translocation of inhaled iridium and carbon nanoparticle aggregates from the lung of rats to the blood and secondary target organs. Inhal Toxicol Suppl 1:55–60
46. Choi HS, Ashitate Y, Lee JH et al (2010) Rapid translocation of nanoparticles from the lung airspaces to the body. Nat Biotechnol 28(12):1300–1304
47. Elder A, Gelein R, Silva V et al (2006) Translocation of inhaled ultrafine manganese oxide particles to the central nervous system. Environ Health Perspect 114(8):1172–1178
48. Abid A D, Anderson D S, Das G K et al (2013) Novel lanthanide-labeled metal oxide nanoparticles improve the measurement of in vivo clearance and translocation. Part Fibre Toxicol 0:1
49. Shinohara N, Osima Y, Kobayashi T et al (2015) Pulmonary clearance kinetics and extrapulmonary translocation of seven titanium dioxide nano- and submicron materials following intratracheal administration in rats. Nanotoxicology 9(8):1050–1058
50. Kreyling WG, Semmler M, Erbe F et al (2002) Translocation of ultrafine insoluble iridium particles from lung epithelium to extrapulmonary organs is size dependent but very low. J Toxicol Environ Health A 65(20):1513–1530
51. Semmler M, Seitz J, Erbe F et al (2004) Long-term clearance kinetics of inhaled ultrafine insoluble iridium particles from the rat lung, including transient translocation into secondary organs. Inhal Toxicol 16(6–7):453–459
52. Lipka J, Semmler-Behnke M, Sperling RA et al (2010) Biodistribution of PEG-modified gold nanoparticles following intratracheal instillation and intravenous injection. Biomaterials 31(25):6574–6581
53. Geiser M, Stoeger T, Casaulta M et al (2014) Biokinetics of nanoparticles and susceptibility to particulate exposure in a murine model of cystic fibrosis. Part Fibre Toxicol 11:19

54. Konduru NV, Jimenez RJ, Swami A et al (2015) Silica coating influences the corona and biokinetics of cerium oxide nanoparticles. Part Fibre Toxicol 12:31
55. Kreyling WG, Hirn S, Möller W et al (2014) Air-blood barrier translocation of tracheally instilled gold nanoparticles inversely depends on particle size. ACS Nano 8(1):222–233
56. Sarlo K, Blackburn KL, Clark ED et al (2009) Tissue distribution of 20 nm, 100 nm and 1000nm fluorecent polystryrene latex nanospheres following acute systemic or acute and repeat aiway exposure in the rat. Toxicology (263):117–126
57. Kreyling WG (2016) Discovery of unique and ENM-specific pathophysiologic pathways: Comparison of the translocation of inhaled iridium nanoparticles from nasal epithelium versus alveolar epithelium towards the brain of rats. Toxicol Appl Pharmacol 299:41–46
58. Oberdorster G, Sharp Z, Atudorei V et al (2004) Translocation of inhaled ultrafine particles to the brain. Inhal Toxicol 16:437–445
59. Garbuzenko OB, Winkler J, Tomassone MS et al (2014) Biodegradable Janus Nanoparticles for local pulmonary delivery of hydrophilic and hydrophobic molecules to the lungs. Langmuir:12941–12949
60. Konduru N, Keller J, Ma-Hock L et al (2014) Biokinetics and effects of barium sulfate nanoparticles. Part Fibre Toxicol 11:55

3 Transmucosal Nanoparticles: Toxicological Overview

Swapnil Talkar, Sagar Dhoble,
Anuradha Majumdar, and Vandana Patravale

Abstract

Nanoparticles have specific physicochemical properties different to bulk materials of the same composition and such properties make them very attractive for commercial and medical applications. Mucoadhesive nanoparticulate dosage forms are designed to enable prolonged retention of these nanoparticles at the site of application, providing a controlled drug release for improved therapeutic outcome. Moreover, drug delivery across the mucosa bypasses the first-pass hepatic metabolism and avoids the degradation by gastrointestinal enzymes. However, like most new technologies, there is a rising debate concerning the possible transmucosal side effects resulting from the use of particles at the nano level. In fact, these nanoparticles on entering the body, deposit in several organs and may cause adverse biological reactions by modifying the physiochemical properties of living matter. Several investigators have found nanoparticles responsible for toxicity in different organs. In addition, the toxicity of nanoparticles also depends on whether they are persistent or cleared from the different organs of entry and whether the host can raise an effective response to sequester or dispose of the particles. In contrast to many efforts aimed at exploiting desirable properties of nanoparticles for medicine, there are limited attempts to evaluate potentially undesirable effects of these particles when administered intentionally for medical purposes. This

S. Talkar · S. Dhoble · V. Patravale (✉)
Department of Pharmaceutical Sciences and Technology, Institute of Chemical Technology, Mumbai, India
e-mail: vbp_muict@yahoo.co.in;
http://www.vbpgroup-ict.in

A. Majumdar
Department of Pharmacology and Toxicology, Bombay College of Pharmacy, Mumbai, India

Q. Saquib et al. (eds.), *Cellular and Molecular Toxicology of Nanoparticles*, Advances in Experimental Medicine and Biology 1048, https://doi.org/10.1007/978-3-319-72041-8_3

chapter focuses on the overview of the mucosal systems, fate of nanoparticles, mechanism of nanoparticle's toxicity and the various toxicity issues associated with nanoparticles through mucosal routes.

Keywords
Nanoparticles · Mucoadhesive · Transmucosal · Toxicity · Fate

Abbreviations

AgNPs	Silver nanoparticles
AP-1	Activator protein-1
ARE	Antioxidant-response element
ATP	Adenosine triphosphate
BBB	Blood brain barrier
BSA	Bovine serum albumin
CBF	Ciliary beat frequency
CeO_2	Cerium dioxide
CNS	Central nervous system
CNTs	Carbon nanotubes
CRP	C-reactive protein
CTAB	Cetyltrimethylammonium bromide
CuO	Copper oxide or cupric oxide
CVM	Cervicovaginal mucus
DCFH-DA	2,7-dichlorofluorescein diacetate
DISC	Death-inducing signalling complex
DMPC	1,2 Dimyristoyl-sn-glycero-3-phosphocholine
DNA	Deoxyribonucleic acid
DPPC	Dipalmitoylphosphatidylcholine
Egg-PC	Egg-phosphatidylcholine
ELISA	Enzyme-linked immunosorbent assay
ERK	Extracellular signal-related kinases
FAS	First apoptosis signal
Fe2O3	Ferric oxide
FeO	Iron oxide or ferrous oxide
GFR	Epidermal growth factor receptor
GIT	Gastrointestinal tract
GRO-alpha	Growth-regulated oncogene-alpha
GSH	Glutathione (reduced)
H2O2	Hydrogen peroxide
Hb	Hemoglobin
HM-EHEC	Hydrophobically modified ethyl hydroxyethyl cellulose
HM-HEC	Hydrophobically modified hydroxyethyl cellulose
HPLC	High pressure liquid chromatography
HSV	Herpes simplex virus
IFN	Interferon
IL	Interleukins
JNK	c-jun NH2-terminal kinases
KD	Kilo Dalton
LDH	Lactate dehydrogenase
MAPK	Mitogen-activated protein kinase
MgO	Magnesium oxide
MHC	Major histocompatibility complex
MNPs	Magnetic nanoparticles
MOCs	Mini organ cultures
mRNA	Messenger ribonucleic acid
MTT	3-(4,5-Dimethylthiazol-2-Yl)-2,5-Diphenyltetrazolium Bromide
MWCNTs	Multi walled carbon nanotubes
NF-κB	Nuclear factor kappa-light-chain-enhancer of activated B cells
NiO	Nickel oxide
NPs	Nanoparticles
Nrf2	LNT induced NF-E2 p45-related factor 2
PbAc	Lead acetate
PCEP	(poly{[(cholesteryl oxocarbonylamido ethyl) methyl bis(ethylene) ammonium iodide] ethyl phosphate})
PCL	Poly ε-caprolactone
PCR	Polymerase chain reaction
PEG	Polyethylene glycol
PEO	Poly(ethylene oxide)
PLGA	Poly(lactic-co-glycolic acid)

PTP	Protein tyrosine phosphatases
QD	Quantum dot
qPCR	Quantitative polymerase chain reaction
RNA	Ribonucleic acid
ROS	Reactive oxygen species
RT-PCR	Reverse transcriptase polymerase chain reaction
SA	Stearic acid
SiO2	Silica dioxide
SOD	Super oxide dismutase
SWCNTs	Single walled carbon nanotubes
TAP	Diacyl-TAP (1,2-diacyl-3-trimethy-lammonium propane
TiO2	Titanium dioxide
TJs	Tight junctions
TNF	Tumor necrosis factor
WST-1	Water-soluble tetrazolium salts
ZnO	Zinc oxide

3.1 Introduction

3.1.1 Overview of Mucosal System

A mucous membrane is an epithelial tissue layer lining various cavity of the body and surrounds internal organs. It is of ectodermal origin and is incessant with the skin at various body openings such as the eyes, ears, inside the mouth, inside the nose, the urogenital tract, digestive tract, and respiratory tract. Mucous membranes are moist due to the presence of glands which secrete a thick fluid known as mucous. It serves many functions like lubrication for the passage of objects, maintenance of a hydrated layer over the epithelium, a barrier to pathogens and noxious substances and as a permeable gel layer for the exchange of gases and nutrients with the underlying epithelium. Mucous is primarily composed of water (95%), but also contains salts, lipids such as fatty acids, phospholipids and cholesterol, proteins which help a defensive purpose with lysozyme, immunoglobulins, defensins and growth factors. However, the glycoprotein mucin is the main component that is responsible for its viscous and elastic gel-like properties. Mucins are large, extracellular glycoproteins with molecular weights ranging from 0.5 to 20 kDa. Both membrane bound mucins, and secreted mucins share many common features. They both are highly glycosylated consisting of 80% carbohydrates primarily N-acetylgalactosamine, N-acetylglucosamine, fucose, galactose, and sialic acid (N-acetylneuraminic acid) and traces of mannose and sulphate. Mucin has been difficult to characterize, due to its large molecular weight, high polydispersity and high degree of glycosylation. The conformation of mucin depends on various factors such as pH and ionic strength though sugars also play important role for maintaining the extended conformation of mucin [1].

3.1.1.1 Mucous and Pharmacology

The physical state of the mucous, change in the concentration of mucin, and the strong dependence of its physicochemical properties on factors such as ionic strength and pH play a significant role in many diseases. For example, many bacteria possess specific adhesins that specifically bind to mucous which helps them to reside within the mucus. This includes pathogenic strains of *Helicobacter*, *Pseudomonas*, *Streptococcus* and *Pneumococcus*. *Helicobacter pylori* particularly resides in the mucus layer of the stomach, and is a common cause of ulcers [2, 3]. Some parasitic organisms also secret their own layers of mucus to escape the immune system. Secondly, overproduction of mucus is involved in cystic fibrosis, bronchitis, asthma and in middle ear infections, and mucus gels serve as the matrix in which gallstones are nucleated and grow. Whilst, mucus underproduction is present in dry eye syndromes and in some forms of ulcer disease. Various drug delivery systems based on mucoadhesive interactions like polyelectrolytic interactions (chitosans, polyacrylic acid, etc.), hydrogen bonds (hydrogels), and disulphide binding (thiomers) have been optimized to increase the residence time. Development of nanoparticles for mucosal DNA vaccines and gene therapy are also being underway [4].

3.1.2 Brief Discussion on Mucosal Routes of Exposure

Currently many of the treatments depend on systemically administered therapies which treat the diseased sites but are toxic to healthy tissues limiting treatment efficiency due to patient noncompliance [5]. The advent of micro-and nano-delivery technologies combined with the non-invasive administration brings new confidence for the treatment of disease. Micro-and nano-delivery technologies overcomes the problem of drugs and genes poor solubility, protect drugs from acid degradation or enzymatic degradation, increase blood circulation, reduce plasma clearance, escape the reticuloendothelial system uptake, and achieve higher cellular interaction. Micro- and nano-carriers must increase retention time in the mucus to improve the diffusion across the mucus barrier, which is a challenge in drug delivery. Positively or negatively charged nanocarriers could prolong the retention time in the mucus by binding forces with negatively or positively charged mucin glycoproteins. Mucoadhesive drug delivery systems in the past have been formulated as powders, compacts, sprays, semisolids, or films [6, 7]. For example, compacts have been used for drug delivery to the oral cavity, and powders and nanoparticles have been used to facilitate drug administration to the nasal mucosa. Recently oral strips were developed for tongue or buccal cavity. Transmucosal routes covered in the review are discussed as follows:

3.1.2.1 Ocular Mucosa

Drug administration to the eye is a challenge because of several clearance mechanisms (tear production, tear flow, and blinking) that protect the eye from harmful agents. The mucus layer, 40 mm, which is secreted by the goblet cells onto the eye surface, is intimately associated with the glycocalyx of the corneal/conjunctival epithelial cells. The mucus layer is very sensitive to hydration and forms a gel-layer with viscoelastic rheological properties. It protects the epithelia from damage and enables movements of the eyelids. The mucus gel entraps bacteria, cell debris, and foreign bodies, forming "mucous threads" consisting of thick fibres arranged in bundles. Due to the composition, physicochemical properties and structure of the tear film, various factors impact the ocular mucoadhesion. Very few ophthalmic formulations containing bioadhesives or penetration enhancers are commercially available in the market. The use of bioadhesives considerably extends the corneal retention time, whereas the absorption promoters increase the rate and amount of drug transport. Combining the two approaches will promise an increase in the bioavailability [8, 9].

3.1.2.2 Nasal Mucosa

The nasal mucous membrane lines the nasal cavities having area of approximately 150 cm^2 with highly dense vascular network and relatively permeable membrane structure, and is adherent to the periosteum or perichondrium of the nasal conchae [10]. From the nasal cavity, it is continuous with the conjunctiva through the nasolacrimal and lacrimal ducts; and with the frontal, ethmoidal, sphenoidal, and maxillary sinuses, through the several openings in the meatuses. The mucus layer is 5–20 mm thick and is divided into two layers, where the outer layer has a high viscosity and a gel-like character, while the layer closest to the cells has a lower viscosity enabling the cilia to move. The turnover time for mucus is usually given as 10–15 min, but it is affected by both environmental conditions and diseases [11]. Nasal mucoadhesive drug delivery has been under active investigation for controlled release dosage forms to deliver drugs directly to the CNS bypassing the BBB. Drugs administered intranasally can travel along the olfactory and trigeminal nerves to reach many regions within the CNS and achieve brain targeting. Although, nasal route of administration has gained substantial interest, it is limited by the rapid mucociliary clearance, resulting in a limited contact period allowed for drug absorption through the nasal mucosa [12].

3.1.2.3 Oral Mucosa

Drug delivery through the oral mucosa (buccal and sublingual) has gained significant attention due to its convenient accessibility. The total surface area of the oral cavity is approximately 100 cm^2, of which the buccal mucosa represents approximately one-third. The epithelium of the oral mucosa consists of a stratified squamous epithelium, the thickness of which varies depending on the site. In the buccal region, the epithelium is around 40–50 cells thick, whereas it is somewhat thinner in the sublingual area. The mucus in the oral cavity is secreted by salivary glands as a component of the saliva and is adsorbed to the surface of the oral mucosa, forming a 0.1-0.7mm thick layer. Sublingual mucosa is more permeable than buccal mucosa, but sublingual administration is difficult for formulations intended to act over a long period of time [11]. Drug delivery through the oral mucosa offers several advantages over other drug delivery systems including bypassing hepatic first-pass metabolism, increasing the bioavailability of drugs, improved patient compliance, excellent accessibility, unidirectional drug flux, and improved permeability [13]. The oral cavity has been used as a site for local and systemic drug delivery in different dosage forms like adhesive gels, tablets, films, patches, ointments, mouth washes, and pastes.

3.1.2.4 Pulmonary Mucosa

The bronchial wall is made up of mucosa, lamina propria, smooth muscle, and submucosa with interspersed cartilage. Submucosal glands are found in the normal human bronchial tree in airways with cartilage in the wall. The glands lie between the epithelium and plates of cartilage and between, and occasionally external to, the plates of cartilage. The secretory tubules in bronchial tree arise directly from the collecting duct and are usually branched. At the end of each, a cluster of short tubules is found. Two types of secretory cells have been recognized in the bronchial submucosal glands lining the tubules, the mucous and serous. Mucous cells line each secretory tubule and its main branches, from the collecting duct to the distal cluster of short tubules. The mucous tubules comprise only columnar mucous cells with goblet cells and basal cells. The density of goblet cells progressively decreases from the periphery and disappears at the level of terminal bronchioles. The presence of mucin, water, and electrolytes contributes to the solubility of bronchial secretions while the cilia present in the luminal facet of epithelial cells are responsible for the rhythmic upward movement of bronchial secretions within the lung to the pharynx [14, 15].

3.1.2.5 Rectal and Vaginal Mucosa

The geometry and morphology of the lower colorectal canal vary with location, and exhibit both macroscopic and microscopic features. On the macroscopic scale, rectal folds create creases and canyons on the mucosal surface. The large folds are found in the relatively short anal canal, which is about 2–4 cm long, ending in the anal verge and contains a stratified squamous epithelium. The rectum has crypts on its surface with a thin columnar epithelium having 40–120 μm in diameter and up to about 1 mm in depth. The characteristic length for transport into the mucosal tissue is only about 1 mm, i.e. the thickness of the mucosa.

The human vaginal walls are lined with stratified squamous epithelium containing numerous folds, or rugae, which permit for distension and increased surface area for absorption. Due to the intra-abdominal pressure that collapses the rugae, high internal surface area, and tortuosity of the vaginal canal, to achieve adequate distribution of a vaginal product is a challenge. Cervicovaginal mucus (CVM) serves as a physical barrier to protect the vagina against infection in addition to the epithelium. Mucus produced at the cervix bathes and coats the vaginal walls, mix with vaginal epithelial cells and vaginal transudate. The CVM is composed mostly of water (~90 to 95%) with gel-forming glycoproteins, lipids, soluble proteins, enzymes, and various immune factors. However, ovulatory mucus is produced in more copious amounts, thus facilitating clearance and impeding drug absorption [10, 11].

3.2 Fate of Nanoparticles via Transmucosal Route

Nanoparticles (NPs), when administered via transmucosal route aid in efficient delivery of the drug without eliciting pain. Drug absorption is higher through a mucosal surface as compared to transdermal delivery due to the absence of stra-

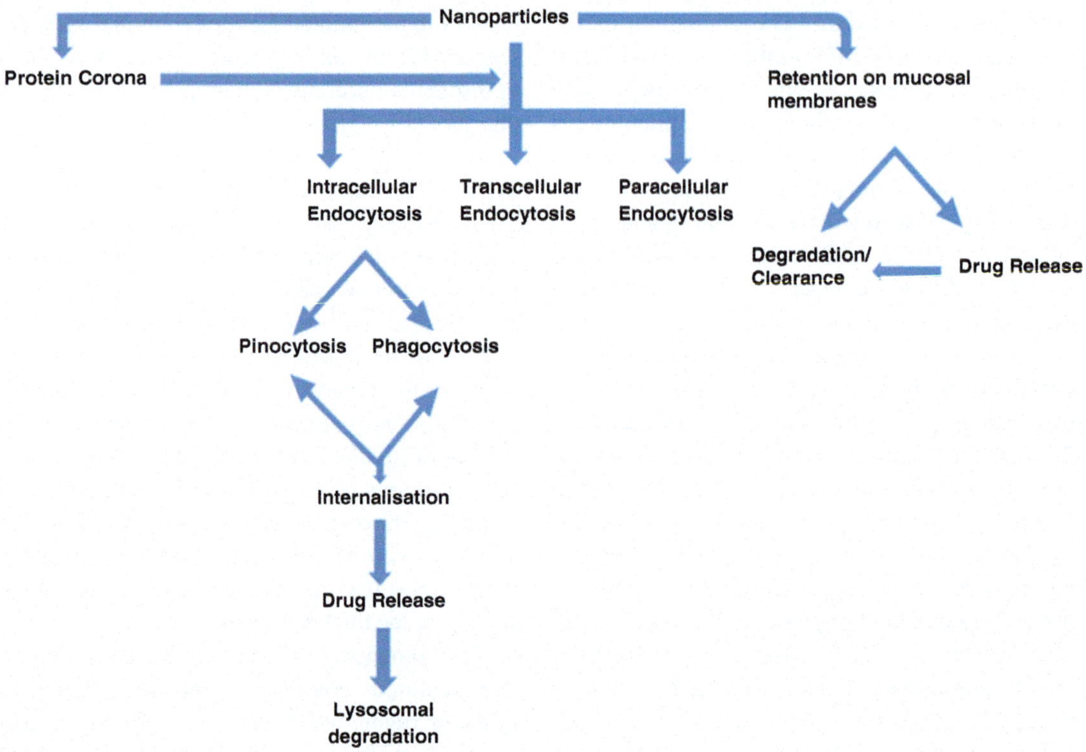

Fig. 3.1 Pathways followed by NPs following administration via transmucosal route

tum corneum. Mucosal surfaces are usually rich in blood supply, providing the means for rapid drug transport to the systemic circulation and avoiding degradation by first-pass hepatic metabolism. However, mucus acts as a barrier to diffusion of lipophilic drugs that interact with the glycoproteins and lipids in the mucus [16]. Nanoparticle (NP) research has proved that NPs cross mucosal barriers and undergo cellular uptake. Properties such as size, surface charge, shape, hydrophobicity, surface chemistry, and protein and ligand conjugates affect the phenomena [17].

3.2.1 Molecular and Cellular Interactions

Following administration, NPs interact with the mucosal membranes in the vicinity and are either internalised into the cell or remain attached to the mucosal lining and are eliminated without any further activity. After internalisation, NPs release their contents inside the cell and are then subjected to degradation. Figure 3.1 represents the pathways followed by Nanoparticles following administration via transmucosal route.

3.2.1.1 Pathways of Internalisation

Internalization occurs through intracellular, paracellular, and transcellular pathways [18].

(a) **Intracellular endocytosis**

Intracellular endocytosis is of two types i.e. pinocytosis and phagocytosis.

Pinocytosis is the ingestion of liquid into a cell by the budding of small vesicles from the cell membrane. Pinocytosis is further divided into Clathrin mediated, Caveolae mediated and macropinocytosis [17]. Clathrin-mediated endocytosis involves clathrin-coated vesicle formation in the presence of adaptor and accessory proteins. Signalling of the NP on the cell surface, aligns surface proteins and aids to begin clathrin-coating on the inner membrane of the cell. An adaptor protein, Epsin, helps pit formation and accessory protein dynamin (GTPase) affects vesicle formation. Thus, a clathrin-coated vesicle with a size of 100–150 nm is formed due to polymerization of the coat complex. The NP containing clathrin-coated vesicle then internally detaches from the donor membrane. Once within the cell, clathrin and adaptor proteins uncoat to allow fusing of the vesicle within the cell to release the endocytosed NPs [19].

In case of Caveole mediated pathway, NPs signalling induces actin reorganization and dynamin recruitment from the cytosol to stimulate membrane invagination and vesicle budding. The caveolae membrane then fuses into the acceptor compartment and releases its contents [17].

Macropinocytosis proceeds by forming protrusions due to actin polymerization from the cell membrane, which then encapsulates the substance to be internalized and once again fuses back with the cell membrane causing internalisation [17].

Phagocytosis is the ingestion of material by phagocytes and amoeboid protozoans. It involves cell surface recognition followed by sequential instigation of receptors leading to internalization by encircling it into triggered cup-shaped cell membrane deformations forming a phagosome [18].

(b) **Transcellular endocytosis**

In this pathway, the NPs to be transported bind to the cell membrane receptors and form a complex. Membrane invagination is then followed by internalization. The endocytosed vesicle is then converted into a transcytotic vesicle to prevent typical endosome degradation. The transcytotic vesicle is then transported to the other end of the cell, where the vesicle membrane fuses with the cell membrane and the content of the vesicle is secreted externally [20].

(c) **Paracellular endocytosis**

Paracellular delivery of hydrophilic drugs occurs through the intercellular space between adjacent cells via tight junctions (TJs).

3.2.1.2 Protein Binding to NPs

It has been established through a study by Fleishcher and Payne [21] that extracellular serum proteins present in blood get adsorbed onto the surface of NPs, forming a "protein corona". When polystyrene NPs functionalized with either amine or carboxylate groups were prepared, serum proteins got adsorbed onto the surface of both the NPs. Bovine serum albumin (BSA)–NP complexes formed from anionic NPs bonded with albumin receptors on the cell surface. BSA–NP complexes formed from cationic NPs were redirected to scavenger receptors. This observation that similar NPs with identical protein corona compositions were bound to different cellular receptors suggested that a difference in the structure of the adsorbed protein may be responsible for the differences in cellular binding of the protein–NP complexes. Similar results were obtained for anionic quantum dots and colloidal gold nanospheres. Protein corona remained bound to the NP throughout the endocytic uptake and transport, however, the NP itself altered the structure of the adsorbed protein [21].

3.2.1.3 Degradation

Particles generally end intracellularly in endosomes or lysosomes followed by degradation. Chemical characteristics such as surface charge determine the fate of NPs in cells.

In the case of macromolecular therapeutics, following intracellular uptake, the contents of the endocytic vesicle are delivered to lysosomes for degradation. Although a therapeutic agent encapsulated in NPs are less susceptible to degradation in the endo-lysosomal compartment, the relatively faster degradation of NPs under acidic conditions in the endo-lysosomal compartment may result in the release of the therapeutic agent, which could then degrade quite rapidly. Thus, NPs are expected to be efficiently internalized into the cells and then deliver their payload into the cytoplasmic compartment rather than be retained in the degradative environment of endo-lysosomal compartment.

A study by Panyam et al. [22], showed that (PLGA) poly (lactic-co-glycolic acid) NPs were internalized through clathrin depended endocytosis. Following their uptake, NPs were localized in the early, recycling endosomes and late endosomes and lysosomes. It was summarised that NPs are either recycled back to the surface from the early endosomes or are transported to the secondary endosomes and lysosomes from which the NPs escape into the cytosol. The early endocytic vesicles possess physiological pH wherein NPs have a net negative charge and hence are repelled by the negatively charged endosomal membrane. The secondary endosomes and lysosomes are predominantly acidic, with pH values ranging from 4–5. In this pH, NPs have a net cationic potential and hence interact with the negatively charged membrane leading to their escape into cytoplasmic compartment. NPs do not open up the endo-lysosomal vesicles but are released by localized destabilization of the endo-lysosomal membrane at the point of contact with NP, followed by extrusion of the NP through the membrane. PLGA NPs are cationic only in the endosomal compartment and do not destabilize the lysosomes. After their escape, NPs deliver their payload in the cytoplasm at a slow rate, leading to a sustained therapeutic effect. Because NPs are biodegradable and biocompatible and are capable of sustained intracellular delivery of multiple classes of cargoes, they are a suitable system for intracytoplasmic delivery of drugs, proteins, or genes [22].

3.3 Mechanism of Toxicity at Cellular and Molecular Level

Nanotoxicity or NP related toxicity implies toxic effects of NPs which are uncommon and not seen with larger particles on the biological system. The significance of nanotoxicity is such that even when the NPs made up of inert materials like gold or silver, they are highly active owing to their nanosized dimension. Nanotoxicity pursues the level or extent to which these properties may cause any threat to environment and living beings. It also intends to quantitatively determine the severity and regularity of toxic effects of the exposure of the NPs on the organism.

The impetus for designing a nano drug delivery system is to reduce the toxicity of a drug and to increase its bioavailability as well as biocompatibility. On the other hand, their exceptional properties like surface area to volume ratio, particle size, solubility, surface coating and shape or structure may pose additional risks to the patients [23]. Toxic manifestations observed with in vitro models are hardly relatable to the effects seen in vivo. Though major entry routes as well as recognised targets have been identified, intense research is still required to demonstrate the pathway and mechanism of toxicity of NPs in the body [24].

It is in general consent that NPs display toxicity through varied mechanisms and can affect in allergy, fibrosis, organ failure, neurotoxicity, hepatological toxicities, nephrotoxicities, haematological toxicities, splenic toxicities, and pulmonary toxicities, among others.

3.3.1 Physiochemical Properties of Nanoparticle and Their Toxic Effects

(a) **Particle size and surface area:** Particle size and surface area of NPs play an important role in their interaction with biological molecules or system. Interestingly, particle size is inversely proportional to the surface area relative to the volume, means decrease in size leads to an increase in surface area to volume ratio. Several biological mechanisms including phagocytosis, endocytosis and passive diffusion as well as endocytic processing (antigen presentation on MHC class molecules) are dependent on size of the material. One of the prime mechanism for toxicity is generation of reactive oxygen species, these free radicals have been known for hazardous impact on biological molecules like DNA, lipids, proteins etc.

Furthermore, surface area also results in some toxic manifestation, i.e. increase in surface area leads to extensive interaction with biomolecules that root more oxidation and

DNA damage abilities as compared to larger particles of same mass [25].

(b) **Particle shape and aspect ratio:** Particle shape dependent toxicity is related to NPs made of gold, silver, carbon nanotube, nickel, titanium etc. Endocytosis and phagocytosis processes are mostly influenced by this property. It has been shown that spherical particles are more prone to endocytosis than any other particle shapes [25]. Studies also report that particle shape can affect the cellular level e.g. blocking of K^+ channel by rod shaped SWNTs were two to three times more efficient than spherical C_{60} fullerenes [26]. Also in another study nanorod ZnO was found to me more cytotoxic than spherical ZnO [27]

Similarly, greater the aspect ratio more will be the toxicity of NPs. It has been observed that asbestos particles which are <2 μm in size caused asbestosis, <5 μm caused mesothelioma and 10 μm caused lung carcinoma [28]. TiO_2 nanofibres having length of 15 mm were more toxic than fibres having length 5 mm, here former induced more inflammatory response by alveolar macrophages in mice than later one. Similarly, in case of Carbon Nanotubes (CNTs), long MWCNTs caused inflammatory response in mice abdominal cavity whereas small MWCNTs did not cause any inflammation at all [29].

(c) **Surface charge:** Surface charge of NPs also play an important role in toxicity. Rather, they have an even greater impact on the biological system. Surface charge on the particles dictate various interactions such as plasma protein binding, selective absorption, blood brain barrier integrity and membrane permeability. Mammalian cell membranes possess negative charge on their surface, thereby promoting association of cationic particles with the cells to a greater extent as compared to the negative or neutral particles. However, higher cationic charge leads to the severe toxicity via haemolysis and platelet aggregation [25]. Positively charged silica NPs have been shown to induce more reactive oxygen species than neutral and negatively charged silica NPs [30].

(d) **Crystalline structure:** Besides the three parameters contributing substantially towards toxicity, crystalline structure may also be responsible for nanotoxicity. Studies have claimed that anatase form of TiO_2 NPs induce higher lipid peroxidation and oxidative DNA damage in presence of light compared to their rutile form [31].

(e) **Aggregation:** Particle aggregation also imparts toxicity. Aggregation is mostly dependent on the surface charge, size and composition. Aggregation of NPs are mostly seen in the case of CNTs, where it has been observed that aggregated CNTs have more cytotoxic effects than dispersed ones [32].

(f) **Surface coating:** Surface coating eventually alters the physiochemical properties of NPs such as surface charge, magnetic, electric, optical and chemical properties. These changes may lead to varied interactions with biomolecules that result into significant toxicity of NPs. It was well acknowledged that the existence of ozone, oxygen radicals along with heavy metals on nanoparticle surface leads to the formation of ROS that induces inflammation in cells.

However, in most cases surface coating could also be employed to abate the toxicity of NP. For example, coating is very essential in the case of quantum dots to render them nontoxic since their metallic core is hydrophobic and composed of heavy toxic metals like cadmium.

While these factors mainly contribute to the toxicity of NPs, concentration is the principle factor dictating the toxicity of macroparticles [25].

3.3.2 Mechanism of Toxicity of NPs

Toxic scenario of NPs in organisms and environment are well established. Unique physiochemical properties of NPs pave way for their application in several fields. On the other hand, they also increase the risk of their exposure to humans and environment.

The mechanisms due to which these properties result in toxicity are discussed in the following section.

(a) **Reactive Oxygen Species (ROS) or Oxidative stress production:** ROS are oxygen containing chemically reactive species that have important roles in cell signalling and homeostasis. They are generally formed as a natural by-product of oxygen metabolism. ROS are generated intrinsically as well as extrinsically within the cell, the pool of ROS constitutes of oxidative species including superoxide anion (O_2^-), hydroxyl radical (OH^-), hydrogen peroxide (H_2O_2), singlet oxygen ($^1O_2$), and hypochlorous acid (HOCl), hypochlorite ion ($-OCl^-$) [33].

NADPH oxidase (NOX) complexes in cell membranes, mitochondria, peroxisomes, and endoplasmic reticulum are responsible for endogenous production of ROS. Whereas, exogenous ROS can be formed from tobacco, pollutants, smoke, engineered NPs, drugs, xenobiotics, or radiation.

Enzymes like catalase, glutathione peroxidase, super oxide dismutase, peroxiredoxins as well as light are responsible for ROS generation. Apart from the deleterious effects of ROS, they also have some positive effects such as programmed cell death i.e. apoptosis and induction of host defence mechanism.

The harmful effects on biomolecules are often seen in the form of:

(i) **DNA or RNA:** Single and double stranded breaks in DNA or RNA
(ii) **Proteins:** Oxidation of amino acids
(iii) **Lipids:** Oxidation of polyunsaturated fatty acids i.e. lipid peroxidation
(iv) **Enzymes:** Inactivation by oxidation of co-factors

NPs may generate ROS via three mechanisms; particle-cell interaction, active redox cycling on surface of NPs (especially in case of transition metals) and oxidative groups functionalized on NPs [23]. Overproduction of ROS leads to the activation of interleukins, cytokines, kinase and tumour necrosis factors which eventually cause the proinflammatory response.

Eom and Choi [34] reported an elevated ROS production when Jurkat T cells that were exposed to AgNPs in contrast to the unexposed ones [34]. Moreover, numerous researchers have established single stranded DNA damage caused due to TiO_2, Carbon black and diesel exhaust particles [31, 35, 36].

(b) **Apoptosis:** It is programmed cell death that occurs in multicellular organisms. This process involves various events such as chromosome condensation, nuclear fragmentation, cell shrinkage, DNA and mRNA decay etc. Apoptosis can be initiated by an intrinsic pathway (cell kills itself because it senses cell stress) as well as an extrinsic pathway (cell kills itself because of signals from other cells). In case of NPs, they are interacting with macrophages which causes the activation extrinsic pathway of apoptosis. Extrinsic pathway is activated via two signals:

(i) **Tumour Necrosis Factor (TNF) signal:** TNF-α is a cytokine produced extensively by activated macrophages. Human cells have two receptors for it i.e. TNFR1 and TNFR2. Binding of TNF-α to these receptors leads to the activation of caspases.

(ii) **First Apoptosis Signal (FAS):** FAS, a transmembrane protein belonging to the TNF family, binds to the FAS ligand. Interaction between them results in the formation of the death-inducing signaling complex (DISC), which contains, caspase-8 and caspase-10. In some cases, caspase-8 is directly activated by interacting with foreign materials and subsequently activate further caspases.

Eom and Choi [34] also reported that 39% of Jurkat T cells underwent apoptosis when exposed to the AgNPs [34]. Zno NPs also triggered cell death via Caspase mediated apoptosis as reported by Wilhelmi et al. [37]. Similarly, high concentration of FeO NPs led to the 35–40% apoptosis [38].

(c) **Genotoxicity:** Nanogenotoxicity is a new term that has emerged in the field of nanotechnology. This term highlighted NP induced genotoxicity and carcinogenesis. Literature showed that, long term inflammation and oxidative stress present in a cell ultimately leads to DNA damage. Continuous production of ROS causes mutagenesis (due to the oxidation, hydrolysis and deamination of nucleic acid bases, substitution etc.) and carcinogenesis (due to the gene deletion, insertion etc.)

The results reported by Eom and Choi also indicate that DNA damage would have resulted into release of the DNA damage marker protein, p-H2AX which increased drastically after the exposure of AgNPs [34].

(d) **Cell surface interaction with proteins:** It is well known that NPs, especially heavy metal NPs tend to interact with proteins and amino acids. These interactions cause the formation of protein corona, protein unfolding, and altered protein function, which finally culminate into protein damage or non-functional. Similarly, interaction of NPs with protein molecules such as serum albumin, human blood protein haemoglobin (Hb), and cytoskeletal proteins result in protein conformational changes or protein damage.

Silica NPs significantly influence the unfolding of RNase, where the enzyme is less stable on NPs surface than in free solution [39].

(e) **Mitochondrial interaction:** Mitochondria is the powerhouse of cell, which provides the energy for vital cell functions. Several toxicological studies shown that NPs have identified mitochondria as a potentially relevant target organelle.

Mitochondrial interaction of NPs lead to following consequences:

(i) Disturbance in proton gradient pumps, voltage-gated channels

(ii) Change in mitochondrial outer membrane permeability, which releases cytochrome C (key event in intrinsic pathway of apoptosis)

(iii) Mitochondrial DNA damage, also releases cytochrome C (results into the necrosis)

Similarly, mitochondrial DNA encodes many structural proteins involved in various pathways. DNA mutagenesis causes the production of defective proteins, thus hampering the pathways like ATP synthesis, Electron Transfer Chain, oxidative phosphorylation etc. [33].

Cationic polystyrene NPs induced mitochondrial damage which eventually resulted into cell death [40]. Ning et al. [41] asserted that ultrafine particles are much potent for induction of mitochondrial damage [41].

(f) **Interaction with cellular signalling pathway:** NPs are inclined to induce the signalling cascade pathways via particle induced release of cytokines, particle to cell interactions and binding to the several cellular receptors as a ligand [33]. Signalling pathways which are activated through NPs; for e.g. MAPK, ERK, EGFR etc. are described in detail in the following section:

(i) **Nuclear factor kappa-light-chain-enhancer of activated B cells (NF-κB):** NF-κB is a group of proteins that controls transcription of DNA, cytokine production, cell proliferation, apoptosis, inflammation and cell survival. These proteins are also responsible for the activation of defence mechanisms as well as cellular responses to stimuli such as stress, cytokine, heavy metals, ultraviolet radiation. ROS activate NF-B via the release of IBs resulting in the nuclear translocation of NF-B [42].

ZnO and CdS NPs were found to induce toxicity via ROS-dependent NF-B activation [43].

(ii) **Activator protein-1 (AP-1):** Activator protein-1 is a transcription factor that regulates gene expression in response to a stimuli; including stress, oxidants, bacterial and viral infection, growth factors, and cytokines. AP-1 regulates cell proliferation, cell differentiation, and apoptosis. This gene is activated via phosphorylation of protooncogene c-jun. Cr, Ni, and Fe NPs have been shown to activate AP-1 via ROS generation [44].

(iii) **Mitogen-activated protein kinase (MAPK):** MAPK is a type of serine/threonine-specific protein kinase. It is involved in various cellular responses to a stimuli; such as proinflammatory cytokines, osmotic stress, heat shock and mitogens. It regulates cell proliferation, cell differentiation, apoptosis, gene expression, cell survival and mitosis. MAPK consist of growth factor regulated extracellular signal-related kinases (ERK) and the stress-activated MAPK, p38MAPK and c-jun NH2-terminal kinases (JNK). MAPK activation is based on the oxidative modification of MAPK signalling proteins (e.g., RTK and MAP3 K). The concentration and kinetics of ROS production and cellular antioxidant pool are mostly important for activation of MAPK signalling pathway. Silver NPs tend to activate JNK pathway and apoptosis whereas CeO_2 NPs trigger p38 MAPK signalling in broncho alveolar cells [45].

(iv) **Protein tyrosine phosphatases (PTP):** PTP is a tyrosine kinase that regulates the phosphorylation of various signalling molecules involved in signal transduction cascades. Signal transduction cascade pathways are involved in oncogenic transformation, mitosis, cell growth and cell differentiation. PTP is highly susceptible to oxidative stress in the form of free radicals and H_2O_2 [42]. Zn^{2+} and V^{4+} NPs are critical in redox regulation of PTP via the inhibition of MAPK and EGFR [46].

(v) **Epidermal growth factor receptor (EGFR):** It acts as an extracellular protein ligand for various members of epidermal growth factor family (EGF family). EGFR dimerization induces intracellular protein-tyrosine kinase activity which results into auto phosphorylation. This auto phosphorylation initiates further downstream signalling pathway such as MAPK, Akt and JNK leading to DNA synthesis and cell proliferation [46].

(vi) **Src family kinase:** Src family belongs to a non-receptor tyrosine kinases family, which is involved in regulation of cell growth, cell differentiation and oncogenic transformation. Oxidative stress is responsible for activation of this family, which later on triggers the signal transduction pathway. Src family kinases interact with many membranes, cytosolic and nuclear proteins by phosphorylation of tyrosine [42]. Low dose of Chromium NPs induced cell death via ROS dependent Src Kinase [47].

3.4 Toxicological Aspects of Nanoparticles via Different Transmucosal Routes

3.4.1 Ocular Mucosa-

In perceiving the advantages of new advances in nanobioadhesives for enhancing topical ocular delivery, the other side of the coin must also be considered. The toxicity literature in this area of research is not as robust as other fields, because most publications focus on the discovery and development of new therapeutic agents. It comes as no surprise that the same properties that make nanosystems attractive for drug delivery applications, may confer reactivity in biological systems and lead to toxicity. For topically ocular administered nanosystems, aggregation and tissue accumulation must be considered. Nanosystem aggregation may block cell metabolism and could impair tissue function. For example, blockage of the lachrymal drainage punctum and decreased tear film recycling can occur due to aggregation of topically applied nanosystems on the ocular surface. Furthermore, indiscriminate ocular nanosystem accumulation results in distortion of the ocular tissue architecture leading to altered function. A very important consideration of toxic effects with nanosystems is, that it may be attributable to actual

approaches that are directed to enhance ocular drug bioavailability; the presence of high concentration of the loaded drug in a non-target tissue.

Chitosan has some unique and important characteristics like being mucoadhesive and cationic. The positively charged chitosan NPs bind to the negatively charged surface of the cornea. Prow et al. [48] evaluated Chitosan, PCEP (poly{[(cholesteryl oxocarbonylamido ethyl) methyl bis(ethylene) ammonium iodide] ethyl phosphate}), and magnetic NPs (MNPs) for the safe gene delivery in the eye. Rabbits were administered with NPs either intravitreally (IV) or subretinally (SR) and sacrificed 7 days later. Eyes were grossly evaluated for retinal pigment epithelium abnormalities, retinal degeneration, and inflammation. IV chitosan showed inflammation in 12/13 eyes, whereas IV PCEP and IV MNPs were not inflammatory and did not induce retinal pathology. Acute inflammation and polymorphonuclear cell infiltrates resulted with injection of these nanoparticle formulations which were grossly visible in the eye cup after 7 days. Histological examination confirmed massive numbers of immune cells at the site of injection. It is possible that the hyalocytes, the sentinel immune cells of the vitreous, are particularly sensitive to polysaccharides leading to the inflammation [48].

Guo et al. [49] synthesized and evaluated a series of positively charged phospholipids and cholesterols as membrane components for liposomes. Selected liposome preparations formulated with these synthetic lipid materials were found to be non-cytotoxic in vitro by using a cell growth inhibition assay, whereas liposomes containing positively charged components (stearylamine and cetyltrimethylammonium bromide) showed considerable cytotoxicity. Their investigations thus, indicate a specific adhesion of the cationic liposomes to the surface of mucosal tissues owing to the presence of negative charge [49].

3.4.2 Nasal Mucosa-

Nasal mucoadhesive drug delivery system offers the advantage of higher residence time as compared to the normal dosage forms. Mucoadhesive polymers are being used to increase the residence time of formulation within the nasal cavity; the polymers possibly interact with the epithelial tight junctions to facilitate drug absorption by some histopathological alterations of the nasal mucosa and effect on ciliary beating. Hence it becomes necessary to evaluate the toxicological effects of such nano systems on the structural alterations of the nasal environment.

Hackenberg et al. [50] evaluated the toxic effect of repeated exposure of ZnO-NPs in three-dimensional (3D) mini organ cultures (MOCs) of human nasal mucosa determined by trypan blue exclusion and caspase-3 activity, respectively. MOCs were exposed once, twice, or three times to 0.1 or 5 µg/ml of ZnO-NPs for 1 h. per exposure and then evaluated for cytotoxicity and genotoxicity. DNA fragmentation augmented after 24 h of regeneration at both concentrations of ZnO-NPs. In contrast, DNA damage induced by the positive control, methyl methanesulfonate, was significantly reduced after 24-h regeneration. Thus, results suggest that repetitive exposure to low concentrations of ZnO-NPs results in persistent DNA damage. Various mechanisms responsible for ZnO-NPs related genotoxicity were proposed by the authors based on the study, such as direct interaction of particles with the DNA in the nucleus, ROS generation, or influence of dissolved zinc ions to DNA damage [50].

Genter et al. [51] assessed the distribution and toxic potential of 25-nm Silver NPs (AgNPs) (100 or 500 mg/kg) following intranasal (IN) exposure in adult male C57BL/6J mice. Histopathology of selected organs was performed, and tissue reduced glutathione (GSH) levels were measured after 1 or 7 days as an indicator of oxidative stress. Aggregated AgNPs were found in the spleen, lung, kidney, and nasal airway by routine light microscopy. Autometallography revealed AgNPs distributed in olfactory bulb and the lateral brain ventricles. Elevated tissue GSH levels was observed in nasal epithelia (both doses at 1 day, 500 mg/kg at 7 days) and blood (500 mg/kg at 7 days). Therefore, intranasal administration of AgNPs permits systemic distribution, produces oxidative stress in the nose and in blood, and develops macrophage-mediated erythrocyte destruction in the spleen [51].

3.4.3 Oral Buccal Mucosa

The buccal area appears to be an attractive site for administration of drugs due to the smooth and relatively immobile surface, good accessibility, the evasion of possible degradation in the gastrointestinal tract, and no first-pass metabolism in the liver. However, continuous salivation and swallowing may lead to a very short residence time in the oral cavity [4]. To overcome this problem, novel bioadhesive dosage forms have been developed; such as bioadhesive tablets, patches, bioadhesive gels and ointments, and medicated chewing gums. Smistad et al. [52] evaluated the toxicity of liposomal formulations on the human buccal cell line TR146. The main objective of this paper was to identify important liposomal formulation factors influencing the toxicity on cells in the buccal region, and identifying significant interactions between the formulation variables. Positively charged liposomes were shown to be toxic compared to the negatively charged liposomes. Diacyl-TAP (1,2-diacyl-3-trimethylammonium propane) was less toxic than SA, and DPPC was less toxic than DMPC. The amount of negatively charged component, the liposome size, and the total lipid concentration did not affect the toxicity within the experimental room. The most toxic combination in this study was egg-PC/positively charged lipid (20 mol%) [52].

Klemetsrud et al. [53] also investigated the in vitro toxicity, mucoadhesive potential and impact on cell permeability of polymer coated liposomes intended for oral use in TR146 cell line. The following polymers were tested: chitosan, low-methoxylated pectin (LM-pectin), high-methoxylated pectin (HM-pectin), amidated pectin (AM-pectin), Eudragit, poly(N-isopropylacrylamide-co-methacrylic acid) (p(NIPAAM-co-MAA)), hydrophobically modified hydroxyethyl cellulose (HM-HEC), and hydrophobically modified ethyl hydroxyethyl cellulose (HM-EHEC). All the systems, except with chitosan, exhibited no significant effect on cell viability and permeability at the considered concentrations. At the lowest chitosan concentration (0.05%) there was no sign of cell toxicity. However, there was a significant reduction of cell viability (65%) when the polymer concentration was increased to 0.1%. At even higher concentrations ($\geq$0.25%), the cell viability was around 9%; representing an almost complete cell death. No significant reductions in the cell viability were observed for the other polymers [53].

Teubl et al. [54] investigated interactions of three different titanium dioxide (TiO_2) particles (i.e. NM 100, NM 101 and NM 105) with oral tissues. Physicochemical properties were addressed in relevant media, and particle penetration was investigated with an ex vivo model using porcine mucosa. Although TiO_2 particles formed large aggregates once dispersed in media, 10–50% remained in the nanoscale range, rapidly interacting with the mucus layer and infecting the epithelium. NM 100 and NM 105 were found in both the upper part and the lower part of the buccal mucosa, while NM 101 (smallest particle sizes) only penetrated the upper parts. Transport studies revealed that TiO_2 NPs were found in vesicles, as well as freely distributed in the cytoplasm. However, NM 105 triggered the production of reactive oxygen species [54].

3.4.4 Pulmonary Mucosa-

Pulmonary route serves as a non-invasive approach for systemic delivery of therapeutic actives like drug, proteins and peptides. Larger surface area of lungs, rich blood supply and a thin mucous layer make the pulmonary route popular for efficient systemic delivery. This route is an effective alternative for the delivery of those therapeutic actives that show less bioavailability via oral route, especially in the case of proteins and peptides. Specialized devices such as dry powder inhaler, metered dose inhaler and nebulizer have been fabricated to facilitate pulmonary delivery by ensuring deep lung deposition. The only significant formulation based parameter to be considered when designing a dosage form for pulmonary delivery is the particle size, the ideal size being 1–5 μ. Besides the use of polymeric or lipidic NPs for relief for pulmonary diseases, human beings are also inadvertently exposed to inorganic NPs. Metal NPs such as iron, silica, nickel, carbon etc. are inhaled into the body via upper respiratory tract in the work place environment, which is a common event [55]. In these conditions, lung primarily acts as a site of accumulation and long term exposure of NPs. Once they enter into the interstitial spaces, they are rapidly taken up by alveolar cells and induce toxic effects. Occupational exposure of these NPs causes hazardous and harmful effects on human being leading to cancer, asthma, fibrosis, and pneumonitis etc. Naturally occurring NPs are well tolerated or adapted by human and the environment. However, the unintentional and intentional inhalation and accumulation of NPs pose a serious threat to human as well as environment [56].

Coating of metallic NPs with polysaccharides can overcome the drawbacks by increasing stability and biocompatibility, improving size distributions and introducing chemical groups that allow for further functionalization of the NPs. Worthington et al. [57] have demonstrated that chitosan coating reduced the toxicity of copper NPs significantly after 24 and 52 h and the generation of ROS. Conversely, inflammatory response of mice exposed to chitosan coated NPs, measured using the number of WBC and cytokines/chemokines in the bronchoalveolar fluid was shown to increase, as was the concentration of copper ions. These results suggest that coating of metal NPs with mucoadhesive polysaccharides (e.g. chitosan) could increase their potential for use in controlled release of copper ions to cells, but will result in a higher inflammatory response if administered via the lung [57]. Tables 3.1 and 3.2 summarises the other in vivo and in vitro research studies pertaining to the possible toxic effects of NPs via pulmonary route.

3.4.5 Rectal/Vaginal Mucosa-

Nanotechnology based drug delivery systems may be an interesting option to advance in the field of microbicides and have been advocated for the delivery of promising microbicide drug candidates such as dapivirine. Among other advantages, nanosystems may be able to enhance drug/virus interaction, penetrate the mucosa, target HIV-susceptible cells, and provide an effective and durable drug barrier along the epithelial lining when administered by the route. In the case of vaginal and rectal administration, only a few studies explored this possibility thus resulting in a substantial lack of data supporting the potential value of nanotechnology-based drug delivery systems. However early methods for assessing toxic effects of NPs are insufficient to detect these toxicities. There is also a paucity of data regarding excipients, components often assumed to be non-toxic on the basis of past experience with commercial vaginal products.

The first report of microbicide toxicity was by Phillips and Zacharopoulos [76] who found that rectal application of N-9 caused rapid exfoliation of sheets of epithelial cells and failed to protect mice against rectal transmission of HSV-2 [76]. das Neves J et al. [77] reported the preparation and characterization of drug-loaded poly ε-caprolactone (PCL), poly(ethylene oxide) (PEO), cetyl trimethylammonium bromide (CTAB) NPs of dapivirine, as well as their influence on the permeation and retention in cervicovaginal and colorectal cell monolayer models, and pig vaginal and rectal mucosa at the cell and tissue level. Dapivirine-loaded NPs of around 175–200 nm and different surface properties were successfully prepared and were readily taken up by different anogenital epithelial cells. NPs were shown able to modulate differently the permeability and monolayer/tissue retention kinetics of dapivirine. PEO-PCL NPs reduced the permeability of dapivirine, while CTAB-PCL NPs increased the diffusion of the drug across studied models. Further, toxicity results in pig vaginal and rectal mucosa showed unacceptable toxicity with CTAB-PCL NPs as compared to free dapivirine where it did not show any toxicity issues [77]. The other toxicological cases of the NP drug delivery via vaginal route are discussed in Table 3.3.

Table 3.1 In vivo NPs toxicity via pulmonary route

No	Types of NPs	Toxicity	Analysis	References
1	Diesel exhaust NPs	Enhanced gene transcription of interleukin-8 (IL-8) in the bronchial tissue and growth-regulated oncogene-alpha (GRO-alpha) protein expression in the bronchial epithelium	RT-PCR and ELISA	[58]
2	Fe_2O_3	Significant enhancement of free radicals and reduction of the GSH in lung tissue and caused pulmonary emphysema, interstitial hyperaemia and inflammation in lungs	Histopathology	[59]
3	Multi-wall carbon nanotubes (MWCNT)	Induced inflammatory and fibrotic reactions by increase production of TNF-α	Histopathology, ELISA and biochemical test	[60]
4	Silver	Dose dependent increase in chronic alveolar inflammation, including thickened alveolar walls and small granulomatous lesions	Histopathology	[61]
5	Silver	Disruption in the blood/alveolar epithelial permeability barrier, oxidative stress and activation of eosinophils, with release of cytokines IL-13 and IgE	Histopathology and PCR	[62]
6	TiO_2 dose >5.0 mg/kg	Developed lung lesions, leakage of cytoplasmic contents, compensatory proliferation and fibrosis of lung tissues	Histopathology	[63]
7	Nickel oxide (NiO)	Pulmonary inflammation accompanied by inflammatory cell infiltration, alveolar proteinosis, and cytokine secretion. Increase ROS production induced IL-1β production	Histopathology. western blotting and immunohistochemistry	[64]
8	Cerium Oxide(CeO_2)	Induced a persistent influx of neutrophils and expression of cytokines such as CINC-1, CINC-2, and HO-1	Histopathology and cytokine assay	[65]
9	Copper oxide	Generation of acute neutrophilic inflammation and cytokine (MCP-1 and IL-6) responses in the lungs	ELISA kit	[66]
10	Barium sulphate	Dose-dependent lung injury and inflammation	Lactate dehydrogenase and myeloperoxidase assay	[67]
11	Silica coated Titanium dioxide	Neutrophilic pulmonary inflammation, elicited increased expression of proinflammatory cytokine TNF-α and neutrophil chemoattractant CXCL1	RT-PCR and ELISA kit	[68]

3.5 Summary and Future Prospects

Nanomaterials are endowed with unique properties, which are responsible for their toxicological manifestations. This had led to concerns over the use of such materials, inspite of their widespread applications. Moreover, rapid developments are viewed in the field of nanotechnology every day, which will only further complicate the toxicological profile of nanomaterials. This necessitates development of a strategy to evaluate the toxic effects of the upcoming nanomaterials and their products before they are being brought into mainstream use. The toxicity of nanomaterials depends on the basic chemistry of interaction of the material with the molecular pathways in an organism. A thorough investigation on the molecular interaction and alteration in the biochemical machinery of an organism upon contact with a nanomaterial is a prerequisite for designing safe and sustainable nanomaterials. There is a multitude of research lit-

Table 3.2 In vitro NP toxicity via pulmonary routes

No	Types of NPs	Toxicity	Cell line	Analysis	References
1	Diesel exhaust NPs	Induces the release of proinflammatory cytokines after triggering transduction pathways, NF-κB activation and MAPK phosphorylation	Human bronchial epithelial cells (16HBE)	Western blotting	[69]
2	Nanoparticulate carbon black	Stimulates proliferation which result into autocrine release of EGF and the subsequent EGF-R transactivation and ERK cascade activation	Human bronchial epithelial cell line	Immunocytochemistry and western blotting	[70]
3	Silica	Dose-dependent increase in Cytotoxicity and oxidative stress	A549 human lung cancer cells	MTT assay	[71]
4	Nickel ferrite	Dose dependent cytotoxicity, Induced apoptosis in A549 cells through ROS generation and oxidative stress via p53, survivin, bax/bcl-2 and caspase pathways	A549 human lung cancer cells	MTT asaay, qPCR	[72]
5	Copper oxide	Dose dependent cytotoxicity, induce oxidative stress in response to ROS production	Hep-2 Human laryngeal epithelial cells	Alamar blue assay	[73]
6	Copper oxide	Dose dependent increase in cytotoxicity	BEAS-2B-human bronchial epithelial and RAW 264.7-murine myeloid cells	MTT assay	[65]
7	Copper oxide	Induction of cytotoxic, genotoxic (upregulation of cell cycle checkpoint protein p53 and DNA damage repair proteins Rad51 and MSH2 expression) and oxidative stress response	A549 human lung cancer cells	MTT assay, neutral red assay, western blotting and Catalse, Superoxide dismutase assay	[74]
8	Aluminium oxide (Al_2O_3)	Dose dependent cytotoxicity	A549 and Hep-2 cell line	MTT assay	[75]

erature now available on the toxicity of nanomaterials to various model organisms from human to ecological models. But majority of the reports focus on acute high dose exposures. Research on the toxicity of other chemicals has shown that dose of a chemical can have a tremendous impact on the pathways that are affected within the organism. Overall, it appears from the literature that many of the current categories of available nanomaterials are not acutely toxic but are most likely to have toxic implications following long-term low dose exposures. There is a significant gap regarding these types of potential impacts.

The most common pathways investigated in nanotoxicity experiments are related to oxidative stress, yet oxidative stress can be a temporary and natural response to an insult without a negative outcome. Currently, a few individual biomarkers are being explored in this capacity and they are often limited to pathways involved in oxidative stress, which is known to be a complicated biomarker. There are a multitude of other potential non-oxidative stress mechanisms that may be triggered in response to sub-lethal exposures of nanoparticles. Testing these materials at low concentrations will allow the development of bio-

Table 3.3 NPs toxicity via vaginal route

No	Types of NPs	Toxicity	Analysis	References
1	Silver	Silver particles accumulates in vaginal tissue which caused ultra-structural pathological changes cause cytotoxic reaction	TEM	[78]
2	Acyclovir loaded polystyrene NPs	Acute inflammation and high neutrophil infiltration into the lumen	Standard hematoxylin and eosin (H&E) staining	[79]
3	Paclitaxel loaded NPs	At 24 h after treatment, changes in nuclear morphology (chromatin clumping, nuclear fragmentation) were evident in a minority of vaginal epithelial cells, increase in the number of apoptotic epithelial cells was noted after 48 h	Histopathology	[80]
4	Melittin NPs	concentrations ≥40 μM mel-NPs resulted in complete loss of vaginal epithelial cell viability	MTT assay	[81]

markers for evaluating nanomaterial toxicity. Also, the ‘unique’ properties demand newer concepts and methodologies to understand and foresee the size (and shape)-specific interaction of nanoparticles within the human body. Recent research has led scientists to identify some discrete physicochemical characteristics of nanoparticles that required on priority basis to assess their toxicological potential. These include particle size and surface area in relevant media the route of administration, physical form, degree of aggregation, surface characteristics, sample purity, etc. Such assessment in the case of nanoparticles poses an additional challenge, since the characteristics of nanoparticles largely depend on their form and chemical composition; both exhibiting a dynamic nature before, during and after administration into in vitro or in vivo systems.

Also, there is a pressing need for the guidance and documentation of nanotoxicity which all researchers come across. The documentation will serve as a reference for further studies involving nanoparticles and nanocarriers. To our understanding, at present data obtained from diverse nanotoxicity testings are neither comparable nor reliable, robust or reproducible. It is a daunting task to extract reliable data from the huge amount of studies in total and to come to comprehensive and valid conclusions that allow us to safely implement nanotechnology in medical and other sophisticated applications. In addition, the strict utilization and realization of validated and standardized intelligent testing policies should become second nature to all ‘nano-scientists/toxicologists’.

Second, nanomaterials clearly demonstrate the need for alternatives assessment methods to consider the intrinsic exposure potential as part of the comparative assessment process because there are distinct physicochemical properties as well as use characteristics that will distinguish from the bulk material. In this respect the role of the peer-reviewing process of manuscripts should be to rigorously demand and enforce these high standards. All studies that address effects of nanomaterials should be reviewed by experts in the field of nanotoxicology. Moreover, these experts ought to closely stick to the given recommendations.

References

1. Dekker J, Rossen JW, Büller HA et al (2002) The MUC family: an obituary. Trends Biochem Sci 27:126–131
2. Scharfman A, Lamblin G, Roussel P (1995) Interactions between human respiratory mucins and pathogens. Biochem Soc Trans 23:836–839
3. Cone RA (2009) Barrier properties of mucus. Adv Drug Deliv Rev 61:75–85
4. Rathbone MJ, Tucker IG (1993) Mechanisms, barriers and pathways of oral mucosal drug permeation. Adv Drug Deliv Rev 12:41–60
5. Wang Y-Y, Lai SK, So C et al (2011) Mucoadhesive nanoparticles may disrupt the protective human

mucus barrier by altering its microstructure. PLoS One 6(6):e21547
6. Patel VF, Liu F, Brown MB (2011) Advances in oral transmucosal drug delivery. J Control Release 153:106–116
7. Sosnik A, Neves JD, Sarmento B (2014) Mucoadhesive polymers in the design of nano-drug delivery systems for administration by non-parenteral routes: a review. Prog Polym Sci 39:2030–2075
8. Ludwig A (2005) The use of mucoadhesive polymers in ocular drug delivery. Adv Drug Deliv Rev 57:1595–1639
9. Kaur IP, Smitha R (2002) Penetration enhancers and ocular bioadhesives: two new avenues for ophthalmic drug delivery. Drug Dev Ind Pharm 28:353–369
10. Donnelly R, Shaikh R, Singh TR et al (2011) Mucoadhesive drug delivery systems. J Pharm Bioallied Sci 3:89
11. Edsman K, Hägerström H (2005) Pharmaceutical applications of mucoadhesion for the non-oral routes. J Pharm Pharmacol 57:3–22
12. Ugwoke M, Agu R, Verbeke N et al (2005) Nasal mucoadhesive drug delivery: background, applications, trends and future perspectives. Adv Drug Deliv Rev 57:1640–1665
13. Hoogstraate JA, Wertz PW, Wertz PW (1998) Drug delivery via the buccal mucosa. Pharm Sci Technol Today 1:309–316
14. Rhodin JA (1966) Ultrastructure and function of the human tracheal mucosa. Am Rev Respir Dis 93(3P2):1–15
15. Meyrick B, Sturgess JM, Reid L (1969) A reconstruction of the duct system and secretory tubules of the human bronchial submucosal gland. Thorax 24:729–736
16. Sigurdsson HH, Kirch J, Lehr C-M (2013) Mucus as a barrier to lipophilic drugs. Int J Pharm 453:56–64
17. Murugan K, Choonara YE, Kumar P et al (2015) Parameters and characteristics governing cellular internalization and trans-barrier trafficking of nanostructures. Int J Nanomedicine 10:2191
18. Kumari S, Mg S, Mayor S (2010) Endocytosis unplugged: multiple ways to enter the cell. Cell Res 20:256–275
19. Xiang S, Tong H, Shi Q et al (2012) Uptake mechanisms of non-viral gene delivery. J Control Release 158:371–378
20. Ferrati S, Mcconnell KI, Mack AC et al (2014) Cellular communication via nanoparticle-transporting biovesicles. Nanomedicine 9:581–592
21. Fleischer CC, Payne CK (2014) Nanoparticle–cell interactions: molecular structure of the protein corona and cellular outcomes. Acc Chem Res 47:2651–2659
22. Panyam J (2002) Rapid endo-lysosomal escape of poly(DL-lactide-co-glycolide) nanoparticles: implications for drug and gene delivery. FASEB J 16:1217–1226
23. Aydın A, Sipahi H, Charehsaz M (2012) Nanoparticles toxicity and their routes of exposures. In: Recent advances in novel drug carrier systems. InTechOpen. Edited by Ali Demir Sezer.
24. Elsaesser A, Howard CV (2012) Toxicology of nanoparticles. Adv Drug Deliv Rev 64:129–137
25. Gatoo MA, Naseem S, Arfat MY et al (2014) Physicochemical properties of nanomaterials: implication in associated toxic manifestations. Biomed Res Int 2014(498420):1–8
26. Park KH, Chhowalla M, Iqbal Z et al (2003) Single-walled carbon nanotubes are a new class of ion channel blockers. J Biol Chem 278:50212–50216
27. Hsiao IL, Huang YJ (2011) Effects of various physicochemical characteristics on the toxicities of ZnO and TiO2 nanoparticles toward human lung epithelial cells. Sci Total Environ 409:1219–1228
28. Lippmann M (1990) Effects of fiber characteristics on lung deposition, retention, and disease. Environ Health Perspect 88:311–317
29. Poland CA, Duffin R, Kinloch I et al (2008) Carbon nanotubes introduced into the abdominal cavity of mice show asbestos-like pathogenicity in a pilot study. Nat Nanotechnol 3:423–428
30. Bhattacharjee S, Haan LHJD, Evers NM et al (2010) Role of surface charge and oxidative stress in cytotoxicity of organic monolayer-coated silicon nanoparticles towards macrophage NR8383 cells. Part Fibre Toxicol 7:25
31. Gurr JR, Wang AS, Chen CH et al (2005) Ultrafine titanium dioxide particles in the absence of photoactivation can induce oxidative damage to human bronchial epithelial cells. Toxicology 213:66–73
32. Wick P, Manser P, Limbach L et al (2007) The degree and kind of agglomeration affect carbon nanotube cytotoxicity. Toxicol Lett 168:121–131
33. Unfried K, Albrecht C, Klotz LO et al (2007) Cellular responses to nanoparticles: target structures and mechanisms. Nanotoxicology 1:52–71
34. Eom H-J, Choi J (2010) p38 MAPK activation, DNA damage, cell cycle arrest and apoptosis as mechanisms of toxicity of silver nanoparticles in Jurkat T cells. Environ Sci Technol 44:8337–8342
35. Carero ADP, Hoet P, Verschaeve L et al (2001) Genotoxic effects of carbon black particles, diesel exhaust particles, and urban air particulates and their extracts on a human alveolar epithelial cell line (A549) and a human monocytic cell line (THP-1). Environ Mol Mutagen 37:155–163
36. Dybdahl M, Risom L, Bornholdt J et al (2004) Inflammatory and genotoxic effects of diesel particles in vitro and in vivo. Mutat Res 562:119–131
37. Wilhelmi V, Fischer U, Weighardt H et al (2013) Zinc oxide nanoparticles induce necrosis and apoptosis in macrophages in a p47phox-and Nrf2-independent manner. PLoS One 8(6):e65704
38. Naqvi S, Samim M, Abdin M et al (2010) Concentration-dependent toxicity of iron oxide nanoparticles mediated by increased oxidative stress. Int J Nanomedicine 5:983

39. Shang W, Nuffer JH, Dordick JS et al (2007) Unfolding of ribonuclease A on silica nanoparticle surfaces. Nano Lett 7:1991–1995
40. Xia T, Kovochich M, Brant J et al (2006) Comparison of the abilities of ambient and manufactured nanoparticles to induce cellular toxicity according to an oxidative stress paradigm. Nano Lett 6:1794–1807
41. Li N, Sioutas C, Cho A et al (2002) Ultrafine particulate pollutants induce oxidative stress and mitochondrial damage. Environ Health Perspect 111:455–460
42. Manke A, Wang L, Rojanasakul Y (2013) Mechanisms of nanoparticle-induced oxidative stress and toxicity. Biomed Res Int 2013:1–15
43. Pujalté I, Passagne I, Brouillaud B et al (2011) Cytotoxicity and oxidative stress induced by different metallic nanoparticles on human kidney cells. Part Fibre Toxicol 8:10
44. Poljak-Blazi M, Jaganjac M, Mustapic M et al (2009) Acute immunomodulatory effects of iron polyisomaltosate in rats. Immunobiology 214:121–128
45. Park E-J, Choi J, Park Y-K et al (2008) Oxidative stress induced by cerium oxide nanoparticles in cultured BEAS-2B cells. Toxicology 245:90–100
46. Kim Y-M, Reed W, Wu W et al (2006) Zn 2+--induced IL-8 expression involves AP-1, JNK, and ERK activities in human airway epithelial cells. Am J Phys Lung Cell Mol Phys 290(5):L1028–L1035
47. Balamurugan K, Rajaram R, Ramasami T et al (2002) Chromium(III)-induced apoptosis of lymphocytes: death decision by ROS and Src-family tyrosine kinases. Free Radic Biol Med 33:1622–1640
48. Prow TW, Bhutto I, Kim SY et al (2008) Ocular nanoparticle toxicity and transfection of the retina and retinal pigment epithelium. Nanomedicine 4:340–349
49. Guo L, Sarris A, Levy M (1989) A safe bioadhesive liposomal formulation for ophthalmic applications. Invest Ophthalmol Vis Sci (Suppl) 29:439
50. Hackenberg S, Zimmermann F-Z, Scherzed A et al (2011) Repetitive exposure to zinc oxide nanoparticles induces dna damage in human nasal mucosa mini organ cultures. Environ Mol Mutagen 52:582–589
51. Genter MB, Newman NC, Shertzer HG et al (2012) Distribution and systemic effects of intranasally administered 25 nm silver nanoparticles in adult mice. Toxicol Pathol 40(7):1004–1013
52. Smistad G, Jacobsen J, Sande SA (2007) Multivariate toxicity screening of liposomal formulations on a human buccal cell line. Int J Pharm 330(1):14–22
53. Klemetsrud T, Kjoniksen AL, Hiorth M et al (2016) Polymer coated liposomes for use in the oral cavity–a study of the in vitro toxicity, effect on cell permeability and interaction with mucin. J Liposome Res:1–12
54. Teubl BJ, Leitinger G, Schneider M et al (2015) The buccal mucosa as a route for TiO2 nanoparticle uptake. Nanotoxicology 9(2):253–261
55. Li JJ, Muralikrishnan S, Ng CT et al (2010) Nanoparticle-induced pulmonary toxicity. Exp Biol Med 235(9):1025–1033
56. Magaye R, Zhao J, Bowman L et al (2012) Genotoxicity and carcinogenicity of cobalt-, nickel-and copper-based nanoparticles. Exp Ther Med 4(4):551–561
57. Worthington KL, Adamcakova-Dodd A, Wongrakpanich A et al (2013) Chitosan coating of copper nanoparticles reduces in vitro toxicity and increases inflammation in the lung. Nanotechnology 24(39):395101
58. Salvi SS, Nordenhall C, Blomberg A et al (2000) Acute exposure to diesel exhaust increases IL-8 and GRO-α production in healthy human airways. Am J Respir Crit Care Med 161(2):550–557
59. Sadeghi L, Yousefi BV, Espanani H (2015) Toxic effects of the Fe2O3 nanoparticles on the liver and lung tissue. Bratisl Lek Listy 116(6):373–378
60. Muller J, Huaux F, Moreau N et al (2005) Respiratory toxicity of multi-wall carbon nanotubes. Toxicol Appl Pharmacol 207(3):221–231
61. Sung JH, Ji JH, Yoon JU et al (2008) Lung function changes in Sprague-Dawley rats after prolonged inhalation exposure to silver nanoparticles. Inhal Toxicol 20(6):567–574
62. Seiffert J, Hussain F, Wiegman C et al (2015) Pulmonary toxicity of instilled silver nanoparticles: influence of size, coating and rat strain. PLoS One 10(3):e0119726
63. Liu R, Yin L, Pu Y et al (2009) Pulmonary toxicity induced by three forms of titanium dioxide nanoparticles via intra-tracheal instillation in rats. Prog Nat Sci 19(5):573–579
64. Cao Z, Fang Y, Lu Y et al (2016) Exposure to nickel oxide nanoparticles induces pulmonary inflammation through NLRP3 inflammasome activation in rats. Int J Nanomedicine 11:3331
65. Morimoto Y, Izumi H, Yoshiura Y et al (2015) Pulmonary toxicity of well-dispersed cerium oxide nanoparticles following intratracheal instillation and inhalation. J Nanopart Res 17(11):442
66. Zhang H, Ji Z, Xia T et al (2012) Use of metal oxide nanoparticle band gap to develop a predictive paradigm for oxidative stress and acute pulmonary inflammation. ACS Nano 6:4349–4368
67. Konduru N, Keller J, Ma-Hock L et al (2014) Biokinetics and effects of barium sulfate nanoparticles. Part Fibre Toxicol 11(1):55
68. Rossi EM, Pylkkänen L, Koivisto AJ et al (2009) Airway exposure to silica-coated TiO2 nanoparticles induces pulmonary neutrophilia in mice. Toxicol Sci 113(2):422–433
69. Bonvallot V, Baeza-Squiban A, Baulig A et al (2001) Organic compounds from diesel exhaust particles elicit a proinflammatory response in human airway epithelial cells and induce cytochrome p450 1A1 expression. Am J Respir Cell Mol Biol 25(4):515–521
70. Tamaoki J, Isono K, Takeyama K et al (2004) Ultrafine carbon black particles stimulate proliferation of human airway epithelium via EGF receptor-mediated signaling pathway. Am J Phys Lung Cell Mol Phys 287(6):L1127–L1133
71. Lin W, Huang YW, Zhou X-D et al (2006) In vitro toxicity of silica nanoparticles in human lung cancer cells. Toxicol Appl Pharmacol 217(3):252–259
72. Ahamed M, Akhtar MJ, Siddiqui MA et al (2011) Oxidative stress mediated apoptosis induced by

nickel ferrite nanoparticles in cultured A549 cells. Toxicology 283(2):101–108
73. Fahmy B, Cormier SA (2009) Copper oxide nanoparticles induce oxidative stress and cytotoxicity in airway epithelial cells. Toxicol in Vitro 23(7):1365–1371
74. Ahamed M, Siddiqui MA, Akhtar MJ et al (2010) Genotoxic potential of copper oxide nanoparticles in human lung epithelial cells. Biochem Biophys Res Commun 396(2):578–583
75. Arul Prakash F, Dushendra Babu G, Lavanya M et al (2011) Toxicity studies of aluminium oxide nanoparticles in cell lines. Int J Nanotechnol Appl 5(2):99–107
76. Phillips DM, Zacharopoulos VR (1998) Nonoxynol-9 enhances rectal infection by herpes simplex virus in mice. Contraception 57(5):341–348
77. das Neves J, Araujo F, Andrade F et al (2013) In vitro and ex vivo evaluation of polymeric nanoparticles for vaginal and rectal delivery of the anti-HIV drug dapivirine. Mol Pharm 10(7):2793–2807
78. Chen D, Yang Z (2015) Tissue toxicity following the vaginal administration of nanosilver particles in rabbits. Regen Biomater 2(4):261–265
79. Ensign LM, Tang BC, Wang YY et al (2012) Mucus-penetrating nanoparticles for vaginal drug delivery protect against herpes simplex virus. Sci Transl Med 4(138):138ra179–138ra179
80. Yang M, Yu T, Wang YY et al (2014) Vaginal delivery of paclitaxel via nanoparticles with non-mucoadhesive surfaces suppresses cervical tumor growth. Adv Healthcare Mat 3(7):1044–1052
81. Jallouk AP, Moley KH, Omurtag K et al (2014) Nanoparticle incorporation of melittin reduces sperm and vaginal epithelium cytotoxicity. PLoS One 9(4):e95411

The Toxicity of Nanoparticles to Human Endothelial Cells

4

Yi Cao

Abstract

The use of nanoparticles (NPs) in commercially available products and as biomedicinal materials could lead to increasing contact of human blood vessels with NPs, and it is necessary to assess the potential adverse effects of NPs to cells lining blood vessels. Of them, endothelial cells (ECs) are of particular relevance as they play a crucial role in the regulation of function of blood vessels. In this book chapter, I discussed studies that used human ECs to study the toxicity and mechanisms of NPs. It has been shown that exposure of human ECs to NPs could lead to cytotoxicity, genotoxicity, endothelial activation and impaired NO signaling. Oxidative stress and inflammation induced by NPs have been suggested as the mechanisms associated with the toxicity of NPs to ECs, and a three-tier model has been proposed to explain the association between NP induced oxidative stress and toxicity. In recent years, dysfunction of autophagy (excessive autophagy induction) has also been suggested as one of the mechanisms associated with the toxicity of NPs to human ECs. In the future, it is necessary to use human ECs to assess the toxicity of NPs to better understand the potential adverse effects of NPs entering circulation.

Keywords

Nanoparticle (NP) · Endothelial cell (EC) · Oxidative stress · Endothelial activation · Genotoxicity

4.1 Introduction

Nanoparticles (NPs) are defined as 'natural, incidental or manufactured material containing particles, in an unbound state or as an aggregate or as an agglomerate and where, for 50% or more of the particles in the number size distribution,

Y. Cao (✉)
Key Laboratory of Environment-Friendly Chemistry and Applications of Ministry Education, Laboratory of Biochemistry, College of Chemistry, Xiangtan University, Xiangtan, People's Republic of China
e-mail: caoyi39@xtu.edu.cn

Q. Saquib et al. (eds.), *Cellular and Molecular Toxicology of Nanoparticles*, Advances in Experimental Medicine and Biology 1048, https://doi.org/10.1007/978-3-319-72041-8_4

one or more external dimensions is in the size range 1 nm–100 nm' by the European Commission in 2011. In specific cases, the threshold between 1% and 50% for the number size distribution may also be used instead of 50% [1]. Nano-sized particles could be generated through combustion, and combustion-derived NPs have been convincingly shown to induce adverse health effects to blood vessels [2, 3]. In recent years, the rapid development of nanotechnologies also leads to increasing use of engineered NPs in many commercial products, marking engineered NPs as one of the major sources for particle exposure in modern society. According to a recent survey, a total of 1814 commercial products sold in global markets contain at least one type of NPs, which can lead to human exposure to NPs in daily life via three major routes, namely dermal (count for 58% of products evaluated), inhalational (25%) and oral (16%) exposure [4]. By the analog with combustion-derived NPs, it has been suggested that engineered NPs added in commercial products may also induce adverse health effects [2, 3].

Besides intended use in commercial products, NPs may also have potential use in medicine. For example, carbon nanotubes (CNTs), including single-walled carbon nanotubes (SWCNTs) and multi-walled carbon nanotubes (MWCNTs), could be used as nanocarriers for drug delivery [5]. Ag NPs are potential anticancer therapeutic agents that could be used in the treatment of a number of cancer, such as leukemia, breast cancer and lung cancer [6]. Metal-based NPs including Ag and ZnO NPs are also effective alternative antimicrobial agents for the treatment of infectious diseases [7]. Iron oxide NPs (IONPs) could be applied as contrast agents in magnetic resonance imaging (MRI) as well as for the treatment of iron deficiency [8]. However, despite the enthusiasm for the development of NPs for medicinal purposes, progress into clinics is relatively slow. One of the concerns is the adverse health effects of NPs, especially to the vasculature, because intravenous injection of NPs in nanomedicine could lead to increasing contact of vasculature to NPs [9]. Thus, it is necessary to evaluate the potential effects of NPs to cells lining blood vessels.

Human blood vessels are covered by a thin layer of endothelial cells (ECs), which act as the first contact and barrier for NPs entering circulation. Moreover, ECs play a crucial role in the regulation of blood vessel tone, thrombogenicity, homeostasis and monocytes recruitment [10]. In normal vascular physiology, the ECs generate NO by endothelial NOS (eNOS), which plays a key role to maintain the quiescent state of the endothelium by inhibition of inflammation, cellular proliferation and thrombosis, and the monocytes adhere poorly onto the endothelium under normal conditions [11]. With the development of cardiovascular diseases, persistent inflammation in blood vessels activates ECs, which in turn express excessive adhesion molecules such as selectins, ICAM-1 and VCAM-1 to recruit monocytes. Monocytes adhere to the activated ECs, migrate through the endothelium to the intima, accumulate lipids and transform to lipid laden foam cells, which result in the earliest visible atherosclerotic lesions as fatty streaks [12]. Meanwhile, the development of cardiovascular diseases was also associated with a gradual loss of NO bioactivity due to the uncoupling of eNOS (eNOS uncoupling; Fig. 4.1). Given the importance of ECs in maintaining the proper function of blood vessels, it is necessary to evaluate the toxicity of NPs to ECs in order to better understand the potential effects of NPs in vivo [9].

In this book chapter, I will discuss about the toxicity of NPs to ECs as well as the possible mechanisms. The discussion will be limited to studies using human ECs, rather than animal ECs, as the results are easier to be extrapolated to human endothelium in vivo. I hope this book chapter will help future studies which consider to evaluate the adverse effects of NPs to endothelium.

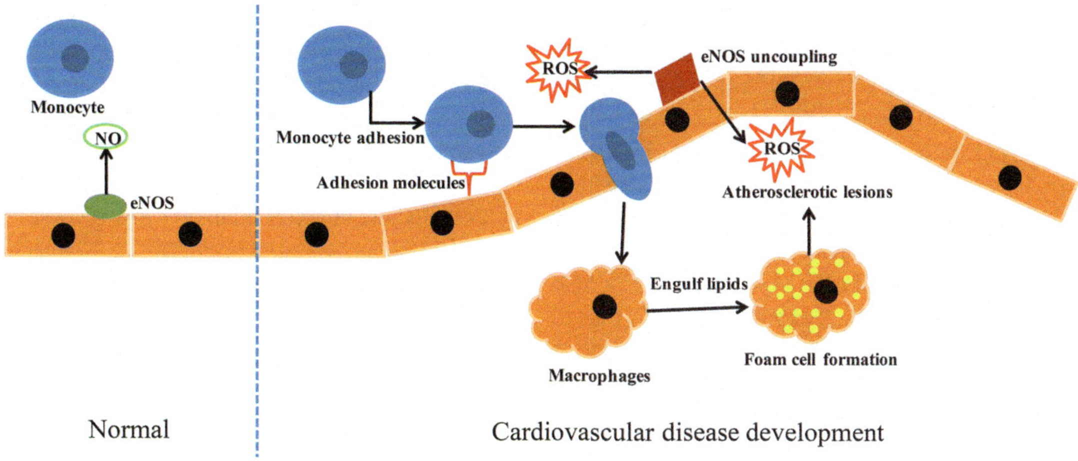

Fig. 4.1 Key events associated with the early development of atherosclerosis. In normal endothelial cells (ECs), endothelial NOS (*eNOS*) generate NO to maintain the quiescent state of the endothelium, and the monocytes adhere poorly onto the endothelium. With the development of cardiovascular diseases, ECs are activated to express excessive adhesion molecules, and monocytes subsequently adhere to the activated ECs, migrate through the endothelium to the intima, accumulate lipids and transform to lipid laden foam cells to form the earliest visible atherosclerotic lesions. Meanwhile, there is also a gradual loss of NO bioactivity due to eNOS uncoupling

4.2 Toxicity of NPs to Human Endothelial Cells

4.2.1 Cytotoxicity

A number of NPs has been shown to induce cytotoxicity to human ECs. Of them, the soluble metal based NPs appear to be the most cytotoxic due to the dissolution of NPs to release metal ions. For example, ZnO NPs are partially soluble to release Zn ions, which has been convincingly shown as one of the mechanisms for ZnO NP induced toxicity to mammalian cells [13]. A pilot study by Sun et al. [14] showed that 24 h exposure to ZnO NPs (primary size 45.3 nm) was associated with significantly decreased mitochondrial activity in human cardiac microvascular endothelial cells (HCMECs), with a threshold as low as 5 μg/mL. Liang et al. [15] showed that 24 h exposure to ZnO NPs (primary size 70 nm) at the concentrations ≥15 μg/mL significantly induced cytotoxicity in human aortic endothelial cells (HAECs) as decreased mitochondrial activity, lactate dehydrogenase (LDH) release and apoptosis. We have also recently shown that 24 h exposure to 32 μg/mL ZnO NPs (primary size 100 nm) induced cytotoxicity in human umbilical vein endothelial cells (HUVECs) as decreased mitochondrial activity, LDH release and lysosomal damages, which was associated with increased intracellular Zn ions [16, 17]. Ag NPs, which may lead to the accumulation of Ag ions, were also cytotoxic to endothelial cells. Guo et al. [18] showed that 24 h incubation of HUVECs with citrated-coated Ag NPs (10, 75, and 110 nm) from 1 to 40 μg/mL significantly induced cytotoxicity associated with increased accumulation of intracellular Ag, with 110 nm Ag NPs being most effective. Meanwhile, $AgNO_3$ at the same concentrations was more cytotoxic compared with Ag NPs, which indicated that $AgNO_3$ induced cells death through a mechanism different from Ag NPs. A comparative study by Danielsen et al. [19] also showed that Ag NPs (<20 nm) induced cytotoxicity in HUVECs, but the threshold of Ag NPs (64 μg/mL) was higher than that of ZnO NPs (32 μg/mL). Another example is Cd based quantum dot (QD), which can release the highly toxic Cd ions. Yan et al. [20] showed that CdTe QD (mercaptosuccinic acid coated; primary size ~4 nm) induced cytotoxicity in HUVECs, with a threshold as low as 0.1 μg/mL. At 10 μg/mL, 24 h incubation of HUVECs with CdTe QD reduced cellular viability to about

50% of control and induced apoptosis about five times higher over control.

Some of the insoluble NPs may also induce cytotoxicity to human ECs. For example, Yang et al. [21] showed that 24 h exposure of HUVECs to 20 μg/mL silica NPs (primary size 56.8 nm) significantly induced cytotoxicity as about 50% decrease of viability and about 6.6 times higher LDH release. Similarly, Guo et al. [22] also showed that 24 h or 48 h exposure of HUVECs to amorphous silica NPs (average diameter 57.66 nm) from 12.5 to 100 μg/mL significantly induced cytotoxicity. Walker et al. [23] compared the effects of single-walled carbon nanotube (SWCNT; surface area 641 m^2/g) and multi-walled carbon nanotube (MWCNT; 56 m^2/g), and results showed that 24 h exposure to both types of CNT (10, 50, 100 μg/10^6 cells) induced cytotoxicity in human aortic endothelial cells (HAEC) to a similar extent, which indicated that surface area is not important in influencing the cytotoxicity of CNT to ECs. Orecna et al. [24] showed that exposure to 100 μg/mL pristine and carboxylated MWCNT (average diameter 60 nm) induced similar effects in mitochondrial viability and LDH release in HUVECs, but the carboxylated MWCNT was associated with higher apoptosis. Ge et al. [25] showed that 6 h and 12 h exposure to 30 μg/mL SWCNT (primary size not provided; with ion impurities) caused over 50% decrease in HUVEC viability, which was attenuated by the presence of proteins due to the formation of protein corona.

In contrast, organic NPs generally showed little to no cytotoxicity. For example, Matuszak et al. [26] showed that exposure to up to 100 μg/mL lipid NPs (50, 80, 120 nm) or liposomes (109, 139 nm) did not significantly affect HUVEC viability (indicated as cell index by real-time measurement), whereas 100 μg/mL micelles (145, 227 nm) significantly decreased the viability. Menard et al. [27] showed that micelles based on diglutamic acid (linked to lithocholic, arachidonic or linoleic acid) did not significantly induce LDH release in HUVECs, and the EC50 to reduce mitochondrial viability was larger than 250 μg/mL. Our recent study also showed that up to 200 μg/mL micelles based on MPEG-PLA (hydrodynamic size ~20 nm) did not significantly induce cytotoxicity in HUVECs [28]. Indeed, coating with organics could be a strategy to improve the biocompatibility of toxic NPs. For example, Su et al. [29] coated Ag NPs with polyurethane micelles with MPEG brush, and the nanocomposite showed relatively low cytotoxicity to HUVECs (viability approximate 72% after 24 h exposure to 20 μg/mL NPs) while efficiently inhibited the growth of bacteria. All of these studies in combination suggested that NPs could induce cytotoxicity to human ECs in vitro, and the cytotoxic potential is probably dependent on the physicochemical properties of NPs.

4.2.2 Genotoxicity

Comet assay, also known as the single cell gel electrophoresis assay, is one of the most popular assays used in particle toxicology studies to detect DNA damage, i.e., DNA strand breaks and oxidative DNA damage [30]. By using this method, it has been shown that exposure of human ECs to TiO_2 NP, silica NP, carbon black (CB) NPs and micelles based on polyethylenimine (PEI) was associated with DNA damage [31–35]. Another popular method is to measure the formation of γH2AX foci, which has been applied to indicate the DNA damage in human ECs induced by MWCNT, multiwall carbon nano-onion (MWCNO) and CdSe/ZnS QD [36–38]. Duan et al. [32] further showed that silica NP (average size 62 nm; concentrations >25 μg/mL) induced DNA damage was associated with G2/M cell cycle checkpoint activation and inhibition of proliferation, which suggested that NP exposure could induce the downstream events associated with DNA damage.

Based on available reports, there appears to be no strong correlation between NP induced DNA damage and cytotoxicity in human ECs. For example, Bayat et al. [31] showed that 24 h exposure to 10 μg/mL TiO_2 NPs (primary size 1–3 nm and 30 nm) significantly induced DNA strand damage without an effect on LDH release in human dermal microvascular endothelial cells (HDMVCs). Guo et al. [38] showed that cytotoxicity was significantly induced in HUVECs after 24 h exposure to 10 μg/mL MWCNT (average diameter 30 nm, length <1 μm), whereas DNA damage was induced after incubation with MWCNT as low as 0.5 μg/mL. Xu et al. [36] showed that the EC50 for MWCNO (hydrodynamic size 31.2 nm) induced cytotoxicity was 44.12 μg/mL, whereas γH2AX foci formation was observed in HUVECs after 6 h, 12 h and 24 h incubation with 5 μg/mL MWCNO. These studies indicated that DNA damage could be used as a sensitive marker to reflect NP induced toxicity.

Cowie et al. [35] compared DNA damage induced by TiO_2 NPs (20 nm), iron oxide NPs (8 nm, with or without oleic acid coating), silica NPs (25 nm and 50 nm; rhodamine-labeled) and polylactic glycolic acid polyethylene oxide polymeric NPs in a number of different cell lines, namely human lymphoblastoid TK6 cells, human kidney HEK293, monkey kidney COS-1 cells, human cerebral endothelial cells (HCECs), bronchial 16HBE14o cells and human BeWo b30 placental cells. The results showed that HCECs were as effective as other types of cells to identify genotoxic and non-genotoxic NPs, however, the TK6 cells, BeWo b30 and kidney cells seem to be the most reliable for detecting a dose-response.

4.2.3 Endothelial Activation

It has been shown that direct exposure of human ECs to different types of NPs could induce endothelial activation in vitro, i.e., expression of adhesion molecules, release of inflammatory markers and monocyte adhesion. For example, Zhu et al. [39] showed that exposure of HAECs to iron oxide NPs (22 nm and 43 nm) at 2, 20, 100 μg/mL significantly promoted the expression of interleukin 8 (IL-8) and ICAM-1 as well as adhesion of U937 to ECs. Montiel-Davalos et al. [40] reported that TiO_2 NPs (average size <50 nm) from 5 to 40 μg/cm^2 induced the expression of adhesion molecules (i.e., E-selectin, P-selectin, ICAM-1, VCAM-1 and PECAM-1) as well as adhesion of U937 cells to HUVECs. Li et al. [41] found that exposure of HUVECs to 0.1, 1 and 10 μg/mL ZnO NPs (50 nm), but not the microparticles (>100 nm), was associated with increased expression of ICAM-1 and monocyte adhesion. Meanwhile, elevated expression of ICAM-1 and foam cell formation were also observed in ZnO NP exposed mice. Shi et al. [42] showed that Ag NPs (hydrodynamic size 65 nm) from 0.5 to 2.0 μg/mL induced the expression of inflammatory cytokines (i.e., IL-6, IL-8 and MCP-1) and adhesion molecules (ICAM-1, VCAM-1 and P-selectin). For carbonaceous NPs, Cao et al. [43] showed that 24 h exposure of HUVECs to MWCNT (primary size 700–3000 nm in length × 5–35 nm in diameter, and 400–4000 nm in length × 6–20 nm in diameter) significantly induced expression of ICAM-1 and VCAM-1 at concentrations from 16 to 64 μg/mL as well as monocyte adhesion at 64 μg/mL, whereas CB NPs (primary size 14 nm) only induced significantly increased expression of ICAM-1 and VCAM-1 without an effect on monocyte adhesion at the concentrations of 50 and 100 μg/mL [44].

Meanwhile, by using co-culture models, it has been shown that NPs may also activate ECs without direct contact with ECs. For example, Napierska et al. [45] incubated EA.hy 926 cells (an immortalized EC cell line) with A549 and THP-1 co-cultures that had been exposed to silica NPs (2 and 60 nm) for 12 h, and found increased release of IL-6, IL-8 and MIP-1 in indirectly exposed endothelial monolayer. Snyder-Talkington et al. [46] showed that apical exposure of human small airway epithelial cells (SAEC) to 1.2 μg/mL MWCNT (length 3.86 μm × diameter 49 nm) for 24 h induced the release of VEGFA, sICAM-1 and sVCAM-1 in human microvascular endothelial cells (HMVEC) cultured in basolateral chamber. In a later study, the same group further showed that 24 h incubation of SAEC with 0.5 and 1.0 μg/mL printer-emitted NPs (aerodynamic diameters ranged from 39 to 122 nm) promoted the release of cytokines and chemokines without the internalization of NPs into HMVEC by using the same model [47]. Cao et al. [48] showed that exposure of A549 and THP-1 cells with 8 μg/mL MWCNT (primary size 700–3000 nm in length × 5–35 nm in diameter, and 400–4000 nm in length × 6–20 nm in diameter) for 24 h modestly promoted the release of IL-6 and IL-8 as well as monocyte adhesion in HUVEC monolayer in basolateral chamber. In combination, it is possible that exposure to NPs could lead to endothelial activation in vitro, either due to the direct effect of NPs or an indirect effect by inflammatory mediators.

4.2.4 Dysfunction of NOS and Impaired NO Signaling

The proper function of blood vessels requires the endothelium to release NO, and dysfunction of NO, particularly diminished NO bioactivity due to eNOS uncoupling, has been implicated in the development of cardiovascular diseases [11]. Some of the NPs have been shown to induce eNOS uncoupling and/or decreased NO production. For example, Duan et al. [49] showed that exposure to silica NPs (diameter 62 nm) dose-dependently (25–100 μg/mL) decreased the NO production and activity of eNOS and NOS but increased the activity of iNOS in HUVECs. Similarly, decreased activity of eNOS and NO production have been observed in human ECs after exposure to magnetic ferroferric oxide NPs (diameter ranges from 10 to 15 nm) [50] and superparamagnetic iron oxide (size ranges from 5 to 75 nm) [51]. A decreased NO/peroxynitrite ratio has been observed in HUVECs after exposure to silica NPs with different sizes (10, 50, 150 and 500 nm), especially 10 nm silica NPs [52]. Exposure to detonation diamond NPs (size grains 2–5 nm) [53] or TiO_2 NPs (1–3 nm) [31] was also associated with decreased NO production. Although these studies did not attempt measuring the activity of NOS, the decreased NO production could be due to decreased activity of NOS, particularly eNOS uncoupling.

In contrast, exposure to NPs may also promote NOS activity and NO over-production. For example, Zhu et al. [39] showed that iron oxide NPs (22 nm and 43 nm) at 2, 20, 100 μg/mL promoted NO over-production due to elevated NOS activity. Similarly, Su et al. [54] showed that magnetic NPs (diameter 15–20 nm) at 400 μg/mL significantly induced NO production and eNOS activity. Furthermore, an increase of serum NO was also observed in mice after 3 days exposure to 20 mg/kg magnetic NPs. Han et al. [55] showed that exposure to 200 μg/mL ZnS nanoarchitectures was associated with increased NO production and NOS activity in HUVECs, whereas exposure to 5 mg/kg and 10 mg/kg ZnS nanoarchitectures promoted NO in serum and damages to endothelium in aortic root of normal mice. NO over-production was also observed in human ECs after exposure to TiO2 NPs (average size <50 nm) [40] and ZnO NPs (diameter 20 nm; not significantly affected by ZnO NPs with diameter of 90–210 nm) [56], although the NOS activity was not further measured. In combination, exposure of human ECs to NPs could lead to impaired NO signaling, showing as diminished NO bioactivity due to eNOS uncoupling or NO over-production due to increased NOS activity.

4.3 Mechanisms

4.3.1 Oxidative Stress

It well known that exposure to NPs could lead to oxidative stress and inflammation, which could be the mechanisms associated with cardiovascular toxicity of NPs [57, 58]. To explain NP induced oxidative stress and toxicity, a three-tier model has been proposed [59, 60]. This model suggested that at low amount of oxidative stress (tier-1 level), NPs could lead to the antioxidant responses mediated by Nrf-2 signaling pathway. At intermediate amount of oxidative stress (tier-2 level), NPs could induced inflammatory responses due to the activation of NF-κB cascades. At high amount of oxidative stress, NPs could result in oxidative damage and eventually apoptosis and necrosis [59, 60]. Over-production of reactive oxygen species (ROS) or NO has been observed in human ECs after exposure to different kinds of NPs, e.g., Ag NPs, silica NPs, TiO_2 NPs and carbonaceous NPs, which may further mediate the oxidative damage of bio-molecules and inhibit the function of antioxidant systems [32, 42–44, 52, 61]. Activation of Nrf-2 signaling pathway has also been observed in human ECs after NP exposure. For example, Lai et al. [62] showed that Au NPs enhanced the level and nuclear translocation of the Nrf-2 protein and Bach1 export/tyrosine phosphorylation, leading to the expression of heme oxygenase-1 (HO-1) protein in HUVECs. Guo et al. [63] also found increased protein expression of Nrf-2 and HO-1 in silica NP exposed HUVECs. Interestingly, it has been shown that the presence of antioxidant may prevent NP induced toxicity in human ECs. For instance, Shi et al. [42] showed that Ag NP induced cytotoxicity, intracellular ROS and inflammatory responses in HUVECs were inhibited by the presence of antioxidant N-acetylcysteine (NAC). Similarly, Guo et al. [22] also found that silica NP induced cytotoxicity and oxidative damage was restored by NAC. In another study, Liang et al. [15] showed that exposure of HAECs to ZnO NPs induced cytotoxicity associated with ROS mediated mitochondria membrane potential decrease, cytochrome C release, activation of caspases 3 and caspases 9 and increase of Bax/Bcl-2 ratio. Moreover, these effects were partially or completely inhibited by the treatment of α-lipoic acid (LA). All of these studies suggested that oxidative stress could be the mechanism associated with the toxicity of NPs to ECs, which is in agreement with the three-tier model [59, 60].

4.3.2 Inflammation

As suggested by the three-tier model, intermediate amount of oxidative stress induced by NP exposure could lead to inflammation due to the activation of NF-κB cascades [59, 60]. Activation of NF-κB cascades has also been observed in NP exposed human ECs. For example, Corbalan et al. [52] showed that exposure of HUVECs to silica NPs was associated with increased NF-κB DNA binding activity as well as up-regulation of a number of inflammatory genes. In addition, suppression of NF-κB activity by proteasome inhibitor significantly prevented the up-regulation of ICAM-1, VCAM-1, SELE, F3, and IL-8 induced by silica NPs. In a later study, Guo et al. [63] also showed that exposure of HUVECs to silica NPs was associated with activation of NF-κB and increased expression of a number of inflammatory markers, including IL-1β, IL-6, IL-8, TNFα, ICAM-1, VCAM-1, and MCP-1. Shi et al. [42] found that exposure of HUVECs to Ag NPs increased the phosphorylation of IKKα/β and IκBα (two key proteins of NF-κB), p65 subunit and the formation of the NF-κB nuclear protein-DNA complex, associated with increased expression of inflammatory cytokines and adhesion molecules. Moreover, all of these effects could be prevented by the treatment of antioxidant NAC, which indicated a role of oxidative stress in mediating NP induced NF-κB cascades. Liu and Sun [64] found increased NF-κB activity as well as expression of inflammatory cytokines and adhesion molecules in hydroxyapatite NP exposed HUVEC+THP-1 co-cultures, but not the HUVEC mono-culture. TiO_2 NPs have also been shown to induce the translocation of NF-κB and IκBα degradation, which could be responsible for NP induced expression of adhesion molecules and U937 adhesion to HUVECs [40]. All of these studies in combination indicated that exposure to NPs could lead to endothelial dysfunction due to the activation of NF-κB cascades, which is in agreement of the three-tier model.

4.3.3 Dysfunction of Autophagy

Autophagy is a process where cytoplasmic components are sequestered in double-membrane vesicles for degradation into lysosomes, and dysfunction of autophagy has been implicated in endothelial dysfunction and development of cardiovascular diseases [65, 66]. In recent years, it has been suggested that exposure to NPs may lead to autophagy dysfunction (defined as excessive autophagy induction) as the mechanism for NP induced toxicity [67]. In human ECs, it has been shown that a number of NPs, such as iron oxide NPs, TiO2 NPs, Ag NPs and silica NPs, could lead to induction of autophagy [49, 61, 68, 69]. In addition, inhibition of autophagy dysfunction has been shown to attenuate NP induced toxicity to ECs. For example, it has been shown that exposure of HUVECs to Fe_3O_4 NPs promoted endothelial dysfunction (eNOS uncoupling, inflammatory responses) associated with an induction of autophagy, and suppression of autophagy induction or stimulation of autophagic flux partially attenuated the toxic effects of NPs [50]. Similarly, exposure of HUVECs to carboxylated MWCNT was also shown to be associated with the profound accumulation of autophagosomes due to blockade of the autophagic flux, and stimulation of the autophagic flux attenuated the cytotoxicity of NPs associated with the extracellular release of NPs in autophagic microvesicles [24]. In another study, it was shown that the presence of antioxidant NAC attenuated silica NP induced endothelial toxicity partially through the inhibition of autophagy [22]. All of these studies suggested that excessive autophagy induced by NPs could be one of the mechanisms responsible for NP induced endothelial toxicity.

4.4 Conclusions

Once entering circulation, it is likely that NPs will interact with the endothelium, and it is necessary to assess the toxicity of NPs to ECs to better understand the potential adverse effects of NPs in vivo [9]. By using human ECs, including primary ECs or immortalized EC cell lines, it has been shown that NPs could induce cytotoxicity, genotoxicity, endothelial activation and impairment of NO signaling (Fig. 4.2). Oxidative stress and inflammatory response induced by NPs have been suggested as the mechanisms, as revealed by the three-tier model (Fig. 4.3). In recent years, dysfunction of autophagy (excessive autophagy induction) has also been suggested as one of the mechanisms responsible for NP induced toxicity to ECs. In the future, it is still necessary to use ECs to assess the potential toxic effects of NPs and to study the mechanisms.

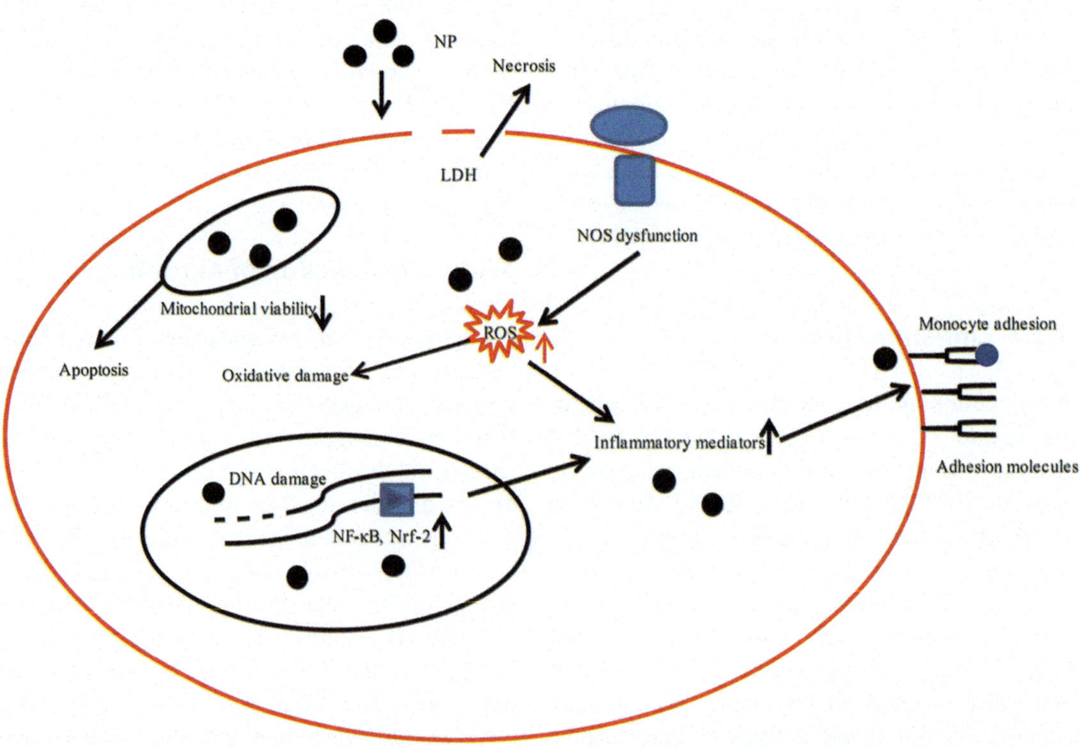

Fig. 4.2 The toxicity of NPs to human ECs. Exposure to NPs may induce cytotoxicity (apoptosis, necrosis), genotoxicity (DNA damage), endothelial activation (adhesion molecules, monocyte adhesion) and dysfunction of NOS (over-production of ROS or NO) in human ECs

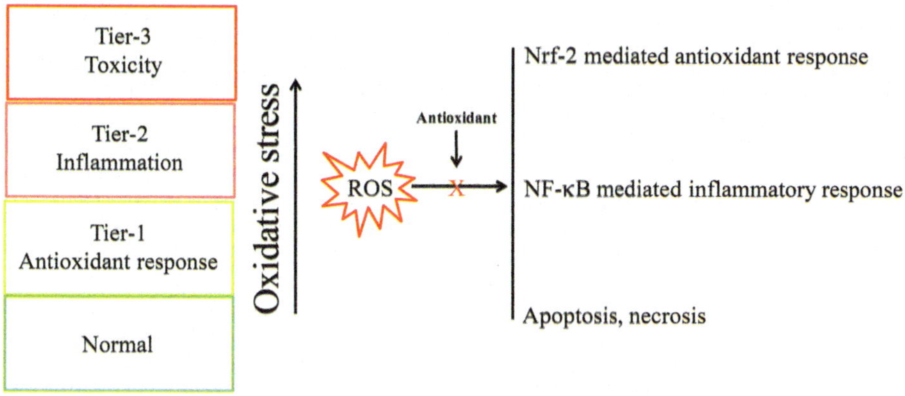

Fig. 4.3 The three-tier model to explain NP induced oxidative stress and toxicity. At low amount of oxidative stress (tier-1 level), NPs could lead to the antioxidant responses mediated by Nrf-2 signaling pathway. At intermediate amount of oxidative stress (tier-2 level), NPs could induced inflammatory responses due to the activation of NF-κB cascades. At high amount of oxidative stress, NPs could result in oxidative damage and eventually apoptosis and necrosis. The presence of antioxidant may prevent NP induced toxicity (Originally proposed by Nel et al. [59])

References

1. European Commission Commission recommendation of 18 October 2011 on the definition of nanomaterial text with EEA relevance. http://eur-lex.europa.eu/legal-content/EN/TXT/?uri=CELEX:32011H0696. Accessed 18 Dec 2011
2. Donaldson K, Tran L, Jimenez LA et al (2005) Combustion-derived nanoparticles: a review of their toxicology following inhalation exposure. Part Fibre Toxicol 2:10
3. Donaldson K, Duffin R, Langrish JP et al (2013) Nanoparticles and the cardiovascular system: a critical review. Nanomedicine (Lond) 8:403–423
4. Vance ME, Kuiken T, Vejerano EP et al (2015) Nanotechnology in the real world: redeveloping the nanomaterial consumer products inventory. Beilstein J Nanotechnol 6:1769–1780
5. Wong BS, Yoong SL, Jagusiak A et al (2013) Carbon nanotubes for delivery of small molecule drugs. Adv Drug Deliv Rev 65:1964–2015
6. Wei L, Lu J, Xu H et al (2015) Silver nanoparticles: synthesis, properties, and therapeutic applications. Drug Discov Today 20:595–601
7. Khan ST, Musarrat J, Al-Khedhairy AA (2016) Countering drug resistance, infectious diseases, and sepsis using metal and metal oxides nanoparticles: current status. Colloids Surf B: Biointerfaces 146:70–83
8. Soenen SJ, Parak WJ, Rejman J et al (2015) (Intra) cellular stability of inorganic nanoparticles: effects on cytotoxicity, particle functionality, and biomedical applications. Chem Rev 115:2109–2135
9. Setyawati MI, Tay CY, Docter D et al (2015) Understanding and exploiting nanoparticles' intimacy with the blood vessel and blood. Chem Soc Rev 44:8174–8199
10. Libby P (2012) Inflammation in atherosclerosis. Arterioscler Thromb Vasc Biol 32:2045–2051
11. Karbach S, Wenzel P, Waisman A et al (2014) eNOS uncoupling in cardiovascular diseases – the role of oxidative stress and inflammation. Curr Pharm Des 20:3579–3594
12. Gimbrone MA Jr, Garcia-Cardena G (2016) Endothelial cell dysfunction and the pathobiology of atherosclerosis. Circ Res 118:620–636
13. Saptarshi SR, Duschl A, Lopata AL (2015) Biological reactivity of zinc oxide nanoparticles with mammalian test systems: an overview. Nanomedicine (Lond) 10:2075–2092
14. Sun J, Wang S, Zhao D et al (2011) Cytotoxicity, permeability, and inflammation of metal oxide nanoparticles in human cardiac microvascular endothelial cells: cytotoxicity, permeability, and inflammation of metal oxide nanoparticles. Cell Biol Toxicol 27:333–342
15. Liang S, Sun K, Wang Y et al (2016) Role of Cyt-C/caspases-9,3, Bax/Bcl-2 and the FAS death receptor pathway in apoptosis induced by zinc oxide nanoparticles in human aortic endothelial cells and the protective effect by alpha-lipoic acid. Chem Biol Interact 258:40–51
16. Gong Y, Ji Y, Liu F et al (2017) Cytotoxicity, oxidative stress and inflammation induced by ZnO nanoparticles in endothelial cells: interaction with palmitate or lipopolysaccharide. J Appl Toxicol 87:895–901
17. Gu Y, Cheng S, Chen G et al (2017) The effects of endoplasmic reticulum stress inducer thapsigargin on the toxicity of ZnO or TiO2 nanoparticles to human endothelial cells. Toxicol Mech Methods 27:191–200
18. Guo H, Zhang J, Boudreau M et al (2016) Intravenous administration of silver nanoparticles causes organ

toxicity through intracellular ROS-related loss of inter-endothelial junction. Part Fibre Toxicol 13:21
19. Danielsen PH, Cao Y, Roursgaard M et al (2015) Endothelial cell activation, oxidative stress and inflammation induced by a panel of metal-based nanomaterials. Nanotoxicology 9:813–824
20. Yan M, Zhang Y, Xu K et al (2011) An in vitro study of vascular endothelial toxicity of CdTe quantum dots. Toxicology 282:94–103
21. Yang L, Yan Q, Zhao J et al (2013) The role of potassium channel in silica nanoparticle-induced inflammatory effect in human vascular endothelial cells in vitro. Toxicol Lett 223:16–24
22. Guo C, Yang M, Jing L et al (2016) Amorphous silica nanoparticles trigger vascular endothelial cell injury through apoptosis and autophagy via reactive oxygen species-mediated MAPK/Bcl-2 and PI3K/Akt/mTOR signaling. Int J Nanomedicine 11:5257–5276
23. Walker VG, Li Z, Hulderman T et al (2009) Potential in vitro effects of carbon nanotubes on human aortic endothelial cells. Toxicol Appl Pharmacol 236:319–328
24. Orecna M, De Paoli SH, Janouskova O et al (2014) Toxicity of carboxylated carbon nanotubes in endothelial cells is attenuated by stimulation of the autophagic flux with the release of nanomaterial in autophagic vesicles. Nanomedicine 10:939–948
25. Ge C, Du J, Zhao L et al (2011) Binding of blood proteins to carbon nanotubes reduces cytotoxicity. Proc Natl Acad Sci U S A 108:16968–16973
26. Matuszak J, Baumgartner J, Zaloga J et al (2016) Nanoparticles for intravascular applications: physicochemical characterization and cytotoxicity testing. Nanomedicine (Lond) 11:597–616
27. Menard N, Tsapis N, Poirier C et al (2012) Novel surfactants with diglutamic acid polar head group: drug solubilization and toxicity studies. Pharm Res 29:1882–1896
28. Liu F, Huang H, Gong Y et al (2017) Evaluation of in vitro toxicity of polymeric micelles to human endothelial cells under different conditions. Chem Biol Interact 263:46–54
29. Su Y, Zhao L, Meng F et al (2017) Silver nanoparticles decorated lipase-sensitive polyurethane micelles for on-demand release of silver nanoparticles. Colloids Surf B: Biointerfaces 152:238–244
30. Moller P, Hemmingsen JG, Jensen DM et al (2015) Applications of the comet assay in particle toxicology: air pollution and engineered nanomaterials exposure. Mutagenesis 30:67–83
31. Bayat N, Lopes VR, Scholermann J et al (2015) Vascular toxicity of ultra-small TiO2 nanoparticles and single walled carbon nanotubes in vitro and in vivo. Biomaterials 63:1–13
32. Duan J, Yu Y, Li Y et al (2013) Toxic effect of silica nanoparticles on endothelial cells through DNA damage response via Chk1-dependent G2/M checkpoint. PLoS One 8:e62087
33. Frikke-Schmidt H, Roursgaard M, Lykkesfeldt J et al (2011) Effect of vitamin C and iron chelation on diesel exhaust particle and carbon black induced oxidative damage and cell adhesion molecule expression in human endothelial cells. Toxicol Lett 203:181–189
34. Calarco A, Bosetti M, Margarucci S et al (2013) The genotoxicity of PEI-based nanoparticles is reduced by acetylation of polyethylenimine amines in human primary cells. Toxicol Lett 218:10–17
35. Cowie H, Magdolenova Z, Saunders M et al (2015) Suitability of human and mammalian cells of different origin for the assessment of genotoxicity of metal and polymeric engineered nanoparticles. Nanotoxicology 9(1):57–65
36. Xu Y, Wang SY, Yang J et al (2013) Multiwall carbon nano-onions induce DNA damage and apoptosis in human umbilical vein endothelial cells. Environ Toxicol 28:442–450
37. Soenen SJ, Montenegro JM, Abdelmonem AM et al (2014) The effect of nanoparticle degradation on amphiphilic polymer-coated quantum dot toxicity: the importance of particle functionality assessment in toxicology [corrected]. Acta Biomater 10:732–741
38. Guo YY, Zhang J, Zheng YF et al (2011) Cytotoxic and genotoxic effects of multi-wall carbon nanotubes on human umbilical vein endothelial cells in vitro. Mutat Res 721:184–191
39. Zhu MT, Wang B, Wang Y et al (2011) Endothelial dysfunction and inflammation induced by iron oxide nanoparticle exposure: risk factors for early atherosclerosis. Toxicol Lett 203:162–171
40. Montiel-Davalos A, Ventura-Gallegos JL, Alfaro-Moreno E et al (2012) TiO(2) nanoparticles induce dysfunction and activation of human endothelial cells. Chem Res Toxicol 25:920–930
41. Li CH, Liao PL, Shyu MK et al (2012) Zinc oxide nanoparticles-induced intercellular adhesion molecule 1 expression requires Rac1/Cdc42, mixed lineage kinase 3, and c-Jun N-terminal kinase activation in endothelial cells. Toxicol Sci 126:162–172
42. Shi J, Sun X, Lin Y et al (2014) Endothelial cell injury and dysfunction induced by silver nanoparticles through oxidative stress via IKK/NF-kappaB pathways. Biomaterials 35:6657–6666
43. Cao Y, Jacobsen NR, Danielsen PH et al (2014) Vascular effects of multiwalled carbon nanotubes in dyslipidemic ApoE−/− mice and cultured endothelial cells. Toxicol Sci 138:104–116
44. Cao Y, Roursgaard M, Danielsen PH et al (2014) Carbon black nanoparticles promote endothelial activation and lipid accumulation in macrophages independently of intracellular ROS production. PLoS One 9:e106711
45. Napierska D, Thomassen LC, Vanaudenaerde B et al (2012) Cytokine production by co-cultures exposed to monodisperse amorphous silica nanoparticles: the role of size and surface area. Toxicol Lett 211:98–104
46. Snyder-Talkington BN, Schwegler-Berry D, Castranova V et al (2013) Multi-walled carbon

nanotubes induce human microvascular endothelial cellular effects in an alveolar-capillary co-culture with small airway epithelial cells. Part Fibre Toxicol 10:35
47. Sisler JD, Pirela SV, Friend S et al (2015) Small airway epithelial cells exposure to printer-emitted engineered nanoparticles induces cellular effects on human microvascular endothelial cells in an alveolar-capillary co-culture model. Nanotoxicology 9:769–779
48. Cao Y, Roursgaard M, Jacobsen NR et al (2016) Monocyte adhesion induced by multi-walled carbon nanotubes and palmitic acid in endothelial cells and alveolar-endothelial co-cultures. Nanotoxicology 10:235–244
49. Duan J, Yu Y, Yu Y et al (2014) Silica nanoparticles induce autophagy and endothelial dysfunction via the PI3K/Akt/mTOR signaling pathway. Int J Nanomedicine 9:5131–5141
50. Zhang L, Wang X, Miao Y et al (2016) Magnetic ferroferric oxide nanoparticles induce vascular endothelial cell dysfunction and inflammation by disturbing autophagy. J Hazard Mater 304:186–195
51. Astanina K, Simon Y, Cavelius C et al (2014) Superparamagnetic iron oxide nanoparticles impair endothelial integrity and inhibit nitric oxide production. Acta Biomater 10:4896–4911
52. Corbalan JJ, Medina C, Jacoby A et al (2011) Amorphous silica nanoparticles trigger nitric oxide/peroxynitrite imbalance in human endothelial cells: inflammatory and cytotoxic effects. Int J Nanomedicine 6:2821–2835
53. Solarska-Sciuk K, Gajewska A, Skolimowski J et al (2013) Stimulation of production of reactive oxygen and nitrogen species in endothelial cells by unmodified and Fenton-modified ultradisperse detonation diamond. Biotechnol Appl Biochem 60:259–265
54. Su L, Han L, Ge F et al (2012) The effect of novel magnetic nanoparticles on vascular endothelial cell function in vitro and in vivo. J Hazard Mater 235–236:316–325
55. Han L, Su L, Chen D et al (2015) ZnS nanoarchitectures induced dysfunction of vascular endothelial cells in vitro and in vivo. Environ Toxicol 30:755–768
56. Chuang KJ, Lee KY, Pan CH et al (2016) Effects of zinc oxide nanoparticles on human coronary artery endothelial cells. Food Chem Toxicol 93:138–144
57. Moller P, Christophersen DV, Jacobsen NR et al (2016) Atherosclerosis and vasomotor dysfunction in arteries of animals after exposure to combustion-derived particulate matter or nanomaterials. Crit Rev Toxicol 46:437–476
58. Cao Y, Long J, Ji Y et al (2016) Foam cell formation by particulate matter (PM) exposure: a review. Inhal Toxicol 28:583–590
59. Nel A, Xia T, Madler L et al (2006) Toxic potential of materials at the nanolevel. Science 311:622–627
60. Nel A, Xia T, Meng H et al (2013) Nanomaterial toxicity testing in the 21st century: use of a predictive toxicological approach and high-throughput screening. Acc Chem Res 46:607–621
61. Halamoda KB, Chapuis BC, Guney-Ayra S et al (2012) Induction of oxidative stress, lysosome activation and autophagy by nanoparticles in human brain-derived endothelial cells. Biochem J 441:813–821
62. Lai TH, Shieh JM, Tsou CJ et al (2015) Gold nanoparticles induce heme oxygenase-1 expression through Nrf2 activation and Bach1 export in human vascular endothelial cells. Int J Nanomedicine 10:5925–5939
63. Guo C, Xia Y, Niu P et al (2015) Silica nanoparticles induce oxidative stress, inflammation, and endothelial dysfunction in vitro via activation of the MAPK/Nrf2 pathway and nuclear factor-kappaB signaling. Int J Nanomedicine 10:1463–1477
64. Liu X, Sun J (2014) Potential proinflammatory effects of hydroxyapatite nanoparticles on endothelial cells in a monocyte-endothelial cell coculture model. Int J Nanomedicine 9:1261–1273
65. Jiang F (2016) Autophagy in vascular endothelial cells. Clin Exp Pharmacol Physiol 43:1021–1028
66. Gatica D, Chiong M, Lavandero S et al (2015) Molecular mechanisms of autophagy in the cardiovascular system. Circ Res 116:456–467
67. Stern ST, Adiseshaiah PP, Crist RM (2012) Autophagy and lysosomal dysfunction as emerging mechanisms of nanomaterial toxicity. Part Fibre Toxicol 9:20
68. Manshian BB, Pfeiffer C, Pelaz B et al (2015) High-content imaging and gene expression approaches to unravel the effect of surface functionality on cellular interactions of silver nanoparticles. ACS Nano 9:10431–10444
69. Yang FY, Yu MX, Zhou Q et al (2012) Effects of iron oxide nanoparticle labeling on human endothelial cells. Cell Transplant 21:1805–1820

The Role of Autophagy in Nanoparticles-Induced Toxicity and Its Related Cellular and Molecular Mechanisms

5

Yubin Li and Dianwen Ju

Abstract

In the past decades, nanoparticles have been widely used in industry and pharmaceutical fields for drug delivery, anti-pathogen, and diagnostic imaging purposes because of their unique physicochemical characteristics such as special ultrastructure, dispersity, and effective cellular uptake properties. But the nanotoxicity has been raised over the extensive applications of nanoparticles. Researchers have elucidated series of mechanisms in nanoparticles-induced toxicity, including apoptosis, necrosis, oxidative stress, and autophagy. Among upon mechanisms, autophagy was recently recognized as an important cell death style in various nanoparticles-induced toxicity, but the role of autophagy and its related cellular and molecular mechanisms during nanoparticles-triggered toxicity were still confusing. In the chapter, we briefly introduced the general process of autophagy, summarized the different roles of autophagy in various nanoparticle-treated different in vitro/in vivo models, and deeply analyzed the physicochemical and biochemical (cellular and molecular) mechanisms of autophagy during nanoparticles-induced toxicity through listing and summarizing representative examples. Physicochemical mechanisms mainly include dispersity, size, charge, and surface chemistry; cellular

Y. Li (✉)
Department of Microbiological and Biochemical Pharmacy & The Key Laboratory of Smart Drug Delivery, Ministry of Education, School of Pharmacy, Fudan University, Shanghai, People's Republic of China

Perelman School of Medicine, University of Pennsylvania, Philadelphia, PA, USA
e-mail: yubinli@upenn.edu

D. Ju (✉)
Department of Microbiological and Biochemical Pharmacy & The Key Laboratory of Smart Drug Delivery, Ministry of Education, School of Pharmacy, Fudan University, Shanghai, People's Republic of China
e-mail: dianwenju@fudan.edu.cn

Q. Saquib et al. (eds.), *Cellular and Molecular Toxicology of Nanoparticles*, Advances in Experimental Medicine and Biology 1048, https://doi.org/10.1007/978-3-319-72041-8_5

mechanisms primarily focus on lysosome impairment, mitochondria dysfunction, mitophagy, endoplasmic reticulum stress and endoplasmic reticulum autophagy; while molecular mechanisms were mainly including autophagy related signaling pathways, hypoxia-inducible factor, and oxidative stress. This chapter highlighted the important role of autophagy as a critical mechanism in nanoparticles-induced toxicity, and the physicochemical and biochemical mechanisms of autophagy triggered by nanoparticles might be useful for establishing a guideline for the evaluation of nanotoxicology, designing and developing new biosafety nanoparticles in the future.

Keywords

Autophagy · Nanoparticles · Nanotoxicology · Mitophagy · Oxidative stress

5.1 Introduction

In the past decades, nanoparticles have been widely used in industry and pharmaceutical fields for drug delivery, anti-pathogen, and diagnostic imaging purposes because of their unique physicochemical characteristics such as special ultrastructure, dispersity, and effective cellular uptake properties [1, 2]. But the nanotoxicity has been raised over the extensive applications of nanoparticles. Researchers have elucidated series of mechanisms in nanoparticles-induced toxicity, including apoptosis, necrosis, oxidative stress, and autophagy [3, 4]. Among upon mechanisms, autophagy was recently recognized as an important cell death style in various nanoparticles-induced toxicity, but the role of autophagy and its related cellular and molecular mechanisms during nanoparticles-triggered toxicity were still confusing.

In the chapter, we briefly introduced the general process of autophagy, summarized the different roles of autophagy in various nanoparticle-treated different in vitro/in vivo models, and deeply analyzed the physicochemical and biochemical (cellular and molecular) mechanisms of autophagy during nanoparticles-induced toxicity through listing and summarizing representative examples. Physicochemical mechanisms mainly include dispersity, size, charge, and surface chemistry; Cellular mechanisms primarily focus on lysosome impairment, mitochondria dysfunction, mitophagy, endoplasmic reticulum stress and endoplasmic reticulum autophagy; while molecular mechanisms were mainly including autophagy related signaling pathways, hypoxia-inducible factor, and oxidative stress. This chapter highlighted the important role of autophagy as a critical mechanism in nanoparticles-induced toxicity, the physicochemical and biochemical mechanisms of autophagy triggered by nanoparticles might be useful for establishing a guideline for the evaluation of nanotoxicology, designing and developing new biosafety nanoparticles in the future.

5.2 Classification of Autophagy

The general term of autophagy was derived from the Greek and meaning for "self-eating". And this evolutionarily conserved process of degradation cytoplasmic components (mis-folded protein and dysfunctional organelles) within lysosomes was first described nearly 50 years ago [5, 6]. Different with endocytosis-mediated degradation and recycle of cytoplasmic components, there are three types of autophagy including microautophagy, macroautophagy, and chaperone-mediated autophagy [7]. Microautophagy is mediated by lysosomal directly engulf cytoplasmic components in nonselective manners. Macroautophagy consists of three steps: cytoplasmic constituents (mis-folded proteins or damaged organelles) are enveloped by double-membraned autophagosome; autophagosomes are fused with lysosome to form an autolysosome; and the cytoplasmic constituents are degraded and recycled in lysosome (Fig. 5.1). Macroautophagy could degrade most specific components, such as mitochondria (mitophagy), endoplasmic reticulum (ERphagy or reticulophagy), ribosomes (ribophagy), and pexophagy (peroxisomes); and is the most popular and well investigated type of autophagy. Chaperone-mediated autophagy is the process that soluble cytosolic proteins directly translocated into lysosomes for degradation in chaperone-dependent selecting manner. Chaperone-mediated autophagy dose not require the formation of additional vesicles [7]. Autophagy can be activated by series of stressful conditions such as nutrient deprivation, oxidative stress, hypoxia, and inflammatory mediates [8]. The internalized nanoparticles were regarded as foreign materials and autophagic cargoed by cells and then triggered autophagy.

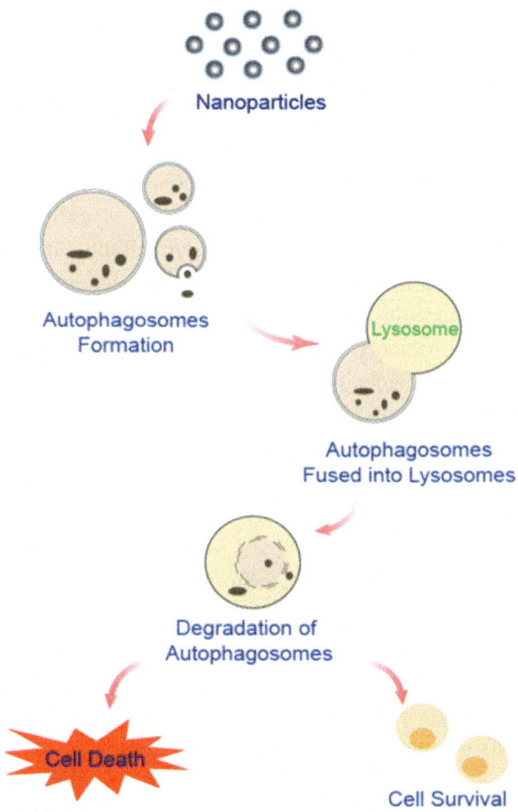

Fig. 5.1 The dual role of autophagy in nanoparticles-induced toxicity. Macroautophagy consists of three steps: cytoplasmic constituents (mis-folded proteins or damaged organelles) are enveloped by double-membraned autophagosome; autophagosomes are fused with lysosome to form an autolysosome; and the cytoplasmic constituents are degraded and recycled in lysosome. Autophagy is likened to the opposite faces of Janus, as it could protect cells to survive under certain severe conditions, as well as inducing cell death when too much autophagy occurs

5.3 The Role of Autophagy in Nanoparticles-Induced Toxicity

Autophagy is likened to the opposite faces of Janus, as it could protect cells to survive under certain severe conditions, as well as inducing cell death when too much autophagy occurs [9]. Similarly, nanoparticles-mediated autophagy during nanotoxicity might be an adaptive cellular response aiding in the clearance of nanoparticles, and also might be harmful in cellular dysfunction (Fig. 5.1).

In the section, the dual role of autophagy in nanoparticles-induced toxicity will be reviewed. Researchers investigated the autophagic effects triggered by non-mental nanoparticles, mental nanoparticles, mental oxide nanoparticles, and polymers in various models, and found most nanoparticles could induce autophagic cell death: Yu et al., reported that silica nanoparticles induced autophagic cell death in hepatoma HepG2 cells, autophagy inhibitor could effectively impair autophagy and cell death triggered by silica nanoparticles [10]. Using different cell models, Liu et al., and Park et al., both confirmed that single-walled carbon nanotubes induced autophagic cell death in A549 human lung cells and BEAS-2B human bronchial epithelial cells [11, 12]. Besides, mental oxide nanoparticles could also trigger autopahgic cell death: Copper oxide nanoparticles induced autophagic cell death in A549 cells, and autophagy inhibitors wortmannin and 3-methyladenine could protect against copper oxide nanoparticles-induced A549 cell death [13]; Yu et al., showed that zinc oxide nanoparticle also induced autophagic cell death and mitochondrial damage via reactive oxygen species generation [14], meanwhile, Johnson et al., also found that acute exposure to zinc oxide nanoparticles could induce autophagic immune cell death [15]. And nano sized neodymium oxide particles could trigger massive vacuolization and induce autophagic cell death in lung cancer cells [16]. Apart from mental and non-mental nanoparticles, polymers could also induce autophagic cell death during their-induced toxicity: Cationic polystyrene nanoparticles induced autophagic cell death through endoplasmic reticulum stress induction in macrophage and lung epithelial cells [17]. And many research groups have confirmed that cationic poly-amidoamine dendrimers promoted liver injury, lung damage and broke neuronal function by inducting autophagic cell death in different cell and mice models [18–20].

Compared with massive studies on nanoparticles-induced autophagic cell death, only a few literatures showed that nanoparticles triggered cyto-protective autophagy in cells. Zhou et al., found a novel nanoparticles paramontroseite VO_2 induced cyto-protective autophagy in HeLa cells, while autophagy inhibitor 3-methyladenine could obviously increase cell death rate in nanoparticels-treated HeLa cells [21]. The same research group also confirmed that silver nanoparticles induced cyto-protective autophagy during their-induced HeLa cell death, suppression of autophagy enhanced the anticancer activity of silver nanoparticles [22].

5.4 Physicochemical Mechanisms of Autophagy in Nanoparticles-Induced Toxicity

The mechanisms of autophagy in nanoparticles-induced toxicity were very complex regarding different physicochemical and biochemical properties of nanoparticles and various interactions between nanoparticles and cells during their-induced toxicity.

5.4.1 Dispersity

Huang and his colleagues explored the role of dispersity in nanoparticles-induced autophagy and found metal oxide Nanoparticles iron oxide nanoparticles could induce significant autophagic effect when in aggregated conditions; when surface modification by dopamine (DA), 3,4-dihydroxyphenylacetic acid (DOPAC), and meso-2,3-dimercaptosuccinicacid (DMSA) or protein adsorption by water dilution or bovine serum albumin (BSA) incubation improved their dispersity, the autophagic effects were also be diminished. Moreover, other kinds of nanoparticles such as mental nanoparticles (gold) and silica nanoparticles also exhibited dispersity-dependent autophagic effects. It was the first and the only study on dispersity of nanoparticles affecting autophagy, and suggested that autophagy triggered by nanoparticles could be modulated through tuning their dispersity [23].

5.4.2 Size

Compared with rare study on the role of dispersity in nanoparticles-activated autophagy, a series of literatures have widely investigated the size-dependent manner of nanoparticles during their-induced autophagy, including quantum dots, silver, gold, and silica nanoparticles.

In the year of 2006, Seleverstov O and colleagues compared the cytotoxicity and intracellular process of two different-sized quantum dots in human mesenchyal stem cells, and then reported for the first time that nanoparticles could trigger autophagy activation in a size-dependent manner [24]. Since then, more and more scientists investigated and found other nanoparticles could also induct autophagy in size-dependent manner.

Mishra and coauthors reported that silver nanoparticle could also activate autophagic-lysosomal interruption in a size-dependent manner; smaller (10 nm) silver nanoparticles exhibited highest uptake, accumulation and strongest autophagy and enhanced lysosomal activity in HepG2 cells when compared with larger (50 nm and 100 nm) silver nanoparticles [25]. Another mental gold nanoparticles could also induce autophagosome accumulation from blocking autophagy flux in a size-dependent nanoparticle uptake manner [26].

Apart from mental nanoparticles, researchers found that non-mental nanoparticles could also activate autophagy in a size-dependent manner during their-induced toxicity. Li et al., compared the difference of autophagy dysfunction triggered by nano-scale size (40 nm and 60 nm) and micro-scale size (200 nm) silica particles during their-induced cytotoxicity in human bronchial epithelial BEAS-2B cells: Nano-scale silica particles, but not micro-scale silica nanoparticles, could induce mitochondrial damage and autophagy via the PI3K/Akt/mTOR signaling pathway in a size- and dose-dependent manner in human bronchial epithelial BEAS-2B cells [27]. Using different cell model, Huang and colleagues also confirmed the size effect of silica submicrospheres in autophagy induction; Cells treated with 0.5–0.7 μm silica particles displayed many GFP-LC3 fluorescent puncta and high expression levels of autophagy related proteins; however, when the particle size was smaller than 0.5 μm or larger than 0.7 μm, autophagic level decreased. Cells treated with 0.1 μm or 2.1 μm silica particles had negligible change in GFP-LC3 puncta or expression levels of autophagy related proteins [28].

5.4.3 Charge

The autophagy triggered by nanoparticles was found not only to be dependent on particles' size, but also to be affected by porosity and surface charge. The polystyrene nanoparticles with neutral, anionic, and cationic surface charges could all activate autophagy and triggered the regulator of autophagy and lysosome biogenesis-transcription factor EB activation. But autophagic cargo clearance was tightly related to the charges of nanoparticles: neutral and anionic surface enhanced clearance of autophagic cargo, while cationic surface caused lysosomal dysfunction, reduced formation of autophagolysosomes, and finally blocked autophagic flux in HeLa cells [29]. There also showed different autophagic effects triggered by cationic or anionic PAMAM dendrimers during their-induced acute lung injury: Cationic PAMAM dendrimers G3 could obviously induce autophagosomes accumulation and enhance the expression of the microtubule-associated protein 1 light chain 3 in human lung A549 cells; while the same concentration of anionic PAMAM dendrimers G5.5 failed to trigger autophagy [18].

5.4.4 Surface Chemistry

Multiwalled carbon nanotubes could alter autophagy via interacting with cell membranes and membrane-associated molecules. While surface ligands modified by combinatorial chemistry on multiwalled carbon nanotubes could tune cell autophagy to various levels in different cell lines. Furthermore, multiwalled carbon nanotubes with different surface chemistries could induce autophagy in mTOR signaling pathway-dependent or -independent manner via binding to different cell surface receptors [30]. Fe_3O_4 nanoparticles extensively impaird lysosomes and led to LC3-positive autophagosomes formation, while PLGA-coated on the surface of Fe_3O_4 nanoparticles reduced this damaging effect on lysosomes. Moreover, Fe_3O_4 nanoparticles also induced mitochondrial dysfunction and endoplasmic reticulum/Golgi body stresses, which triggered autophagy, while PLGA-coated Fe_3O_4 nanoparticles reduced the devastating effect on these organelles. In vivo experimental results suggested that Fe_3O_4 nanoparticles led to massive autophagosomes accumulation in the kidney and spleen of mice when compared with the PLGA-coated Fe_3O_4 and PLGA nanoparticles [31]. Besides, N-alkyl-PEI-lactobionic acid wrapped superparamagnetic iron oxide nanocomposites showed better cell viability in RAW 264.7 cells when compared with the unsubstituted ones, and PEI induced cell autophagy can be reduced via lactose modification [32], confirmed that surface chemistry played an important role in nanoparticles-induced autophagy.

5.5 Cellular Mechanisms of Autophagy in Nanoparticles-Induced Toxicity

5.5.1 Lysosome Impairment

Lysosome is a membrane-bound organelle that consists of hydrolytic enzymes containing vesicles, and is the most important organelle in the final step of autophagy steps. Lysosomal dysfunction is the major mechanism by which nanoparticles trigger autophagy, and mainly including lysosomal ultrastructures damage, changes of lysosome pH, and disactivities of lysosomal proteases (Fig. 5.2) [33]. Wang et al., reported that silica nanoparticles induced autophagy dysfunction via lysosomal impairment and inhibition of autophagosome degradation in hepatocytes. They found that silica nanoparticles could trigger autophagy formation in two kinds of hepatocytes even at the noncytotoxic level and suppress the autophagic flux at high concentration. Silica nanoparticles impaired the lysosomal function through damaging lysosomal ultrastructures, increasing membrane permeability, and downregulating the expression of lysosomal proteases, cathepsin B [34]. Meanwhile, Ji and coauthors found that graphene oxide quantum dots induced autophagosome accumulation but blocked autophagic flux by decreasing the amount and enzymatic activity of cathepsin B and inhibiting lysosome proteolytic capacity in GC-2 and TM4 cells [35]. Besides, Fe_3O_4 nanoparticles extensively impaired lysosomes and led to the accumulation of LC3-positive autophagosomes, while PLGA-coated Fe_3O_4 nanoparticles reduced this destructive effect on lysosomes [31].

Apart from lysosmal ultrastructures damage and disactivaties of lysosomal proteases, changes of pH in lysosomes were also involved in lysosome impairment. Ma et al., investigated the

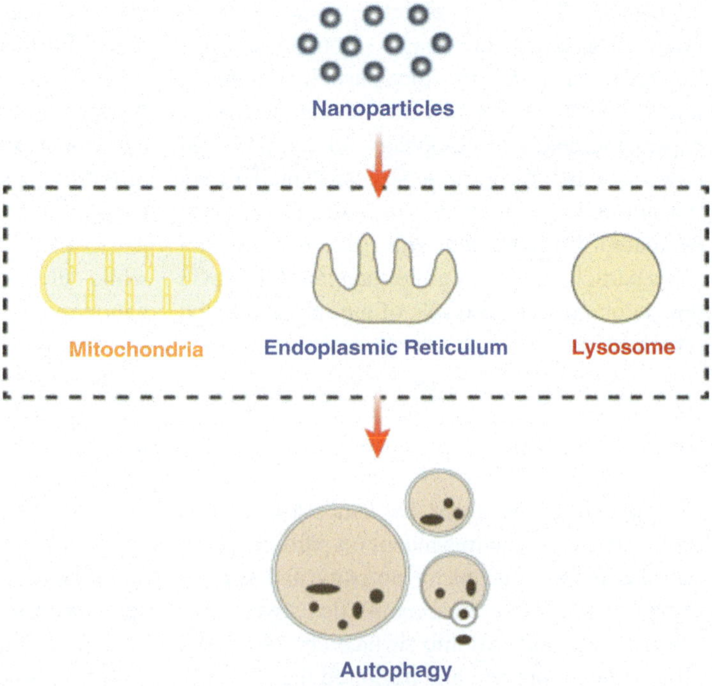

Fig. 5.2 Cellular mechanisms of autophagy in nanoparticles-induced toxicity. (1) lysosome impairment; (2) mitochondria dysfunction and mitophagy; (3) endoplasmic reticulum stress and endoplasmic reticulum autophagy

mechanisms of gold nanoparticles-triggered autophagy in cells and found that the internalized gold nanoparticles was finally accumulated in lysosomes and caused impaired lysosome degradation ability through alkalinization of lysosomal pH during their-induced autophagosome accumulation from blocking autophagy flux [26]. While using another nanoparticles, Schutz et al., reported that internalized silica nanoparticles accumulated in lysosomes resulted in suppression of autophagy-mediated protein recycle and impaired degradation of internalized epidermal growth factor, while endosomal recycling proceeds was not disturbed. The phenotype was caused by perturbed delivery of cargo via autophagosomes and late endosomes to silica nanoparticles-filled cathepsin B/L-containing lysosomes, rather than elevated lysosomal pH or altered mTOR activity [36].

5.5.2 Mitochondria Dysfunction and Mitophagy

Mitochondria are major places for cells to producing energy and operating oxidative reaction; besides, mitochondria were also involved in many cell activities such as cell proliferation, differentiation, and apoptosis [37, 38]. In the year of 2010, Johnson-Lyles et al., investigated that renal proximal tubule cells exposed to fullerenol showed cytoskeletion disruption, autophagic vacuole accumulation, loss of cellular mitochondrial membrane potential and ATP depletion; moreover, fullerenol-induced ATP depletion and loss of mitochondrial potential were partially ameliorated by co-treatment with the autophagy inhibitor 3-MA, which confirmed the critical relation between mitochondrial dysfunction and autophagy [39]. Wu et al., investigate that iron core and gold shell nanoparticles could cause an irreversible membrane-potential loss in the mitochondria of cancer cells. Iron elements, before oxidation, triggered mitochondria-mediated autophagy was the key factor responsible for the differential cytotoxicity observed between cancerous and healthy cells [40]. Similarly, Fe_3O_4 nanoparticles could also cause mitochondrial damage and ER and Golgi body stresses, which induce autophagy, while PLGA-coated Fe_3O_4 nanoparticles reduce the destructive effect on these organelles [31].

Apart from mitochondrial dysfunction in nanoparticles-induced autophagy during their toxicity, mitophagy is one of the most direct mechanisms linking mitochondria and autophagy. Mitophagy is a kind of autophagy, which is the process of removing abnormal mitochondria through autophagy in cells [41]. The detailed process were: abnormal conditions such as reactive oxygen species, nutrition lacking could induce depolarized and damaged mitochondrion. Dysfunctional mitochondria were packaged into autophagosome and then fused and degraded in lysosomes [42].

Zhang et al., developed polyoxometalates nanoparticle-peptide conjugates targeting mitochondria to explore the interactions between nanoparticles and cells. Autophagy of Mitochondria induced by polyoxometalates nanoparticles-peptide conjugates was the cell response for the damaged organelles recycle via mitochondrial membrane potential-related molecular mediation [43]. Besides, Nano-alumina (Al_2O_3) could induce autophagy and mitochondria damage in primary cortical neuronal cells while the damaged mitochondria were removed by mitophagy [44].

5.5.3 Endoplasmic Reticulum Stress and Endoplasmic Reticulum Autophagy

Endoplasmic reticulum plays several common functions, including protein molecules folding and transporting synthesized proteins from vesicles to Golgi apparatus [45]. Abnormal conditions in redox and calcium regulation, glucose deficiency or viral infection can trigger endoplasmic reticulum stress response [46]. An accumulation of unfolded or misfolded proteins in the endoplasmic reticulum leads to stress conditions. Recent studies have revealed that endoplasmic reticulum stress can either stimulate or inhibit autophagy [47].

Titanium dioxide nanoparticles could induce endoplasmic reticulum stress inhuman bronchial epithelial cells and disrupted the mitochondria-associated endoplasmic reticulum membranes and calcium ion balance, thereby increasing autophagy. Tauroursodeoxycholic acid, inhibitor of endoplasmic reticulum stress, could significantly mitigated titanium dioxide nanoparticles-induced cellular toxic response [48]. In vivo experimental results also showed that endoplasmic reticulum and mitochondria were disrupted and dysfunctional in the TiO_2 nanoparticles-exposed lung leading to abnormal autophagy [49]. Another mental oxide nanoparticles magnetic iron oxide nanoparticles could induce autophagy preceding apoptosis through mitochondrial damage and endoplasmic reticulum stress in RAW264.7 cells: Blocking of autophagosome formation may accelerate apoptotic cell death and endoplasmic reticulum stress [50].

Besides mental oxide nanoparticles, cationic polystyrene nanospheres could also induce autophagic cell death through the induction of endoplasmic reticulum stress. Cationic polystyrene nanospheres were highly toxic with enhanced uptake in RAW 264.7 macrophage and BEAS-2B lung epithelial cells. The nanoparticles could induce autophagic cell death, and the increased autophagic flux triggered by reactive oxygen species generation and endoplasmic reticulum stress was caused by misfolded protein accumulation. The inhibition of endoplasmic reticulum stress could impair cytotoxicity and autophagy in cationic polystyrene-treated cells [17].

The role of endoplasmic reticulum stress in nanoparticle-induced autophagy has been extensively studied, however, real time information about the endoplasmic reticulum involved autophagic process (endoplasmic reticulum autophagy) induced by nanoparticles remains confusing. Wei et al., reported that silica nanoparticles could be captured, accumulated in endoplasmic reticulum, and triggered autophagy in HCT-116 human colon cancer cells. The co-location of cells between endoplasmic reticulum, lysosomes and autophagic vacuoles confirmed that silica nanoparticles-induced endoplasmic reticulum autophagy. These findings enable us to know more about endoplasmic reticulum autophagy [51].

5.6 Molecular Mechanisms of Autophagy in Nanoparticles-Induced Toxicity

5.6.1 PI3K/Akt/mTOR Signaling Pathway

The phosphatidylinositol 3-kinase (PI3K)/Akt/mammalian target of rapamycin (mTOR) pathway plays critical role in multiple cellular functions, and is a major regulator of autophagy [52]. Upstream of mTOR PI3K/Akt signaling pathway could regulate mTOR activity [53]. Scientists have found that autophagy is negatively adjusted by the activation of mTOR as mTORC1 could regulate autophagy under unusual stressful conditions [54]. When dephosphorylated, mTOR will be inhibited and then trigger autophagy [55]. Besides, the ribosomal protein S6, one substrate of mTOR, was also involved in autophagy processes (Fig. 5.3).

Liu et al., reported that Akt-TSC2-mTOR signaling pathway played the key role in single-walled carbon nanotube-induced autophagy in A549 cells. Single-walled carbon nanotube impaired phosphorylated Akt, mTOR and its substrate ribosomal protein S6. Using siRNA to

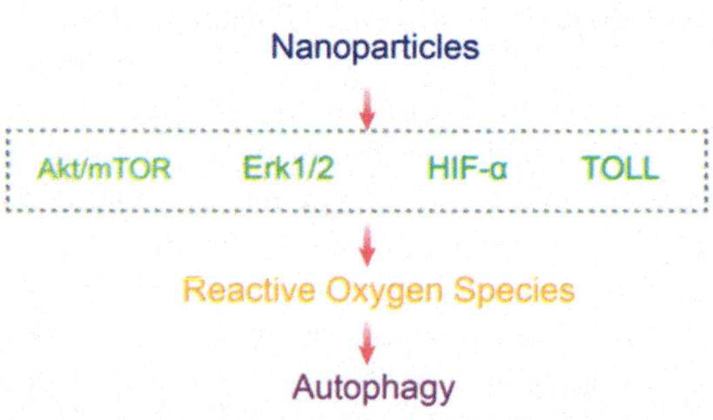

Fig. 5.3 Molecular mechanisms of autophagy in nanoparticles-induced toxicity. (1) PI3K/Akt/mTOR signaling pathway; (2) MAPK/ERK signaling pathway; (3) Toll-like receptor signaling pathways; (4) Hypoxia-inducible factor; (5) Oxidative stress

knockdown TSC2 could obviously enhance cell viability treated by carbon nanotube [11]. Using another non-mental nanoparticles in different cell model, Duan et al., confirmed that phosphorylated PI3K, Akt and mTOR were obviously suppressed in endothelial cells treated by silica nanoparticles in a dose-dependent manner [56]. For mental nanoparticles-induced autophagy, Roy et al., investigated that zinc oxide nanoparticles induced apoptosis by enhancement of autophagy via PI3K/Akt/mTOR inhibition. The phosphorylated levels of PI3K, Akt, and mTOR were significantly decreased during zinc oxide nanoparticles exposing on macrophage [57]. Apart from mental and non-mental nanoparticles, polymers could also induce autophagy via PI3K/Akt/mTOR signaling pathway: Chiu et al., found that Akt/mTOR signaling pathway was involved in cationic polystyrene nanospheres-induced autophagic cell death in macrophage and lung epithelial cells [17]. Meanwhile, Li et al., and Wang et al., both confirmed that cationic polyamidoamine dendrimers could inhibit phosphorylation of Akt/mTOR during their-induced autophagy in hepatocyte and neuronal cells [19, 20].

5.6.2 MAPK/ERK Signaling Pathway

The mitogen-activated protein kinases (MAPKs) involving extracellular signaling-regulated kinase (ERK), was tightly related to reactive oxygen species induction [3]. Previous studies have suggested that ERK activation could contribute to autophagic effects and promote cell survival [58, 59]. Researchers also found that MAPK/ERK signaling pathway was involved in nanoparticles-induced autophagy during their-induced toxicity (Fig. 5.3).

Park et al., explored the underlying mechanism of iron oxide nanoparticles-induced autophagy in RAW 264.7 macrophage, and found that the autophagy related protein increased in a dose-dependent manner together with phosphorylated ERK [60]. Activated phosphorylated ERK was also involved in copper oxide nanoparticle-induced cytotoxicity in HaCaT human keratinocytes and mouse embryonic fibroblasts [61]. Besides, ERK activation also played important roles in the radio-sensitivity enhancement of silver nanoparticles; suppression of ERK could reduce autophagy levels triggered by silver nanoparticles [62]. Rinna et al., also explored effects of silver nanoparticles on mitogen-activated protein kinases activation, and confirmed the role of reactive oxygen species and implication in DNA damage during silver nanoparticles-induced toxicity [63].

5.6.3 Toll-Like Receptor Signaling Pathways

Chen et al., reported that grapheme oxide nanosheets could simultaneously induce autophagy, provoke the toll-like receptor signaling cascades, and trigger ensuing cytokine responses in RAW 264.7 macrophage cells. Grapheme oxide nanoparticles-induced autophagy was regulated by toll-like receptor 4 and toll-like receptor 9, suggested that autophagy was partly regulated by the toll-like receptors pathway in grapheme oxide nanoparticles-treated immune cells [64]. Using CT26 colon cancer cell model and mice model, they also confirmed that grapheme oxide nanoparticles could induce the toll-like receptors response and autophagy in cancer cells and showed antitumor effects. The grapheme oxide nanoparticles-triggered autophagy was regulated through the myeloid differentiation primary response gene 88- and tumor necrosis factor receptor associated factor 6-associated toll-like receptor-4/9 signaling pathways [65].

5.6.4 Hypoxia-Inducible Factor-1α

Hypoxia-inducible factor-1 is a heterodimer composed of α and β subunits and is the transcription factor, which mediates adaptive responses to hypoxia [66]. Hypoxia-inducible factor-1 is mainly regulated by oxygen-dependent changes and could regulate autophagy and other hypoxia-responses [67]. Lin et al., explored the role of hypoxia-inducible factor-1α in zinc oxide nanoparticle-induced nephrotoxicity in vitro and in vivo, and found that zinc oxide nanoparticles could enhance reactive oxygen species generation, apoptosis, autophagy, and hypoxia-inducible factor-1α signaling pathway in HEK-293 human embryonic kidney cells and mouse kidney tissues. Hypoxia-inducible factor -1α knockdown resulted in significantly decreased levels of autophagy and increased cytotoxicity in HEK-293 cells [68]. Silver nanoparticles induced reactive oxygen species generation in lung cancer cells could also trigger high susceptibility to oxidative stress, whereas pre-exposure to hypoxia blocked silver nanoparticles-induced oxidative stress. Hypoxia-inducible factor-1α inhibited silver nanoparticles-induced mitochondria-mediated apoptosis by regulating autophagic flux through the regulation of ATG5, LC3-II, and p62 [69].

5.6.5 Oxidative Stress

Reactive oxygen species are highly reactive molecules containing an oxygen atom, and are mainly produced by the mitochondrial and cytoplasmic oxidation processes under physiological conditions [70]. Reactive oxygen species could induce membrane damage that could influence inner proteins, lipid denaturation, and DNA structures [71]. An increasing amount of evidence indicated that nanoparticles could induce autophagy because of their large surface and positive charges, which is capable of inducing reactive oxygen species, during their-induced toxicity (Fig. 5.4) [72]. Cadmium-based quantum dot increased intracellular reactive oxygen species levels, affected mitochondrial function and induced autophagy, led to subsequent apoptosis in mouse renal adenocarcinoma cells. Antioxidant agent N-Acetylcysteine, reduced intracellular reactive oxygen species levels and impaired quantum dots-induced autophagy but enhanced cell death. Autophagic inhibitor 3-MA also reduced cell viability in quantum dots-treated cells, suggested that oxidative stress-induced autophagy played a survival mechanism against the cytotoxicity of quantum dots [73]. On the contrary, Fan and coworkers found that another kind of cadmium-based quantum dots CdTe/CdS 655 could induce autophagic cell death in vitro and in vivo. Suppress autophagy could attenuate the toxicity induced by cadmium-based quantum dots CdTe/CdS 655 [74].

Apart from quantum dots, mental oxide nanoparticles could also trigger autophagy via reactive oxygen species induction during their-induced toxicity. Yu et al., investigated the toxicity of zinc oxide nanoparticles and explored the underlying molecular mechanisms in normal skin cell model, and found that zinc oxide nanoparticles led to cell death through autophagic vacuole accumulation and mitochondria damage in normal skin cells via reactive oxygen species induction [14]. Apart from skin cells, acute exposure of immune cells to ZnO nanoparticles resulted in autophagic death and increased levels of LC3. Accordingly, ZnO nanoparticles-mediated upregulation of LC3 and induction of immune cell death were inhibited by blocking autophagy and reactive oxygen species production. Release of Zn^{2+} from ZnO nanoparticles triggered excessive intracellular reactive oxygen species production, and resulted in autophagic death of immune cells [15].

Besides, reactive oxygen species were also involved in non-mental nanoparticles and polymers-induced autophagy during their-triggered nanotoxicity. Silica nanoparticles could induce autophagy in a dose-dependent manner in HepG2 cells. The elevated reactive oxygen species level was in line with the increasing of autophagy activation, while both the autophagic inhibitor (3-MA) and reactive oxygen species inhibitor (N-Acetylcysteine) effectively suppressed the autophagy and cell death induced by silica nanoparticles [10]. The role of reactive oxygen species during silica nanoparticles-induced autophagy was also been explored in two other cell models: SiO_2 nanoparticles induced reactive oxygen species -mediated autophagy in MRC-5 human lung fibroblast cells as a possible mechanism of cell survival [75]. And amorphous silica nanoparticles trigger vascular endothelial

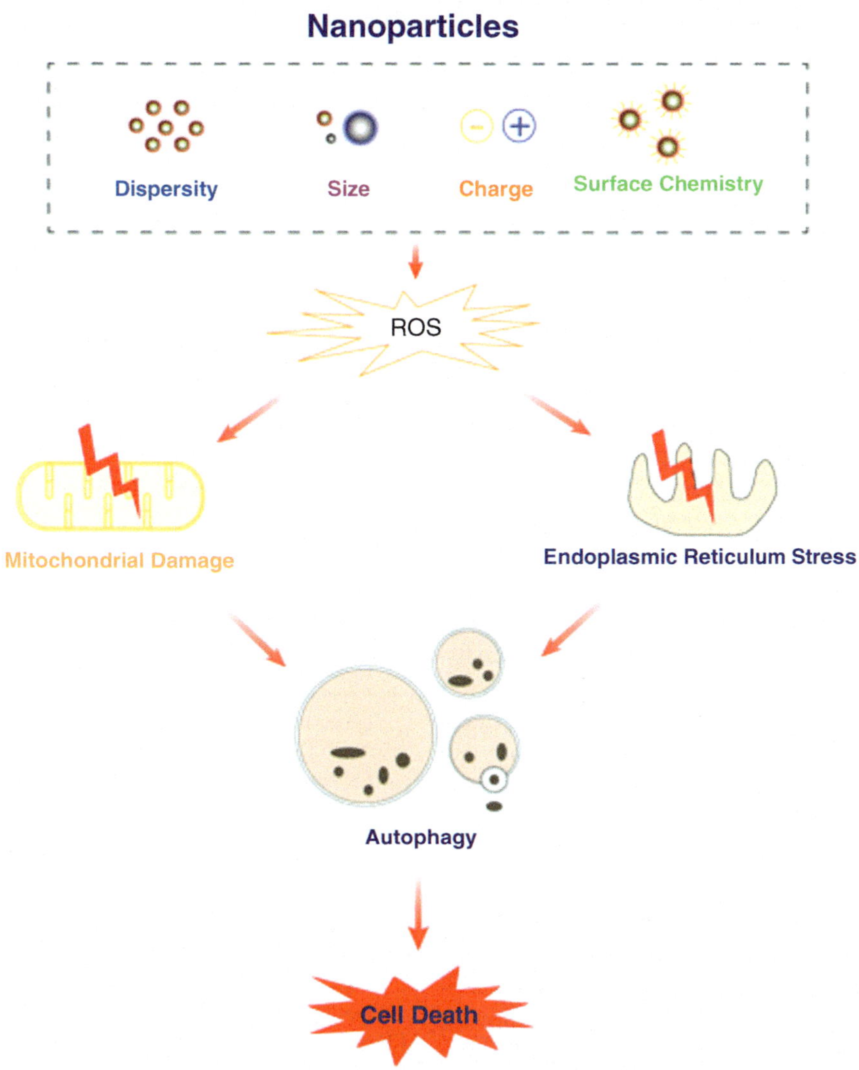

Fig. 5.4 Crosstalk of physicochemical and biochemical mechanisms of autophagy in nanoparticles-induced toxicity. (1) Physicochemical mechanisms including dispersity, size, charge, and surface chemistry of nanoparticles; (2) Nanoparticles triggered reactive oxygen species induced mitochondria dysfunction and endoplasmic reticulum stress. (3) Mitochondria dysfunction and endoplasmic reticulum stress was involved in nanoparticles-induced autophagic cell death

cell injury through apoptosis and autophagy via reactive oxygen species-mediated MAPK/Bcl-2 and PI3K/Akt/mTOR signaling [76].

Poly-amidoamine dendrimers induced reactive oxygen species and autophagy flux in PC-12 and SH-SY5Y neuronal cells and glioma cells [20]. Interestingly, autophagy might be triggered by the formation of reactive oxygen species induced by poly-amidoamine dendrimers. Suppression of reactive oxygen species could not only impair poly-amidoamine dendrimers-induced autophagic effects, but also reduce poly-amidoamine dendrimers-induced neuronal cell death [77] (Fig. 5.4).

5.7 Conclusion

As nanoparticles possess unique physical structures and specific properties, series of nanoparticles have been developed in the past few decades for wide industry or medicinal application. Because nanoparticles could highly interact with cells after they entering into bodies, they may cause nanotoxicity and induce body damage, which could seriously limit their application. Although the underlying mechanisms of nanotoxicity triggered by nanoparticles have been widely investigated, unfortunately, our understanding of the role of autophagy and its related mechanisms in nanotoxicity is still poor. Thus, it is an instant need for researchers to address the questions about the role and mechanism of autophagy in nanoparticles-induced toxicity. While evaluating the mechanisms of autophagy in nanoparticles-induced toxicity, the dispersity, size, surface charge, and surface modification of nanoparticles were expected to present critical roles in nanoparticles-triggered autophagy, and should be cautiously considered. Besides, the research initiatives highlighted in the chapter showed cellular and molecular mechanisms of autophagy during nanoparticles-induced toxicity, including lysosome impairment, mitochondria dysfunction, endoplasmic reticulum stress, PI3K/Akt/mTOR signaling pathway, MAPK/ERK signaling pathway, toll-like receptor pathway, Hypoxia-inducible factor, and oxidative stress. These mechanisms of autophagy triggered by nanoparticles might be help for evaluating nanotoxicity, and might offer a foundation for deeply design safe nanoparticles in the future.

Acknowledgments This work was supported by grants from the National Key Basic Research Program of China (2015CB931800, 2013CB932502), and the National Natural Science Foundation of China (81573332, 81773620). Yubin Li would like to acknowledge his baby girl Charlotte (Qianxun) Li.

Conflict of Interests All the Figures in "The Role of Autophagy in Nanoparticles-Induced Toxicity and Its Related Cellular and Molecular Mechanisms" are original, unpublished materials designed and prepared by Yubin Li and Dianwen Ju. The authors declared that there's no conflict of interests.

References

1. Li J, Fan C, Pei H et al (2013) Smart drug delivery nanocarriers with self-assembled DNA nanostructures. Adv Mater 25:4386–4396
2. Linko V, Ora A, Kostiainen MA (2015) DNA nanostructures as smart drug-delivery vehicles and molecular devices. Trends Biotechnol 33:586–594
3. Fu PP, Xia Q, Hwang HM et al (2014) Mechanisms of nanotoxicity: generation of reactive oxygen species. J Food Drug Anal 22:64–75
4. Liu Y, Liang J, Wang Q et al (2016) Copper nanoclusters trigger muscle cell apoptosis and atrophy in vitro and in vivo. J Appl Toxicol 36:454–463
5. Deter RL, De Duve C (1967) Influence of glucagon, an inducer of cellular autophagy, on some physical properties of rat liver lysosomes. J Cell Biol 33:437–449
6. Glick D, Barth S, Macleod KF (2010) Autophagy: cellular and molecular mechanisms. J Pathol 221:3–12
7. Mizushima N (2007) Autophagy: process and function. Genes Dev 21:2861–2873
8. Zeng X, Zhao H, Li Y et al (2015) Targeting Hedgehog signaling pathway and autophagy overcomes drug resistance of BCR-ABL-positive chronic myeloid leukemia. Autophagy 11:355–372
9. Levine B, Kroemer G (2008) Autophagy in the pathogenesis of disease. Cell 132:27–42
10. Yu Y, Duan J, Yu Y et al (2014) Silica nanoparticles induce autophagy and autophagic cell death in HepG2 cells triggered by reactive oxygen species. J Hazard Mater 270:176–186
11. Liu HL, Zhang YL, Yang N et al (2011) A functionalized single-walled carbon nanotube-induced autophagic cell death in human lung cells through Akt-TSC2-mTOR signaling. Cell Death Dis 2:e159
12. Park EJ, Zahari NE, Lee EW et al (2014c) SWCNTs induced autophagic cell death in human bronchial epithelial cells. Toxicol In Vitro 28:442–450
13. Sun T, Yan Y, Zhao Y et al (2012) Copper oxide nanoparticles induce autophagic cell death in A549 cells. PLoS One 7:e43442
14. Yu KN, Yoon TJ, Minai-Tehrani A et al (2013) Zinc oxide nanoparticle induced autophagic cell death and mitochondrial damage via reactive oxygen species generation. Toxicol In Vitro 27:1187–1195
15. Johnson BM, Fraietta JA, Gracias DT et al (2015) Acute exposure to ZnO nanoparticles induces autophagic immune cell death. Nanotoxicology 9:737–748
16. Chen Y, Yang L, Feng C et al (2005) Nano neodymium oxide induces massive vacuolization and autophagic cell death in non-small cell lung cancer NCI-H460 cells. Biochem Biophys Res Commun 337:52–60
17. Chiu HW, Xia T, Lee YH et al (2015) Cationic polystyrene nanospheres induce autophagic cell death through the induction of endoplasmic reticulum stress. Nanoscale 7:736–746
18. Li C, Liu H, Sun Y et al (2009) PAMAM nanoparticles promote acute lung injury by inducing autophagic cell death through the Akt-TSC2-mTOR signaling pathway. J Mol Cell Biol 1:37–45

19. Li Y, Zeng X, Wang S et al (2015a) Inhibition of autophagy protects against PAMAM dendrimers-induced hepatotoxicity. Nanotoxicology 9:344–355
20. Wang S, Li Y, Fan J et al (2014) The role of autophagy in the neurotoxicity of cationic PAMAM dendrimers. Biomaterials 35:7588–7597
21. Zhou W, Miao Y, Zhang Y et al (2013) Induction of cyto-protective autophagy by paramontroseite VO2 nanocrystals. Nanotechnology 24:165102
22. Lin J, Huang Z, Wu H et al (2014) Inhibition of autophagy enhances the anticancer activity of silver nanoparticles. Autophagy 10:2006–2020
23. Huang D, Zhou H, Gao J (2015) Nanoparticles modulate autophagic effect in a dispersity-dependent manner. Sci Rep 5:14361
24. Seleverstov O, Zabirnyk O, Zscharnack M et al (2006) Quantum dots for human mesenchymal stem cells labeling. A size-dependent autophagy activation. Nano Lett 6:2826–2832
25. Mishra AR, Zheng J, Tang X et al (2016) Silver nanoparticle-induced autophagic-lysosomal disruption and NLRP3-inflammasome activation in HepG2 cells is size-dependent. Toxicol Sci 150:473–487
26. Ma X, Wu Y, Jin S et al (2011) Gold nanoparticles induce autophagosome accumulation through size-dependent nanoparticle uptake and lysosome impairment. ACS Nano 5:8629–8639
27. Li Q, Hu H, Jiang L et al (2016) Cytotoxicity and autophagy dysfunction induced by different sizes of silica particles in human bronchial epithelial BEAS-2B cells. Toxicol Res 5:1216–1228
28. Huang D, Zhou H, Gong X et al (2017) CSilica sub-microspheres induce autophagy in an endocytosis dependent manner. RSC Adv 7:12496–12502
29. Song W, Popp L, Yang J et al (2015) The autophagic response to polystyrene nanoparticles is mediated by transcription factor EB and depends on surface charge. J Nanobiotechnol 13:87
30. Wu L, Zhang Y, Zhang C et al (2014) Tuning cell autophagy by diversifying carbon nanotube surface chemistry. ACS Nano 8:2087–2099
31. Zhang X, Zhang H, Liang X et al (2016) Iron oxide nanoparticles induce autophagosome accumulation through multiple mechanisms: lysosome impairment, mitochondrial damage, and ER stress. Mol Pharm 13:2578–2587
32. Du J, Zhu W, Yang L et al (2016) Reduction of polyethylenimine-coated iron oxide nanoparticles induced autophagy and cytotoxicity by lactosylation. Regen Biomater 3:223–229
33. Stern ST, Adiseshaiah PP, Crist RM (2012) Autophagy and lysosomal dysfunction as emerging mechanisms of nanomaterial toxicity. Part Fibre Toxicol 9:20
34. Wang J, Yu Y, Lu K et al (2017) Silica nanoparticles induce autophagy dysfunction via lysosomal impairment and inhibition of autophagosome degradation in hepatocytes. Int J Nanomedicine 12:809–825
35. Ji X, Xu B, Yao M et al (2016) Graphene oxide quantum dots disrupt autophagic flux by inhibiting lysosome activity in GC-2 and TM4 cell lines. Toxicology 374:10–17
36. Schutz I, Lopez-Hernandez T, Gao Q et al (2016) Lysosomal dysfunction caused by cellular accumulation of silica nanoparticles. J Biol Chem 291:14170–14184
37. Miettinen TP, Bjorklund M (2017) Mitochondrial function and cell size: an allometric relationship. Trends Cell Biol 27:393–402
38. Altshuler-Keylin S, Kajimura S (2017) Mitochondrial homeostasis in adipose tissue remodeling. Sci Signal 10:eaai9248
39. Johnson-Lyles DN, Peifley K, Lockett S et al (2010) Fullerenol cytotoxicity in kidney cells is associated with cytoskeleton disruption, autophagic vacuole accumulation, and mitochondrial dysfunction. Toxicol Appl Pharmacol 248:249–258
40. Wu YN, Yang LX, Shi XY et al (2011) The selective growth inhibition of oral cancer by iron core-gold shell nanoparticles through mitochondria-mediated autophagy. Biomaterials 32:4565–4573
41. Fivenson EM, Lautrup S, Sun N et al (2017) Mitophagy in neurodegeneration and aging. Neurochem Int 109:202–209
42. Kim MJ, Yoon JH, Ryu JH (2016) Mitophagy: a balance regulator of NLRP3 inflammasome activation. BMB Rep 49:529–535
43. Zhang Z, Zhou L, Zhou Y et al (2015) Mitophagy induced by nanoparticle-peptide conjugates enabling an alternative intracellular trafficking route. Biomaterials 65:56–65
44. Chang L, Guo W, Ge C et al (2014) Effect of nano-alumina on mitophagy in primary cortical neuronal cells from Wistar newborn rats. Chin J Pharmacol Toxicol 5:737–742
45. Preston GM, Brodsky JL (2017) The evolving role of ubiquitin modification in endoplasmic reticulum-associated degradation. Biochem J 474:445–469
46. Lee WS, Yoo WH, Chae HJ (2015) ER stress and autophagy. Curr Mol Med 15:735–745
47. Rashid HO, Yadav RK, Kim HR et al (2015) ER stress: autophagy induction, inhibition and selection. Autophagy 11:1956–1977
48. Yu KN, Chang SH, Park SJ et al (2015a) Titanium dioxide nanoparticles induce endoplasmic reticulum stress-mediated autophagic cell death via mitochondria-associated endoplasmic reticulum membrane disruption in normal lung cells. PLoS One 10:e0131208
49. Yu KN, Sung JH, Lee S et al (2015b) Inhalation of titanium dioxide induces endoplasmic reticulum stress-mediated autophagy and inflammation in mice. Food Chem Toxicol 85:106–113
50. Park EJ, Choi DH, Kim Y et al (2014a) Magnetic iron oxide nanoparticles induce autophagy preceding apoptosis through mitochondrial damage and ER stress in RAW264.7 cells. Toxicol In Vitro 28:1402–1412
51. Wei F, Wang Y, Luo Z et al (2017) New findings of silica nanoparticles induced ER autophagy in human colon cancer cell. Sci Rep 7:42591

52. McAuliffe PF, Meric-Bernstam F, Mills GB et al (2010) Deciphering the role of PI3K/Akt/mTOR pathway in breast cancer biology and pathogenesis. Clin Breast Cancer 10(Suppl 3):S59–S65
53. Heras-Sandoval D, Perez-Rojas JM, Hernandez-Damian J et al (2014) The role of PI3K/AKT/mTOR pathway in the modulation of autophagy and the clearance of protein aggregates in neurodegeneration. Cell Signal 26:2694–2701
54. Jung CH, Ro SH, Cao J et al (2010) mTOR regulation of autophagy. FEBS Lett 584:1287–1295
55. Yu X, Long YC, Shen HM (2015c) Differential regulatory functions of three classes of phosphatidylinositol and phosphoinositide 3-kinases in autophagy. Autophagy 11:1711–1728
56. Duan J, Yu Y, Yu Y et al (2014) Silica nanoparticles induce autophagy and endothelial dysfunction via the PI3K/Akt/mTOR signaling pathway. Int J Nanomedicine 9:5131–5141
57. Roy R, Singh SK, Chauhan LK et al (2014) Zinc oxide nanoparticles induce apoptosis by enhancement of autophagy via PI3K/Akt/mTOR inhibition. Toxicol Lett 227:29–40
58. Ogier-Denis E, Pattingre S, El Benna J et al (2000) Erk1/2-dependent phosphorylation of Galpha-interacting protein stimulates its GTPase accelerating activity and autophagy in human colon cancer cells. J Biol Chem 275:39090–39095
59. Cagnol S, Chambard JC (2010) ERK and cell death: mechanisms of ERK-induced cell death – apoptosis, autophagy and senescence. FEBS J 277:2–21
60. Park EJ, Umh HN, Kim SW et al (2014b) ERK pathway is activated in bare-FeNPs-induced autophagy. Arch Toxicol 88:323–336
61. Luo C, Li Y, Yang L et al (2014) Activation of Erk and p53 regulates copper oxide nanoparticle-induced cytotoxicity in keratinocytes and fibroblasts. Int J Nanomedicine 9:4763–4772
62. Wu H, Lin J, Liu P et al (2015) Is the autophagy a friend or foe in the silver nanoparticles associated radiotherapy for glioma? Biomaterials 62:47–57
63. Rinna A, Magdolenova Z, Hudecova A et al (2015) Effect of silver nanoparticles on mitogen-activated protein kinases activation: role of reactive oxygen species and implication in DNA damage. Mutagenesis 30:59–66
64. Chen GY, Yang HJ, Lu CH et al (2012) Simultaneous induction of autophagy and toll-like receptor signaling pathways by graphene oxide. Biomaterials 33:6559–6569
65. Chen GY, Chen CL, Tuan HY et al (2014) Graphene oxide triggers toll-like receptors/autophagy responses in vitro and inhibits tumor growth in vivo. Adv Healthc Mater 3:1486–1495
66. Semenza GL, Wang GL (1992) A nuclear factor induced by hypoxia via de novo protein synthesis binds to the human erythropoietin gene enhancer at a site required for transcriptional activation. Mol Cell Biol 12:5447–5454
67. Bruick RK, McKnight SL (2001) A conserved family of prolyl-4-hydroxylases that modify HIF. Science 294:1337–1340
68. Lin YF, Chiu IJ, Cheng FY et al (2016) The role of hypoxia-inducible factor-1alpha in zinc oxide nanoparticle-induced nephrotoxicity in vitro and in vivo. Part Fibre Toxicol 13:52
69. Jeong JK, Gurunathan S, Kang MH et al (2016) Hypoxia-mediated autophagic flux inhibits silver nanoparticle-triggered apoptosis in human lung cancer cells. Sci Rep 6:21688
70. Indo HP, Hawkins CL, Nakanishi I et al (2017) Role of mitochondrial reactive oxygen species in the activation of cellular signals, molecules, and function. Handb Exp Pharmacol 240:439–456
71. Chen Y, Gibson SB (2008) Is mitochondrial generation of reactive oxygen species a trigger for autophagy? Autophagy 4:246–248
72. Anozie UC, Dalhaimer P (2017) Molecular links among non-biodegradable nanoparticles, reactive oxygen species, and autophagy. Adv Drug Deliv Rev 17:30001–30007
73. Luo YH, Wu SB, Wei YH et al (2013) Cadmium-based quantum dot induced autophagy formation for cell survival via oxidative stress. Chem Res Toxicol 26:662–673
74. Fan J, Sun Y, Wang S et al (2016) Inhibition of autophagy overcomes the nanotoxicity elicited by cadmium-based quantum dots. Biomaterials 78:102–114
75. Petrache Voicu SN, Dinu D, Sima C et al (2015) Silica nanoparticles induce oxidative stress and autophagy but not apoptosis in the MRC-5 cell line. Int J Mol Sci 16:29398–29416
76. Guo C, Yang M, Jing L et al (2016) Amorphous silica nanoparticles trigger vascular endothelial cell injury through apoptosis and autophagy via reactive oxygen species-mediated MAPK/Bcl-2 and PI3K/Akt/mTOR signaling. Int J Nanomedicine 11:5257–5276
77. Li Y, Zhu H, Wang S et al (2015b) Interplay of oxidative stress and autophagy in PAMAM dendrimers-induced neuronal cell death. Theranostics 5:1363–1377

Check for updates

Nanoparticles-Caused Oxidative Imbalance

6

Mariusz Zuberek and Agnieszka Grzelak

Abstract

Application of nanomaterials in nearly every single branch of industry results in their accumulation in both abiotic environment and tissues of living organisms. Despite the common use of nanomaterials, we are not able to precisely define their toxicity towards humans and surrounding biota. Although we were able to determine final effects of chronic exposure to nanoparticles which consist of many pathologies such as respiratory diseases, allergies, diseases of cardiovascular system, disorders in embryonic life differentiation and growth disorders, toxic effects on the immune system and cancers. The most predominantly investigated feature of most nanoparticles is their ability to induce oxidative stress on cellular level. Imbalance in redox state of cells can lead to various malfunctions in their internal metabolism, which in turn can lead to mentioned pathologies on the organismal level if the exposure is persistent and spread wide enough. Imbalance in redox state translate into production of reactive oxygen species in amounts impossible to be scavenged in given time. Many reactive oxygen species play crucial role in physiological processes in properly functioning cells. It was proven on numerous occasions that abundance of ROS, aside from oxidative damage, can lead to more subtle adverse effects tied to disturbances in intra- and intercellular signaling pathways. In this chapter we would like to address the nanoparticle-induced redox imbalance in cells and its effects.

Keywords

Nanoparticles · Oxidative stress · Signal transduction · ROS · DNA damage · Nanotoxicity

M. Zuberek (✉) · A. Grzelak
Department of Molecular Biophysics, Faculty of Biology and Environmental Protection, University of Lodz, Lodz, Poland
e-mail: zuberekmariusz@gmail.com

Q. Saquib et al. (eds.), *Cellular and Molecular Toxicology of Nanoparticles*, Advances in Experimental Medicine and Biology 1048, https://doi.org/10.1007/978-3-319-72041-8_6

6.1 Redox Status - Background

Oxidative stress is a phenomenon concerning organisms living in aerobic conditions. It can be described as imbalance between removal and production of reactive oxygen species. In effect of prolonged exposure to redox imbalance, cell structures and biomolecules can undergo oxidative damage, resulting in amassing of damaged DNA, proteins and lipids, which may result in alterations of cellular metabolism and eventually cause cell death.

Free oxygen radicals are natural products of cellular metabolism and function as signaling molecules activating various signaling pathways, such as NFκB and AP1 [1]. Term free radical was created in 1785 by Louis-Bernard Guyton de Morveau and was also used by Antoine Lavosidier in 1789 in his *Traité Élémentaire de Chimie*, although not in the current meaning. Modern free radical term defines particles with unpaired electron in valence shell. This kind of chemical entity was firstly described by Moses Gomberg in 1900 and concerned organic radical created during the synthesis of tetraphenylmethane [2]. Free radicals most common in biological systems have unpaired electron on the oxygen atom.

Primary free radicals produced in cells are transformed into reactive oxygen species by low molecular weight antioxidants and antioxidant enzymes. Free radicals are created locally in cell compartments and have no ability to cross intracellular membranes and their damaging effects are limited to biomolecules in direct vicinity of the radical. Reactive oxygen species created in following reactions can diffuse at considerable distances in the cytosol and translocate through membranes.

The source of oxygen free radicals in cells are enzymatic reactions of oxidoreductases i.a. NADPH oxidase [3] and active mitochondrial electron transport chain [4], which was pointed out as organelle producing superoxide radical in 1966 [5].

Reactive nitrogen species, as well as reactive oxygen species play crucial roles in cellular metabolism. Nitric oxide is the precursor of cellular reactive nitrogen species and it is generated in the course of enzymatic reaction of nitric oxide synthase in many types of cells e.g. endothelial cells [6], hepatocytes [7] and cells associated with immune system [8]. Nitric oxide is a molecule extensively studied due to its signaling role [9], resulting in regulation of blood pressure on organismal level [10], influencing apoptosis [11] and activity of enzymes [12] on cellular level.

Nitric oxide reaction with superoxide anion produces peroxynitrite, which is a molecule of great reactivity towards DNA, proteins and lipids, which contributes to the damage of cell structures [13]. Proteins can undergo various modifications in reaction with peroxynitrite, but the most characteristic one is formation of 3-nitrotyrosine [14].

6.2 Redox Balance as Guarantee of Properly Functioning Signal Transmission

Cells functioning aerobically developed mechanisms enabling them to cope with conditions of temporary increased levels of oxygen free radicals and other reactive oxygen species, which may be produced under the influence of external stimuli such as radiation, xenobiotics and toxins, including nanoparticles. The most important regulating aspects of reactive oxygen species are redox sensitive signal cascades activation of which leads to increase in antioxidant enzymes and glutathione levels. Enzymes closely tied to maintaining redox homeostasis are pointed out in Fig. 6.1 and they consist of glutathione peroxidase, glutathione reductase, thioredoxins, glutaredoxins, superoxide dismutase and catalase. These enzymes utilize low molecular weights antioxidants: glutathione and NADPH to maintain proper redox state in cells. Superoxide dismutase and catalase are responsible for detoxification of, respectively, superoxide radical and hydrogen peroxide.

Under physiological conditions oxygen free radicals and reactive oxygen species are crucial for intracellular signal transduction. Many transcription factors have cysteine residue in DNA binding sites, which is extremely prone to oxida-

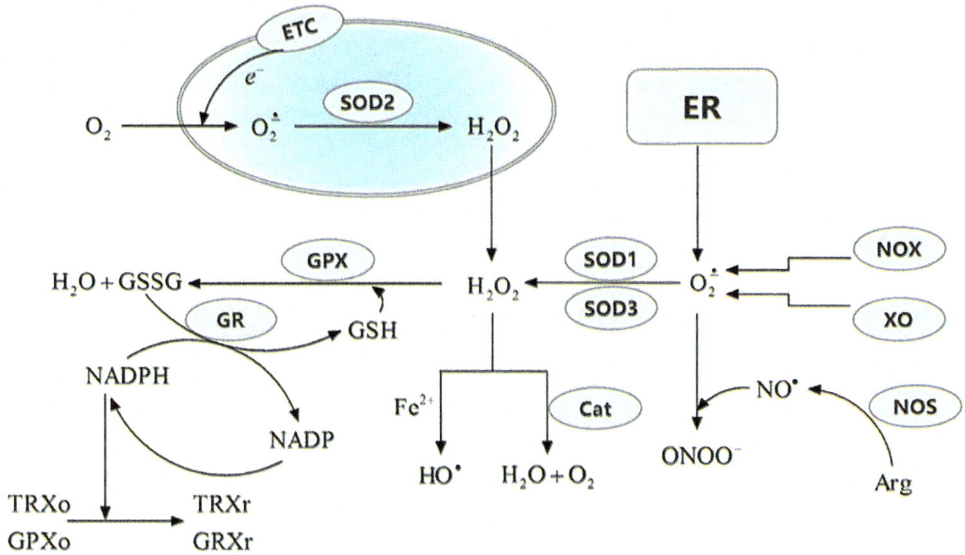

Fig. 6.1 Redox homeostasis. *ETC* electron transport chain, *SOD* superoxide dismutase, H_2O_2 hydrogen peroxide, *GPX* glutathione peroxidase, *GSSG* glutathione disulfide, *GSH* glutathione, *GR* glutathione reductase, *NADP/NADPH* nicotinamide adenine dinucleotide phosphate (oxidized and reduced), *TRXo/TRXr* thioredoxin (oxidized and reduced), *GRXo/GRXr* glutaredoxin (oxidized and reduced), *HO.* hydroxyl radical, *Cat* catalase, $O_2^{\cdot-}$ - superoxide radical, *NOX* NADPH oxidase, *XO* xanthine oxidase, $ONOO^-$ peroxynitrite, *NO* nitric oxide, *Arg* arginine, *NOS* nitric oxide synthase

tion [15]. Exemplary transcription factors whose DNA binding is regulated by oxidation of thiols are NFκB, AP-1, HIF-1 and p53 [16]. Intranuclear level of glutathione is one of the key factors in maintaining proper redox state and ensures proper gene activation [17]. Oxidative modifications of proteins resulting from reactions with oxygen free radicals are one of the crucial post-translational modifications of proteins, which influence their activity and functions [18].

Direct oxidation of amino acid residues is possible in reactions with hydroxyl radical and nitric oxide. The most frequently oxidized amino acids are those containing the thiol group, methionine and cysteine. Oxidation of these amino acids can lead to alterations in protein conformation and result in protein denaturation and fragmentation. Mild oxidative stress can promote number of cysteine modifications: formation of disulfide bonds [19], reversible glutathionylation [20] and S-nitrosylation [21]. Mentioned modifications are important in the process of regulation of protein activity, such as Trx, p53, Ras, and Akt. Oxidative modification of methionine is crucial for calmodulin, a protein involved in calcium metabolism [22]. Another mechanism regulating protein activity is the formation of nitrotyrosine residues. This mechanism mainly refers to kinases e.g. JNK, p38, and PKC. It has been found that formation of nitrotyrosine inhibits their activity [23]. Structure and activity of many proteins is stabilized by protein-protein interactions. Examples of proteins whose functionality is dependent on this mechanism are ASK1-TRX, JNK-GST, p53-JNK, NRF2-Keap1 [24–28].

Particular attention should be paid to the influence of free radicals on the protein phosphorylation process, which is considered the key in intracellular homeostasis. It can lead to protein activation or be a signal for targeted protein degradation [29].

Turnover of proteins is also regulated by redox mechanisms. Proteins are degraded by proteasome and under physiological conditions, ubiquitin and 26S proteasome are responsible for the removal of damaged or poorly folded proteins [30]. In the occurrence of oxidative stress proteasome activity might be inhibited [31], due to inactivation of ubiquitin-activating enzymes and 26S proteasome, which can lead to amassing of damaged proteins and effect in cell death.

6.3 Reactive Oxygen Species Generated in Response to Nanoparticles

Reactive oxygen species produced as outcome of nanoparticles biological activity are studied by various methods ranging from detection via fluorescence measurement of oxidized fluorescent probes [32–35] to indirect methods such as analysis of expression of gene profiles [36] and determination of antioxidant inhibition of nanoparticle cytotoxic effects [37]. Summary of assessment methods of various parameters defining NP-induced oxidative stress is given in Table 6.1. Prefix nano- refer to any particle of diameter of less than a 1 μm, despite the range of diameters of each studied metallic nanoparticle researchers are consistent in their findings – despite the shifts of size from study to study. If the properties of metallic nanoparticle allow it to generate oxidative stress it will be detectable despite the differences in size from study to study. Imbalance in redox state of cells after exposure to nanoparticles is insensitive to experimental system. Similar effects are observed in variety of mammalian cell lines and microorganisms (Table 6.1). Paradigm of oxidative stress

Table 6.1 Summary of detection methods along with experimental systems utilized in studies upon nanoparticle induced oxidative stress, cl stands for cell line

No.	Method of detection	Detection of	Nanoparticle type	Experimental system
1	Fluorescence of oxidized fluorescein derivatives	ROS	Ag [38, 50], Fe_2O_3 [51], SiO_2 [51], CuO [51], TiO_2[39, 52], ZnO[38, 52, 53], CdS [52], USPIO [39], Si[39, 40], SWCNT [46], NiO [54]	*E.coli* [38, 50], HEp-2 cl[51, 54], IP14 cl [52], HK-2 cl [52], HCEC cl [39], Hs68 cl [53], HaCaT cl [53], A549 [53], *S. aureus* [38], *P. aeruginosa* [38], RL 65 cl [46], MCF-7 cl [54], NR8383 [40]
2	Fluorescence of oxidized CellROX Deep Red Reagent	ROS	Fe_3O_4 [43]	A549 cl [43], AA8 cl [43]
3	Fluorescence of oxidized DHE	$O_2^{\cdot-}$	Si [41]	RAW264.7[41]
4	Comet assay	ROS mediated DNA damage	Au [55], USPIO [39], TiO_2 [39], Ag [56]	BALB/c 3T3 cl [55], HCEC cl [39], HepG2 cl [56], A549 cl [56], HT29 cl [56]
5	Ellman's test, MBB assay	Thiol content	Fe_2O_3[51, 57], Fe_3O_4 [58] SiO_2 [51], CuO [51], TiO_2[5], ZnO [52], CdS [52], NiO [54], USPIO [39], PLGA-PEO [39], silica [39], TiO_2 [39]	HEp-2 cl[51, 54], IP14 cl [52], HK-2 cl [52], Jb6 P+ cl [57], MCF-7[54], CHSE-214 [58], HCEC cl [39]
6	Fluorescence of oxidized DHR123	Cellular H_2O_2	Ag [56]	HepG2 cl [56], A549 cl [56]
7	Fluorescence of oxidized MitoSOX RED	Mitochondrial $O_2^{\cdot-}$	UFP* [48], NH_2-PS	RAW 264.7 cl [48]
8	MDA assays and related	Lipid peroxidation	NiO [54], Au [59], Fe_3O_4 [58], Fe_2O_3 [51], SiO_2 [51], CuO [51]	HEp-2 cl[54; 51], MCF-7[54], CHSE-214 cl [58], MRC-5 cl [59]

Nanoparticle abbreviations in table: USPIO – ultrasmall superparamagnetic iron oxide, UFP* – Ultra Fine Particles obtained in the Los Angeles basin through the use of particle concentrator

mediated cell death after exposure to nanoparticles is ingrained so strong in nanotoxicology that many researchers do not assay the oxidative stress itself, but make literature-based assumption that cell death in given experimental system was induced by it. Extensively studied silver nanoparticles were proven to have strong antimicrobial activity due to induction of ROS in bacteria, presenting similar biological activity to many commonly used antibiotics [38]. This characteristic is extremely important due to multidrug resistance phenomenon rendering many overused antibiotics obsolete.

Uncoated metal nanoparticles prone to agglomeration exhibit greater prooxidative potency than their coated counterparts. Nanoparticles such as USPIO (ultrasmall supermagnetic iron oxide) were proven to exhibit prooxidative properties, which effected in DNA damage and cell death, however oleic acid coating strongly inhibited their biological activity [39]. Coating of silica nanoparticles seems crucial in terms of their ability to induce oxidative stress, while functionalization with positively charged and neutral groups like $-NH_2$ and $-N_3$ promotes induction of ROS in cells, negatively charged –COOH group diminishes the effect to non-detectable level [40]. Although charge of the coating is only one of the factors in silica nanoparticles, as functionalization with aminopropyltriethoxysilane giving positive charge increased or decreased ROS producing potency of porous and nonporous silica nanoparticles respectively [41]. ROS in cells can be produced (aside of mitochondrial electron transport) by one-electron reaction of transition metals [42]. Coating of iron nanoparticles limited reaction-available iron resulting in less ROS production and less oxidative damage to cell components.

Nanoparticles of varying sizes are formed from iron on different degrees of oxidation, the common denominator for mentioned Fe_2O_3 and Fe_3O_4 [43] is the production of ROS in studied experimental systems. All forms of iron (Fe^0, Fe(II), Fe(III)) utilized in nanoparticles have the ability to promote generation of reactive oxygen species in aqueous solutions in the course of Fenton, Fenton-like, Haber-Weiss and heterogeneous redox reactions [44]. Many non-ferrous metals were investigated in terms of their catalytic input in Fenton-like reactions at neutral pH. Cu(II), Cr(III), Co(II), Mn(II), all present catalytic activity, while Cd(II), ZN(II) and Ni(II) did not produce any oxidizing species at pH 7 in cell free system [45]. Nevertheless nanoaprticles composed of zinc, cadmium and nickel are reported to induce generation of reactive oxygen species in various cellular experimental systems (Table 6.1). Those findings suggest that the mechanism of induction of ROS in cells consists of Fenton-like reactions but is not limited to them. Experiments on carbon nanoparticles further enhance this concept as single-walled carbon nanotubes were proven to increase ROS levels in cells and decrease cellular glutathione level [46], which are effects common after treatment with metal nanoparticles. Reactive oxygen species can be generated by nanoparticles in photoelectric phenomena in materials where photon absorption is accompanied by electron transfer to the conduction band. Migrating electrons can form superoxide radical in reaction with oxygen and hydroxyl radicals in reaction with water [47]; this effect would be enhanced in nanomaterials due to their high surface to mass ratio. Reactive oxygen species generation induced by semiconductors like TiO_2 is not explicitly confirmed in all studied experimental systems involving mammalian cells due to adaptation processes or passivation of the nanoparticle surface by factors contained in culture medium [48]. Metalloid nanoparticles in their nature are not prone to produce ROS in Fenton chemistry, mechanism of their direct induction of ROS was proposed, implicating that the cleavage of strained three-membered rings is responsible for silica surface ROS production [49] (Fig. 6.2).

6.4 ROS Generation in Course of Action of Nanoparticles Is Strongly Dependent on Particle Size

Toxic effect of AgNPs is strictly tied to size of the nanoparticle and to its bioavailability. Bulk of toxicity related to silver nanoparticles is in fact toxicity of ionic Ag. The effect of nanopar-

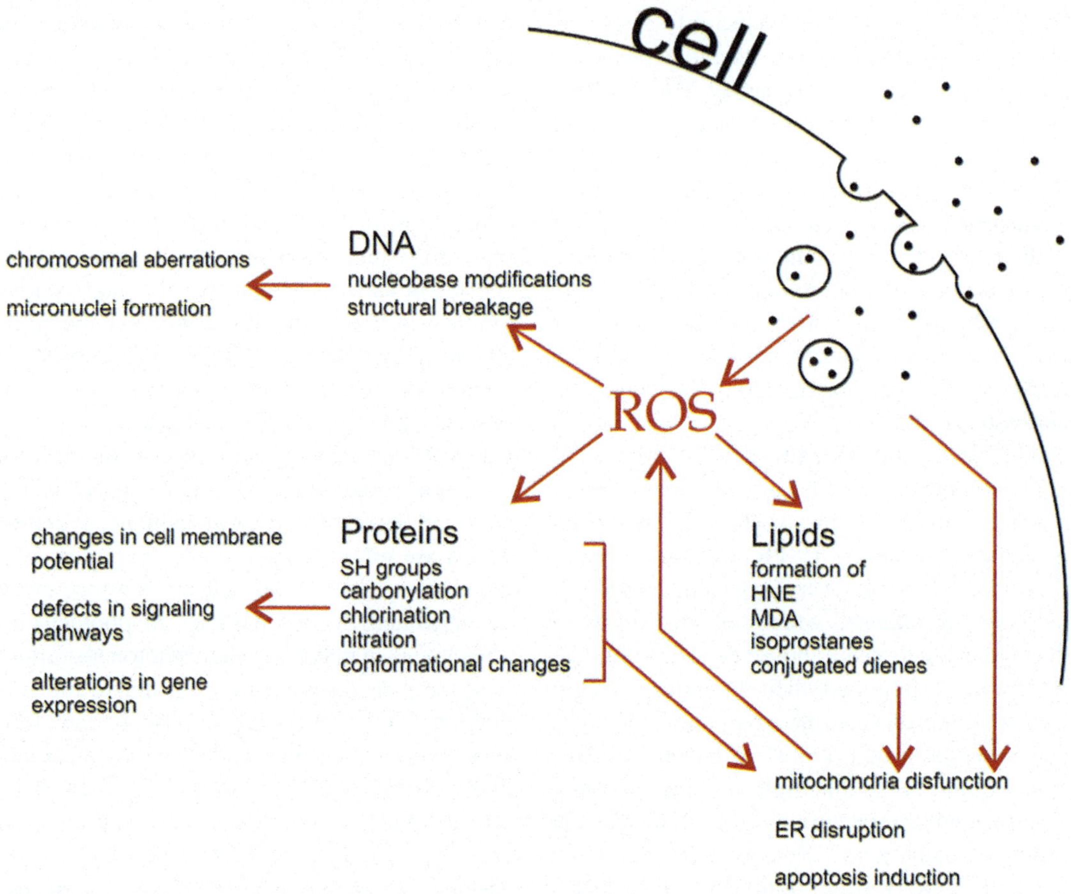

Fig. 6.2 Direct biological effects caused by nanoparticle induced oxidative stress and their outcomes on cellular level

ticles can be distinguished from toxicity of ionic silver for particles of diameter ≤10 nm, suggesting that toxicity of silver nanoparticles consists of two independent mechanisms [60]. Similar effect was observed in regard to iron nanoparticles [44]. Studies of non-metal nanoparticles further enhance this hypothesis, as silica particles of ⌀ = 25 nm were shown to induce ROS detectable through standard DCF fluorescence assay, while those of ⌀ = 50 nm did not procure such effect [39]. Similar findings were published in regard to gold nanoparticles [61] suggesting common denominator for nanoparticles in form of inverse relationship between amount of ROS induced and surface area to diameter ratio.

The toxicity of nanoparticles is closely tied with reactive oxygen species. Aside of direct detection of ROS, their follow-up products and assessment of ROS-induced damage to cellular macromolecules there are many studies providing an insight in mechanism of nanoparticle action by abolition of their biological activity with antioxidants [54]. Those findings further enhance the thesis of reactive oxygen species mediated nanoparticles toxicity. Among many mechanisms that would allow nanoparticles to play direct role in formation of reactive oxygen species there are other, indirect paths for nanoparticles to promote redox imbalance in cells. Those consist of nanoparticle-cell interactions influencing intracellular signal transduction pathways.

6.5 Reactive Oxygen Species Generated by Nanoparticles: Influence of Both on Signal Transduction Pathways

Many studies indicate interaction between nanoparticles and cells in in vitro and in vivo research. In vitro studies provide important information on the accumulation and toxicity of nanomaterials (e.g. generation of oxidative stress, apoptosis, proinflammatory cytokines production, protein and nucleic acid damage, lipid peroxidation). Although data on many types of nanoparticles and cell lines are sometimes contradictory [62, 63], the in-depth analysis allows us to draw some general conclusions and allows to predict the effect of nanoparticles in in vivo systems. Cells in response to nanoparticles biological activity show changes in expression of proteins (fibronectin, cadherin, enzymes connected with oxidative homeostasis and inflammation), disturbances in regulation of cell cycle and increase in apoptosis markers level [64–68]. It was found that nanoparticles cause various types of damage to cells [69–77], increase level of proinflammatory cytokines [64, 78, 79] and reactive oxygen species [63, 78, 80, 81], and abnormalities in cell adhesion [66]. However, there are also studies reporting absence of any impact of nanomaterials on organ tissues or in vitro cell cultures [82, 83]. Nanoparticle toxicity occurs after nanoparticles penetrate into cells or organisms. There are many research papers where the size of nanoparticles is investigated. It has been found, inter alia, that it is possible for non-phagocytic cells to internalize particles of 500 nm diameter [84]. The size of 30–50 nm seems to be optimal for nanoparticles to translocate into cells, which has been confirmed for many types of nanomaterials (e.g., gold NPs, silica NPs, single-walled carbon nanotubes, and quantum dots) [85–87]. Nanoparticles of diameter less than 100 nm can penetrate the cell membrane. Presence of nanoparticles of diameter less than 40 nm was confirmed in the nucleus, the ability to cross the blood/brain barrier was found for 35 nm nanoparticles [88]. Nanoparticles penetrate to higher organisms mainly by inhalation or *per os*. Translocation of nanoparticles to cells depends on their size, surface charge and magnetic properties. As nanoparticles vary in diameter from several to several hundred nanometers, the way in which they enter cells also varies significantly. It was found that it is possible for nanoparticles to cross cell membrane due to pinocytosis, endocytosis and receptor-mediated endocytosis [89], while large nanoparticles may be phagocytosed. It is possible to modulate the ability of nanoparticles to enter cells by changing the charge of the nanoparticle surface [90].

Studies confirm the contribution of the microtubule network and oxygenases in the uptake and intracellular transfer of positively charged nanoparticles. It has been found that the uptake of this type of nanoparticles is inhibited by nocodazole, indomethacin and chlorpromazine (recognized endocytosis inhibitors) [90]. The intensity of nanoparticle internalization into cells is determined mainly through two factors: surface modification and its charge. Unambiguous data has been obtained during research on many cell lines showing that the surface modification of nanoparticles via the carboxyl and amino groups allows for better penetration of the nanoparticles into cells [91, 92]. Endocytosis of positively charged particles is more efficient than uncharged or negatively charged particles, which is closely tied with the negative charge on cell surface [93]. Despite the hindrance of electrostatic force in interactions of negatively charged nanoparticles and cell membrane, there are many studies confirming their internalization [94–96]. This phenomenon is explained by non-specific binding and endocytosis of negatively charged clusters of nanoparticles mediated by scarce fragments of cationic cell membrane [97]. Furthermore, charged nanoaprticles are prone to interact with serum proteins and formed corona can enhance their internalization [98]. Phagocytic cells internalize negatively charged nanoparticles preferably in comparison to positively charged variants, due to similarity to negatively charged bacteria cells [99]. In some studies it is hypothesized that positively charged nanoparticles affect cell membrane fluidity, whereas inert nanoparticles cause local gelation of membranes [100,

101]. Alterations in local membrane compositions mediated by nanoparticle interactions with membrane domains such as rafts and surface proteins are the basis for nanoparticles influence on intracellular signal transduction [65, 102].

From the medical science point of view the ability to predict the nanoparticle-induced activation of intracellular transduction pathways is extremely significant. Nanoparticles and their coatings are tailored to procure effects along specific intracellular transcription pathways, unfortunately due to the complexity of biological systems the outcome can be different than intended [103]. There is evidence that many types of nanoparticles have the ability to interact with cells by activating the signaling pathways directly through receptor activation [104]. Apart from direct activation, nanoparticles can activate signaling pathways through induced ROS.

It has been observed that nanoparticles can inflict damage to DNA without any evidence of their crossing the barrier tissue itself [72, 105]. Unlike cells that were exposed to nanoparticles directly, DNA damage caused in cells behind the barrier did not cause cell death. That indirect damage is associated with the production of ROS by the mitochondria, which is the result of operation of purine nucleotides being secondary relays between barrier cells connected with gap junctions [106]. Another example of the action of nanomaterials through receptors is induction of lung epithelial cell proliferation via activation of the epidermal growth factor receptor (EGFR) and β1-integrins [107]. Integrins are transmembrane proteins that participate in cell adhesion and in communication between cells and the extracellular environment, and are activated by extracellular ligands and control such processes as angiogenesis, differentiation and cell migration [108]. Number of studies [109, 110] evidences that in human bronchial epithelial cells exposed to PM2.5 or diesel exhaust particles ERK is activated, which in turn leads to expression and secretion of the epidermal growth factor – amphiregulin creating autocrine loop with EGFR leading to chronic inflammation. Studies on small negatively charged supermagnetic iron oxide nanoparticles (snSPIONs) reveled activation of ERK in ROS-independent manner [111]. It was also observed that cell having Ras mutation, which enhances cell proliferation, promotes proliferation more effectively in cells treated with snSPIONs, implicating that some nanoaprticles can mimic growth factors (Fig. 6.3). It is estimated that 20% of human Ras tumors mutate, it is crucial to take into account possible influence of certain classes of nanoparticles on cancer cells proliferation, as Ras is a part of EGFR/ERK signaling pathway

AP-1 is one of the transcription factors activated by metallic nanoparticles [112]. It is responsible for regulation of cell proliferation and differentiation and it is also implicated in the process of apoptosis and carcinogenesis [113]. Activation of AP-1 was confirmed for TiO_2 nanoparticles, which were stimulating MAPK cascade including extracellular signal-regulated protein kinases, p38 kinase and C-Jun N-terminal kinase [114].The use of specific kinase inhibitors has shown that AP-1 activation is mediated through the generation of the hydroxyl radical and mediates activation of p38 kinase and the ERK pathway.

In many works, the MAPK pathway is identified as one of the main pathways activated by nanoparticles. MAPKs are serine-threonine protein kinases that are essential for normal cellular metabolism with particular focus on cell division, differentiation, gene expression and cell death. It has been found that silver nanoparticles have the ability to activate the MAPK pathway [115] and it has been shown how its activation translate into nanoparticle toxicity indicating free radical mechanism as key in cell death induction [116].

Changes in the intracellular redox state are also connected with NFκB and Nrf-2 pathways. NFκB is crucial in cell proliferation, inflammation, immune response and apoptosis. The earliest known way of activating this pathway is via hydrogen peroxide through the classical IKK-dependent pathway. There is evidence of expression of genes coding proinflamatory proteins such as TNF-α, IL-8, IL-2 and IL-6 under the action of nanoparticles in variety of cell lines. Activation of these proinflammatory pro-

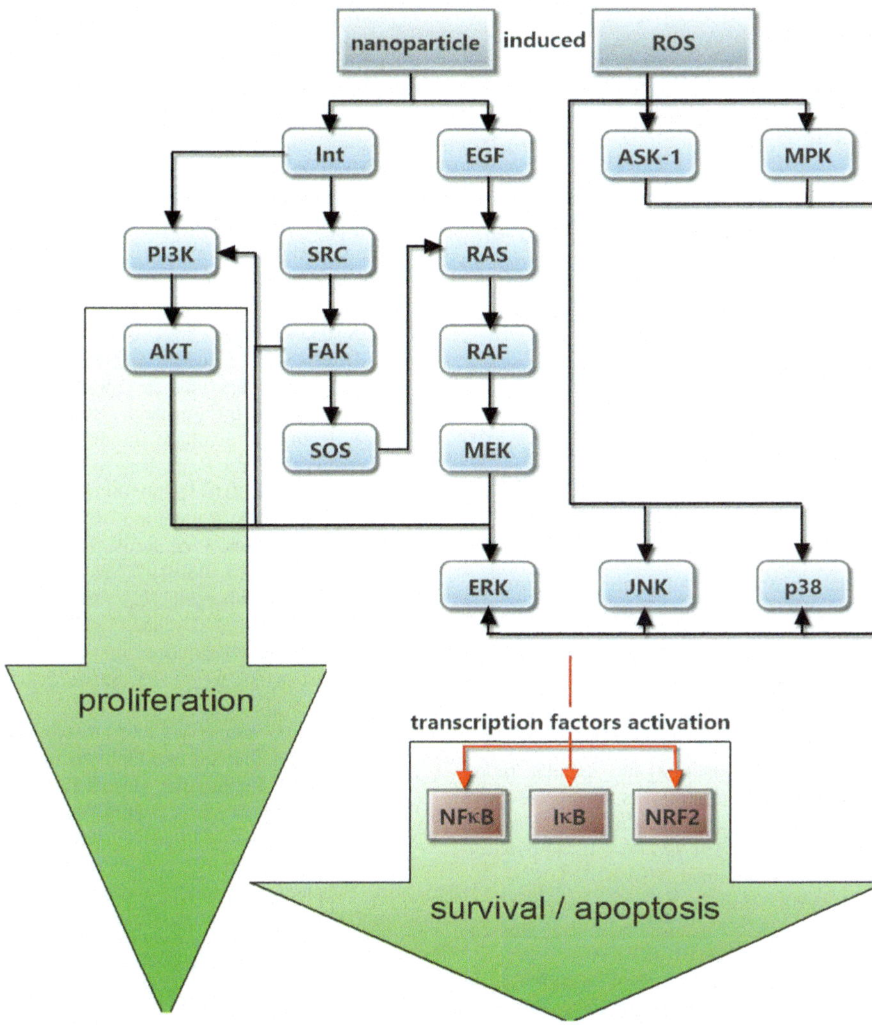

Fig. 6.3 Summary of signaling pathways activated by nanoparticles and nanoparticle – induced ROS

teins is closely tied with activation of the NFκB pathway [117]. It has been shown that nanoparticle induced free oxygen radicals and other reactive oxygen species activate NFκB in variety of cell lines, which suggests that this is a common phenomenon [52, 118, 119]. The Nrf-2 pathway is a well-known pathway for cellular response to mild oxidative stress. It is indicated as key factor in cell adaptation to oxidative stress as it regulates the expressions of proteins involved in antioxidant defense [120]. Nanoparticles activate the Nrf-2 pathway in human keratinocytes resulting in modulation of hemoxygenase I [121], which suggests involvement of Nrf-2 pathway in response to nanoparticle biological activity.

6.6 Summary

Biological and chemical sciences work together to create new types of nanomaterials that can be the answer to the growing need of medicine and industry. Full understanding of the toxicity mechanisms of nanoparticles and how they can modify the intracellular metabolism of higher organisms will improve the application properties of newly synthesized nanomaterials.

References

1. Droge W (2002) Free radicals in the physiological control of cell function. Physiol Rev 82(1):47–95. https://doi.org/10.1152/physrev.00018.2001
2. M. G (1897) Tetraphenylmethan. Ber Dtsch Chem Ges 30((2)):2043–2047. https://doi.org/10.1002/cber.189703002177
3. Panday A, Sahoo MK, Osorio D et al (2015) NADPH oxidases: an overview from structure to innate immunity-associated pathologies. Cell Mol Immunol 12(1):5–23. https://doi.org/10.1038/cmi.2014.89
4. Cadenas E, Davies KJ (2000) Mitochondrial free radical generation, oxidative stress, and aging. Free Radic Biol Med 29(3–4):222–230
5. Jensen PK (1966) Antimycin-insensitive oxidation of succinate and reduced nicotinamide-adenine dinucleotide in electron-transport particles. II Steroid effects Biochim Biophys Acta 122(2):167–174
6. Sohn HY, Krotz F, Zahler S et al (2003) Crucial role of local peroxynitrite formation in neutrophil-induced endothelial cell activation. Cardiovasc Res 57(3):804–815
7. Taylor BS, Alarcon LH, Billiar TR (1998) Inducible nitric oxide synthase in the liver: regulation and function. Biochemistry (Mosc) 63(7):766–781
8. McNeill E, Crabtree MJ, Sahgal N et al (2015) Regulation of iNOS function and cellular redox state by macrophage Gch1 reveals specific requirements for tetrahydrobiopterin in NRF2 activation. Free Radic Biol Med 79:206–216. https://doi.org/10.1016/j.freeradbiomed.2014.10.575
9. Ignarro LJ (1990) Nitric oxide. A novel signal transduction mechanism for transcellular communication. Hypertension 16(5):477–483
10. Hermann M, Flammer A, Luscher TF (2006) Nitric oxide in hypertension. J Clin Hypertens (Greenwich) 8(12 Suppl 4):17–29
11. Manderscheid M, Messmer UK, Franzen R et al (2001) Regulation of inhibitor of apoptosis expression by nitric oxide and cytokines: relation to apoptosis induction in rat mesangial cells and raw 264.7 macrophages. J Am Soc Nephrol 12(6):1151–1163
12. Cuzzocrea S, Salvemini D (2007) Molecular mechanisms involved in the reciprocal regulation of cyclooxygenase and nitric oxide synthase enzymes. Kidney Int 71(4):290–297. https://doi.org/10.1038/sj.ki.5002058
13. Szabo C, Ischiropoulos H, Radi R (2007) Peroxynitrite: biochemistry, pathophysiology and development of therapeutics. Nat Rev Drug Discov 6(8):662–680. https://doi.org/10.1038/nrd2222
14. Ischiropoulos H, Al-Mehdi AB (1995) Peroxynitrite-mediated oxidative protein modifications. FEBS Lett 364(3):279–282
15. Haddad JJ (2002) Antioxidant and prooxidant mechanisms in the regulation of redox(y)-sensitive transcription factors. Cell Signal 14(11):879–897
16. Turpaev KT (2002) Reactive oxygen species and regulation of gene expression. Biochemistry (Mosc) 67(3):281–292
17. Green RM, Graham M, O'Donovan MR et al (2006) Subcellular compartmentalization of glutathione: correlations with parameters of oxidative stress related to genotoxicity. Mutagenesis 21(6):383–390. https://doi.org/10.1093/mutage/gel043
18. England K, Cotter TG (2005) Direct oxidative modifications of signalling proteins in mammalian cells and their effects on apoptosis. Redox Rep 10(5):237–245. https://doi.org/10.1179/135100005X70224
19. O'Brian CA, Chu F (2005) Post-translational disulfide modifications in cell signaling – role of inter-protein, intra-protein, S-glutathionyl, and S-cysteaminyl disulfide modifications in signal transmission. Free Radic Res 39(5):471–480. https://doi.org/10.1080/10715760500073931
20. Ghezzi P (2005) Regulation of protein function by glutathionylation. Free Radic Res 39(6):573–580. https://doi.org/10.1080/10715760500072172
21. Sun J, Steenbergen C, Murphy E (2006) S-nitrosylation: NO-related redox signaling to protect against oxidative stress. Antioxid Redox Signal 8(9–10):1693–1705. https://doi.org/10.1089/ars.2006.8.1693
22. Bigelow DJ, Squier TC (2005) Redox modulation of cellular signaling and metabolism through reversible oxidation of methionine sensors in calcium regulatory proteins. Biochim Biophys Acta 1703(2):121–134. https://doi.org/10.1016/j.bbapap.2004.09.012
23. Schopfer FJ, Baker PR, Freeman BA (2003) NO-dependent protein nitration: a cell signaling event or an oxidative inflammatory response? Trends Biochem Sci 28(12):646–654. https://doi.org/10.1016/j.tibs.2003.10.006
24. Liu H, Nishitoh H, Ichijo H et al (2000) Activation of apoptosis signal-regulating kinase 1 (ASK1) by tumor necrosis factor receptor-associated factor 2 requires prior dissociation of the ASK1 inhibitor thioredoxin. Mol Cell Biol 20(6):2198–2208
25. Thevenin AF, Zony CL, Bahnson BJ et al (2011) GST pi modulates JNK activity through a direct interaction with JNK substrate, ATF2. Protein Sci 20(5):834–848. https://doi.org/10.1002/pro.609
26. Lee S, Das HK (2008) Inhibition of basal activity of c-jun-NH2-terminal kinase (JNK) represses the expression of presenilin-1 by a p53-dependent mechanism. Brain Res 1207:19–31. https://doi.org/10.1016/j.brainres.2008.02.016
27. Lu MC, Ji JA, Jiang YL et al (2016) An inhibitor of the Keap1-Nrf2 protein-protein interaction protects NCM460 colonic cells and alleviates experimental colitis. Sci Rep 6:26585. https://doi.org/10.1038/srep26585
28. Abed DA, Goldstein M, Albanyan H et al (2015) Discovery of direct inhibitors of Keap1-Nrf2 protein-protein interaction as potential therapeutic and preventive agents. Acta Pharm Sin B 5(4):285–299. https://doi.org/10.1016/j.apsb.2015.05.008

29. Johnson L (2007) Protein kinases and their therapeutic exploitation. Biochem Soc Trans 35(1):7–11. https://doi.org/10.1042/BST0350007
30. Reinstein E, Ciechanover A (2006) Narrative review: protein degradation and human diseases: the ubiquitin connection. Ann Intern Med 145(9):676–684
31. Bhaskar PT, Hay N (2007) The two TORCs and Akt. Dev Cell 12(4):487–502. https://doi.org/10.1016/j.devcel.2007.03.020
32. Wardman P (2007) Fluorescent and luminescent probes for measurement of oxidative and nitrosative species in cells and tissues: progress, pitfalls, and prospects. Free Radic Biol Med 43(7):995–1022. https://doi.org/10.1016/j.freeradbiomed.2007.06.026
33. Yapici NB, Mandalapu S, Gibson KM, Bi L (2015) Targeted fluorescent probes for detection of oxidative stress in the mitochondria. Bioorg Med Chem Lett 25(17):3476–3480. https://doi.org/10.1016/j.bmcl.2015.07.011
34. Chen J, Zeng L, Xia T et al (2015) Toward a biomarker of oxidative stress: a fluorescent probe for exogenous and endogenous malondialdehyde in living cells. Anal Chem 87(16):8052–8056. https://doi.org/10.1021/acs.analchem.5b02032
35. Gomes A, Fernandes E, Lima JL (2006) Use of fluorescence probes for detection of reactive nitrogen species: a review. J Fluoresc 16(1):119–139. https://doi.org/10.1007/s10895-005-0030-3
36. Zuberek M, Wojciechowska D, Krzyzanowski D et al (2015) Glucose availability determines silver nanoparticles toxicity in HepG2. J Nanobiotechnol 13:72. https://doi.org/10.1186/s12951-015-0132-2
37. Kim JS, Kuk E, Yu KN et al (2007) Antimicrobial effects of silver nanoparticles. Nanomedicine 3(1):95–101. https://doi.org/10.1016/j.nano.2006.12.001
38. Quinteros MA, Cano Aristizabal V, Dalmasso PR et al (2016) Oxidative stress generation of silver nanoparticles in three bacterial genera and its relationship with the antimicrobial activity. Toxicol In Vitro 36:216–223. https://doi.org/10.1016/j.tiv.2016.08.007
39. Halamoda Kenzaoui B, Chapuis Bernasconi C, Guney-Ayra S et al (2012) Induction of oxidative stress, lysosome activation and autophagy by nanoparticles in human brain-derived endothelial cells. Biochem J 441(3):813–821. https://doi.org/10.1042/BJ20111252
40. Bhattacharjee S, de Haan LH, Evers NM et al (2010) Role of surface charge and oxidative stress in cytotoxicity of organic monolayer-coated silicon nanoparticles towards macrophage NR8383 cells. Part Fibre Toxicol 7:25. https://doi.org/10.1186/1743-8977-7-25
41. Lehman SE, Morris AS, Mueller PS et al (2016) Silica nanoparticle-generated ROS as a predictor of cellular toxicity: mechanistic insights and safety by design. Environ Sci Nano 3(1):56–66. https://doi.org/10.1039/C5EN00179J
42. Manke A, Wang L, Rojanasakul Y (2013) Mechanisms of nanoparticle-induced oxidative stress and toxicity. Biomed Res Int 2013:942916. https://doi.org/10.1155/2013/942916
43. Kawanishi M, Ogo S, Ikemoto M et al (2013) Genotoxicity and reactive oxygen species production induced by magnetite nanoparticles in mammalian cells. J Toxicol Sci 38(3):503–511
44. Wu H, Yin JJ, Wamer WG et al (2014) Reactive oxygen species-related activities of nano-iron metal and nano-iron oxides. J Food Drug Anal 22(1):86–94. https://doi.org/10.1016/j.jfda.2014.01.007
45. Strlic M, Kolar J, Selih VS et al (2003) A comparative study of several transition metals in Fenton-like reaction systems at circum-neutral pH. Acta Chim Slov 50(4):619–632
46. Sharma CS, Sarkar S, Periyakaruppan A et al (2007) Single-walled carbon nanotubes induces oxidative stress in rat lung epithelial cells. J Nanosci Nanotechnol 7(7):2466–2472
47. Hoffmann MR, Martin ST, Choi WY et al (1995) Environmental applications of semiconductor photocatalysis. Chem Rev 95(1):69–96. https://doi.org/10.1021/cr00033a004
48. Xia T, Kovochich M, Brant J et al (2006) Comparison of the abilities of ambient and manufactured nanoparticles to induce cellular toxicity according to an oxidative stress paradigm. Nano Lett 6(8):1794–1807. https://doi.org/10.1021/nl061025k
49. Zhang H, Dunphy DR, Jiang X et al (2012) Processing pathway dependence of amorphous silica nanoparticle toxicity: colloidal vs pyrolytic. J Am Chem Soc 134(38):15790–15804. https://doi.org/10.1021/ja304907c
50. Choi O, Hu Z (2008) Size dependent and reactive oxygen species related nanosilver toxicity to nitrifying bacteria. Environ Sci Technol 42(12):4583–4588
51. Fahmy B, Cormier SA (2009) Copper oxide nanoparticles induce oxidative stress and cytotoxicity in airway epithelial cells. Toxicol In Vitro 23(7):1365–1371. https://doi.org/10.1016/j.tiv.2009.08.005
52. Pujalte I, Passagne I, Brouillaud B et al (2011) Cytotoxicity and oxidative stress induced by different metallic nanoparticles on human kidney cells. Part Fibre Toxicol 8:10. https://doi.org/10.1186/1743-8977-8-10
53. Chiang HM, Xia Q, Zou X et al (2012) Nanoscale ZnO induces cytotoxicity and DNA damage in human cell lines and rat primary neuronal cells. J Nanosci Nanotechnol 12(3):2126–2135
54. Siddiqui MA, Ahamed M, Ahmad J et al (2012) Nickel oxide nanoparticles induce cytotoxicity, oxidative stress and apoptosis in cultured human cells that is abrogated by the dietary antioxidant curcumin. Food Chem Toxicol 50(3–4):641–647. https://doi.org/10.1016/j.fct.2012.01.017
55. Di Guglielmo C, De Lapuente J, Porredon C et al (2012) In vitro safety toxicology data for evaluation of gold nanoparticles-chronic cytotoxicity,

genotoxicity and uptake. J Nanosci Nanotechnol 12(8):6185–6191
56. Kruszewski M, Gradzka I, Bartlomiejczyk T et al (2013) Oxidative DNA damage corresponds to the long term survival of human cells treated with silver nanoparticles. Toxicol Lett 219(2):151–159. https://doi.org/10.1016/j.toxlet.2013.03.006
57. Murray AR, Kisin E, Inman A et al (2013) Oxidative stress and dermal toxicity of iron oxide nanoparticles in vitro. Cell Biochem Biophys 67(2):461–476. https://doi.org/10.1007/s12013-012-9367-9
58. Srikanth K, Trindade T, Duarte AC et al (2017) Cytotoxicity and oxidative stress responses of silica-coated iron oxide nanoparticles in CHSE-214 cells. Environ Sci Pollut Res Int 24(2):2055–2064. https://doi.org/10.1007/s11356-016-7870-z
59. Li JJ, Hartono D, Ong CN et al (2010) Autophagy and oxidative stress associated with gold nanoparticles. Biomaterials 31(23):5996–6003. https://doi.org/10.1016/j.biomaterials.2010.04.014
60. Ivask A, Kurvet I, Kasemets K et al (2014) Size-dependent toxicity of silver nanoparticles to bacteria, yeast, algae, crustaceans and mammalian cells in vitro. PLoS One 9(7):e102108. https://doi.org/10.1371/journal.pone.0102108
61. Khaing Oo MK, Yang Y, Hu Y et al (2012) Gold nanoparticle-enhanced and size-dependent generation of reactive oxygen species from protoporphyrin IX. ACS Nano 6(3):1939–1947. https://doi.org/10.1021/nn300327c
62. Laurent S, Burtea C, Thirifays C et al (2012) Crucial ignored parameters on nanotoxicology: the importance of toxicity assay modifications and "cell vision". PLoS One 7(1):e29997. https://doi.org/10.1371/journal.pone.0029997
63. Mahmoudi M, Laurent S, Shokrgozar MA et al (2011) Toxicity evaluations of superparamagnetic iron oxide nanoparticles: cell "vision" versus physicochemical properties of nanoparticles. ACS Nano 5(9):7263–7276. https://doi.org/10.1021/nn2021088
64. Nel A, Xia T, Madler L et al (2006) Toxic potential of materials at the nanolevel. Science 311(5761):622–627. https://doi.org/10.1126/science.1114397
65. Nel AE, Madler L, Velegol D et al (2009) Understanding biophysicochemical interactions at the nano-bio interface. Nat Mater 8(7):543–557. https://doi.org/10.1038/nmat2442
66. Cui D, Tian F, Ozkan CS et al (2005) Effect of single wall carbon nanotubes on human HEK293 cells. Toxicol Lett 155(1):73–85. https://doi.org/10.1016/j.toxlet.2004.08.015
67. Sayes CM, Gobin AM, Ausman KD et al (2005) Nano-C60 cytotoxicity is due to lipid peroxidation. Biomaterials 26(36):7587–7595. https://doi.org/10.1016/j.biomaterials.2005.05.027
68. Kim JA, Aberg C, Salvati A et al (2011) Role of cell cycle on the cellular uptake and dilution of nanoparticles in a cell population. Nat Nanotechnol 7(1):62–68. https://doi.org/10.1038/nnano.2011.191
69. Chou LY, Chan WC (2012) Nanotoxicology. No signs of illness. Nat Nanotechnol 7(7):416–417. https://doi.org/10.1038/nnano.2012.110
70. Resch-Genger U, Grabolle M, Cavaliere-Jaricot S et al (2008) Quantum dots versus organic dyes as fluorescent labels. Nat Methods 5(9):763–775. https://doi.org/10.1038/nmeth.1248
71. Deng ZJ, Liang M, Monteiro M et al (2011) Nanoparticle-induced unfolding of fibrinogen promotes Mac-1 receptor activation and inflammation. Nat Nanotechnol 6(1):39–44. https://doi.org/10.1038/nnano.2010.250
72. Bhabra G, Sood A, Fisher B et al (2009) Nanoparticles can cause DNA damage across a cellular barrier. Nat Nanotechnol 4(12):876–883. https://doi.org/10.1038/nnano.2009.313
73. Peeters S, Kitz M, Preisser S et al (2012) Mechanisms of nanoparticle-mediated photomechanical cell damage. Biomed Opt Express 3(3):435–446. https://doi.org/10.1364/BOE.3.000435
74. Stone V, Donaldson K (2006) Nanotoxicology: signs of stress. Nat Nanotechnol 1(1):23–24. https://doi.org/10.1038/nnano.2006.69
75. Dobrovolskaia MA, Germolec DR, Weaver JL (2009) Evaluation of nanoparticle immunotoxicity. Nat Nanotechnol 4(7):411–414. https://doi.org/10.1038/nnano.2009.175
76. Dobrovolskaia MA, McNeil SE (2007) Immunological properties of engineered nanomaterials. Nat Nanotechnol 2(8):469–478. https://doi.org/10.1038/nnano.2007.223
77. Kendall M, Holgate S (2012) Health impact and toxicological effects of nanomaterials in the lung. Respirology 17(5):743–758. https://doi.org/10.1111/j.1440-1843.2012.02171.x
78. Nel A (2005) Atmosphere. Air pollution-related illness: effects of particles. Science 308(5723):804–806. https://doi.org/10.1126/science.1108752
79. Xiao GG, Wang M, Li N et al (2003) Use of proteomics to demonstrate a hierarchical oxidative stress response to diesel exhaust particle chemicals in a macrophage cell line. J Biol Chem 278(50):50781–50790. https://doi.org/10.1074/jbc.M306423200
80. Mahmoudi M, Sant S, Wang B et al (2011) Superparamagnetic iron oxide nanoparticles (SPIONs): development, surface modification and applications in chemotherapy. Adv Drug Deliv Rev 63(1–2):24–46. https://doi.org/10.1016/j.addr.2010.05.006
81. Kagan VE, Bayir H, Shvedova AA (2005) Nanomedicine and nanotoxicology: two sides of the same coin. Nanomedicine 1(4):313–316. https://doi.org/10.1016/j.nano.2005.10.003
82. Lidke DS, Arndt-Jovin DJ (2004) Imaging takes a quantum leap. Physiology (Bethesda) 19:322–325. https://doi.org/10.1152/physiol.00030.2004
83. Lidke DS, Nagy P, Heintzmann R et al (2004) Quantum dot ligands provide new insights into erbB/HER receptor-mediated signal transduction. Nat

Biotechnol 22(2):198–203. https://doi.org/10.1038/nbt929
84. Rejman J, Oberle V, Zuhorn IS et al (2004) Size-dependent internalization of particles via the pathways of clathrin- and caveolae-mediated endocytosis. Biochem J 377(Pt 1):159–169. https://doi.org/10.1042/BJ20031253
85. Chithrani BD, Chan WC (2007) Elucidating the mechanism of cellular uptake and removal of protein-coated gold nanoparticles of different sizes and shapes. Nano Lett 7(6):1542–1550. https://doi.org/10.1021/nl070363y
86. Jin H, Heller DA, Sharma R et al (2009) Size-dependent cellular uptake and expulsion of single-walled carbon nanotubes: single particle tracking and a generic uptake model for nanoparticles. ACS Nano 3(1):149–158. https://doi.org/10.1021/nn800532m
87. Lu F, Wu SH, Hung Y et al (2009) Size effect on cell uptake in well-suspended, uniform mesoporous silica nanoparticles. Small 5(12):1408–1413. https://doi.org/10.1002/smll.200900005
88. Dawson KA, Salvati A, Lynch I (2009) Nanotoxicology: nanoparticles reconstruct lipids. Nat Nanotechnol 4(2):84–85. https://doi.org/10.1038/nnano.2008.426
89. Pratten MK, Lloyd JB (1986) Pinocytosis and phagocytosis: the effect of size of a particulate substrate on its mode of capture by rat peritoneal macrophages cultured in vitro. Biochim Biophys Acta 881(3):307–313
90. Dausend J, Musyanovych A, Dass M et al (2008) Uptake mechanism of oppositely charged fluorescent nanoparticles in HeLa cells. Macromol Biosci 8(12):1135–1143. https://doi.org/10.1002/mabi.200800123
91. Lorenz MR, Holzapfel V, Musyanovych A et al (2006) Uptake of functionalized, fluorescent-labeled polymeric particles in different cell lines and stem cells. Biomaterials 27(14):2820–2828. https://doi.org/10.1016/j.biomaterials.2005.12.022
92. Mailander V, Landfester K (2009) Interaction of nanoparticles with cells. Biomacromolecules 10(9):2379–2400. https://doi.org/10.1021/bm900266r
93. Marano F, Hussain S, Rodrigues-Lima F et al (2011) Nanoparticles: molecular targets and cell signalling. Arch Toxicol 85(7):733–741. https://doi.org/10.1007/s00204-010-0546-4
94. Bannunah AM, Vllasaliu D, Lord J et al (2014) Mechanisms of nanoparticle internalization and transport across an intestinal epithelial cell model: effect of size and surface charge. Mol Pharm 11(12):4363–4373. https://doi.org/10.1021/mp500439c
95. Lundqvist M, Stigler J, Elia G et al (2008) Nanoparticle size and surface properties determine the protein corona with possible implications for biological impacts. Proc Natl Acad Sci U S A 105(38):14265–14270. https://doi.org/10.1073/pnas.0805135105
96. Asati A, Santra S, Kaittanis C et al (2010) Surface-charge-dependent cell localization and cytotoxicity of cerium oxide nanoparticles. ACS Nano 4(9):5321–5331. https://doi.org/10.1021/nn100816s
97. Verma A, Stellacci F (2010) Effect of surface properties on nanoparticle-cell interactions. Small 6(1):12–21. https://doi.org/10.1002/smll.200901158
98. Schweiger C, Hartmann R, Zhang F et al (2012) Quantification of the internalization patterns of superparamagnetic iron oxide nanoparticles with opposite charge. J Nanobiotechnol 10:28. https://doi.org/10.1186/1477-3155-10-28
99. Tomita Y, Rikimaru-Kaneko A, Hashiguchi K et al (2011) Effect of anionic and cationic n-butylcyanoacrylate nanoparticles on NO and cytokine production in Raw264.7 cells. Immunopharmacol Immunotoxicol 33(4):730–737. https://doi.org/10.3109/08923973.2011.565345
100. Arvizo RR, Miranda OR, Thompson MA et al (2010) Effect of nanoparticle surface charge at the plasma membrane and beyond. Nano Lett 10(7):2543–2548. https://doi.org/10.1021/nl101140t
101. Wang B, Zhang L, Bae SC et al (2008) Nanoparticle-induced surface reconstruction of phospholipid membranes. Proc Natl Acad Sci U S A 105(47):18171–18175. https://doi.org/10.1073/pnas.0807296105
102. Grecco HE, Schmick M, Bastiaens PI (2011) Signaling from the living plasma membrane. Cell 144(6):897–909. https://doi.org/10.1016/j.cell.2011.01.029
103. Wang N, Butler JP, Ingber DE (1993) Mechanotransduction across the cell surface and through the cytoskeleton. Science 260(5111):1124–1127
104. Jiang W, Kim BY, Rutka JT et al (2008) Nanoparticle-mediated cellular response is size-dependent. Nat Nanotechnol 3(3):145–150. https://doi.org/10.1038/nnano.2008.30
105. Parry MC, Bhabra G, Sood A et al (2010) Thresholds for indirect DNA damage across cellular barriers for orthopaedic biomaterials. Biomaterials 31(16):4477–4483. https://doi.org/10.1016/j.biomaterials.2010.02.038
106. Sood A, Salih S, Roh D et al (2011) Signalling of DNA damage and cytokines across cell barriers exposed to nanoparticles depends on barrier thickness. Nat Nanotechnol 6(12):824–833. https://doi.org/10.1038/nnano.2011.188
107. Unfried K, Sydlik U, Bierhals K et al (2008) Carbon nanoparticle-induced lung epithelial cell proliferation is mediated by receptor-dependent Akt activation. Am J Physiol Lung Cell Mol Physiol 294(2):L358–L367. https://doi.org/10.1152/ajplung.00323.2007
108. Harburger DS, Calderwood DA (2009) Integrin signalling at a glance. J Cell Sci 122(Pt 2):159–163. https://doi.org/10.1242/jcs.018093

109. Auger F, Gendron MC, Chamot C et al (2006) Responses of well-differentiated nasal epithelial cells exposed to particles: role of the epithelium in airway inflammation. Toxicol Appl Pharmacol 215(3):285–294. https://doi.org/10.1016/j.taap.2006.03.002
110. Blanchet S, Ramgolam K, Baulig A et al (2004) Fine particulate matter induces amphiregulin secretion by bronchial epithelial cells. Am J Respir Cell Mol Biol 30(4):421–427. https://doi.org/10.1165/rcmb.2003-0281RC
111. Rauch J, Kolch W, Mahmoudi M (2012) Cell type-specific activation of AKT and ERK signaling pathways by small negatively-charged magnetic nanoparticles. Sci Rep 2:868. https://doi.org/10.1038/srep00868
112. Poljak-Blazi M, Jaganjac M, Mustapic M et al (2009) Acute immunomodulatory effects of iron polyisomaltosate in rats. Immunobiology 214(2):121–128. https://doi.org/10.1016/j.imbio.2008.07.005
113. Shaulian E, Karin M (2001) AP-1 in cell proliferation and survival. Oncogene 20(19):2390–2400. https://doi.org/10.1038/sj.onc.1204383
114. Ding M, Bowman L, Leonard S et al (2015) ROS generation is involved in titanium dioxide nanoparticle induced MAPKs/AP-1 activation in JB6 cells. FASEB J 29(1):1046. https://doi.org/10.1096/fj.1530-6860
115. Eom HJ, Choi J (2010) p38 MAPK activation, DNA damage, cell cycle arrest and apoptosis as mechanisms of toxicity of silver nanoparticles in Jurkat T cells. Environ Sci Technol 44(21):8337–8342. https://doi.org/10.1021/es1020668
116. Rinna A, Magdolenova Z, Hudecova A et al (2015) Effect of silver nanoparticles on mitogen-activated protein kinases activation: role of reactive oxygen species and implication in DNA damage. Mutagenesis 30(1):59–66. https://doi.org/10.1093/mutage/geu057
117. Byrne JD, Baugh JA (2008) The significance of nanoparticles in particle-induced pulmonary fibrosis. Mcgill J Med 11(1):43–50
118. Hubbard AK, Timblin CR, Shukla A et al (2002) Activation of NF-kappaB-dependent gene expression by silica in lungs of luciferase reporter mice. Am J Physiol Lung Cell Mol Physiol 282(5):L968–L975. https://doi.org/10.1152/ajplung.00327.2001
119. Murray AR, Kisin ER, Tkach AV et al (2012) Factoring-in agglomeration of carbon nanotubes and nanofibers for better prediction of their toxicity versus asbestos. Part Fibre Toxicol 9:10. https://doi.org/10.1186/1743-8977-9-10
120. Ma Q (2013) Role of nrf2 in oxidative stress and toxicity. Annu Rev Pharmacol Toxicol 53:401–426. https://doi.org/10.1146/annurev-pharmtox-011112-140320
121. Lai TH, Shieh JM, Tsou CJ et al (2015) Gold nanoparticles induce heme oxygenase-1 expression through Nrf2 activation and Bach1 export in human vascular endothelial cells. Int J Nanomedicine 10:5925–5939. https://doi.org/10.2147/IJN.S88514

Toxicity of Metal Oxide Nanoparticles

7

Koyeli Girigoswami

Abstract

In the recent times, nanomaterials are used in many sectors of science, medicine and industry, without revealing its toxic effects. Thus, it is in urgent need for exploring the toxicity along with the application of such useful nanomaterials. Nanomaterials are categorized with a particle size of 1–100 nm. They have gained increasing attention because of their novel properties, including a large specific surface area and high reaction activity. The various fundamental and practical applications of nanomaterials include drug delivery, cell imaging, and cancer therapy. Nanosized semiconductors have their versatile applications in different areas such as catalysts, sensors, photoelectronic devices, highly functional and effective devices etc. Metal oxides contribute in many areas of chemistry, physics and materials science. Mechanism of toxicity of metal oxide nanoparticles can occur by different methods like oxidative stress, co-ordination effects, non-homeostasis effects, genotoxicity and others. Factors that affect the metal oxide nanoparticles were size, dissolution and exposure routes. This chapter will explain elaborately the toxicity of metal oxide nano structures in living beings and their effect in ecosystem.

Keywords

Metal oxide nanoparticles · Nanoparticle toxicity · Apoptosis · Cell death mechanisms · Genotoxicity

7.1 Introduction

Toxicology is an important branch of science that deals with toxins and poisons that may be harmful to body. According to the father of toxicology, Paracelsus (1493–1541) "all things are poison and nothing is without poison, only the dose

K. Girigoswami (✉)
Department of Medical Bionanotechnology, Chettinad Hospital and Research Institute, Chettinad Academy of Research and Education, Rajiv Gandhi Salai, Kelambakkam,
Kanchipuram, Tamil Nadu, India
e-mail: koyelig@gmail.com

Q. Saquib et al. (eds.), *Cellular and Molecular Toxicology of Nanoparticles*, Advances in Experimental Medicine and Biology 1048, https://doi.org/10.1007/978-3-319-72041-8_7

permits something not to be poisonous" [1]. Toxicants can be classified based on the mode of action and the rate of exposure. Toxic substances may be present in food, air, water and soil and may lead to death or giving some adverse effects [2]. As the field of nanoscience is progressing, researchers tend to make nanostructures, nanoparticles, nanoarrays which have typical characteristics compared to their bulk counterpart [3–7]. These nanomaterials have a particle size usually between 1 and 100 nm, with huge surface area to volume ratio thereby exhibiting high reactivity and different electronic and optical properties [8]. The nanomaterials commonly have low solubility in water and are prone to aggregation. The various applications of nanomaterials include development of biosensors, diagnostic kits, biomedical imaging and theranostics, cancer therapy, drug delivery, etc. [9–11]. Among the most common type of nanostructures, metal oxide nanoparticles are used profoundly like zinc oxide (ZnO) and titanium dioxide (TiO_2) in sunscreens and paints, iron oxide in biomedical applications and bioremediation of ground water, cerium oxide in diesel additives, copper oxide in wood preservatives, antimicrobial agents, silicon dioxide in skin care, textiles and therapeutics, aluminium oxide for polishing and abrasive agents etc. [12]. There is a versatile application of metal oxides in chemistry, physics and materials science [13–18]. These nanoparticles are either directly or indirectly released to surface water, ground water, rivers, lakes, seas, air and soil. The bioaccumulation of these nanoparticles may lead to potential hazards in aquatic life, animals, humans and plants. Thus, it is warranted to explore the various aspects of metal oxide nanoparticle toxicity.

7.1.1 Metal Oxide Nanoparticles

Metal oxides can be found in several geometrical structures having electronic structure that enables them to have metallic, semiconductor or insulating character. The band gap of the nanoparticles enable them to possess a semiconducting nature. The band gap diagram (Fig. 7.1) shows the mechanism of its semiconducting nature. The bands consisting of orbitals are the characteristics of the electronic structure of a semiconductor. These energy bands are separated from one another by gaps in the energy without any orbital. When light falls on the band of minimum energy, it is absorbed by the valance band electron (e^-) which gets excited to the conduction band creating a hole in the valance band (h^+) [19].

The particle size of the oxide nanoparticles can affect three types of basic properties of a material. Firstly, the lattice symmetry and cell parameters, secondly the electronic properties and third is the quantum confinement related to its small size

Fig. 7.1 Diagram showing band gap of a semiconducting material

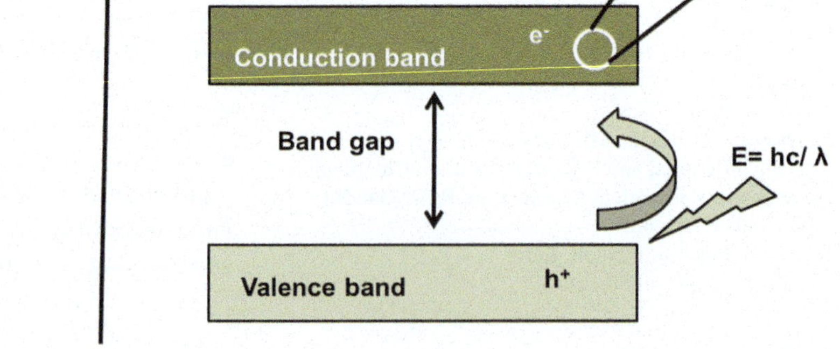

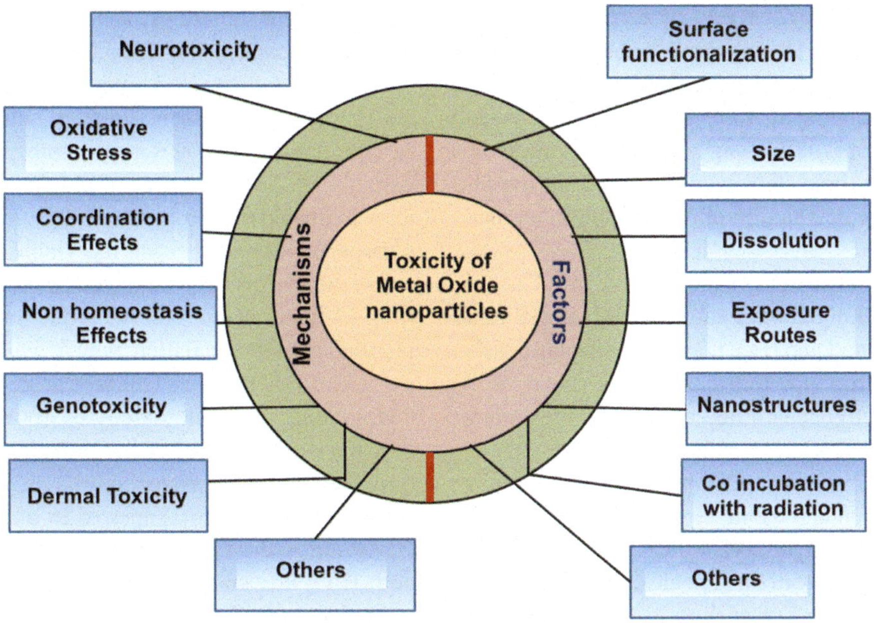

Fig. 7.2 The toxicity mechanisms of metal oxide nanoparticles

[20].These properties make them to stay at different agglomeration states and their dispersion in biological fluids may lead to differential effects. Metal oxides can attribute to toxicity mediated by several routes like size, dissolution and exposure (Fig. 7.2) [21]. The most common exposure routes of nanoparticles are by ingestion, dermal exposure or inhalation. We shall discuss the toxicity of different nanoparticles in different living system in the following sections.

7.1.2 Zinc Oxide (ZnO) Nanoparticles

ZnO is one of the most used semiconductor metal oxide which exhibits several applications in bulk and nanosized form like ultra violet light (UV) blocking property, photocatalytic and antibacterial properties [21–23]. ZnO occurs in two crystalline forms: wurtzite and zinc blend, wurtzite being more stable with a refractive index between 2.3 and 2.0. The band gap energy for wurtzite and zinc blend detected at 77 K is 3.22 eV and 3.32 eV respectively [24].Due to this band gap width, ZnO does not absorb visible light (which is scattered and reflected) but absorbs in the UV-range, (apart from scattering), specially UVA [25]. ZnO at its microsize (0.1–10.0 μm) along with TiO_2 was used as particulate sunscreen ingredients for a long time [26]. Wurtzite ZnO particles, coated and uncoated have been used as sunblockers to attenuate UV from both reflection and scattering of UV radiation and visible light as well as from UV absorption. Among the UV absorption compound, ZnO has gained more interest because of its UVA absorption efficiency [25]. As particles become smaller than 100 nm, the optical characteristics also alter giving rise to transparency of the material and transmitting visible light. For both ZnO and TiO_2, UVB absorption increased rather than UVA1 on reduced particle size making UV protection unbalanced [27, 28]. This shift in UV protection region recommended a balance combination of materials that can protect both UVA and UVB light. Researchers proved that a balanced UVA/ UVB protection was found by a combination of aggregated ZnO particles of 130 nm rather than its primary particles of 20 nm diameter [28]. Although the control of aggregation size was limited, that may lead to larger aggregates.

7.1.2.1 Photocatalysis of ZnO

ZnO loses its UV absorption capacity as it is phtocatalysed by the visible and UV light, thereby making it more suitable to be used in paints for destroying organic compounds. Generally three steps were involved for the photocatalytic destruction of organic compounds:

(i) As the energy of the photon (hν) becomes either equal or greater than the semiconductor band gap (E_g), there is an excitation of an electron to the conduction band (CB), thereby generating a hole in the valance band (VB);
(ii) The excited electrons are then trapped by oxygen and the holes by surface hydroxyl to produce hydroxyl radical (•OH), called as primary oxidizing species; and
(iii) Mineralization of the adsorbed organic matters is done by these hydroxyl radicals.

However, there could be a recombination of the photoexited electrons and holes thereby reducing the semiconductor photocatalytic activity which can be circumvented using noble metals for semiconductor modification [29]. To be a potential UV protector, the ZnO photocatalysis was not desired, as it may lead to production of oxidative stress to the skin. To slow down the photocatalytic activity and oxidation of ZnO, the nanoparticle was coated with a zinc aluminate layer, although the biological acceptability was not studied [30].

7.1.2.2 Toxicity of ZnO Nanoparticles

Bare ZnO was generally recognized as safe by the International Agency for Research on Cancer (IARC), but the size dependent toxicity was found in many research studies [31, 32]. ZnO and TiO_2 nanoparticles (20–30 nm) have been used widely in cosmetics, sunscreen and topical skin care products. Sunscreens or sun blockers protect the skin against harmful ultraviolet (UV) radiation from sunlight. Although the ozone layer around earth partially blocks UVB (290–320 nm) rays, but the remaining UVB, UVA (UVA-320–340 nm; and UVA-1, 340–400 nm) reaches our skin with sunlight exposure and causes biological and metabolic reactions. A thorough investigation on the capacity of TiO_2 and ZnO nanoparticles to penetrate the skin outer layer till the viable cells present in the deeper layer of skin was done in vitro and in vivo by a large number of investigators [33]. The outcomes of most of these studies indicated the neither TiO_2 nor ZnO nanopartticles could reach the viable cells present in the dermis. The other routes of administration of nano ZnO can be oral, intratracheal instillation and inhalation which was also used by researchers to study acute toxicity. Cytotoxicity and genotoxicity of ZnO nanoparticles were often associated with their photo catalytic activity. On photocatalysis of ZnO there was a production of superoxide anion radical ($O_2^{\bullet-}$) and hydroxyl radicals ($OH^{\bullet}$), singlet oxygen, H_2O_2 etc. and other species contributing to oxidative stress [24]. In an attempt to block the production of reactive oxygen species (ROS), the nanoparticle (NP)- containing sunscreen manufacturers have either coated the NPs or added some antioxidant compounds to their product. The material used for coating included polymers, aluminium hydroxide ($Al[OH]_3$), inert oxides of silica and vitamins (A, E, C) were added as antioxidants [34]. The removal of $Al[OH]_3$ surface layer by any means like swimming pool chlorines, can cause the photo-active TiO_2 to react with water and form such intermediates which can further cause skin damage or cancer [35]. One of the reasons of ZnO toxicity, thus, may be attributed to the oxidative stress exerted by the photocatalysis of ZnO. Aggregation of nanoparticles has also been found to be toxic. TiO_2 and ZnO nanoparticles can exist in three states: primary particles (5–20 nm), aggregates (30–150 nm) and agglomerates (1–100 μm). When added in sunscreens the primary particles cluster to form aggregates and these aggregates were found as the smallest unit present in the sunscreen final formulation [36–38]. Agglomeration of ZnO nanoparticles may occur because of two reasons.

1. It had the tendency to produce high surface energy which might occur when they were dispersed in organic solvents.
2. Bond formation of Zn-O- [39].

This phenomenon of aggregation reduced their high surface area to volume ratio and subsequently reduced the effectiveness. The toxicity of ZnO nanoparticles could also be lowered by polymer coating as there were many advantages that included less expensive in cost, prolonged circulation, less hazardous to environment, less expensive than conventional coatings, control released rates, decreased electrostatic interactions with plasma proteins. The polymeric additives get adsorbed on the metal oxide surface by a combination of chemical and electrostatic interactions, hydrogen bonding and van der Waals force [40]. To reduce the toxicity and to prevent the aggregation, surface coating of ZnO nanoparticles with two types of polymer-a synthetic polymer – polyethylene glycol (PEG) and a natural polymer (chitosan) was done and zebrafish embryo model studies showed chitosan coating to be more biocompatible and efficient in UV light scavenging [41].

Many nanoparticles created toxicity both in zebrafish and embryos [42–44]. Toxicity caused by the nanoparticles was dependent on their concentration and the time of exposure. Higher concentrations caused body deformations in 30–100% of adults and embryos. Some studies revealed that ZnO nanoparticles created tail malfunction and delayed hatching of zebrafish embryos [45]. The bioaccumulation of ZnO nanoparticles were observed in the gill, liver, intestine and brain of the fish. The toxicity was attributed to the increase in the ROS generation and occurrence of oxidative stress. Researchers showed that the exposure of ZnO nanoparticles caused some neuro and behavioural changes in the adult fish [46]. Growth rate of marine chlorophyte *Dunaliella tertiolecta* was also altered after exposure to ZnO nanoparticles [47]. In the liver of rats exposed to different doses of ZnO nanoparticles, hepatocyte swelling was observed along with significant increase in plasma AST, ALP and ALT at a dose of 5 mg/kg body weight [48]. Many in vitro studies have shown the toxic effects of ZnO nanoparticles caused due to ROS production, oxidative stress, apoptotic induction, inflammatory responses etc. In addition to these adverse effects, ZnO nanoparticles can also induce some positive effects like anti-inflammatory, anti-diabetic, and pro-coagulant effects at low doses may be due to the low concentration of Zn^{+2} ions required for cell homeostasis [49].

The above studies indicated a potential hazard of ZnO nanoparticles depending on its size and mode of exposure. Another nanoparticle, TiO_2 is profoundly used with nano ZnO in cosmetic products like sunscreens. We shall discuss in elaborate the possible toxicity routes of nano TiO_2 in the next section.

7.1.3 Titanium Dioxide (TiO_2) Nanoparticle Toxicity

Naturally occurring TiO_2 exists in three crystalline structures: rutile, brokite and anatase. The most stable and commonly used form is rutile which is birefringent and has refractive indices in UV and visible wavelength. The most common form of this pigment is the rutile form and its refractive index is in the UV and visible range making this birefringent crystal optically important. Rutile polycrystalline and epitaxial films has an average refractive index (n) of 4.0 and anatase films have 3.6. The band gap energy is found to be ~3.03 eV and ~3.2 eV for rutile bulk and anatase bulk TiO_2 respectively [50].

Researchers have primarily seen the effect of TiO_2 nanoparticles ($nTiO_2$) in vitro for different cell lines. NRF2 and BRCA1 gene expression studies after exposure to $nTiO_2$ in A549 alveolar carcinoma and BEAS-2B normal bronchial lung cell lines showed that it causes genotoxicity by inhibiting the DNA repair pathway [51]. The viability of Sertoli cells from male germ cells is reduced and morphology changed upon exposure to $nTiO_2$. $nTiO_2$ exposure also induced the expression of different immune mediator and cytokines [52]. CeO_2 and TiO_2 nanoparticles and microparticles were exposed to primary cultures of rat cortex and did not show any toxicity at the doses used ($\leq$50 µg/ml) but exhibited changed in spontaneous or GABA receptor mediated activity [53]. $nTiO_2$ and TiO_2 nanoparticles immobilized on nanokaolin (clay) were exposed to HepG2 cells in the presence and absence of serum and toxicity was evaluated. DNA damage was observed in short term exposure without serum and long term exposure with serum in these cells indicating nanokaolin to be unsuitable for immobilizing $nTiO_2$ [54]. These results indicated that TiO_2 can cause genotoxicity in normal and cancer cells, can affect the morphology of male germ cells, can alter neuronal responses as well as immobilization genotoxicity was observed. TiO_2 is widely used in dentistry as whitening agent, but its toxicity is not explored very extensively. Human gingival fibroblasts (HGFs) were exposed to white mineral trioxide aggregate (WMTA) and $nTiO_2$ showed decreased cell viability upon incubation from 24 to 48 h [55]. Dental fillings or carries treatment is a prolonged process and the use of TiO_2 nanoparticles for treatment is to be considered with time of exposure. The incorporation of TiO_2 nanoparticles in our body may induce certain immune responses which may lead to some other health problems. Fibrinogen (FG) is a protein in our body that responds to foreign body reactions. The effect of FG on toxicity of $nTiO_2$, carbon and SiO_2 were studied on alveolar macrophages reporting increase in cytotoxicity and NO production [56]. TiO_2 is widely used in cosmetic products and its toxicity on keratinocytes (HaCaT cells) were exploited showing increase in apoptosis via caspase 8/Fas dependent pathway for TiO_2 nanoparticles. Although the smaller sized nanoparticles of TiO_2 did not exhibit any harmful effect on these cells [57]. Rutile and anatase $nTiO_2$ had differential effect on HaCaT cells before and after UVA light exposure. Before exposure to UVA light, rutile $nTiO_2$ inhibited cellular growth and caused *HO-1* gene expression which was attributed to its hydrophobic nature. Although the inhibitory effect of rutile $nTiO_2$ was not observed after exposure to UVA light and phototoxic effect of anatse $nTiO_2$ was explained to be due to internalization of the particles [58]. TiO_2 nanoparticles can thus exert different effect in the presence and absence of non ionizing radiation like UVA light. The effect of ionizing radiation like X rays were studied in pancreatic cancer cell model after exposure to titanium peroxide nanoparticles (TiO_XNPs) and polyacrylic acid-modified TiO_xNPs (PAA-TiO_xNPs). The combination of X- rays with $nTiO_2$ and PAA-TiO_xNPs produced ROS in a dose dependent manner, whereas, only $nTiO_2$ did not produce ROS in pancreatic cancer model (MIAPaCa-2 cells) [59]. The exposure to nanoparticles causes inflammation of endothelial cells leading to atherosclerosis. Two cell lines – human aorta (HAECs) and human umbilical vein (HUVECs), were exposed to cobalt (CoNPs) and TiO_2 nanoparticles (TiNPs) showing rapid internalization, increased mRNA and protein levels of adhesion molecules (ICAM-1, VCAM-1, E-selectin), monocyte chemoattractant protein-1 (MCP-1) and interleukin 8 (IL-8). Although TiO_2 did not induce any oxidative stress but CoNPs induced ROS generation [60].

TiO_2 nanoparticles get washed off from surfaces and get mixed directly with surface water, lakes, rivers and sea. Thus the aquatic life has a higher risk of exposure to TiO_2 nanoparticles. A fish cell line (BF-2) and zebrafish embryos were exposed to silver (Ag) and titanium dioxide (TiO_2) nanoparticles under mimicking condition of sunlight exposure (Simulated Solar Light (SSL) exposure). The results of this study indicated that solar light can cause physicochemical changes in both the nanoparticles which can further enhance their toxic potential [61]. There were several reports on the effect of TiO_2 nanoparticles alone or along with other nanoparticles on aquatic models which is summarised in Table 7.1.

The studies on the aquatic animals upon $nTiO_2$ exposure did not give any conclusive remarks about its safety, but their up-regulation of damaging effect along with other pollutants is established. The studies for $nTiO_2$ were further extrapolated to higher animals like rat or mouse model to better understand the phenomena in a mammalian animal model. Rat liver cells (BRL-3A) and Sprague–Dawley (SD) rat livers were exposed to H_2O_2 to induce oxidative stress and further exposed to $nTiO_2$ to observe the effect on the stress condition. There was a synergistic increment in the oxidative stress in both in vivo and in vitro upon exposure to $nTiO_2$, indicating a critical screening of oxidative stress in the individual is warranted before using $nTiO_2$ in tooth-

Table 7.1 Reports on toxicity studies of TiO_2 nanoparticles in aquatic animals

S. No.	Nanoparticles used	Aquatic animal used for study	Gross effect	Reference
1.	TiO_2 nanoparticles	Rainbow trout fish (*Oncorhynchus mykiss*)	Oral administration of TiO_2 at 10 and 100 mg/kg body weight. No change in WBC, RBC counts, hematocrit value, hemoglobin, plasma Na^+. Accumulation of Ti was seen in gill, gut, liver, brain and spleen. Brain did not clear Ti and exhibited disturbances in Cu and Zn levels	Ramsden et al. [62]
2.	TiO_2 and CuO	Carp (*Cyprinus carpio*)	Exposure to a mixture of TiO_2 and CuO nanoparticles induced similar histopathological anomalies compared to individual nanoparticle exposure. But the injury to gills, intestine, liver and kidney was higher after exposure to TiO_2 and CuO nanoparticle mixture compared to individual nanoparticle exposure	Mansouri et al. [63]
3.	TiO_2 nanoparticles	Rainbow trout fish (*Oncorhynchus mykiss*)	No major disturbances in haematological values, but intestinal lining injury and brain injury related to oxidative stress was reported	Federici et al. [64]
4.	TiO_2 nanoparticles	Zebrafish (*Danio rerio*)	Exposure to nano TiO_2 did not affect the whole body electrolyte levels but reproduction was affected leading to production of non viable eggs	Ramsden et al. [65]
5.	Chromium Cr(VI), TiO_2 and Al_2O_3	Freshwater micro algae, *Scenedesmus obliquus*	The toxicity of Cr(VI) decreased upon nano TiO_2 addition but unaffected upon nano Al_2O_3 addition. There was oxidative stress induced toxicity	Dalai et al. [66]
6.	TiO_2	Fish cell lines (RTG-2, PLHC-1, RTH-149, RTL-W1) and rainbow trout primary hepatocytes)	Negligible effect of TiO_2 nanoparticles were observed in these cell lines. No significant effect on mitochondrial metabolic activity, lysosome function and plasma membrane integrity was observed	Bermejo-Nogales et al. [67]
7.	TiO_2 and $TiCl_4$	Microalgae (*Phaeodactylumtricornutum*) and crustacean (*Artemia franciscana*)	$nTiO_2$ is more harmful to the microalgae whereas $TiCl_4$ exhibited higher toxicity to the crustacean. The dark conditions lowered the toxicity of $nTiO_2$, but starvation increased the toxicity of both nanoparticles	Minetto et al. [68]
8.	TiO_2	Zebrafish (*Danio rerio*) larvae	BDE-209 (a polybrominated diphenyl ether congener) bioconcentration, its metabolism and effect on thyroid function in the presence of nano TiO_2 was minotored in zebrafish larvae. Nano TiO_2 increased the BDE-209 bioavailability and metabolism, decreased locomotive activity, upregulated genes like *tshβ, tg, dio,* indicating a disruption in endocrine function and developmental neurotoxicity	Wang et al. [69]

(continued)

Table 7.1 (continued)

S. No.	Nanoparticles used	Aquatic animal used for study	Gross effect	Reference
9.	nTiO_2; (fresh, or aged) and bulk TiO_2	Marine mussels (*Mytilus galloprovincialis*)	Higher accumulation of Ti in digestive tissue compared to gills with bulk TiO_2 to be more toxic. Metallothionein gene expression, histology and histochemical studies, comet assay for DNA damage also supported bulk TiO_2 to be more toxic	D'Agata et al. [70]
10.	nZnO, nTiO_2 and nAl_2O_3	Zebrafish embryo and larva (*Danio rerio*)	nZnO and bulk ZnO exposure showed developmental delay in larva whereas TiO_2 and Al_2O_3 did not exhibit any developmental toxicity	Zhu et al. [71]
11.	nTiO_2	Zebrafish (*Danio rerio*) embryos or larvae	The toxicity of nTiO_2 was compared with pentachlorophenol (PCP). Only nTiO_2 did not show oxidative stress, DNA damage or lipid peroxidation whereas along with PCP, nTiO_2 caused lipid peroxidation, DNA damage and ROS generation. Thus, nTiO_2 upregulated the deleterious effect of organic pollutants	Zhu et al. [72]
12.	Mn, Fe, Ni, and Cu doped TiO_2 nanoparticles	Zebrafish embryos	Fe-TiO_2 NPs showed highest toxicity along with high photocatalytic activity. Mn-TiO_2 NPs exhibited lowest toxic effect with improvement in photocatalytic activity	Park et al. [73]
13.	nTiO_2 and bisphenol A (BPA)	Zebrafish embryos	Decreased survival, increased morphological abnormalities, and delayed embryo hatching were observed upon co-exposure to nTiO_2 and BPA	Yan et al. [74]

pastes, cosmetics and potential treatment of diseases [75]. Intraperitonial injection of anatase nTiO_2 in mice resulted in heavy accumulation of nTiO_2 in lungs and liver tissues, impaired DNA and interrupted metabolic homeostasis in liver tissue. Lung tissues were observed to have oxidative stress and inflammatory responses suggesting the adverse effect of nTiO_2 [76]. The oral administration of nTiO_2 in mouse model also disrupted the liver metabolic function causing hepatotoxicity [77]. Although in some studies it has been found that hairless mouse exposed to sunscreen containing nTiO_2 and nZnO in the presence and absence of UV radiation, did not show any apparent toxicity even after long term exposure [78]. The impact on seminal vesicles, thyroid gland and androgen receptor after exposure to nTiO_2 in rats demonstrated abnormal development of seminal vesicles which was further reduced by administration of aged garlic extract [79]. A role of thymoquinone and avenanthramides in protection of nTiO_2 induced toxicity in Sprague-Dawley rats was reported through their antioxidative and anti-inflammatory effect [80]. Therefore, it can be summarized that some studies have shown the effect of nTiO_2 to be deleterious but there are also some natural products that can partially nullify such effect.

7.1.4 Iron Oxide Nanoparticle Toxicity

Superparamagnetic iron oxide nanoparticles (SPIONS) consists of (maghemite) γ-Fe_2O_3 and magnetite (Fe_3O_4) which are considered to be biocompatible, non toxic, biodegradable with a variety of biomedical applications like targeted drug delivery, tumor imaging, contrast agents in magnetic resonance imaging (MRI) etc. [81].

The magnetic moments of the SPIONS change with the interaction with different biomolecules resulting in decrease in its size. The biological effect of the SPIONS although depends on size, shape concentration and mode of incorporation of these particles [82]. The toxicity of SPIONS, in vitro and in vivo, has been extensively studied and reviewed [83–85]. In some cases it has been found to be mutagenic [86], or exhibited long term exposure effects [87] and in some studies it is additionally found to induce genotoxicity, developmental toxicity and neurotoxicity [88]. The redox state of ultrafine SPIONS altered the cellular uptake and induced DNA damage in human lymphoblastoid cell line (MCL-5) [89]. The imaging applications of SPIONS are more related to human brain and nervous system and researchers have exploited the in vitro effect in many neuronal cells. The uptake and metabolism of SPIONS have been studied in cultured brain cells in serum starved and serum containing media. The localized deposition of nanoparticles was seen but the toxicity was only exerted in microglial cells but not in astrocytes, oligodendroglial cells and neurons [90]. Retinoic acid induced neuronal differentiation of mouse embryonic stem cells was reduced with reduced viability due to induction of ROS mediated oxidative stress [91]. Human neuroblastoma SHSY5Y and glioblastoma A172 cells were exposed to different concentrations of silica and oleic acid coated SPIONS in serum free and normal serum containing medium. Significant decrease in cell viability was observed with silica coated SPION to be more toxic [92]. Studies on acute toxicity of iron oxide nanoparticles in human and murine neuroblastoma cell line, neural progenitor cell line and neural stem cells revealed cell type and species specific response in ROS generation calcium homeostasis, mitochondrial integrity and cell morphology, indicating that cellular homeostasis was impaired in distinct ways [93]. Short term and long term exposure of human CNS cells astrocytes (D384) and neurons (SH-SY5Y) to magnetite nanoparticles caused susceptibility of the astrocytes to low and high doses of iron oxide nanoparticles (IONPs) whereas neurons did not show any effect. Ion overload in cells can trigger adverse effects in brain thereby compromising the CNS function [94]. These studies focus on the safety concerns regarding the use of SPIONS in different biomedical applications. The long term effect that these nanoparticle can cause if not cleared from the body can become a major concern on its use. Although many surface functionalizations are being made to make it safe to be used in humans, but the conjugations of such coatings after incorporation into the system is needed to be studied.

Breast cancer cell lines are also used for studying the toxicity of IONPs. Human breast cancer cell line MCF 7 undergoes oxidative stress, DNA damage and caspase dependent apoptosis after exposure to IONPs [95]. Dimercaptosuccinic acid-coated superparamagnetic iron oxide nanoparticles did not show any effect on cell morphology, cytoskeleton organization, cell cycle distribution, ROS generation and cell viability in MCF 7 compared to untreated cells showing a safe mode of drug delivery to breast cancer cells [96]. Human cell lines for leukaemia (Jurkat cells), breast cancer (MCF-7 cells), cervical cancer (HeLa cells), and liver cancer (HepG2 cells) were treated with IONPs green synthesized using brown seaweed (*Sargassum muticum*) aqueous extract. Flow cytometry analysis and gene expression for caspase 3 and caspase 9 concluded the increase in early and late apoptosis indicating a cytotoxic effect in cancer cell lines [97]. Coating IONPs with PEG decreased the cellular uptake and hence decreased cytotoxicity in Chinese Hamster ovaries (CHO-K1 cells) [98]. HepG2 cells also showed oxidative stress induced toxicity after exposure to IONPs [99]. Although the primary concern, after IONPs exposure, should be the blood as it serves as a common platform for any nanoparticle incorporation into the body. Reports exists that the magnetite nanoparticles exert cytotoxicity (MTT and LDH assay), genotoxicity (micronuclei formation and sister chromatid exchange assay) and oxidative damage in human whole blood cells [100]. Secondary concern is our immune system which may be compromised after long term exposure or secondary effect of iron stress in the cells.

Macrophage cells were used by researchers to assess the activity of IONPs on cell viability. SPIONS coated with dimercaptosuccinic acid, aminopropyl silane or aminodextran were added to murine primary IL-4-activated bone marrow-derived macrophages and human M2-like differentiated THP-1 cells. There was internalization of the SPIONS and no significant effect on cell survival, but generation of ROS was observed along with activation of extracellular signal regulated kinase and AKT pathways. The iron metabolism of M2 macrophages were switched to iron-replete state [101]. The contribution of SPIONS towards cellular toxicity also depends on the shape of the nanoparticles. Micro-sized Fe_2O_3 (MFe_2O_3), nano-sized Fe_2O_3 (N-Fe_2O_3), and rod-shaped Fe_2O_3 (R-Fe_2O_3) induced differential toxicity in mouse macrophage cells (RAW 264.7). LDH assay, tumour necrosis factor-α production and ROS generation revealed that rod shaped was more toxic than sphere shaped IONPs [102]. The sensitivity of H9c2 cardiomyocyte cells embedded in acrolein along with IONPs was increased causing acrolyn induced dysfunction [103]. Exposure of IONPs in skin epithelial A431 and lung epithelial A549 cell line induced ROS generation and genotoxicity. Caspase 3 and caspase 9 levels were elevated at the mRNA level as well as protein level [104]. In some cases, SPIONS were also found to be non toxic to mouse embryonic stem cells (mESCs) with respect to growth, morphology, viability, proliferation and differentiation [105]. IONPS are also functionalized with VEGF for targeting the choroid layer of the eyes to yield a safe drug delivery carrier [106].

Another application of IONPs is for remediation of groundwater [107]. Removal of heavy metals using IONPs and functionalised IONPs has made the aquatic animals to be exposed to IONPs. Studies involving exposure of zebrafish embryos to greater than 10 mg/L caused developmental toxicity through mortality, malformation and delayed hatching of the embryos [108]. A short term exposure (72 h) of zebrafish embryos to polyacrylic coated IONPs, $nTiO_2$, nZnO and $nCeO_2$ at different doses neither showed any toxicity nor much free metal ions were present in the growth medium. Only $nCeO_2$ could traverse the chorion showing the protective role of functionalized nanoparticles [109]. The exposure to aquatic animals to IONPs can also affect the lake fishes which are used for human consumption. *Labeo rohita* is a very common kind of carp fish which is widely cultured in lakes and ponds in India and used for human consumption. Researchers have focused on the toxicity of IONPs to *Labeo rohita* for a long term (25 days) showing a significant increase in haemoglobin, RBC count and hematocrit value. Gill Na^+/K^+-ATPase activity, WBC count increased and there were evidences of hyponatremia (Na^+), hypochloremia (Cl^-) and hypokalemia (K^+) [110]. Although a short term exposure (96 h) showed decrease in Haemoglobin, hematocrit, sodium, potassium, chloride and gill Na^+/K^+- ATPase values [111]. The physiological parameters of *Labeo rohita* also changed after exposure to sublethal doses of IONPs exhibiting bottom resting, movements in jerks and surface respiration [112]. Mussels (*Mytilus galloprovincialis)* after exposure to IONPs and zeolite incorporated IONPs developed toxicity through ROS production, ubiquitin conjugation, DNA damage, protein carbonylation, and lipid peroxidation [113]. Thus, it is in urgent need to investigate the consumability of *L. rohita* and other aqueous animals by humans.

The potential application of IONPs in biomedical field may threaten the human health as evidenced from the in vitro studies. Some of the studies have explained the toxicity of IONPs, whereas no effect was found in some cell lines after exposure to IONPs. It becomes important to extrapolate the toxicity studies of IONPs towards in vivo models. The studies on the effect of IONPs on different types of rat and mouse models are listed in Table 7.2

The studies in vivo with IONPs in mice and rat models could conclude that the dose of administration plays a critical role in inducing different types of toxicity irrespective of mode of administration. Fictionalization of IONPs either by coating with a polymer, doping or encapsulation may exert a safer option for the use of IONPs in medical application.

Table 7.2 Reports on toxicity studies of IONPs in rat and mouse models

S. No.	Nanoparticles used	Animal model used for study	Gross effect	Reference
1.	IONPs	Rats	Long term chronic exposure caused increase in lipid peroxidation, reduction in reduced glutathione levels in liver, kidney and brain in a dose dependent manner. Accumulation of IONPs in organs was also observed	Reddy et al. [114]
2.	IONPs	Male mice	Decrease in food intake, water consumption, and iron content elevation in urine, faeces, seminal vesicle and prostate gland. Increase in LDH, total protein and fructose level. Heat shock proteins HSP60, HSP70, HSP90, Caltrin, PSP94, and SSLP1 expression also altered	Sundarraj et al. [115]
3.	IONPs	Male mice	Alanine aminotransfere (ALT), aspartate aminotransferase (AST) and alkaline phosphatase (ALP)] levels were elevated and liver tissues were damaged after administration of high dose of IONPs	Babadi et al. [116]
4.	IONPs	Wistar rats (Female)	Inhibition of acetylcholinesterase in red blood cells and brain cells and inhibition of Na^+-K^+, Mg^{2+}, and Ca^{2+}-ATPase activities in brain were found at acute doses	Kumari et al. [117]
5.	IONPs, positively-charged polyethyleneimine-Fe_2O_3-NPs (PEI-NPs), or negatively-charged poly(acrylic acid)-Fe_2O_3-NPs (PAA-NPs)	Pregnant female mice	At low doses there was no toxic effect on the development of the foetus. High dose exposure induced foetal mortality, charge based toxicity of the uterine lining, testes of the developing offspring revealing PEI-NPs to be more toxic	Di Bona et al. [118]
6.	IONPs	BALB/c mice	At higher doses liver tissues were damaged (150 and 300 μg/gr body weight)	Parivar et al. [119]
7.	IONPs	Mice	Repeated intraperitonial injection of IONPs could enable the particles to cross the blood testis barrier. In testicular cells, they caused increased ROS generation, lipid peroxidation, protein carbonylation, nitric oxide and decreased the antioxidant levels. Bax, caspase 3 expression elevation confirmed apoptosis affecting male fertility	Sunderraj et al. [120]
8.	IONPs	Wistar rats	Intravenous injections of IONPs once a week for 28 days caused alterations in haematological factors, inhibition of antioxidant enzymes, lipid peroxidation increased but no genotoxicity induced as assessed by comet assay	Gaharwar and Paulraj [121]

(continued)

Table 7.2 (continued)

S. No.	Nanoparticles used	Animal model used for study	Gross effect	Reference
9.	IONPs	Wistar rats	Oral administration of IONPs for 28 days did not cause any effect at low and medium doses. At high doses IONPs caused decrease in Na^{+}, K^{+}, Ca^{+2}-ATPases and Mg^{+2} in brain. Dilated central vein, liver damage and kidney damage. No moratality was observed	Kumari et al. [117, 122]
10.	IONPs	Wistar rats	Intratracheal instillation of IONPs did not induce any toxicity except development of pulmonary fibrosis in later stage	Szalay et al. [123]
11.	PVA coated SPIONs	naïve Toll-like receptor-4 deficient (TLR4−/−) or wild-type C57Bl/6 mice, or C57Bl/6 mice with induced arthritis	Intravenous injection of PVA coated SPIONs induced synovial inflammation in joints and genes responsible pro-inflammatory response increased in liver during the initial period of injections. No detrimental effect was found after 14 days of injection	Vermeij et al. [124]

7.1.5 Copper Oxide Nanoparticle (CuO and Cu_2O) Toxicity

Copper oxide nanoparticles (CuO NPs) have many potential physical properties like electron correlation effects, electron spin dynamics and high temperature superconductivity. CuO NPs are semiconductor in nature with monoclinic structure and have photovoltaic and photoconductive properties due to its narrow band gap. It could enhance the viscosity of fluids, increased thermal conductivity thereby enabled them as improved energy saving material for energy conversion. The industrial applications of CuO NPs include batteries, high temperature superconductors, catalysis, gas sensors, field emission emitters, solar energy converters etc. [125]. Biological applications of CuO NPs are antimicrobial agents for preservation of wood, antimicrobial textiles, antifouling paints, agricultural biocides etc. The high surface area and nanoscle size of CuO NPs gets attached to the negatively charged cell wall of the microorganisms and get internalized thereby eliciting various damaging effects after interaction with DNA, membranes of other organelles etc. [126]. The indiscriminate use of CuO NPs may possess potential health concerns to humans, animals as well as aquatic biota. It becomes essential to exploit the toxicity of CuO NPs in different living systems.

CuO NPs induced ROS in HepG2 cells which was reversed by use of N acetyl cyteine (NAC) [127]. Air liquid interface exposure of CuO NPs to human bronchial epithelial cells (HBEC) and lung adenocarcinoma cells (A549 cells) decreased the cell viability, increased LDH leakage and ROS generation and increased IL-8 expression in a dose dependent manner. The effect was more pronounced in A549 cells and could be reverted by the use of antioxidants [128, 129]. In silico the toxicity of CuO NPs was estimated using untargeted metabolomics profiling tool [130]. There were several studies on CuO NP toxicity exerted on neuronal cell lines as it was projected earlier that deposition of Cu^{+2} in brain cells may enhance amyloid plaque formation associated with Alzheimer's disease. In vitro it has been observed that amyloid beta 42 (Aβ 42) can cause aggregation of amyloids in the presence of metal ions like copper, zinc and iron when incubated alone or co-incubated with each other [131]. Dimercaptosuccinic acid and bovine serum albumin coated CuO NPs exposure to C6 glial cells reduced the cell viability but the effect could be improved by the used of copper chelators ensuring the role of free copper ions that escaped from coating towards toxicity [132]. The primary astrocytes isolated from rats were exposed to CuO NPs and showed significant copper accumulation in the cells, reduced viability at higher doses and altered the glycolytic flux, synthesis of glutathione and metallothioneins [133]. These studied can conclude that CuO NPs can be exposed to cells but the dose of exposure plays a pivotal role in induced toxicity. Proper encapsulation of CuO NPs with minimum release of Cu^{+2} ions can become a safe option to be used.

The safety of aquatic animals is also to be regarded, because the consumption of fishes and other aquatic animals by humans may cause bioaccumulation and biomagnification of the nanoparticles in humans and other animals. Macrophytic algae cells of *Nitellopsis obtuse*, microphytic algae *Chlorella*, shrimp *Thamnocephalus platyurus*, and rotifer *Brachionus calyciflorus* were exposed to sonicated and non sonicated CuO NPs and exhibited lethality due to accumulation of CuO not due to Cu^{+2} ions released in water. There was delayed membrane depolarization and the thick charophyte cell wall was also affected in *N. obtuse* [134]. The feeding behavior and growth pattern of the shredder of *Allogamus ligonifer* (Trichoptera, Limnephilidae) was influenced due to accumulation of Cu^{+2} ions produced by the leaching effect after exposure to CuO NPs [135]. Artificial freshwater (AFW) and natural water was used to study the toxicity of CuO NPs and ZnO NPs in crustaceans *Daphnia magna* and *Thamnocephalus platyurus* and protozoan *Tetrahymena thermophila*. AFW showed higher toxicity for the crustaceans for both CuO NPs and ZnO NPs, but in natural water the toxicity decreased. The toxicity was attributed to the solubilised ions liberated from the nanoparticles into the water [136]. Similar mode of toxicity by the dissolution of metal ions in water was observed for microalgae *Pseudokirchneriella subcapitata*, after exposure to CuO NPs, ZnO NPs and TiO_2 NPs [137]. Tight epithelial cells were cultured from *Xenopus laevis* and the toxicity of CuO NPs and Cu ions were estimated. Results showed decreased cell viability upon CuO NPs for a long term exposure, whereas, Cu ions could produce the same amount of toxicity in a short period [138]. A lower RBC count, hematocrit value and elevated WBC count was observed in fish (*R. rutilus)* after treatment with CuO NPs and the LC 50 value for 96 h is found to be 2.19 ± 0.003 mg/l of water [139]. The effect of Cu NPs and copper sulfate was observed for the neuronal cells of *c. elegans* and a 10% degeneration of dopaminergic neurons were found after CuO NPs exposure. There was loss of feeding pattern and average body length of the nematode and elevation in stress response genes after CuO NPs exposure [140]. Male wistar rats were injected intraperitonially to 100, 200 and 400 ppm of CuO NPs and the blood parameters aspartate aminotransferase, alanine transaminase, alkaline phosphatase, creatinine, fasting blood sugar, albumin, blood urea nitrogen, total protein were evaluated. The results showed that at 400 ppm CuO NPs exposure, all the biochemical parameters were alarming indicating that higher doses of CuO NPs can be detrimental in wistar rats [141]. The studies on CuO NPs on different biological systems show that the aquatic animals like fish, crustaceans, algae, are very much affected by decreased viability, altered blood parameters and in some cases neurological disturbances at high dose exposure. The indirect incorporation of CuO NPs toxicity through food chain could magnify and extrapolate the effect in humans.

7.1.6 Cerium Oxide Nanoparticle Toxicity

Cerium oxide nanoparticles, also known as nanoceria is a rare earth metal oxide nanoparticle with versatile application in catalysis, gas sensors, solar cells, UV absorbents and glass polishing material. Nanoceria also has biomedical applications in biosensors, treatment of retinopathy and cancer. It has multi-enzymatic biomimetic properties of superoxide oxidase, catalase and oxidase [142]. With increase in usage there are occupational exposures to nanoceria which makes a concern to test its toxicity. Conflicting reports were found regarding the extent of toxicity of nanoceria and the valance state of Ce atoms in the nano preparation may be the cause of differential toxicity. The valance state of Ce can be both +3 and +4 in CeO_2 nanoparticle and enable to take and release oxygen respectively. The transition of Ce^{+4} to Ce^{+3} produces subsequent oxygen vacancies, producing superoxide anions which further produces hydroxyl ion that are toxic to cells; whereas, Ce^{+3} can react with hydroxyl ions and behave as an antioxidant [143–146]. Thus, the oxidation state of Ce in CeO_2 can decide whether it will exert toxic effect or protective effect to the cells [147].

Nanoceria can be used for treatment of cancer as the cancer cell environment is acidic (pH = 6.7) compared to normal tissues (pH = 7.), whereas the cytosol of cancer cells are alkaline. A selective killing of cancer cells on nanoceria exposure is hypothesized due to the acidic environment of the cancer tumours resulting in the inhibition of catalase like activity of nanoceria. Along with radiation, the superoxide radical generation capacity of nanoceria in acidic environment further enhances cancer cell toxicity together with radiation [148, 149]. The selective killing of cancer cells by nanoceria by inducing oxidative stress and apoptosis after radiation exposure is also observed [150–152]. The radical scavenging effect of nanoceria can be compared with the activity of the enzyme superoxide dismutase (SOD). SOD removes superoxide ($O_2^{\bullet-}$) from the system in the following reaction steps.

$$O_2^{\bullet-} + 2H^+ + (Cu^+) - SOD \rightarrow H_2O_2 + (Cu^{2+}) - SOD$$

$$O_2^{\bullet-} + (Cu^{2+}) - SOD \rightarrow O_2 + (Cu^+) - SOD$$

$$2O_2^{\bullet-} + 2H^+ \rightarrow H_2O_2 + O_2$$

This H_2O_2 produced by SOD is reduced to water by the enzyme catalase (CAT).

$$2H_2O_2 \xrightarrow{\text{CAT}} 2H_2O + O_2$$

A similar process of superoxide radical scavenging is reported for nanoceria as follows [153–155]:

$$O_2^{\bullet-} + Ce^{3+} + 2H^+ \rightarrow H_2O_2 + Ce^{4+} \left(Ce^{3+} \text{is oxidized}\right)$$

$$O_2^{\bullet-} + Ce^{4+} \rightarrow O_2 + Ce^{3+} \left(Ce^{4+} \text{is reduced}\right)$$

To understand the effect of low dose exposure to nanoceria, AL-PEG-600 coated nanoceria was exposed at a very low dose (0.01 μg/ml) to the cancer cell lines- human hepatocellular carcinoma HepG2, Huh7, and SMMC-7721 cell lines, the human osteogenic sarcoma U2OS cell line, the human breast cancer MCF-7 cell line, and the human colon carcinoma HCT116 cell line. The AL-PEG-600 coating was done to make the nanoparticles stable and biocompatible. The results of this study showed that polymer coated nanoceria could promote cell proliferation in hepatoma cells by reducing apoptosis and can activate AKT/ERK signaling pathways [156]. Inhalation of fresh and aged nanoceria by Sprague Dawley rats showed a predominant presence of these nanopartciles in lungs and feces. There was no significant difference in the biodistribution of fresh and aged nanoceria in the organs and with increase in time the amount of nanoceria increased in the feces indicating a clearance of the nanoparticles. Alveolar phagocytosis was also observed for nanoceria clearance but no toxicity was observed [157]. There was no genotoxicity in the blood cells observed after inhalation of nanoceria for 3–6 months in female rats [158]. Lung lavage fluids were collected from nanoceria and NiO nanoparticles exposed male C57BL/6J mice to assay inflammation, cytotoxicity and inflammasome activation at 1 and 7 day post exposure. A pH dependent radical scavenging activity was observed for both nanoceria (acidic environment mimicking the lysosome) and NiO-NP (at physiological pH mimicking the cell cytosol) [159]. After the inhalation studies of nanoceria, studies involving neurological disorders are also warranted as the size of the nanoparticles make them to penetrate the blood brain barrier and may influence the activity of brain. In a mouse model for amyotrophic lateral sclerosis (ALS), $SOD1^{G93A,}$ nanoceria was injected intravenously twice a week and it was observed that there was induction of muscle weakness in the mouse but their life span increased significantly. It was hypothesized that the citrate stabilized nanoceria could act like catalase and oxidase and exerted its protective role by extending the lifespan of $SOD1^{G93A}$ mice [160]. The neuroprotective role of nanoceria was studied in Parkinson's disease induced by 6-hydroxy dopamine (6-OHDA) in rats. The results showed partial decrease in oxidative stress and apoptosis but no alteration in striatal dopamine level. Thus, nanoceria could have a neuroprotective level exerted through its antioxidant properties [161]. Reduction in neuronal death and calcium dysregulation caused by mild brain injury was an after effect of intraperitonial nanoceria treatment in rats [162]. Application of nanceria as a sensor for determination of low concentration of vitamin C makes this nanoparticle a promising agent for biosensor development [163]. Although in some cases nanoceria is found to be toxic by generating ROS [143, 164] most of the studies concluded nanoceria to be protective rather than harmful.

Contribution of nano TiO_2 towards induction of oxidative stress and nervous system is well documented [165, 166]. The toxicity of other metal oxide nanoparticles like Bi_2O_3, Co_3O_4, CuO, Cu_2O, Fe_2O_3, Fe_3O_4, Sb_2O_3, SiO_2, TiO_2, UO_2, ZnO, ZrO_2, SnO_2, Ag, Au, has already been discussed by other researchers [167, 168]. Several cell lines [169] as well as embryonic zebrafish are used for the study of metal oxide toxicity [170] because of its transparent eggs. Zebrafish has a low cost of rearing and is a vertebrate system. Moreover, the transparency of the zebrafish eggs makes the researchers to monitor all the developmental stages microscopically.

7.1.7 Mechanisms of Metal Oxide Nanoparticle Toxicity

The mechanism of toxicity of different metal oxide nanoparticles are attributed mainly due to dissociation of the metal ion from the nanoparticle when exposed to cells or biological fluids. The general toxicity mechanism induced by NPs when internalized inside the cells is depicted in Fig. 7.3.

The internalization by endocytosis can cause much severe damage to the cell compared to non endocytosis entry pathway. The cells carry the metal oxide nanoparticles through the lysosome and several metal ions are released in the cytoplasm due to the acidic pH of the lysosome that acts like a Trojan Horse [171]. These ions interact with the proteins in the cytoplasm thereby generating superoxide radicals. These radicals are converted to hydrogen peroxide (H_2O_2) by the action of our antioxidant defense system, the enzyme called superoxide dismutase (SOD). The released H_2O_2 is converted to water in the presence of another antioxidant enzyme called catalase (CAT). But when the superoxide radical load is higher, some of the H_2O_2 escape through CAT decomposition and enter the nucleus. In the nucleus, H_2O_2 undergoes Fenton type reaction leading to the generation of hydroxyl radicals and other adducts generating ROS that can cause DNA damage [172].

When the free radical enter the mitochondria they depolarize the membrane and trigger apoptosis by the release of cytochrome c. Cytochrome c is released from mitochondria to the cytoplasm leads to the cooperative action of three protein factors designated as apoptotic protease activating factors (Apafs). Released cytochrome c binds to Apaf-1 (apoptotic protease activating factor-1) and promotes its oligomerization to form the so called 'apoptosome' [173]. Apaf-1 contains a sequence homologous to the C.elegans CED-4, and while Apaf-2 was identified as Cytochrome c, recruitment of procaspase-9 to this active apoptosome results in its autoactivation. Apaf-3 was shown to be identical with Caspase-9. It also was demonstrated that, in the presence of dATP, Caspase-9 is directly activated by Apaf-1 and Cytochrome c. Active Caspase-9 activates Caspase-3 and by this the apoptotic machinery gets triggered that leads to DNA fragmentation and cell death [173] (Fig. 7.3).

One of the most used hallmarks of apoptosis is internucleosomal DNA fragmentation, brought

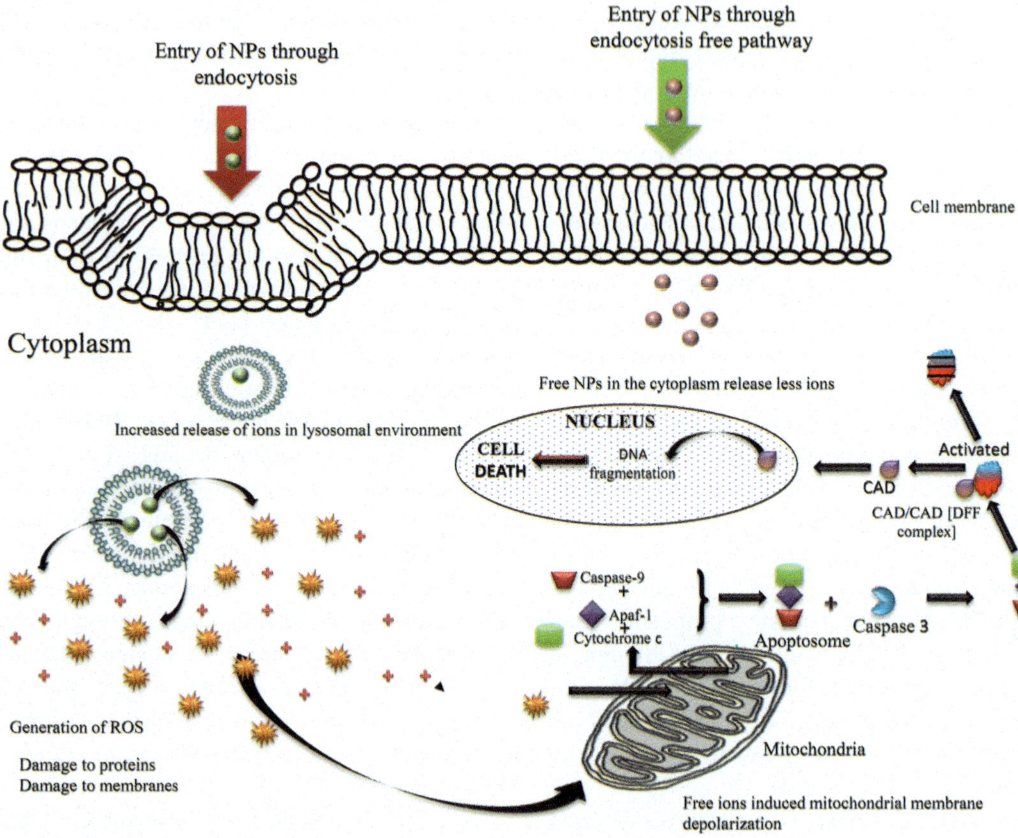

Fig. 7.3 The nanoparticles entry into the cells via endocytosis and endocytosis free pathway. The acidic environment of lysosome causes burst release of the nanoparticles which interact with the cytosolic proteins and other molecules generating ROS. These ROS generate oxidative stress into the mitochondria and cytochrome c is released which triggers the caspase cascade leading to caspase dependent apoptosis

about by an endonuclease, generating DNA fragments with lengths corresponding to multiple integers of approximately 180 basepairs [174]. The responsible endonuclease (called CAD for Caspase Activated DNase) pre-exists in living cells as an inactive complex with an inhibitory subunit, called ICAD (Inhibitor of Caspase-Activated DNase). Activation of CAD occurs by caspase-3-mediated cleavage of the inhibitory subunit, resulting in the release and activation of the catalytic subunit. Caspase-mediated cleavage of specific substrates also explains several other characteristic features of apoptosis, like cleavage of the nuclear lamins, required for nuclear shrinking and budding and cleavage of cytoskeletal proteins such as fodrin and gelsolin, causing loss of overall cell shape [175].

The Fenton reaction involving iron ions generated due to iron oxide internalization by the endothelial cells of the lumen of the blood vessels is given below:

$$Fe^{2+} + H_2O_2 \rightarrow Fe^{3+} + OH^{\bullet} + OH^-$$

$$Fe^{3+} + H_2O_2 \rightarrow Fe^{2+} + OOH^{\bullet} + OH^+$$

Iron oxide nanoparticles are injected intravascularly for imaging applications and the fate of the nanoparticles is shown in Fig. 7.4.

As the maximum nanoparticles get cleared through phagocytosis, the remaining of the particles interact with the endothelial cells leading to increased level of cytokines, adhesion molecule expression, monocyte recruitment and adhesion, nitric oxide level elevation and decreased cell

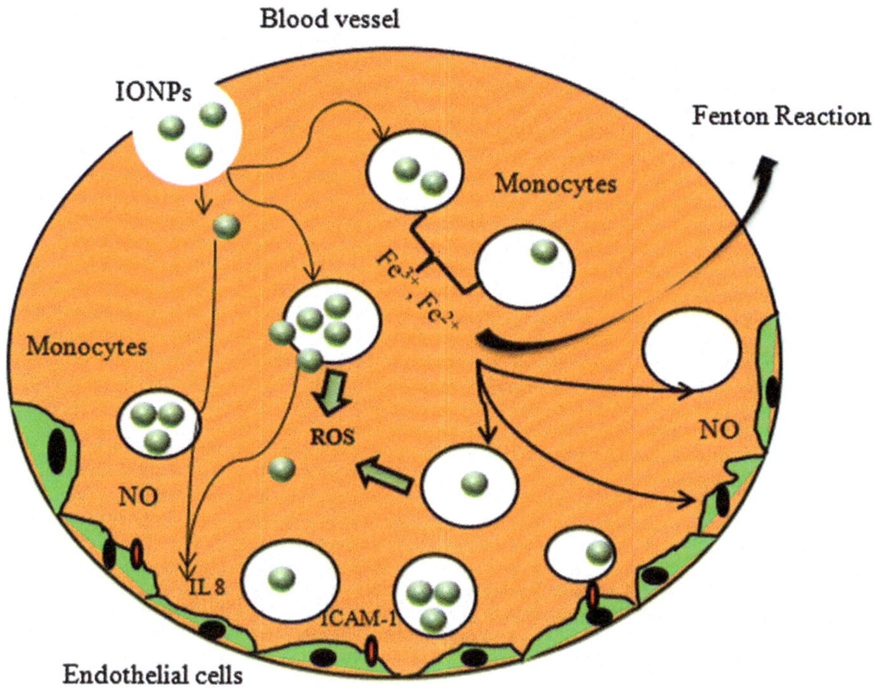

Fig. 7.4 The internalization of IONPs into the endothelial cells and interaction generating various responses

viability. These effects collectively may lead to endothelial inflammation and dysfunction leading to early atherosclerosis [176].

ZnO nanoparticles induced cytotoxicity was compared with Zn^{2+} ions and CeO_2 nanoparticles in HUVEC cells and it was found that the dissolved Zn^{2+} ions exerted toxicity through endoplasmic reticulum (ER) stress. At non toxic concentrations ZnO nanoparticles could induce significant ER stress response showing elevated levels of spliced xbp-1, chop, and caspase-12 at the mRNA level, and associated ER marker proteins including BiP, Chop, GADD34, p-PERK, p-eIF2R, and cleaved Caspase-12 at the protein levels [177]. Expression of these genes and proteins can effectively turn on the apoptotic mechanism leading to apoptotic cell death. The ZnO nanoparticle and CuO NP based toxicity towards eukaryotic cells was discussed through three mechanisms based on oxidative stress, coordination effects, and nonhomeostasis.

Chang et al., illustrated that the NPs can cross the small pores of the cell membrane due to their small size or can enter through the ion channels and transporter proteins present in the plasma membrane or can enter through endocytosis [21]. The mechanism of toxicity is explained earlier in Fig. 7.3. There is a direct interaction of the nanoparticles with mitochondria and the redox active proteins then stimulate ROS production in cells. Cu^{2+} ions that are exerted by the nanoparticles can produce ROS by different chemical reactions and ROS can further induce DNA strand breaks, affect gene expression. Cu^{2+} ions can form chelates after interacting with biomolecules or can dislodge metal ions in some metalloproteins which can inactivate such proteins. Moreover, the Cu^{2+} that are released from CuO NPs can caused a local enhanced concentration of this metal; and can disrupt the cation metal homeostasis in cells leading to cellular toxicity [178].

7.2 Conclusion

Engineered nanoparticles are rapidly capturing the market of nanotechnology for designing nano sensors, biosensors, solar cells, catalyst, paints, cosmetics, diagnostic kits, drug loading and targeted drug delivery and so on. The health issues concerned with the profound use of these nanostructures are also explored very well, but the studies could not end up with any particular conclusion. There are certain factors that can control the toxicity exerted by metal oxide nanoparticles like, size, mode of administration, aggregation state, surface functionalization, oxidation state and also the in vivo environment where it is accumulated.

Acknowledgement I am grateful to Chettinad Academy of Research and Education (CARE), Kelambakkam for providing me the infrastructure. My sincere thanks to Mr. Sanjay K Metkar and Ms. Ramalakshmi M, CARE for their immense help and support.

References

1. Hunter P (2008) A toxic brew we cannot live without. EMBO Rep 9(1):15–18
2. Cope WG (2004) Exposure classes, toxicants in air, water, soil, domestic and occupational settings. In: A textbook of modern toxicology, p 33
3. Gleiter H (1995) Nanostructured materials: state of the art and perspectives. Nanostruc Mater 6(1–4):3–14
4. Valden M, Lai X, Goodman DW (1998) Onset of catalytic activity of gold clusters on titania with the appearance of nonmetallic properties. Science 281(5383):1647–1650
5. Rodriguez JA, Liu G, Jirsak T et al (2002) Activation of gold on titania: adsorption and reaction of SO_2 on Au/TiO_2 (110). J Am Chem Soc 124:5242–5247
6. Bäumer M, Freund H-J (1999) Metal deposits on well-ordered oxide films. Prog Surf Sci 61(7):127–198
7. Trudeau ML, Ying JY (1996) Nanocrystalline materials in catalysis and electrocatalysis: structure tailoring and surface reactivity. Nanostruct Mater 7(1):245–258
8. Tachikawa S, Noguchi A, Tsuge T et al (2011) Structures and optical properties of ZnO nanoparticles capped with polyethylene glycol. Dent Mater 4:1132–1143
9. Amsaveni G, Farook AS, Haribabu V et al (2013) Engineered multifunctional nanoparticles for DLA cancer cells targeting, sorting, MR imaging and drug delivery. Adv Sci Eng Med 5(12):1340–1348
10. Thurn KT, Brown EMB, Wu A et al (2007) Nanoparticles for applications in cellular imaging. Nanoscale Res Lett 2(9):430–441
11. Brannon-Peppas L, Blanchette JO (2012) Nanoparticle and targeted systems for cancer therapy. Adv Drug Deliv Rev 64:206–212
12. Kühnel D, Marquardt C, Nau K et al (2014) Environmental impacts of nanomaterials: providing comprehensive information on exposure, transport and ecotoxicity-the project DaNa2. 0. Environ Sci Eur 26(1):21
13. Noguera C (1996) Physics and chemistry at oxide surfaces. Cambridge University Press, New York
14. Kung HH (1989) Transition metal oxides: surface chemistry and catalysis, vol 45. Elsevier, Amsterdam
15. Henrich VE, Cox PA (1996) The surface science of metal oxides. Cambridge University Press, Cambridge
16. Wells AF (1987) Structural inorganic chemistry, 6th edn. Oxford University Press, New York
17. Rodríguez JA, Fernández-García M (eds) (2007) Synthesis, properties and applications of oxide nanoparticles. Nanomater: inorg & bioinorg perspectives, chapter # 14. Wiley, New Jersey
18. Fernández-García M, Wang X, Belver C et al (2007) Anatase-TiO_2 nanomaterials: morphological/size dependence of the crystallization and phase behavior phenomena. J Phys Chem C 111(2):674–682
19. Smijs TG, Pavel S (2011) Titanium dioxide and zinc oxide nanoparticles in sunscreens: focus on their safety and effectiveness. Nanotechnol Sci Appl 4(1):95–112
20. Fernández-Garcíaa M, Rodriguez JA (2009) Metal oxide nanoparticles. In: Encyclopedia of inorganic chemistry. Wiley Online Library, Hoboken
21. Chang Y-N, Zhang M, Zia L et al (2012) The toxic effects and mechanisms of CuO and ZnO nanoparticles. Dent Mater 5(12):2850–2871
22. Becheri A, Durr M, Nostro PL et al (2008) Synthesis and characterization of zinc oxide nanoparticles: application to textiles as UV-absorbers. J Nanopart Res 10(4):679–689
23. Meruvu H, Vengalapati M, Chippada SM et al (2011) Synthesis and characterization of zinc oxide nanoparticles and its antimicrobial activity against *bacillus subtilis* and *Escherichia coli*. Rasayan J Chem 4:217–222
24. Lee GH, Kawazoe T, Ohtsu M (2002) Difference in optical bandgap between zinc-blende and wurtzite ZnO structure formed on sapphire (0001) substrate. Solid State Commun 124(5):163–165
25. Popov AP, Zvyagin AV, Lademann J et al (2010) Designing inorganic light-protective skin nanotechnology products. J Biomed Nanotechnol 6(5):432–451
26. Gasparro FP, Mitchnick M, Nash JF (1998) A review of sunscreen safety and efficacy. Photochem Photobiol 68(3):243–256
27. Stamatakis P, Palmer BR, Salzman GC et al (1990) Optimum particle size of titanium dioxide and zinc oxide for attenuation of ultraviolet radiation. JCT J Coat Technol 62(789):95–98

28. Using TiO_2 and ZnO for balanced UV protection (2009) Jul. http://www.personalcaremagazine.com/Story.aspx?Story=5243
29. Perez-Lopez OW, Farias AC, Marcilio NR et al (2005) The catalytic behavior of zinc oxide prepared from various precursors and by different methods. Mater Res Bull 40(12):2089–2099
30. Fangli Y, Peng H, Chunlei Y et al (2003) Preparation and properties of zinc oxide nanoparticles coated with zinc aluminate. J Mater Chem 13(3):634–637
31. Sahu D, Kannan GM, Vijayaraghavan R (2014) Size-dependent effect of zinc oxide on toxicity and inflammatory potential of human monocytes. J Toxicol Env Health Part A 77(4):177–191
32. Lopes S, Ribeiro F, Wojnarowicz J et al (2014) Zinc oxide nanoparticles toxicity to Daphnia magna: size-dependent effects and dissolution. Environ Toxicol Chem 33(1):190–198
33. Ose, MAG (2013) Literature review on the safety of titanium dioxide and zinc oxide nanoparticles in sunscreens, pp 1–24 https://www.tga.gov.au/literature-review-safety-titanium-dioxide-and-zinc-oxide-nanoparticles-sunscreens
34. Tran DT, Salmon R (2011) Potential photocarcinogenic effects of nanoparticle sunscreens. Australas J Dermatol 52(1):1–6
35. Virkutyte J, Al-Abed SR, Dionysiou DD (2012) Depletion of the protective aluminum hydroxide coating in TiO_2-based sunscreens by swimming pool water ingredients. Chem Eng J 191:95–103
36. Butler MK, Prow TW, Guo YN et al (2012) High-pressure freezing/freeze substitution and transmission electron microscopy for characterization of metal oxide nanoparticles within sunscreens. Nanomedicine 7(4):541–551
37. Schilling K, Bradford B, Castelli D et al (2010) Human safety review of "nano" titanium dioxide and zinc oxide. Photochem Photobiol Sci 9(4):495–509
38. Wang SQ, Tooley IR (2011) Photoprotection in the era of nanotechnology. Seminars Cutaneous Med Surg 30(4):210–213
39. Yıldırım ÖA, Durucan C (2010) Synthesis of zinc oxide nanoparticles elaborated by microemulsion method. J Alloys Compd 506(2):944–949
40. Zhang YW, Tang M, Jin X (2003) Polymeric adsorption behavior of nanoparticulate yttria stabilized zirconia and the deposition of as-formed suspensions on dense α-Al_2O_3 substrates. Solid State Sci 5(3):435–440
41. Girigoswami K, Meenakshi V, Murugesan R et al (2015) Studies on polymer-coated zinc oxide nanoparticles: UV blocking efficacy and in vivo toxicity. Mater Sci Eng C 56:501–510
42. Kovrižnych JA, Sotnikova R, Zeljenkova D et al (2013) Acute toxicity of 31 different nanoparticles to zebrafish (Danio rerio) tested in adulthood and in early life stages–comparative study. Interdiscip Toxicol 6(2):67–73
43. Xiong D, Fang T, Yu L et al (2011) Effects of nano-scale TiO_2, ZnO and their bulk counterparts on zebrafish: acute toxicity, oxidative stress and oxidative damage. Sci Total Environ 409(8):1444–1452
44. Van Aerle R, Lange A, Moorhouse A et al (2013) Molecular mechanisms of toxicity of silver nanoparticles in zebrafish embryos. Environ Sci Technol 47(14):8005–8014
45. Bai W, Zhang Z, Tian W et al (2010) Toxicity of zinc oxide nanoparticles to zebrafish embryo: a physicochemical study of toxicity mechanism. J Nanopart Res 12:1645–1654
46. Chen TH, Ling CC, Meng PJ (2014) Zinc oxide nanoparticles alter hatching and larval locomotor activity in zebrafish (*Danio rerio*). J Hazard Mater 277:134–140
47. Manzo S, Miglietta ML, Rametta G et al (2013) Toxic effects of ZnO nanoparticles towards marine algae *Dunaliella tertiolecta*. Sci Total Environ 445–446:371–376
48. Mansouri E, Khorsandi L, Orazizadeh M et al (2015) Dose dependent hepatotoxicity effects of zinc oxide nanoparticles. Nanomed J 2(4):273–282
49. Liu J, Feng X, Wei L et al (2016) The toxicity of ion-shedding zinc oxide nanoparticles. Crit Rev Toxicol 46(4):348–384
50. Tanemura S, Miao L, Wunderlich W et al (2005) Fabrication and characterization of anatase/rutile–TiO_2 thin films by magnetron sputtering: a review. Sci Technol Adv Mater 6(1):11–17
51. Biola-Cleir M, Beal D, Caillat S et al (2017) Comparison of the DNA damage response in BEAS-2B and A549 cells exposed to titanium dioxide nanoparticles. Mutagenesis 32(1):161–172
52. Wu N, Hong F, Zhou Y et al (2017) Exacerbation of innate immune response in mouse primary cultured sertoli cells caused by nanoparticulate TiO_2 involves the TAM/TLR3 signal pathway. J Biomed Mater Res Part A 105(1):198–208
53. Strickland JD, Lefew WR, Crooks J et al (2016) In vitro screening of metal oxide nanoparticles for effects on neural function using cortical networks on microelectrode arrays. Nanotoxicology 10(5):619–628
54. Bessa MJ, Costa C, Reinosa J et al (2016) Moving into advanced nanomaterials. Toxicity of rutile TiO_2 nanoparticles immobilized in nanokaolin nanocomposites on HepG2 cell line. Toxicol Appl Pharmacol 316:114–122
55. Samiei M, Ghasemi N, Aghazadeh M et al (2017) Biocompatibility of mineral trioxide aggregate with TiO_2 nanoparticles on human gingival fibroblasts. J Clin Exp Dent 9(2):e182–e185
56. Marucco A, Gazzano E, Ghigo D et al (2016) Fibrinogen enhances the inflammatory response of alveolar macrophages to TiO_2, SiO_2 and carbon nanomaterials. J Nanotoxicol 10(1):1–9
57. Wright C, Iyer AK, Wang L et al (2017) Effects of titanium dioxide nanoparticles on human keratinocytes. J Drug Chem Toxicol 40:90–100
58. Horie M, Sugino S, Kato H et al (2016) Does photocatalytic activity of TiO_2 nanoparticles correspond to

photo-cytotoxicity? Cellular uptake of TiO_2 nanoparticles is important in their photo-cytotoxicity. J Toxicol Mech Methods 26(4):284–294
59. Nakayama M, Sasaki R, Ogino C et al (2016) Titanium peroxide nanoparticles enhanced cytotoxic effects of X-ray irradiation against pancreatic cancer model through reactive oxygen species generation in vitro and in vivo. Radiat Oncol 11:91. https://doi.org/10.1186/s13014-016-0666-y
60. Alinovi R, Goldoni M, Pinelli S et al (2015) Oxidative and pro-inflammatory effects of cobalt and titanium oxide nanoparticles on aortic and venous endothelial cells. Toxicol In Vitro 29(3):426–437
61. George S, Gardner H, Seng EK et al (2014) Differential effect of solar light in increasing the toxicity of silver and titanium dioxide nanoparticles to a fish cell line and zebrafish embryos. Environ Sci Technol 48(11):6374–6382
62. Ramsden CS, Smith TJ, Shaw BJ et al (2009) Dietary exposure to titanium dioxide nanoparticles in rainbow trout, (*Oncorhynchus mykiss*): no effect on growth, but subtle biochemical disturbances in the brain. Ecotoxicology 18(7):939–951
63. Mansouri B, Maleki A, Davari B et al (2016) Histopathological effects following short-term coexposure of *Cyprinus carpio* to nanoparticles of TiO_2 and CuO. Environ Monit Assess 188(10):575
64. Federici G, Shaw BJ, Handy RD (2007) Toxicity of titanium dioxide nanoparticles to rainbow trout (*Oncorhynchus mykiss*): gill injury, oxidative stress, and other physiological effects. Aquat Toxicol 84(4):415–430
65. Ramsden CS, Henry TB, Handy RD (2013) Sublethal effects of titanium dioxide nanoparticles on the physiology and reproduction of zebrafish. Aquat Toxicol 126:404–413
66. Dalai S, Pakrashi S, Bhuvaneshwari M et al (2014) Toxic effect of Cr (VI) in presence of n-TiO_2 and n-Al_2O_3 particles towards freshwater microalgae. Aquat Toxicol 146:28–37
67. Bermejo-Nogales A, Connolly M, Rosenkranz P et al (2017) Negligible toxicity induced by different titanium dioxide nanoparticles in fish cell lines. Ecotoxicol Environ Saf 138:309–319
68. Minetto D, Libralato G, Marcomini A et al (2017) Potential effects of TiO_2 nanoparticles and $TiCl_4$ in saltwater to *Phaeodactylum tricornutum* and *Artemia franciscana*. Sci Total Environ 579:1379–1386
69. Wang Q, Chen Q, Zhou P et al (2014) Bioconcentration and metabolism of BDE-209 in the presence of titanium dioxide nanoparticles and impact on the thyroid endocrine system and neuronal development in zebrafish larvae. Nanotoxicology 8(l):196–207
70. D'Agata A, Fasulo S, Dallas LJ et al (2014) Enhanced toxicity of "bulk" titanium dioxide compared to "fresh" and "aged" nano-TiO_2 in marine mussels (*Mytilus galloprovincialis*). Nanotoxicology 8(5):549–558
71. Zhu X, Zhu L, Duan Z et al (2008) Comparative toxicity of several metal oxide nanoparticle aqueous suspensions to Zebrafish (*Danio rerio*) early developmental stage. J Environ Sci Health A Tox Hazard Subs Environ Eng 43(3):278–284
72. Fang Q, Shi X, Zhang L et al (2015) Effect of titanium dioxide nanoparticles on the bioavailability, metabolism, and toxicity of pentachlorophenol in zebrafish larvae. J Hazard Mater 283:897–904
73. Park H-G, Kim JI, Kang M et al (2014) The effect of metal-doped TiO_2 nanoparticles on zebrafish embryogenesis. Mol Cell Toxicol 10(3):293–301
74. Yan J, Lin B, Hu C et al (2014) The combined toxicological effects of titanium dioxide nanoparticles and bisphenol A on zebrafish embryos. Nanoscale Res Lett 9(1):406
75. Sha B, Gao W, Wang S et al (2014) Oxidative stress increased hepatotoxicity induced by nano-titanium dioxide in BRL-3A cells and Sprague-Dawley rats. J Appl Toxicol 34(4):345–356
76. Li Y, Yan J, Ding W et al (2017) Genotoxicity and gene expression analyses of liver and lung tissues of mice treated with titanium dioxide nanoparticles. Mutagenesis 32(1):33–46
77. Yang J, Luo M, Tan Z et al (2017) Oral administration of nano-titanium dioxide particle disrupts hepatic metabolic functions in a mouse model. Environ Toxicol Pharmacol 49:112–118
78. Osmond-McLeod MJ, Oytam Y, Rowe A et al (2016) Long-term exposure to commercially available sunscreens containing nanoparticles of TiO_2 and ZnO revealed no biological impact in a hairless mouse model. Part Fibre Toxicol 13(1):44
79. Abu-Zeid EH, Alam RTM, Abd El-Hameed NE (2017) Impact of titanium dioxide on androgen receptors, seminal vesicles and thyroid hormones of male rats: possible protective trial with aged garlic extract. Andrologia 49:e12651. https://doi.org/10.1111/and.12651
80. Hassanein KM, El-Amir YO (2016) Protective effects of thymoquinone and avenanthramides on titanium dioxide nanoparticles induced toxicity in Sprague-Dawley rats. Pathol Res Pract. https://doi.org/10.1016/j.prp.2016.08.002
81. Rosen JE, Chan L, Shieh D-B et al (2012) Iron oxide nanoparticles for targeted cancer imaging and diagnostics. Nanomedicine 8(3):275–290
82. Siddiqi KS, Ur Rahman A, Tajuddin et al (2016) Biogenic fabrication of iron/iron oxide nanoparticles and their application. Nanoscale Res Lett 11(1):498
83. Lei L, Ling-Ling J, Yun Z et al (2013) Toxicity of superparamagnetic iron oxide nanoparticles: research strategies and implications for nanomedicine. Chin Phys B 22(12):127503
84. Mahmoudi M, Hofmann H, Rothen-Rutishauser B et al (2011) Assessing the in vitro and in vivo toxicity of superparamagnetic iron oxide nanoparticles. Chem Rev 112(4):2323–2338
85. Singh N, Jenkins GJ, Asadi R et al (2010) Potential toxicity of superparamagnetic iron oxide nanopar-

ticles (SPION). Nanotechnol Rev. https://doi.org/10.3402/nano.v1i0.5358
86. Dissanayake NM, Current KM, Obare SO (2015) Mutagenic effects of iron oxide nanoparticles on biological cells. Int J Mol Sci 16(10):23482–23516
87. Arami H, Khandhar A, Liggitt D et al (2015) In vivo delivery, pharmacokinetics, biodistribution and toxicity of iron oxide nanoparticles. Chem Soc Rev 44(23):8576–8607
88. Valdiglesias V, Kilic G, Costa C et al (2015) Effects of iron oxide nanoparticles: cytotoxicity, genotoxicity, developmental toxicity, and neurotoxicity. Environ Mol Mutagen 56(2):125–148
89. Singh N, Jenkins GJ, Nelson BC et al (2012) The role of iron redox state in the genotoxicity of ultrafine superparamagnetic iron oxide nanoparticles. Biomaterials 33:163–170
90. Petters C (2015) Uptake and metabolism of iron oxide nanoparticles in cultured brain cells. Dissertation, doi:elib.suub.uni-bremen.de/edocs/00104261-1.pdf
91. Rostami AA, Mohseni Kouchesfahani H, Kiani S et al (2015) Iron oxide nanoparticles reduced retinoic acid induced-neuronal differentiation of mouse embryonic stem cells by ROS generation. Arch Iran Med 18(9):586–590
92. Costa C, Brandão S, Bessa MJ et al (2016) In vitro cytotoxicity of superparamagnetic iron oxide nanoparticles on neuronal and glial cells. Evaluation of nanoparticle interference with viability tests. J Appl Toxicol 36(3):361–372
93. Joris F, Valdeperez D, Pelaz B et al (2016) The impact of species and cell type on the nanosafety profile of iron oxide nanoparticles in neural cells. J Nanobiotechnology 14(1):69
94. Coccini T, Caloni F, Ramirez Cando LJ et al (2016) Cytotoxicity and proliferative capacity impairment induced on human brain cell cultures after short- and long-term exposure to magnetite nanoparticles. J Appl Toxicol 37(3):361–373
95. Alarifi S, Ali D, Alkahtani S et al (2014) Iron oxide nanoparticles induce oxidative stress, DNA damage, and caspase activation in the human breast cancer cell line. Biol Trace Elem Res 159(1–3):416–424
96. Calero M, Chiappi M, Lazaro-Carrillo A et al (2015) Characterization of interaction of magnetic nanoparticles with breast cancer cells. J Nanobiotechnology 13(1):16
97. Namvar F, Rahman HS, Mohamad R et al (2014) Cytotoxic effect of magnetic iron oxide nanoparticles synthesized via seaweed aqueous extract. Int J Nanomedicine 9:2479–2488
98. Hanot CC, Choi YS, Anani TB et al (2015) Effects of iron-oxide nanoparticle surface chemistry on uptake kinetics and cytotoxicity in CHO-K1 cells. Int J Mol Sci 17(1):54
99. Sadeghi L, Tanwir F, Babadi VY (2015) In vitro toxicity of iron oxide nanoparticle: oxidative damages on Hep G2 cells. Exp Toxicol Pathol 67(2):197–203
100. Sonmez E, Aydin E, Turkez H et al (2016) Cytotoxicity and genotoxicity of iron oxide nanoparticles: an in vitro biosafety study. Arch Biol Sci 68(1):41–50
101. Rojas JM, Sanz-Ortega L, Mulens-Arias V et al (2015) Superparamagnetic iron oxide nanoparticle uptake alters M2 macrophage phenotype, iron metabolism, migration and invasion. Nanomedicine 12(4):1127–1138
102. Lee JH, Ju JE, Kim BI et al (2014) Rod-shaped iron oxide nanoparticles are more toxic than sphere-shaped nanoparticles to murine macrophage cells. Environ Toxicol Chem 33(12):2759–2766
103. Luo C, Li Y, Yang L et al (2014) Superparamagnetic iron oxide nanoparticles exacerbate the risks of reactive oxygen species-mediated external stresses. Arch Toxicol 89(3):357–369
104. Ahamed M, Alhadlaq HA, Alam J et al (2013) Iron oxide nanoparticle-induced oxidative stress and genotoxicity in human skin epithelial and lung epithelial cell lines. Curr Pharm Des 19(37):6681–6690
105. Parsa H, Shamsasenjan K, Movassaghpour A et al (2015) Effect of superparamagnetic iron oxide nanoparticles-labeling on mouse embryonic stem cells. Cell J 17(2):221–230
106. Giannaccini M, Pedicini L, De Matienzo G et al (2017) Magnetic nanoparticles: a strategy to target the choroidal layer in the posterior segment of the eye. Sci Rep 7:43092
107. Dave PN, Chopda LV (2014) Application of iron oxide nanomaterials for the removal of heavy metals. J Nanotech Article ID 398569, https://doi.org/10.1155/2014/398569
108. Zhu X, Tian S, Cai Z (2012) Toxicity assessment of iron oxide nanoparticles in zebrafish (*Danio rerio*) early life stages. PLoS One 7(9):e46286
109. Felix LC, Ortega VA, Ede JD et al (2013) Physicochemical characteristics of polymer-coated metal-oxide nanoparticles and their toxicological effects on zebrafish (*Danio rerio*) development. Environ Sci Technol 47(12):6589–6596
110. Remya AS, Ramesh M, Saravanan M et al (2015) Iron oxide nanoparticles to an Indian major carp, *Labeo rohita*: impacts on hematology, iono regulation and gill Na^+/K^+ ATPase activity. J King Saud Univ Sci 27:151–160
111. Saravanan M, Suganya R, Ramesh M et al (2015) Iron oxide nanoparticles induced alterations in haematological, biochemical and ionoregulatory responses of an Indian major carp *Labeo rohita*. J Nanopart Res 17:274
112. Keerthika V, Ramesh R, Rajan MR (2016) Impact of iron oxide nanoparticles on behavioural changes in fresh water fish *Labeo Rohita*. Paripex Ind J Res 5(8):158–160
113. Taze C, Panetas I, Kalogiannis S et al (2015) Toxicity assessment and comparison between two types of iron oxide nanoparticles in *Mytilus galloprovincialis*. Aquat Toxicol 172:9–20

114. Reddy UA, Prabhakar PV, Mahboob M (2015) Biomarkers of oxidative stress for in vivo assessment of toxicological effects of iron oxide nanoparticles. Saudi J Biol Sci (in press) https://doi.org/10.1016/j.sjbs.2015.09.029
115. Sundarraj K, Raghunath A, Paneerselvam L et al (2017) Iron oxide nanoparticles modulate heat shock proteins and organ specific markers expression in mice male accessory organs. Toxicol Appl Pharmacol 317:12–24
116. Babadi VY, Najafi L, Najafi A et al (2012) Evaluation of iron oxide nanoparticles effects on tissue and enzymes of liver in rats. J Pharm Biomed Sci 23(23):1–4
117. Kumari M, Rajak S, Singh SP et al (2013) Biochemical alterations induced by acute oral doses of iron oxide nanoparticles in Wistar rats. Drug Chem Toxicol 36(3):296–305
118. Di Bona KR, Xu Y, Gray M et al (2015) Short-and long-term effects of prenatal exposure to iron oxide nanoparticles: influence of surface charge and dose on developmental and reproductive toxicity. Int J Mol Sci 16(12):30251–30268
119. Parivar K, Malekvand Fard F, Bayat M et al (2016) Evaluation of iron oxide nanoparticles toxicity on liver cells of BALB/c rats. Iran Red Crescent Med J 18(1):e28939
120. Sundarraj K, Manickam V, Raghunath A et al (2017) Repeated exposure to iron oxide nanoparticles causes testicular toxicity in mice. Environ Toxicol 32(2):594–608
121. Gaharwar US, Paulraj R (2015) Iron oxide nanoparticles induced oxidative damage in peripheral blood cells of rat. J Biomed Sci Eng 8(4):274–286
122. Kumari M, Rajak S, Singh SP et al (2012) Repeated oral dose toxicity of iron oxide nanoparticles: biochemical and histopathological alterations in different tissues of rats. J Nanosci Nanotechnol 12(3):2149–2159
123. Szalay B, Tátrai E, Nýirő G et al (2012) Potential toxic effects of iron oxide nanoparticles in in vivo and in vitro experiments. J Appl Toxicol 32(6):446–453
124. Vermeij EA, Koenders MI, Bennink MB et al (2015) The in-vivo use of superparamagnetic iron oxide nanoparticles to detect inflammation elicits a cytokine response but does not aggravate experimental arthritis. PLoS One 10(5):e0126687
125. Gawande MB, Goswami A, Felpin F-X et al (2016) Cu and Cu-based nanoparticles: synthesis and applications in catalysis. Chem Rev 116(6):3722–3811
126. Hou J, Wang X, Hayat T et al (2017) Ecotoxicological effects and mechanism of CuO nanoparticles to individual organisms. Environ Pollut 221:209–217
127. Fu X (2015) Oxidative stress induced by CuO nanoparticles (CuO NPs) to human hepatocarcinoma (HepG2) cells. J Cancer Ther 6:889–895
128. Jing X, Park JH, Peters TM et al (2015) Toxicity of copper oxide nanoparticles in lung epithelial cells exposed at the air-liquid interface compared with in vivo assessment. Toxicol In Vitro 29(3):502–511
129. Karlsson HL, Gustafsson J, Cronholm P et al (2009) Size-dependent toxicity of metal oxide particles—a comparison between nano-and micrometer size. Toxicol Lett 188(2):112–118
130. Boyles MSP, Ranninger C, Reischl R et al (2016) Copper oxide nanoparticle toxicity profiling using untargeted metabolomics. Part Fibre Toxicol 13(1):49
131. Ryu J, Girigoswami K, Ha C et al (2008) Influence of multiple metal ions on β-amyloid aggregation and dissociation on a solid surface. Biochemistry 47(19):5328–5335
132. Joshi A, Rastedt W, Faber K et al (2016) Uptake and toxicity of copper oxide nanoparticles in C6 Glioma cells. Neurochem Res 41(11):3004–3019
133. Bulcke F, Dringen R (2015) Copper oxide nanoparticles stimulate glycolytic flux and increase the cellular contents of glutathione and metallothioneins in cultured astrocytes. Neurochem Res 40(1):15–26
134. Manusadzianas L, Caillet C, Fachetti L et al (2012) Toxicity of copper oxide nanoparticle suspensions to aquatic biota. Environ Toxicol Chem 31(1):108–114
135. Pradhan A, Seena S, Pascoal C et al (2012) Copper oxide nanoparticles can induce toxicity to the freshwater shredder Allogamus ligonifer. Chemosphere 89(9):1142–1150
136. Blinova I, Ivask A, Heinlaan M et al (2010) Ecotoxicity of nanoparticles of CuO and ZnO in natural water. Environ Pollut 158(1):41–47
137. Aruoja V, Dubourguier H-C, Kasemets K et al (2009) Toxicity of nanoparticles of CuO, ZnO and TiO_2 to microalgae *Pseudokirchneriella subcapitata*. Sci Total Environ 407(4):1461–1468
138. Thit A, Selck H, Bjerregaard HF (2013) Toxicity of CuO nanoparticles and Cu ions to tight epithelial cells from Xenopus laevis (A6): effects on proliferation, cell cycle progression and cell death. Toxicol In Vitro 27(5):1596–1601
139. Jahanbakhshi A, Hedayati A, Pirbeigi A et al (2015) Determination of acute toxicity and the effects of sub-acute concentrations of CuO nanoparticles on blood parameters in *Rutilus rutilus*. Nanomedicine J 2(3):195–202
140. Mashock MJ, Zanon T, Kappell AD et al (2016) Copper oxide nanoparticles impact several toxicological endpoints and cause neurodegeneration in *Caenorhabditis elegans*. PLoS One 11(12):e0167613. doi:https://doi.org/10.1371/journal.pone.0167613
141. Mohammadyari A, Razavipour ST, Mohammadbeigi M et al (2014) Explore in-vivo toxicity assessment of copper oxide nanoparticle in Wistar rats. J Biol Today's World 3:124–128
142. Xu C, Qu X (2014) Cerium oxide nanoparticle: a remarkably versatile rare earth nanomaterial for biological applications. NPG Asia Mater 6(3). https://doi.org/10.1038/am.2013.88

143. Park E-J, Choi J, Park Y-K et al (2008) Oxidative stress induced by cerium oxide nanoparticles in cultured BEAS-2B cells. Toxicology 245(1–2):90–100
144. Celardo I, De Nicola M, Mandoli C et al (2011) Ce^{3+} ions determine redox-dependent anti-apoptotic effect of cerium oxide nanoparticles. ACS Nano 5(6):4537–4549
145. Das M, Patil S, Bhargava N et al (2007) Auto-catalytic ceria nanoparticles offer neuroprotection to adult rat spinal cord neurons. Biomaterials 28(10):1918–1925
146. Korsvik C, Patil S, Seal S et al (2007) Superoxide dismutase mimetic properties exhibited by vacancy engineered ceria nanoparticles. Chem Commun 10:1056–1058
147. Dunnick KM, Pillai R, Pisane KL et al (2015) The effect of cerium oxide nanoparticle valence state on reactive oxygen species and toxicity. Biol Trace Elem Res 166(1):96–107
148. Wason MS, Colon J, Das S et al (2013) Sensitization of pancreatic cancer cells to radiation by cerium oxide nanoparticle-induced ROS production. Nanomedicine 4:558–569
149. Ali D, Alarifi S, Alkahtani S, AlKahtane AA et al (2015) Cerium oxide nanoparticles induce oxidative stress and genotoxicity in human skin melanoma cells. Cell Biochem Biophys 71(3):1643–1651
150. Colon J, Hsieh N, Ferguson A et al (2010) Cerium oxide nanoparticles protect gastrointestinal epithelium from radiation-induced damage by reduction of reactive oxygen species and upregulation of superoxide dismutase 2. Nanomedicine 6(5):698–705
151. Tarnuzzer RW, Colon J, Patil S et al (2005) Vacancy engineered ceria nanostructures for protection from radiation-induced cellular damage. Nano Lett 5(12):2573–2577
152. Vinardell MP, Mitjans M (2015) Antitumor activities of metal oxide nanoparticles. Nanomaterials 5(2):1004–1021
153. Reed K, Cormack A, Kulkarni A et al (2014) Exploring the properties and applications of nanoceria: is there still plenty of room at the bottom? Environ Sci Nano 1(5):390–405
154. Ivanov VK, Shcherbakov AB, Usatenko AV (2009) Structure-sensitive properties and biomedical applications of nanodispersed cerium dioxide. Russ Chem Rev 78(9):855
155. Nelson BC, Johnson ME, Walker ML et al (2016) Antioxidant cerium oxide nanoparticles in biology and medicine. Antioxidants 5(2):15
156. Cheng H, Liao Z-L, Ning L-H et al (2017) Alendronate-anchored PEGylation of ceria nanoparticles promotes human hepatoma cell proliferation via AKT/ERK signaling pathways. Cancer Med 6(2):374–381
157. Li D, Morishita M, James G et al (2016) In vivo biodistribution and physiologically based pharmacokinetic modeling of inhaled fresh and aged cerium oxide nanoparticles in rats. Part Fibre Toxicol 13(1):45
158. Cordelli E, Keller J, Eleuteri P et al (2017) No genotoxicity in rat blood cells upon 3- or 6-month inhalation exposure to CeO_2 or $BaSO_4$ nanomaterials. Mutagenesis 32(1):13–22
159. Sager TM, Wolfarth M, Leonard SS et al (2016) Role of engineered metal oxide nanoparticle agglomeration in reactive oxygen species generation and cathepsin B release in NLRP3 inflammasome activation and pulmonary toxicity. Inhal Toxicol 228(14):686–697
160. DeCoteau W, Heckman KL, Estevez AY et al (2016) Cerium oxide nanoparticles with antioxidant properties ameliorate strength and prolong life in mouse model of amyotrophic lateral sclerosis. Nanomedicine 12(8):2311–2320
161. Hegazy MAE, Maklad HM, Abd Elmonsif DA et al (2016) The possible role of cerium oxide (CeO2) nanoparticles in prevention of neurobehavioral and neurochemical changes in 6-hydroxydopamine-induced parkinsonian disease. Alexandria J Med. doi:https://doi.org/10.1016/j.ajme.2016.12.006
162. Bailey ZS, Nilson E, Bates JA et al (2016) Cerium oxide nanoparticles improve outcome after in vitro and in vivo mild traumatic brain injury. J Neurotrauma 33:1–11
163. Krishnan A, Sreeremya TS, Ghosh S (2016) Size-tunable hydrophilic cerium oxide nanoparticles as a 'turn-on' fluorescent sensor for the rapid detection of ultralow concentration vitamin C. RSC Adv 6:53550–53559
164. Lin W, Huang YW, Zhou XD et al (2006) Toxicity of cerium oxide nanoparticles in human lung cancer cells. Int J Toxicol 25(6):451–457
165. Song B, Zhou T, Yang W et al (2016) Contribution of oxidative stress to TiO_2 nanoparticle-induced toxicity. Environ Toxicol Pharmacol 48:130–140
166. Czajka M, Sawicki K, Sikorska K et al (2015) Toxicity of titanium dioxide nanoparticles in central nervous system. Toxicol In Vitro 29(5):1042–1052
167. Seabra AB, Durán N (2015) Nanotoxicology of metal oxide nanoparticles. Metals 5:934–975
168. Sarkar A, Ghosh M, Sil PC (2014) Nanotoxicity: oxidative stress mediated toxicity of metal and metal oxide nanoparticles. J Nanosci Nanotechnol 14:730–743
169. Ivask A, Titma T, Visnapuu M et al (2015) Toxicity of 11 metal oxide nanoparticles to three mammalian cell types in vitro. Curr Top Med Chem 15:1914–1929
170. Wehmas LC, Anders C, Chess J et al (2015) Comparative metal oxide nanoparticle toxicity using embryonic zebrafish. Toxicol Rep 2:702–715
171. Sabella S, Carney RP, Brunetti V et al (2014) A general mechanism for intracellular toxicity of metal-containing nanoparticles. Nanoscale 6:7052–7061
172. Huang G, Chen H, Dong Y et al (2013) Superparamagnetic iron oxide nanoparticles: amplifying ROS stress to improve anticancer drug efficacy. Theranostics 3(2):116–126

173. Li P, Nijhawan D, Budihardjo I et al (1997) Cytochrome c and dATP-dependent formation of Apaf-1/caspase-9 complex initiates an apoptotic protease cascade. Cell 91:479–489
174. Wyllie AH (1980) Glucocorticoid-induced thymocyte apoptosis is associated with endogenous endonuclease activation. Nature 284:555–556
175. Hengartner MO (2000) The biochemistry of apoptosis. Nature 407:770–776
176. Zhu M-T, Wang B, Wang Y et al (2011) Endothelial dysfunction and inflammation induced by iron oxide nanoparticle exposure: risk factors for early atherosclerosis. Toxicol Lett 203:162–171
177. Chen R, Huo L, Shi X et al (2014) Endoplasmic reticulum stress induced by zinc oxide nanoparticles is an earlier biomarker for nanotoxicological evaluation. ACS Nano 8(3):2562–2574
178. Ingle AP, Duran N, Rai M (2014) Bioactivity, mechanism of action, and cytotoxicity of copper-based nanoparticles: a review. Appl Microbiol Biotechnol 98:1001–1009

8 Relevance of Physicochemical Characterization of Nanomaterials for Understanding Nano-cellular Interactions

Henriqueta Louro

Abstract

The manufactured nanomaterials (NMs) have specific physicochemical properties that confer unique mechanical, optical, electrical and magnetic characteristics that are beneficial for biomedical and industrial applications. However, recent studies have suggested that such specific physicochemical properties of the NMs may define nano-bio interactions thereby determining their toxic potential.

One of the major concerns about NMs is the potential to induce cancer, suggested by some experimental studies, as seen for titanium dioxide nanomaterials or carbon nanotubes. To analyze in a short term the carcinogenic properties of a compound, genotoxicity assays in mammalian cell lines or animal models are frequently used. However, the investigation of the genotoxic properties of NMs has been inconclusive, up to date, since divergent results have been reported throughout the literature. While trying to understand how the NMs' characteristics may encompass increased toxicological effects that harbor uncertainties for public health, the use of correlation analysis highlights some physicochemical properties that influence the genotoxic potential of these NM.

In this chapter, it is hypothesized that the different genotoxicity observed in closely related NMs may be due to subtle differences in their physicochemical characteristics. The present work provides an overview of the studies exploring the correlation between physicochemical properties of nanomaterials and their genotoxic effects in human cells, with focus

H. Louro (✉)
Department of Human Genetics, National Institute of Health Dr. Ricardo Jorge (INSA), Lisbon, Portugal

Toxicogenomics and Human Health (ToxOmics), Nova Medical School/Faculdade de Ciências Médicas, Universidade Nova de Lisboa, Lisbon, Portugal
e-mail: henriqueta.louro@insa.min-saude.pt

Q. Saquib et al. (eds.), *Cellular and Molecular Toxicology of Nanoparticles*, Advances in Experimental Medicine and Biology 1048, https://doi.org/10.1007/978-3-319-72041-8_8

on the toxicity of two groups of NMs, titanium dioxide nanomaterials and multiwalled-carbon nanotubes. It is suggested that, for tackling NMs' uncertainties, the in-depth investigation of the nano-bio interactions must be foreseen, where in vitro research must be integrated with in vivo and biomonitoring approaches, to cope with the complex dynamic behaviour of nanoscale materials.

Keywords
Nanomaterials · Public health · Genotoxicity · Cytotoxicity · Titanium dioxide · Multiwalled carbon nanotubes · Physicochemical properties

8.1 Introduction

The manufactured nanomaterials (NMs) have specific physicochemical properties that confer unique mechanical, optical, electrical and magnetic characteristics that are beneficial for biomedical and industrial applications. However, the exponential developments of nanotechnologies contrast with the still insufficient risk assessment for human health and the environment, leading to uncertainties in terms of public health. The concerns about the potential risks of NMs for public health arise mainly from epidemiologic studies in humans exposed to nanomaterials generated as by-products from human activity and pollution.

One of the major concerns about NMs is the potential to induce cancer, suggested by some experimental studies, as seen for titanium dioxide nanomaterials (TiO_2) or carbon nanotubes (CNTs). Based on the increased incidence of cancer in the respiratory tract of rodents exposed to high doses of ultrafine (<100 nm) TiO_2 dust, these NMs were classified by IARC as possibly carcinogenic to humans, group 2B substances [1]. However, such finding was not supported by epidemiological studies [2]. To analyse in a short term the carcinogenic properties of a compound, genotoxicity assays in mammalian cell lines or animal models are frequently used. Nevertheless, the investigation of the genotoxic properties of NMs has been inconclusive, up to date, since many of the studies reported in vivo and in vitro describe conflicting results in several biological systems [1, 3]. While several studies in bacteria revealed no genotoxicity, positive results were obtained in eukaryotic cells and animals [4].

Regarding carbon nanotubes (CNTs), the IARC working group published recently a paper that included the assessment of the carcinogenicity of single-walled carbon nanotubes (SWCNTs) and multi-walled carbon nanotubes (MWCNTs) [5]. The main conclusions were that the lack of coherent evidence across the various distinct CNTs precluded generalization to other types of CNTs. Only one MWCNT, Mitsui MWCNT-7, was classified as possibly carcinogenic to humans, group 2B [5]. Other CNTs were categorized as not classifiable as to their carcinogenicity to humans, Group 3 [5]. This classification mirrors the contradictory results that have been reported in the scientific literature about the genotoxicity of CNTs, either showing induction of DNA damage, gene mutations, micronuclei, and chromosomal aberrations in different types of cells [6] or irrelevant effects [7] for several CNTs. Further research suggested that, besides Mitsui MWCNT-7, other types of MWCNTs may be carcinogenic as well, and that straight MWCNTs appear to have a greater potency to induce mesothelioma than tangled MWCNTs [8, 9].

The inconsistent results found in the literature may be related to variables inherent to the test systems and exposure conditions, including dispersion of the nanoparticles. The analysis of the extensive literature existent on NMs toxicity evidences the fact that NMs with the same chemistry can greatly differ by size, surface area, shape, stability, rigidness, coating or electrical charge and these characteristics affect nano-bio interactions, leading to different toxic potential [10]. In fact, recent studies have suggested that specific physicochemical properties of the NM may define their nano-bio interactions, thereby determining their toxic potential [11]. In this chapter, it is hypothesized that the different genotoxicity observed in closely related NMs may be due to subtle differences in their physicochemical characteristics. The present work provides an overview of the studies exploring the correlation between physicochemical properties of nanomaterials and their genotoxic effects in human cells, with focus on the toxicity of two groups of NMs, titanium dioxide nanomaterials and multiwalled-carbon nanotubes.

8.2 The Physicochemical Characterization of Nanomaterials

Recent studies have suggested that specific physicochemical properties of the NM may define their nano-bio interactions, thereby determining their toxic potential. The identification of such determinant of toxicity may enable grouping strategies valuable for risk assessment. In addition, this nanotoxicology paradigm implies that, once the "toxic" characteristic is identified, it will be possible to develop/synthetize new NMs devoid of toxicity, leading to a "safe-by-design" approach to nanotechnology. In this way, early identification of uncertainties and potential risks along the NM life-cycle (production, use and end-of-life process) will also enable safety to keep pace with innovation [12].

Some authors suggested that the intrinsic physicochemical properties may be viewed as the 'synthetic identity' of the nanomaterial [13]. This identity, however, is not simple to ascribe due to the fact that NMs, unlike their constituent chemicals, are defined by statistical parameters such as geometric mean particle diameter and geometric standard deviation [14]. In Europe, major efforts have been engaged by the Joint Research Centre in Ispra, Italy, to characterize representative benchmark NMs that could be used as reference. It is, however, visible that the JRC reports still describe lot, inter- and intra-vial differences in the parameters recorded, that reflect such variation associated with each type of NM (see for example [15, 16]). The intrinsic properties can change with method of production, preparation process, storage, etc., reflecting the dynamic behavior of NMs [17]. Furthermore, many physicochemical properties of NMs are inter-related and thus cannot be varied systematically in isolation from others, e.g. increasing surface charge may impact on hydrophobicity, or changing the shape of a NM may introduce defects or alter the atomic configuration of the surface [12].

Due to their reactive surface, agglomeration and aggregation is a natural behavior of NMs, both in liquids or in air, and determines the actual particle size that will interact with cells and organisms [17]. In addition, following the introduction of NMs into a biological system a "biological identity" is formed, in part, by the adsorption of surrounding biomolecules such as proteins or lipids, onto the surface of the nanoparticles [13, 18]. The formation of a "corona" upon the adsorption of different serum/plasma biomolecules onto NMs has long been recognized [17, 19], being the composition dependent on the portal of entry into the body and on the particular fluid that the NMs come across with (e.g., blood, lung fluid, gastro-intestinal fluid, etc.). Additional dynamic changes can affect the "corona" as the nanoparticle crosses from one biological compartment to another [13].

Having realized such impact, it has been recommended that toxicologists adequately characterize physicochemical properties of the nanoparticles evaluated for hazard testing. For this assignment, many tools are available and can effectively aid in characterizing NMs, including microscopy, spectroscopy, spectrometry, and light scattering devices (reviewed in [20]). Some are depicted in Table 8.1.

The primary characterization end-points include average size/size-distribution including agglomeration/aggregation state, shape/morphology, surface area and porosity, atomic structure, chemical composition, surface chemistry [21].

Table 8.1 Some of the tools available for the characterization of the nano-cellular interface

Property	Tool/methodology
Size, morphology, composition	Electron microscopy (TEM), Energy dispersive spectroscopy (EDS), Small angle X-ray scattering (SAXS)
Crystalline phase	X-ray diffraction (XRD)
Surface Area	Nitrogen adsorption using the Brunauer, Emmett and Teller (BET) method
Surface charge	Zetametry (zeta-potential)
Hydrodynamic size	Dynamic light scattering (DLS)
Deposition efficiency	Inductively coupled plasma- mass spectrometry (ICP-MS)
Internalization	ICP-MS, TEM
Cellular localization	Fluorescence microscopy, TEM
Intracellular agglomeration	TEM, hyperspectral imaging
NM- binding properties	Microscopy
Protein corona	MS, SDS-PAGE

Among the methodologies available, electron microscopy is the most widely used instrument for chemical and structural investigation of nanomaterials [21], allowing the study not only of the primary characteristics but also of their biological identity in cells or tissues. For the use in biological assays, TEM is often coupled with energy dispersive X-ray spectroscopy (EDS) analysis for determining elemental constituents [22]. However, it is a complex and time-consuming technique that requires expensive equipment and experienced technicians [23].

The characteristics and properties of each nanomaterial to be assessed depend strongly on chemical type and chemical-structural complexity of the nanomaterials and on the purpose of their use [21]. According to these authors, one may discriminate between the end-points required to identify the material versus the end-points required to characterize its dispersibility into a given matrix versus characterization required for chemicals registration and finally for full risk assessment, a view that meets the concept of several levels of identity for one nanomaterial above described.

A major drawback is that most biology laboratories do not have expertise in these methods, and therefore a multidisciplinary approach is mandatory in nanotoxicology. Nevertheless, user-friendly methodologies such as dynamic light scattering (DLS) can be easily implemented in each laboratory for an in situ, real time, evaluation of the NMs to be used in biological testing. In the next sections some examples are described, arising from the author's experience, that illustrate the importance of including physicochemical characterization for interpreting the toxicological studies of NMs.

8.3 Closely Related NMs Reveal Distinct Genotoxic Effects

In earlier studies, the in vitro genotoxicity of benchmark NMs has been addressed using cytokinesis-blocked micronucleus assay in human lymphocytes [11]. Two groups of NMs were used, namely, titanium dioxide NMs (TiO_2) and multiwalled carbon nanotubes (MWCNTs), obtained from the Joint Research Center (JRC, Ispra, Italy), except NRCWE-006 (Mitsui&Co., Ltd., Ibaraki, Japan) and NRCWE-007 (Cheap Tubes Inc., Brattleboro, VT, USA) that were provided as sub-samples by the National Research Centre for the Working and Environment (NRCWE) within Nanogenotox Joint Action. The NMs were prepared under Good Laboratory Practices (GLP), allowing their application as international benchmarks (https://ec.europa.eu/jrc/en/scientific-tool/jrc-nanomaterials-repository). Their physicochemical characteristics were previously determined and are summarized in Table 8.1 [11, 16]. The NM dispersion for the biological assays included the preparation of a 2.56 mg/ml stock dispersion of each NM by prewetting powder in 0.5 vol% ethanol (96%) followed by addition of sterile-filtered 0.05 wt% BSA-water and 16 min of probe sonication of the sample, cooled in an ice-water bath, according to a standardized protocol [24]. The cytokinesis-blocked micronucleus assay was used for the investigation of genotoxic effects of the NMs in human lymphocytes, mainly as recommended by

OECD [25] with the modifications suggested for NMs [26]. Such modifications include the exposure of cells to the NMs some hours (typically 6 h) before the addition of cytochalasin-B, since this reagent may impair NM uptake [27]. Considering the micronucleus assay results, the NMs were classified as having no effects or genotoxic effects or equivocal genotoxic effects, when some positive results were observed occasionally. In addition, for the present analysis, a ranking index of the genotoxicity effects was determined, based on the number of concentrations that showed significant increases in the micronucleus frequency. Such classification is depicted in Table 8.2.

It is evident from Table 8.2 that within groups of closely related NMs, displaying a similar core chemical composition, different genotoxic effects are observed, ranging from positive to negative results [11, 28]. This creates a difficulty regarding safety assessment, since it leads to the need of investigating the toxic potential of each NM individually, instead of assuming a common mechanism and equal genotoxic effects for a set of similar NMs [11]. In view of this reality, grouping strategies for risk assessment cannot be based simply in the chemical core of the NMs. The subsequent analysis tries to overcome this problem by trying to discriminate the feature more relevant for genotoxicity of the NMs.

Among TiO_2, which are widely used in a broad variety of final products, e.g. as a pigment in paints, varnishes and plastics, as an additive to food (colorant E171), or as UV-filter in cosmetic products [29], the rutile NMs presented the most genotoxic effects. On the contrary, in general the literature describing in vitro genotoxicity testing using human cell lines, shows the induction of micronuclei by anatase TiO_2 and negative results for rutiles [30, 31]. Nevertheless, recently it was reported that rutile, and not anatase, induces toxic effects in Balb/3T3 mouse fibroblasts, whereas the internalization is dependent on the particle size [32]. In addition, one study used NM-103 TiO_2 (hydrophobic, dimethicone-coated) and those authors reported a slight increase in micronucleus formation for at the lower doses tested (1 and 5 μg/ml) in Beas-2B cells [33]. No comparable results were reported specifically for NM-104 TiO_2 (hydrophilic, glycerine-coated). Furthermore, NM-105 TiO_2, which consisted of a mixture of rutile and anatase (15–85%), was not genotoxic under similar experimental conditions [33].

Using correlation analysis, here we try to understand further the characteristics that may underlie the genotoxicity of these TiO_2. For the correlation analysis, the micronucleus frequency obtained in each concentration, was plotted against each of the characteristics presented in Table 8.2, to check linear relations among the test variables. Three replicate values were used for each NM. Whenever a linear relation was found, Pearson correlation analysis was used to test if there was a significant correlation.

The results of this analysis show that for the lower concentrations of TiO_2 (5 and 15 μg/cm^2), there is a strong positive correlation of the micronucleus frequency with the agglomerate size (correlation coefficients: $r = 0.595$ and 0.683; $p = 0.054$ and 0.02, respectively; Pearson correlation test), meaning that NMs forming agglomerates of larger sizes yielded more elevated micronucleus frequencies. Such correlation was also observed at the highest concentrations (250 μg/cm^2) ($r = 0.573$, $p = 0.052$; Pearson correlation test) and is represented in Fig. 8.1.

Pooling together the micronucleus frequencies for all concentrations of TiO_2, the strongest correlation was also with the agglomerate size ($r = 0.479$, $p = 0.0002$; Fig. 8.2).

In fact, the agglomeration behaviour is expected under physiological conditions [17], and may play an important role in explaining the high levels of variation found in genotoxicity testing. Also the absence of dose-response curves upon NM exposure may be explained by the time- and concentration-dependent agglomeration and subsequent sedimentation of NMs [11]. Accordingly, sedimentation has been studied for this set of NMs. In NM-102, sedimentation occurs in combination of fining, suggesting that preferentially the coarsest particles drop out of suspension, while in NM-103 and NM-104 there is a strong component of total sedimentation [11]. This extensive sedimentation may be

Table 8.2 Main physicochemical characteristics of the NMs used for biological assessment and the outcome in micronucleus assay

	NM	Chrystal phase/ morphology [15, 16]	Specific surface area (m^2/g) [15, 16]	Average diameter (nm) [15, 16]	Average length (nm) [15, 16]	Aspect ratio ± SD [11]	Agglomerates mean size (nm) [15, 16]	Genotoxicity Ranking Index	Outcome
TiO_2	NM-102	Anatase	78	21	22	1.5 ± 1.3	442.6	1	(+)
	NM-103	Rutile (hydrophobic)	51	22	26	1.7 ± 1.3	113.8	2	+
	NM-104	Rutile (hydrophilic)	56	23	27	1.4 ± 1.3	128.6	2	+
	NM-105	Rutile-anatase	46	15	24	1.4 ± 1.2	124.5	0	–
MWCNT	NM-400	Highly bend	254	11	846	67.3 ± 1.8	NA	1	(+)
	NM-401	Straight-wall	140	67	4048	53.6 ± 2.0	NA	0	–
	NM-402	Highly bend	226	11	1372	107.1 ± 1.9	NA	1	(+)
	NM-403	Highly bend	189	12	443	35.6 ± 1.8	NA	5	+
	NRCWE-006[11]	Straight-wall	26	70	4424	63.7 ± 2.4	NA	2	+
	NRCWE-007[11]	Highly bend	233	15.3	369	24.1 ± 1.9	NA	0	–

Genotoxicity Ranking Index – number of concentrations with significant increases in micronucleus frequency ($p < 0,05$; 2-sided Fisher); Outcome: + genotoxic effects; (+) equivocal genotoxic effects; – no effects

NA not applicable

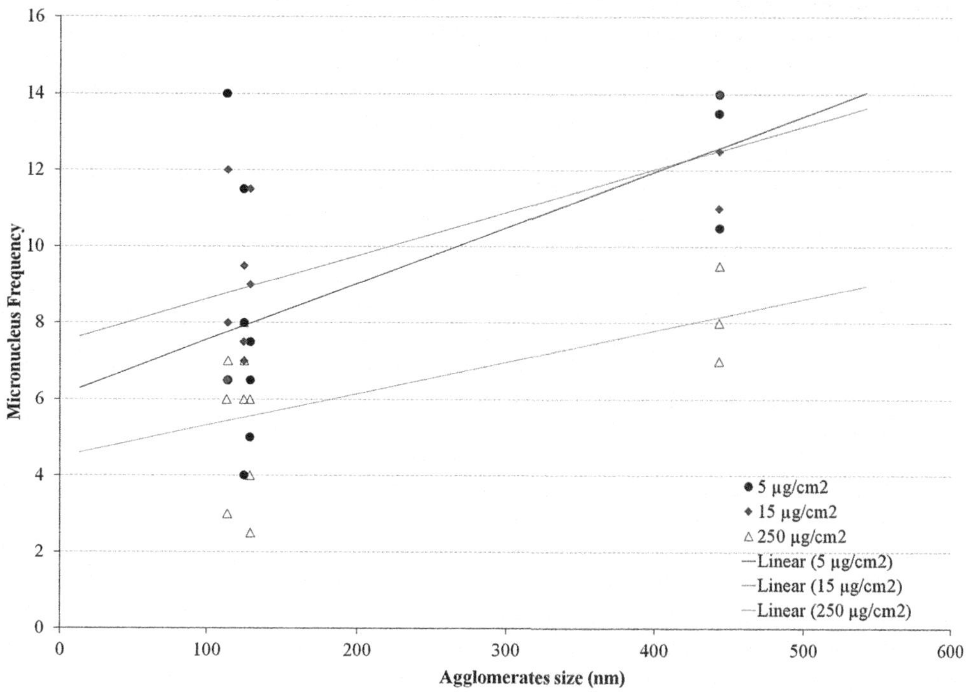

Fig. 8.1 Correlation of the micronucleus frequencies at the concentrations of 5, 15 and 250 μg/cm² TiO_2 and the agglomerates size

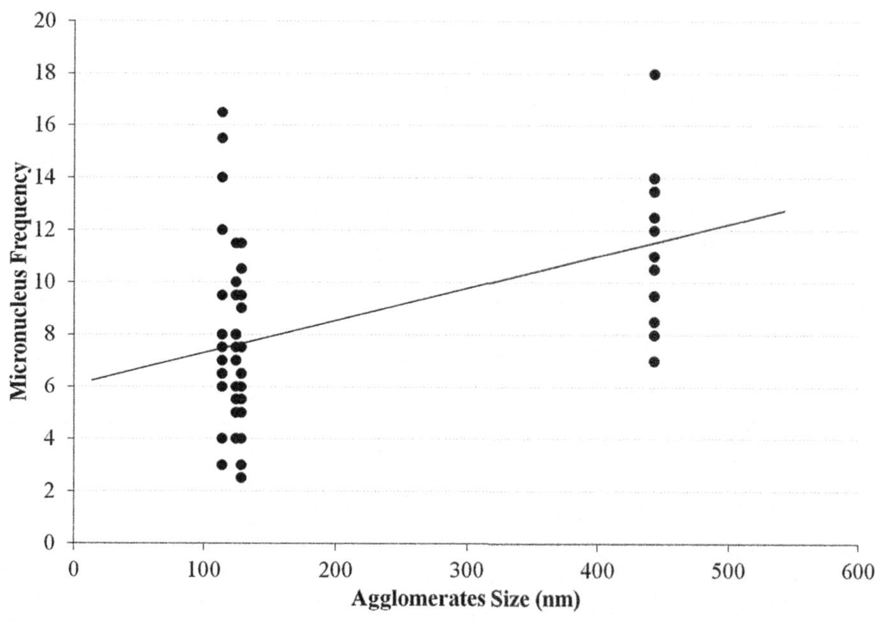

Fig. 8.2 Correlation of the micronucleus frequency at all concentrations of TiO_2 and the agglomerates mean size

responsible for increased contact/interaction of the NMs with the cells, namely particles and small agglomerates, leading to increased genotoxic effects in the case of NM-103 and NM-104. For NM-105, sedimentation studies showed a slow deposition and accumulation at the vial bottom resulting in concentration-induced agglomeration [11].

Considering one of the mechanisms usually proposed to explain the genotoxic effects of NMs [34], a direct action of the small NMs on the genome would require them to enter the cells, bypassing cellular barriers and reaching the nucleus where they could bind to DNA molecule or interfere with replication process. This mechanism would imply that smaller NMs would reach cell nucleus easily, thus increasing agglomerate size would decrease their genotoxic effect, on the contrary of our observations. However, if the genotoxic effects observed are originated from the indirect/secondary action of NMs in cellular metabolism, for example by the increased production of reactive oxygen species or mediating inflammation processes [34], the increasing size of agglomerates might favour increased biological effects, that may explain our findings. Accordingly, a previous work described that TiO_2 dispersions with large agglomerates induced DNA damage in all three cell lines, while the TiO_2 dispersed with agglomerates less than 200 nm had no effect on genotoxicity [35].

It cannot be excluded that larger agglomerates formed might be less stable and thus NMs can be released and reach cells as individual particles, as suggested in the literature using rutile/anatase mixture for toxicity testing in vitro [35].

Although the present study points to the agglomerates size as the major determinant of TiO_2 genotoxicity independently of the concentration, for the concentration of 15 μg/cm^2, a correlation was detected with specific surface area ($r = 0.737$, $p = 0.010$, Pearson correlation test; Fig. 8.3).

It is possible that, at lower concentrations when agglomeration/sedimentation is not yet massive, primary/intrinsic characteristics of the TiO_2 are of relevance. An increased SSA determines increased surface reactivity, possibly leading to more pronounced cellular effects.

In addition, at the highest concentration investigated, 250 μg/cm^2, assuming a high rate of sedimentation, this would mean a higher deposit of

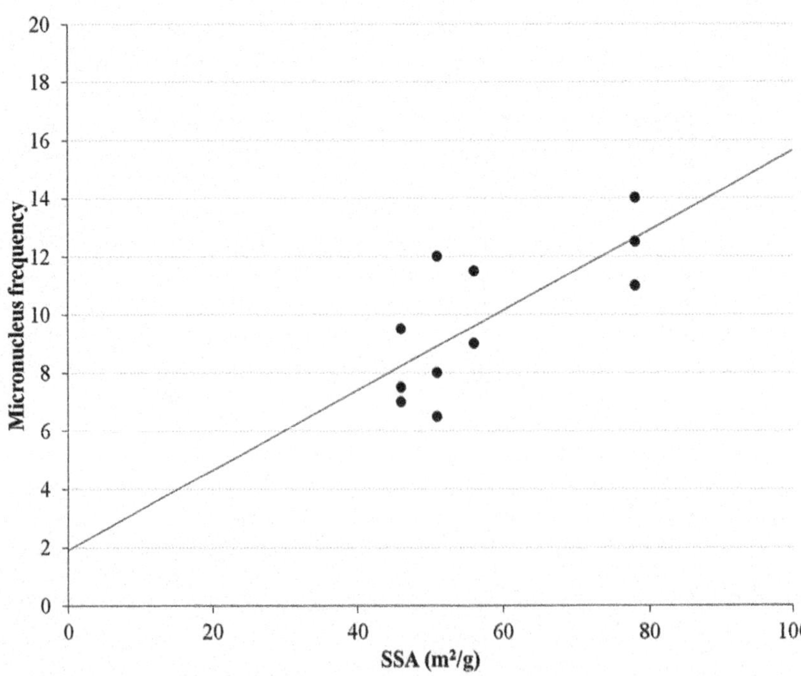

Fig. 8.3 Correlation of the micronucleus frequency at the concentration of 15 μg/cm^2 TiO_2 and the specific surface area (SSA)

NMs on cells, increasing nano-bio interactions. At this concentration, a negative strong correlation between the length of the NM and the micronucleus frequency was detected ($r = -0.758$, $p = 0.004$). However, it must be noted that the length of the NMs did not vary significantly, ranging from 15 to 23 nm, thus this correlation has limited value.

Although the present findings need to be confirmed in studies involving a larger panel of NMs and other cell media, it is proposed that the agglomerate size is a major predictor of TiO_2 genotoxicity in human lymphocytes. However, we cannot exclude the possibility that the major determinant is a combination of characteristics, such as SSA plus agglomerate size, which are interdependent, but in the future further studies should investigate this issue.

Within the panel of NMs hereby presented, the most genotoxic NM, based on the genotoxicity ranking index obtained in the micronucleus assay (Table 8.2), was NM-403, a MWCNT presenting a highly bend morphology, followed by NRCWE-006, which is a more rigid MWCNT. The latter is the same MWCNT previously shown to induce mesothelioma in mice and rats [36–39]. Amongst MWCNTs, the thin (diameter 12 nm) and short (443 nm) NM-403 was positive, but also the thick (diameter 70 nm) and long (4424 nm) NRCWE-006 had positive results. However, NRCWE-007 and NM-400, two other thin (15 and 11 nm, respectively) and short (369 and 846 nm, respectively) MWCNTs, as well as NM-402 (11 nm diameter and 1372 nm length) presented negative outcomes in the micronucleus assay (Table 8.2).

Focusing on MWCNTs, no linear relations were observed between the physicochemical characteristics and the micronucleus frequency and thus no correlation analysis could be envisaged. This fact may reflect the impact of cell culture medium in the NMs size, an issue that is addressed in the next section.

In addition, in previous studies, correlation analysis was used in respiratory cells, to test the association of the micronucleus frequency in A549 cells at the maximum (128 mg/cm^2) and intermediate (64 mg/cm^2) concentration of MWCNTs with the NMs physicochemical characteristics [28]. No correlation was found between micronucleus frequency and the surface area, thickness, length, aspect ratio, or with the size of NMs in the batch dispersions and the dilutions in cell culture medium [28]. A recent work analysed a panel of 15 MWCNTs that included NM-401, NM-402 and NM-402, and the tested CNTs were weakly genotoxic (using comet assay) after 24 h exposure to doses up to 200 mg/ml [40]. Even using powerful statistical methodologies such as principal component analysis, no association was identified between physicochemical parameters or the dispersion quality of stock dispersion and the observed biological effects [40].

In accordance, other authors have reported that the differences observed in genotoxicity among MWCNTs may not be simply explained by variation in length and diameter. Other structural differences, including surface activity or transition metals present as impurities, might be implicated [29, 41] and the influence of the environmental context may be determinant for the observed effects.

8.4 The Influence of Cell Environment in the Biological Effects of NMs

Recent work was dedicated to the investigation of the cytotoxic and genotoxic effects of four benchmark MWCNTs in relation to their physicochemical characteristics in human cells [28]. Considering that inhalation is the most probable route of human exposure to these NMs, two types of human respiratory cells were selected for in vitro complementary analysis: a bronchial epithelial cell line (Beas-2B) and a lung adenocarcinoma epithelial cell line (A549). The Beas-2B cell line is derived from normal human bronchial epithelium obtained from non-cancerous individuals and is able to generate and release mediators of inflammation [42] while, A549 cells mimics function of Type II pneumocytes, retaining the endocytic ability of the pulmonary epithelium and localization of cytochrome P450 systems [43]. Both cell lines were exposed to NM-400, NM-401, NM-402 and NM-403 and their cytotoxic and genotoxic effects were analysed concomitantly to the hydrodynamic sizes in cell culture medium for each cell type, using DLS [28].

The DLS analysis revealed that, for each MWCNT, different hydrodynamic sizes were observed after the dispersion of NMs in aqueous solutions (batch dispersions), using the standardized dispersion protocol [24], or in the dilution of the batch in each cell culture medium (Table 8.3).

While it is apparent that the dilution of the batch dispersion in the cell culture medium leads to increased agglomeration of the MWCNTs, there is major impact when using the Beas-2B cell line culture medium, as can be clearly seen in Fig. 8.4.

A shift to higher hydrodynamic size is seen upon the dilution of MWCNTs in cell culture medium, and, in the case of Beas-2B cell medium, this also results in a multimodal size-distribution. This effect is probably related to differences in the components of the solutions used. While batch dispersion is prepared in aqueous solution containing 0.5 vol% ethanol (96%) and sterile-filtered 0.05 wt% BSA-water [24], the A549 cell culture medium contains DMEM with 1% Penicillin/Streptomycin, HEPES buffer, 10% fetal bovine serum inactivated and 1% fungizone. Conversely, Beas-2B medium is rich in biomolecules and contains Serum-free Bronchial Epithelium Growth Medium (BEGM, Lonza) supplemented with bovine pituitary extract, Hydrocortisone, Human Epidermal Growth Factor, Epinephrine, Transferrin, Insulin, Retinoic Acid and Triiodothyronin. Apparently, the Beas-2B medium components affect more significantly the MWCNTs size-distribution, which may have impact in the resulting biological effects. Our findings are in agreement with the work by Magdolenova et al. [35] showing that the level of agglomeration/aggregation of NMs and size distribution depends on medium components and the dispersion procedure.

In fact, the results obtained in Beas-2B were negative for all the genotoxicity endpoints while in A549 some positive findings were reported (Table 8.4 and Fig. 8.5).

The two longest MWCNTs, NM-401 and NM-402, showed genotoxic effects in A549 cells but not in Beas-2B cells and the remaining MWCNTs did not induce micronucleus formation in any of the cell lines. Several reports showed the ability of some MWCNTs to increase micronucleus formation in respiratory cells [6, 44–46] but others described negative results

Table 8.3 Hydrodynamic sizes of the batch dispersion of the MWCNTS in aqueous solution or its dilution in A549 or Beas-2B cell culture medium used for the genotoxicity assays

	Concentration	Zav, nm (mean±sd)		
	(μg/ml)	Batch	A549 medium	Beas-2B medium
NM-400	2560	182.0 ± 2.1	–	–
	64	–	428.5 ± 34.8	2192.0 ± 604.3
	256	–	774.9 ± 47.9	6817.0 ± 1457.0
NM-401	2560	1019.0 ± 78.8	–	–
	64	–	939.0 ± 79.7	1670.0 ± 168.5
	256	–	1009.0 ± 31.4	2855.0 ± 473.1
NM-402	2560	202.0 ± 3.8	–	–
	64	–	533.8 ± 9.6	1308.0 ± 135.6
	256	–	584.0 ± 26.5	4772.0 ± 1359.0
NM-403	2560	198.1 ± 3.9	–	–
	64	–	428.8 ± 13.7	1391.0 ± 87.0
	256	–	562.9 ± 11.3	4336.0 ± 751.4

Zav- mean hydrodynamic size

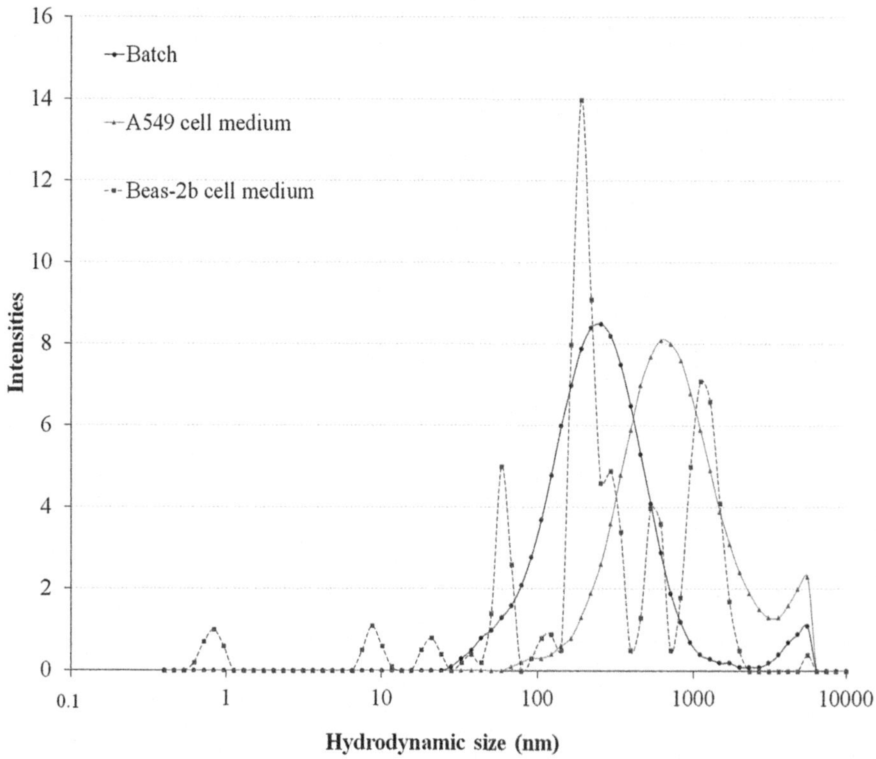

Fig. 8.4 Size-distribution of the batch dispersion of NM-402 (2560 μg/ml) or its dilution in A549 or Beas-2B cell culture medium (at a concentration of 256 μg/ml)

Table 8.4 Main results and the outcome in micronucleus assay in respiratory cells exposed to MWCNTs

	A549 cells		Beas-2B cells	
MWCNT	Genotoxicity ranking index	Outcome	Genotoxicity ranking index	Outcome
NM-400	0	–	0	–
NM-401	1	+	0	–
NM-402	2	+	0	–
NM-403	0	–	0	–

Genotoxicity Ranking Index – number of concentrations with significant increases in micronucleus frequency ($p < 0{,}05$; 2-sided Fisher); Outcome: + genotoxic effects; (+) equivocal genotoxic effects; – no effects.

either after straight or tangled MWCNTs in the same assay [47]. Such inconsistencies have prompted the search further investigation focused on the properties that may underlie toxicity. However, in the work with respiratory cell lines, the correlation analysis did not reveal any significant association between a given physicochemical property of the MWCNTs and the frequency of micronucleated A549 cells [28], similarly to what was previously observed in human lymphocytes [11].

Adding the results of micronucleus assay in human lymphocytes using the same standardized procedures (Table 8.2) to this comparison in respiratory cells, the conclusions become more unclear. In that case, RPMI medium was used (Gibco-Invitrogen, Carlsbad, CA) supplemented with fetal calf serum (15–20%, Gibco-Invitrogen)

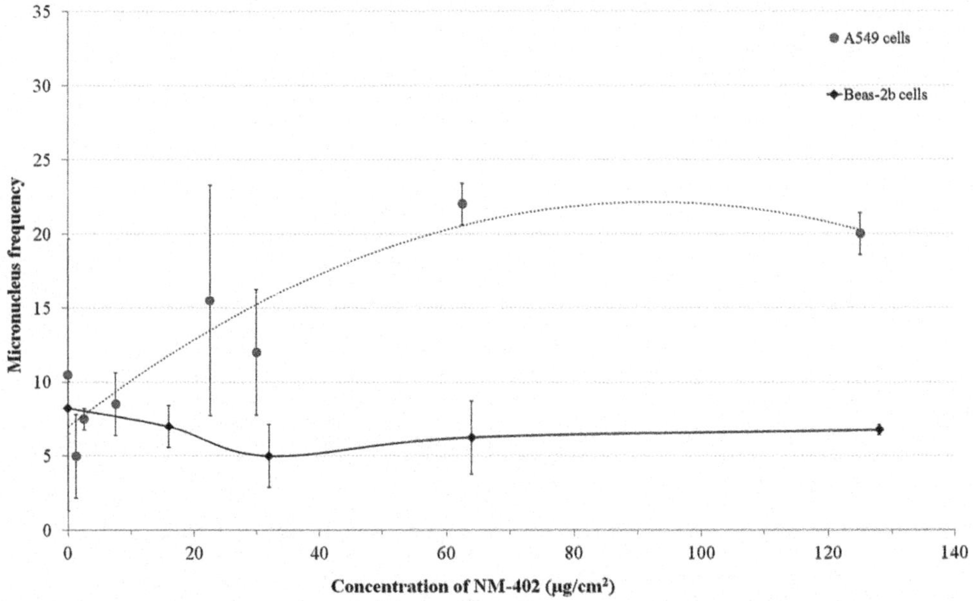

Fig. 8.5 Comparison of the micronucleus assay results following NM-402 exposure of A549 or Beas-2B cells [27]

and phytohemaglutinin A (2.5%, Gibco-Invitrogen) and positive results were obtained for different MWCNTs as compared with A549. These discrepancies may reflect the intrinsic characteristics of the MWCNTs analysed in each report, their preparation for toxicological assays, or other factors related to experimental systems, such as the one here illustrated.

The question remains whether the differences observed in the several cell lines are due to specificities of the cell type, such as uptake capacity, or due to changes in cell culture medium and consequent impact on NMs mean size. While A549 are cancerous cells that may have undergone substantial modifications in their DNA damage response network, Beas-2B are non-cancerous cells. In addition, A549 were reported to have endocytic ability (Foster [43, 48]), and such mechanism may have contributed to a more efficient internalization/ uptake of CNTs. However, in one study MWCNT exerted adverse metabolic effects without being internalized by human epithelial and mesothelial pulmonary cell lines, not changing cell membrane permeability or apoptosis [49]. Besides, other authors showed that human normal bronchial epithelial cells and mesothelium cells endocytosed MWCNTs [50]. Concerning human lymphocytes, their uptake capacity has been discussed also [11].

The second hypothesis, of the changes in cell culture medium affecting the outcome, seems reasonable when considering the impact in hydrodynamic sizes of MWCNTs shown by DLS. The JRC considered that the size of the agglomerates formed was an endpoint not relevant for characterizing MWCNTs [16]. Due to the morphological nature of MWCNTs, there is not a good understanding of what the size-spectra in DLS analyses represent. Even though, the corresponding sizes may be understood as hydrodynamic equivalent size-distributions, which may be due to signals from more than one particle dimension as well as the catalyst contaminants [11]. Apparently, aggregated CNTs might produce a different effect because of the resulting change in bioavailability, width, length, surface area, shape, and other factors [9]. It is possible that the coarsest dispersions in Beas-2B cell medium may have negatively influenced the cell uptake of the CNTs, leading to mostly negative results in this cell line. The fact that the same MWCNT showed distinct genotoxic effects on different cell types, may also reflect, as suggested by some authors [12], a toxicological behaviour that is "context dependent", being affected by the surrounding matrix (pH, ionic strength, biomolecules or macromolecules etc.). In agreement, a study described that culture medium type affects endocytosis of MWCNTs in Beas-2B cells and subsequent biological response, showing that cells cultured in serum-free medium did not internalize MWCNTs and the IC_{50} was increased by more than tenfold under these conditions [44].

In this way, the dynamic behaviour of NMs in physiological conditions, determining the agglomeration status of the suspension used, seems to lead to different genotoxic effects. Overall, our findings reporting different size-distributions depending on cell culture medium suggest that the primary physicochemical descriptors of NMs may not be the most appropriate to predict their toxicological behaviour. Therefore, to obtain results relevant for risk assessment, in vitro assays can be of limited value and one may suggest that only in vivo studies mimicking the route of exposure and physiological interactions can add relevant information.

8.5 Shifting from In Vitro Toxicology Assays to In Vivo Evidence

The previous section has illustrated the difficulty of using in vitro assays to obtain straightforward information relevant for hazard assessment, since within in vitro environment the NMs may not exhibit the same secondary characteristics occurring in complex tissues or organisms, under physiological conditions. Eukaryotic cells possess complex mechanisms and DNA damaging surveillance molecules, or damage sensors, that activate and coordinate many aspects of the DNA damage response, including delay of cell cycle progression, activation of DNA repair pathways and induction of cell death, all of them conjugated in the same effort of maintaining the genomic stability [51], many times dependent on the interaction of several cell types within a tissue. Once DNA damage occurs, a complex network of repair functions is triggered to protect cells against the deleterious consequences of mutations, in order to maintain genomic stability. Although in vitro assays are of utmost importance for large-scale screening and for hazard identification, only in vivo methodologies account for all these events while in vitro methodologies do not reflect the complexity of an organism concerning bioavailability, toxicokinetics and immune responses. In this view, data from suitable in vivo assays are foreseen as useful to evaluate the performance of in vitro assays and to strengthen the knowledge about the genotoxicity of chemicals and even more importantly, for nanomaterials, using several routes of exposure, at a whole organism level [52]. It is expected that the application of in vivo assays in which in vitro results have been confirmed may validate the use of the in vitro methodologies in safety assessment and, in a near future, may provide alternative testing strategies.

For these reasons, in vivo genotoxicity assays have been recognized as important tools to evaluate the carcinogenic potential of pharmaceuticals and environmental agents, while accounting for the complexity of an organism, including the route of exposure, variability among tissues and organs in physiological responses, metabolic and DNA repair capacities, etc., that can greatly influence the genotoxic effects in somatic and germinal cells [52]. Therefore, in particular gene mutation assays in vivo can be used as follow-up tests to develop weight of evidence in assessing results of in vitro or in vivo assays [53, 54].

In the recent studies focused in understanding the genotoxicity of TiO_2, no clear results were obtained for NM-102 in the in vitro assays above described, either in human lymphocytes [11] or in respiratory cell lines [55]. Subsequent in vivo investigation was the next option to try to clarify the safety of this NM and the LacZ-plasmid based transgenic mouse was selected as the model system for an integrated approach involving multiple endpoints, following repeated exposure to NM-102 [57]. The LacZ-transgenic mouse line derives from the C57Bl/6 mouse and harbours the pUR288 plasmid shuttle vector (containing the *LacZ* reporter gene) that has the double role of being a target for mutations and a

Table 8.5 Main results and the outcome in cells or LacZ mice exposed to the titanium dioxide nanomaterial, NM-102

	Human lymphocytes [11]	A549 cells [55]		Beas-2B cells [55]		LacZ mice [57]		
	Micronucleus	Micronucleus	Comet	Micronucleus	Comet	Micronucleus	Comet	Mutation
Genotox ranking index	1	0	2	0	3	0	0	0
Outcome	(+)	–	+	–	+	–	–	–

Genotoxicity Ranking Index – number of concentrations with significant increases in the endpoint. Outcome: + genotoxic effects; (+) equivocal genotoxic effects; – no effects

reporter for their detection [56]. Mutation analysis in the LacZ-transgene was used to access genotoxicity in the mouse liver and spleen tissues following daily doses of 15 mg/kg body weight of NM-102 for two consecutive days and tissues were sampled 28 days after the last treatment. In addition, micronucleus assay and the comet assay were used in vivo [57]. The latter methodology allows the detection of primary DNA lesions such as DNA single and double-strand breaks and provides complementary information on genotoxicity [58].

The results obtained for all the endpoints in the mouse model, together with the in vitro data obtained in the several cell types [55] are summarized in Table 8.5.

In vitro, it was observed that the same NM, prepared using standardized procedures and dispersion protocols [24], yields different outcomes in blood or respiratory cell line. As described above, cell culture conditions may have different effects in secondary characteristics of the NMs, leading to the discrepant observations. Likewise, a Pubmed search reveals around 115 papers on this issue, showing contradictory information on genotoxicity testing of titanium dioxide NMs in vitro.

The in vivo exposure did not evidence any genotoxic effects in mouse liver, spleen or lymphocytes, under the experimental conditions used, either in micronucleus, comet or mutation assays [57]. However, the histopathological and cytological analyses performed in liver samples evidenced a moderate inflammatory response in the liver, together with the bioaccumulation of NM-102 in liver cells, as observed by TEM [57]. Using TEM analysis of liver samples, it was observed the uptake of the NMs in liver cells and that they were not eliminated by the organism (Fig. 8.6), as seen in the form of large clusters of NM-102 inside vesicles within the Kupffer cells of these mice [57].

Apparently, as discussed by Louro et al. [57], the NM overload induced an increase in the phagocytosis of particles by macrophages, probably as an attempt to eliminate them. Indeed, the accumulation of TiO_2 particles inside Kupffer cells has been already described by other authors as well as in in organ parenchyma [59, 60] and the role of these macrophages in the removal of NMs from the blood has already been reported [61].

The issue of cellular uptake of NMs has long been discussed in the literature. Size-dependent cellular uptake of nanoparticles has been extensively investigated in various cell lines because the nanoparticle size has been known to be a key determinant of the uptake pathways (reviewed in [62]). In theory, the penetration of NMs in cells is larger than the bulk materials, allowing crossing cellular membranes, even reaching the nucleus, where they can act directly on DNA. It has been described that anatase TiO_2 accumulated human lung cells and forms DNA adducts [63]. Alternatively, NMs will remain in cells and will have access to nucleus during cell division. In fact, this type of events has been described for TiO_2, that enter the nucleus and lead to the formation of aggregates with intranuclear proteins, leading to inhibition of replication, transcription and cell proliferation (reviewed in [34, 64]). Conversely, the NMs may lead to DNA lesions indirectly, through the induction of oxidative stress and inflammatory responses.

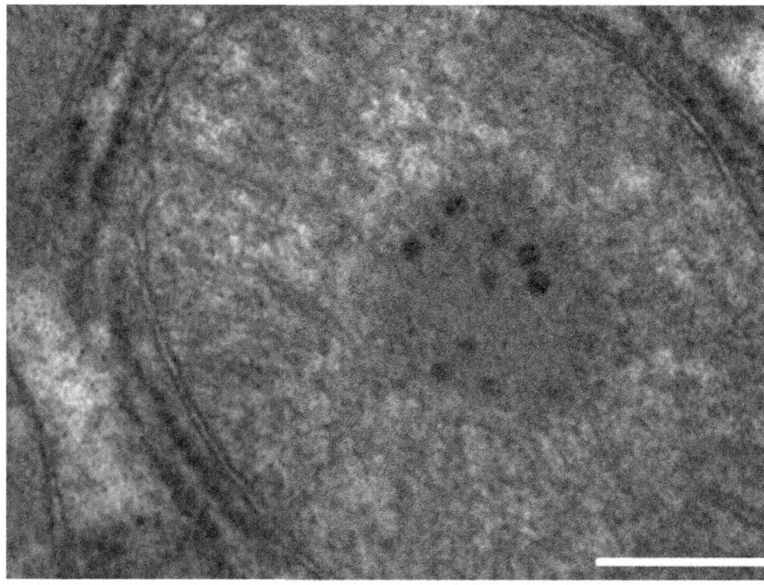

Fig. 8.6 TEM microphotograph of hepatocyte showing NMs inside a mitochondrion. Mice were exposed to daily doses of 15 mg/kg body weight of NM-102 for two consecutive days and tissues were sampled 28 days after the last treatment. Scale bar 200 nm (Image credits: Elsa Alverca, INSA)

In spite of the negative results in genotoxicity endpoints, the findings in vivo further suggested that a prolonged exposure and accumulation of NMs in vivo may lead to long-term toxicological consequences, thus in the future this hypothesis should be clarified. The negative results in the comet assay agree with the absence of a DNA damaging effect in rat lung cells [65] when the small sized anatase TiO_2 (5 nm) was instilled intratracheally at low doses (1 or 5 mg/kg, single dose and 0.2 or 1 mg/kg once a week, for 5 weeks). In addition, in lung cells from broncho-alveolar fluid of Wistar rats exposed by inhalation to rutile TiO_2, negative data were reported [66]. Male gpt Delta C57BL/6J mice administered TiO_2 (2, 10 or 50 mg/kg body weight per week) for four consecutive weeks did not indicate genotoxic effects on mouse liver or bone marrow, in spite TiO_2 were found in the liver in the sinuses and inside Kupffer cells [67]. However, recently the 10 nm anatase- exposed mice (repeated dose of 50 mg/kg/day for 3 days via intraperitoneal injection) induced genotoxicity and genomic change in mouse, concomitantly with NMs accumulation in liver and lung tissues [68]. Furthermore, anatase TiO_2 showed genotoxic effects both in vitro in lung cells (V79 cells) and in vivo in Sprague-Dawley rats via intragastric administration (0, 10, 50 and 200 mg/kg body weight every day for 30 days) [69].

The overview of the research carried out for understanding NM-102 safety does not show a correlation between the results obtained in vitro and in vivo, and the extrapolation of data from in vitro to in vivo seems unfeasible. The moderate genotoxicity detected in vitro in human lymphocytes or respiratory cells was not seen in vivo in blood, spleen or liver cells. Even though, in vitro studies based on standardized testing procedures are valued to contribute for the knowledge of cell type-specific mechanisms and may also allow a first hazard identification to guide the risk assessment process. Recognizing the gap between in vitro and in vivo data that hampers the formulation of unambiguous conclusions on NM toxicity, some authors suggest the need of more reliable in vitro models with a higher predictive power, mimicking the in vivo environment more closely, such as air-liquid interface cellular models or 3D-cell cultures [70]. It is recognized that cells react differently to NM exposure when they are cultured in mono- or co-cultures and culturing cells in a 3D setup has a major influence on the cellular phenotype and function and therefore causes cells to react in a drastically different way to NM exposure [70]. In this way, such models are promising tools to allow extrapolation from in vitro to in vivo genotoxicity.

For risk assessment, in vivo results remain mandatory. While moving from in vitro studies to in vivo, the need to standardize the preparation of the samples to use for exposure remains, and the selection of the target organs to analyse and the dose-range selection is a challenge to allow the clear interpretation of the results. Shifting from in vitro studies to in vivo models harbours several constrains. The most relevant is the world-wide recognized need to reduce animal experimentation to the least possible, in a context where new NMs are expanding and need to be tested in a time- and cost-effectively way. A more realistic approach, which would require less testing and thus reduce the number of animals, is to first assess exposure levels and routes relevant for human health [71].

8.6 New Avenues for Understanding Nano-cellular Interactions

With the expansion of the NMs being developed and used throughout the world, human exposure is already a reality, since NMs are now used in many consumer products and in biomedicine and will continue to have a number of innovative applications in the future. Therefore, for understanding nano-cellular interactions, relevant for human health, a biomonitoring strategy needs to be implemented.

In fact, some measurements in occupational settings suggest that workers may be already exposed to significant amounts of carbon nanotubes [72–76], TiO_2 and iron NMs [77, 78]. The latter study provided evidence of several elevated oxidative stress markers in exhaled breath condensate of workers exposed to NMs, showing an early biological effect that can have impact on human health.

Other exposure and effect biomarkers can be exploited using human cells obtained from exposed populations if a biomonitoring strategy would be designed. Three main types of biomarkers are usually considered: biomarkers of exposure, biomarkers of effect and biomarkers of susceptibility [79].

The quantification of human exposure to nanomaterials is hampered by the difficulty to differentiate manufactured nanomaterials from other nanomaterials (incidental to processes or naturally occurring) and the lack of a single metric that can be used for health risk assessment [80]. The assessment of consumers' exposure to nanomaterials is even more challenging, since the materials contained in the products are mixed with other chemicals that affect the release and transfer and their bioavailability [80].

In addtion to environment determination of the NMs, using many of the methodologies depitect in Table 8.1 it is possible to determine the level of exposure in human cells or tissues. Since the extent of NM uptake by cells of the target tissue is dependent on the interactions at the nano-bio interface [79] the measurements obtained in samples of human blood or other cells can provide a realistic scenario of internal exposure. Concerning genotoxicity, biomarkers of early genetic effect such as chromosomal aberrations, micronucleus and comet assay can be used. While the chromosome aberration assay is time- and resource-requiring, the micronucleus assay using human blood seems a feasible option for detection of early genotoxic effects and has the advantage of detecting both direct and indirect genotoxic effects [79]. In addition, this assay has been shown to be predictive for cancer risk [81]. Newer versions on of the micronucleus assay allow using non-invasive methods for biomonitoring, while collecting for example exfoliated buccal mucosa cells that can be easily obtained [82]. In fact, in vitro, human oral mucosa buccal epithelial cells (TR146) have been successfully used to issue the toxicological effects of TiO_2, allowing to discriminate NMs subcellular localization, oxidative species production, inflammatory responses, and induction of apoptosis accordingly to their chemical nature [83].

The use of a biomonitoring approach to NMs would mean a reversal of the actual in vitro-in vivo strategy for the safety assessment of NMs that could provide relevant in-depth information up to the molecular mechanisms of action of

NMs. Such an approach would require the identification of exposed populations, only possible if a registry of NMs use was available. However, the EU Member States did not agree on the creation of EU-wide registration of nanomaterials and only three Member States, namely France, Denmark and Belgium, have decided to adopt their own national register for nanomaterials. Other national legislation framework does not require it, thus future biomonitoring research may become compromised by the difficulty in mapping exposed populations.

8.7 Final Remarks

Correlating seemingly homogeneous group of NMs, which is separated into sub-groups according to measurable characteristics, arise as a promising tool for risk assessment [12, 84]. This implies testing a large number of NMs from the same class, systematically varying a single property, which will be feasible only by applying high-throughput technologies. Therefore, developing high throughput technologies focused on the study of these NMs remains a priority concern for addressing potential human risks.

Although further investigation beyond the data here presented with larger panels of NMs is necessary, this work identified some correlations among the NMs properties that contribute for their genotoxicity. The dynamic behaviour of NMs in physiological conditions, determining the agglomeration status of the suspension used, seems to lead to different genotoxic effects. According to the data here presented, the main determinant of in vitro genotoxicity is then thought to be the size-distribution of the NMs that is dependent on the cellular environment. This major determinant, leading to variation of the results reported worldwide in in vitro studies, may undercover other primary properties that significantly influence the genotoxic outcome. Therefore, an in vivo approach based on realistic dose ranges and exposure routes might provide information more relevant for the safety assessment of NMs. To identify the mentioned realistic exposure conditions, biomonitoring studies are of utmost importance and should be foreseen. For such, registry or monitoring, high quality information on exposure is essential.

In conclusion, tackling the uncertainties of the innovative nanomaterials to public health implies the in-depth investigation of the nano-bio interactions, where in vitro research must be integrated with in vivo and biomonitoring approaches, to cope with the complex dynamic behaviour of nanoscale materials.

Acknowledgments The author wishes to thank Dr. Maria João Silva and all the team from the Genetic Toxicology Laboratory (INSA, Portugal). HL acknowledges the support of the partners of the Nanogenotox Joint Action (Health Programme under Grant Agreement no. 2009 21), NANOREG Project (A common European approach to the regulatory testing of nanomaterials, Grant Agreement 310584) and Dr. José Catita (Paralab, Portugal) for the help with the DLS analysis.

References

1. IARC – International Agency for Research on Cancer (2010) Volume 93: carbon black, titanium dioxide and talc. In: World Health Organization (ed) IARC monographs on the evaluation of carcinogenic risks to humans. World Health Organization, Lyon
2. NIOSH - National Institute for Occupational Safety and Health. Department Health and Human Services. Centers for Disease Control and Prevention. 2011. Current intelligence bulletin 63: occupational exposure to titanium dioxide. Available at: https://www.cdc.gov/niosh/docs/2011-160/pdfs/2011-160.pdf
3. Sycheva LP, Zhurkov VS, Iurchenko VV et al (2011) Investigation of genotoxic and cytotoxic effects of micro- and nanosized titanium dioxide in six organs of mice in vivo. Mutat Res 726:8–14
4. SCCS - Scientific Committee on Consumer Safety (2013) Opinion on titanium dioxide (nano form). COLIPA n° S75
5. Grosse Y, Loomis D, Guyton KZ et al (2014) Carcinogenicity of fluoro-edenite, silicon carbide fibres and whiskers, and carbon nanotubes. Lancet Oncol 15:1427–1428
6. Cveticanin J, Joksic G, Leskovac A et al (2010) Using carbon nanotubes to induce micronuclei and double strand breaks of the DNA in human cells. Nanotechnology 21:015102
7. Szendi K, Varga C (2008) Lack of genotoxicity of carbon nanotubes in a pilot study. Anticancer Res 28:349–352
8. Donaldson K, Murphy FA, Duffin R et al (2010) Asbestos, carbon nanotubes and the pleural mesothelium: a review of the hypothesis regarding the role of

long fibre retention in the parietal pleura, inflammation and mesothelioma. Part Fibre Toxicol 7:5
9. Rittinghausen S, Hackbarth A, Creutzenberg O et al (2014) The carcinogenic effect of various multi-walled carbon nanotubes (MWCNTs) after intraperitoneal injection in rats. Part Fibre Toxicol 11:59
10. Pelaz B, Charron G, Pfeiffer C et al (2012) Interfacing engineered nanoparticles with biological systems: anticipating adverse nano-bio interactions. Small 9(9–10):1573–1584
11. Tavares AM, Louro H, Antunes S et al (2014) Genotoxicity evaluation of nanosized titanium dioxide, synthetic amorphous silica and multi-walled carbon nanotubes in human lymphocytes. Toxicol In Vitro 28(1):60–69
12. Lynch I, Weiss C, Valsami-Jones E (2014) A strategy for grouping of nanomaterials based on key physico-chemical descriptors as a basis for safer-by-design NMs. Nano Today 9(3):266–270
13. Fadeel B, Fornara A, Toprak MS et al (2015) Keeping it real: the importance of material characterization in nanotoxicology. Biochem Biophys Res Commun 468(3):498–503
14. Maynard AD (2016) Navigating the risk landscape. Nat Nanotechnol 11(3):211–212
15. Rasmussen K, De Temmerman PI, Verleysen E et al (2014) Titanium dioxide, NM-100, NM-101, NM-102, NM-103, NM-104, NM-105: characterisation and physicoChemical properties. NM-series of Representative Manufactured Nanomaterials. JRC, Joint Research Centre. Available at: https://ec.europa.eu/jrc/en
16. Rasmussen K, Mast J, De Temmerman P-J et al (2014). Multi-walled carbon nanotubes, NM-400, N,M-401, NM-402, NM-403: characterisation and physico-chemical properties. JRC Repository: NM-series of Representative Manufactured Nanomaterials European Commission- Joint Research Centre, Institute for Health and Consumer Protection. Available at: http://publications.jrc.ec.europa.eu/repository/handle/JRC91205
17. Oberdorster G (2010) Safety assessment for nanotechnology and nanomedicine: concepts of nanotoxicology. J Intern Med 267(1):89–105
18. Chen R, Riviere JE (2017) Biological surface adsorption index of nanomaterials: modelling surface interactions of nanomaterials with biomolecules. Adv Exp Med Biol 947:207–253
19. Monopoli MP, Aberg C, Salvati A et al (2012) Biomolecular coronas provide the biological identity of nanosized materials. Nat Nanotechnol 7(12):779–786
20. Hussain SM, Warheit DB, Ng SP et al (2015) At the crossroads of nanotoxicology in vitro: past achievements and current challenges. Toxicol Sci 147:5–16
21. Atluri R, Jensen KA (2017) Engineered nanomaterials: their physicochemical characteristics and how to measure them. Adv Exp Med Biol 947:3–23
22. Ren G, Hu D, Cheng EW et al (2009) Characterisation of copper oxide nanoparticles for antimicrobial applications. Int J Antimicrob Agents 33(6):587–590
23. Seoane JR, Llovet X (2012) Handbook of instrumental techniques for materials, chemical and biosciences research. Centres Cientfics i Tecnològics, Barcelona, Spain, Barcelona
24. Jensen KA, Kembouche Y, Christiansen E et al (2011) The generic NANOGENOTOX dispersion protocol – Standard Operation Procedure (SOP). Available at: https://www.anses.fr/en/system/files/nanogenotox_deliverable_5.pdf
25. OECD – Organisation for Economic Co-operation and Development (2010) OECD guideline for the testing of chemicals – in vitro mammalian cell micronucleus test, Test Guideline 487. OECD, Paris
26. Magdolenova Z, Lorenzo Y, Collins A et al (2012) Can standard genotoxicity tests be applied to nanoparticles? J Toxicol Environ Health A 75:800–806
27. Gonzalez L, Sanderson BJ, Kirsch-Volders M (2011) Adaptations of the in vitro MN assay for the genotoxicity assessment of nanomaterials. Mutagenesis 26(1):185–191
28. Louro H, Pinhão M, Santos J et al (2016) Evaluation of the cytotoxic and genotoxic effects of benchmark multi-walled carbon nanotubes in relation to their physicochemical properties. Toxicol Lett 262:123–134
29. JRC – Joint Research Center (2011) Impact of engineered nanomaterials on health: considerations for benefit-risk assessment. EASAC Policy Report - JRC Reference Report. Ispra
30. Osman IF, Baumgartner A, Cemeli E et al (2010) Genotoxicity and cytotoxicity of zinc oxide and titanium dioxide in HEp-2 cells. Nanomedicine 5:1193–1203
31. Shukla RK, Sharma V, Pandey AK et al (2011) ROS-mediated genotoxicity induced by titanium dioxide nanoparticles in human epidermal cells. Toxicol In Vitro 25:231–241
32. Uboldi C, Urban P, Gilliland D et al (2016) Role of the crystalline form of titanium dioxide nanoparticles: rutile, and not anatase, induces toxic effects in Balb/3T3 mouse fibroblasts. Toxicol In Vitro 31:137–145
33. Di Bucchianico S, Cappellini F, Le Bihanic F et al (2017) Genotoxicity of TiO2 nanoparticles assessed by mini-gel comet assay and micronucleus scoring with flow cytometry. Mutagenesis 32(1):127–137
34. Magdolenova Z, Collins A, Kumar A et al (2014) Mechanisms of genotoxicity. A review of in vitro and in vivo studies with engineered nanoparticles. Nanotoxicology 8(3):233–278
35. Magdolenova Z, Bilanicova D, Pojana G et al (2012) Impact of agglomeration and different dispersions of titanium dioxide nanoparticles on the human related in vitro cytotoxicity and genotoxicity. J Environ Monit 14(2):455–464
36. Nagai H, Okazaki Y, Chew SH et al (2011) Diameter and rigidity of multiwalled carbon nanotubes are critical factors in mesothelial injury and carcinogenesis. Proc Natl Acad Sci U S A 108:E1330–E1338
37. Takagi A, Hirose A, Nishimura T et al (2008) Induction of mesothelioma in p53+/− mouse by intra-

peritoneal application of multi-wall carbon nanotube. J Toxicol Sci 33:105–116
38. Takagi A, Hirose A, Futakuchi M et al (2012) Dose-dependent mesothelioma induction by intraperitoneal administration of multi-wall carbon nanotubes in p53 heterozygous mice. Cancer Sci 103:1440–1444
39. Sakamoto Y, Nakae D, Fukumori N et al (2009) Induction of mesothelioma by a single intrascrotal administration of multi-wall carbon nanotube in intact male Fischer 344 rats. J Toxicol Sci 34:65–76
40. Jackson P, Kling K, Jensen KA et al (2015) Characterization of genotoxic response to 15 multi-walled carbon nanotubes with variable physicochemical properties including surface functionalizations in the FE1-Muta(TM) mouse lung epithelial cell line. Environ Mol Mutagen 56(2):183–203
41. Lindberg HK, Falck GCM, Suhonen S et al (2009) Genotoxicity of nanomaterials: DNA damage and micronuclei induced by carbon nanotubes and graphite nanofibres in human bronchial epithelial cells in vitro. Toxicol Lett 186(3):166–173
42. Atsuta J, Sterbinsky SA, Plitt J et al (1997) Phenotyping and cytokine regulation of the BEAS-2B human bronchial epithelial cell: demonstration of inducible expression of the adhesion molecules VCAM-1 and ICAM-1. Am J Respir Cell Mol Biol 17(5):571–582
43. Foster KA, Oster C, Mayer M et al (1998) Characterization of the A549 cell line as a type II pulmonary epithelial cell model for drug metabolism. Exp Cell Res 243(2):359–366
44. Haniu H, Saito N, Matsuda Y, Tsukahara T et al (2013) Culture medium type affects endocytosis of multi-walled carbon nanotubes in BEAS-2B cells and subsequent biological response. Toxicol In Vitro 27(6):1679–1685
45. Asakura M, Sasaki T, Sugiyama T et al (2010) Genotoxicity and cytotoxicity of multi-wall carbon nanotubes in cultured Chinese hamster lung cells in comparison with chrysotile fibers. J Occup Health 52:155–166
46. Migliore L, Saracino D, Bonelli A et al (2010) Carbon nanotubes induce oxidative DNA damage in RAW 264.7 cells. Environ Mol Mutagen 51:294–303
47. Catalan J, Siivola KM, Nymark P et al (2016) In vitro and in vivo genotoxic effects of straight versus tangled multi-walled carbon nanotubes. Nanotoxicology 10:794–806
48. Stearns RC, Paulauskis JD, Godleski JJ (2001) Endocytosis of ultrafine particles by A549 cells. Am J Respir Cell Mol Biol 24(2):108–115
49. Tabet L, Bussy C, Amara N et al (2009) Adverse effects of industrial multiwalled carbon nanotubes on human pulmonary cells. J Toxicol Environ Health A 72(2):60–73
50. Maruyama K, Haniu H, Saito N et al (2015) Endocytosis of multiwalled carbon nanotubes in bronchial epithelial and mesothelial cells. Biomed Res Int 2015:793186. https://doi.org/10.1155/2015/793186
51. Louro H, Silva MJ (2011) Cost/benefit of mutation induction under PARP1 deficiency: from genomic instability to therapy. In: Urbano KV (ed) Advances in genetics research. Nova Science Publishers, New York, pp 109–134
52. Louro H, Silva MJ (2010) In vivo mutagenic effects of alkylating agents eliciting different DNA-adducts. In: Emerson A, Cunha R (eds) DNA adducts: formation, detection and mutagenesis. Nova Science Publishers, New York, pp 39–60
53. ICH- International Conference on Harmonisation of Technical Requirements for Registration of Pharmaceuticals for Human Use (2012) S2(R1) genotoxicity testing and data interpretation for pharmaceuticals intended for human use. Food and Drug Administration, HHS. (Ed)
54. OECD – Organisation for Economic Co-operation and Development (2011) OECD guidelines for the testing of chemicals: transgenic rodent somatic and germ cell gene mutation assays, Test Guideline 488. OECD, Paris
55. Louro H (2013) Nanomateriais manufaturados: avaliação de segurança através da caracterização dos seus efeitos genéticos. Ph.D. Dissertation, Nova University of Lisbon
56. Louro H, Pinto M, Vital N et al (2014) The LacZ plasmid-based transgenic mouse model: an integrative approach to study the genotoxicity of nanomaterials. In: Sierra LM, Gaivão I (eds) Genotoxicity and DNA repair – a practical approach. Methods in pharmacology and toxicology. Humana Press, Springer, New York, pp 451–477
57. Louro H, Tavares A, Vital N et al (2014) Integrated approach to the in vivo genotoxic effects of a titanium dioxide nanomaterial using LacZ plasmid-based transgenic mice. Environ Mol Mutagen 55(6):500–509
58. Tice RR, Agurell E, Anderson D et al (2000) Single cell gel/comet assay: Guidelines for in vitro and in vivo genetic toxicology testing. Environ Mol Mutagen 35(3):206–221
59. Bruno ME, Tasat DR, Ramos E et al (2013) Impact through time of different sized titanium dioxide particles on biochemical and histopathological parameters. J Biomed Mater Res A 102(5):1439–1448
60. Olmedo DG, Tasat DR, Evelson P et al (2008) Biological response of tissues with macrophagic activity to titanium dioxide. J Biomed Mater Res A 84(4):1087–1093
61. Sadauskas E, Wallin H, Stoltenberg M et al (2007) Kupffer cells are central in the removal of nanoparticles from the organism. Part Fibre Toxicol 4:10
62. Oh N, Park JH (2014) Endocytosis and exocytosis of nanoparticles in mammalian cells. Int J Nanomedicine 9(Suppl 1):51–63
63. Bhattacharya K, Davoren M, Boertz J et al (2009) Titanium dioxide nanoparticles induce oxidative stress and DNA-adduct formation but not DNA-breakage in human lung cells. Part Fibre Toxicol 2009:6–17

64. Singh N, Manshian B, Jenkins GJ et al (2009) NanoGenotoxicology: the DNA damaging potential of engineered nanomaterials. Biomaterials 30(23–24):3891–3914
65. Naya M, Kobayashi N, Ema M, Kasamoto S, Fukumuro M, Takami S, Nakajima M, Hayashi M, Nakanishi J (2012) In vivo genotoxicity study of titanium dioxide nanoparticles using comet assay following intratracheal instillation in rats. Regul Toxicol Pharmacol 62:1–6
66. Landsiedel R, Ma-Hock L, Van Ravenzwaay B et al (2010) Gene toxicity studies on titanium dioxide and zinc oxide nanomaterials used for UV-protection in cosmetic formulations. Nanotoxicology 4:364–381
67. Suzuki T, Miura N, Hojo R et al (2016) Genotoxicity assessment of intravenously injected titanium dioxide nanoparticles in gpt delta transgenic mice. Mutat Res 802:30–37
68. Li Y, Yan J, Ding W et al (2017) Genotoxicity and gene expression analyses of liver and lung tissues of mice treated with titanium dioxide nanoparticles. Mutagenesis 32:33–46
69. Chen Z, Wang Y, Ba T et al (2014) Genotoxic evaluation of titanium dioxide nanoparticles in vivo and in vitro. Toxicol Lett 226:314–319
70. Joris F, Manshian BB, Peynshaert K et al (2013) Assessing nanoparticle toxicity in cell-based assays: influence of cell culture parameters and optimized models for bridging the in vitro-in vivo gap. Chem Soc Rev 42(21):8339–8359
71. De Boo J, Hendriksen C (2005) Reduction strategies in animal research: a review of scientific approaches at the intra-experimental, supra-experimental and extra-experimental levels. Altern Lab Anim 33(4):369–377
72. Bello D, Hart AJ, Ahn K et al (2008) Particle exposure levels during CVD growth and subsequent handling of vertically-aligned carbon nanotube films. Carbon 46(6):974–977
73. Bello D, Wardle B, Yamamoto N et al (2009) Exposure to nanoscale particles and fibers during machining of hybrid advanced composites containing carbon nanotubes. J Nanopart Res 11(1):231–249
74. Han JH, Lee EJ, Lee JH et al (2008) Monitoring multiwalled carbon nanotube exposure in carbon nanotube research facility. Inhal Toxicol 20(8):741–749
75. Lee JH, Lee S-B, Bae GN et al (2010) Exposure assessment of carbon nanotube manufacturing workplaces. Inhal Toxicol, 2010 22(5):369–381
76. Tsai SJ, Hofmann M, Hallock M et al (2009) Characterization and evaluation of nanoparticle release during the synthesis of single-walled and multiwalled carbon nanotubes by chemical vapor deposition. Environ Sci Technol 43(15):6017–6023
77. Pelclova D, Barosova H, Kukutschova J et al (2015) Raman microspectroscopy of exhaled breath condensate and urine in workers exposed to fine and nano TiO2 particles: a cross-sectional study. J Breath Res 9(3):036008
78. Pelclova D, Zdimal V, Kacer P et al (2016) Oxidative stress markers are elevated in exhaled breath condensate of workers exposed to nanoparticles during iron oxide pigment production. J Breath Res 10(1):016004
79. Gonzalez L, Kirsch-Volders M (2016) Biomonitoring of genotoxic effects for human exposure to nanomaterials: the challenge ahead. Mutat Res 768:14–26
80. Jimenez AS, van Tongeren M (2017) Assessment of human exposure to ENMs. Adv Exp Med Biol 947:27–40
81. Bonassi S, Znaor A, Ceppi M et al (2007) An increased micronucleus frequency in peripheral blood lymphocytes predicts the risk of cancer in humans. Carcinogenesis 28(3):625–631
82. Holland N, Bolognesi C, Kirsch-Volders M et al (2008) The micronucleus assay in human buccal cells as a tool for biomonitoring DNA damage: the HUMN project perspective on current status and knowledge gaps. Mutat Res 659(1–2):93–108
83. Tay CY, Fang W, Setyawati MI et al (2014) Nano-hydroxyapatite and nano-titanium dioxide exhibit different subcellular distribution and apoptotic profile in human oral epithelium. ACS Appl Mater Interfaces 6(9):6248–6256
84. Sayes CM, Smith PA, Ivanov I (2013) A framework for grouping nanoparticles based on their measurable characteristics. Int J Nanomedicine 8:45–56

Toxicogenomics: A New Paradigm for Nanotoxicity Evaluation

9

Sourabh Dwivedi, Quaiser Saquib, Bilal Ahmad, Sabiha M. Ansari, Ameer Azam, and Javed Musarrat

Abstract

The wider applications of nanoparticles (NPs) has evoked a world-wide concern due to their possible risk of toxicity in humans and other organisms. Aggregation and accumulation of NPs into cell leads to their interaction with biological macromolecules including proteins, nucleic acids and cellular organelles, which eventually induce toxicological effects. Application of toxicogenomics to investigate molecular pathway-based toxicological consequences has opened new vistas in nanotoxicology research. Indeed, genomic approaches appeared as a new paradigm in terms of providing information at molecular levels and have been proven to be as a powerful tool for identification and quantification of global shifts in gene expression. Toxicological responses of NPs have been discussed in this chapter with the aim to provide a clear understanding of the molecular mechanism of NPs induced toxicity both in in vivo and in vitro test models.

Keywords

Nanoparticles · Toxicogenomics · Oxidative stress · RNA-Seq · Microarray

S. Dwivedi · A. Azam
Department of Applied Physics, Faculty of Engineering and Technology, Aligarh Muslim University, Aligarh, Uttar Pradesh, India

Q. Saquib
Zoology Department, College of Sciences, King Saud University, P.O. Box 2455, Riyadh 11451, Saudi Arabia

B. Ahmad
Department of Agricultural Microbiology, Faculty of Agricultural Sciences, Aligarh Muslim University, Aligarh, Uttar Pradesh, India

S. M. Ansari
Department of Botany and Microbiology, College of Sciences, King Saud University, P.O. Box 2455, Riyadh 11451, Saudi Arabia

J. Musarrat (✉)
Department of Agricultural Microbiology, Faculty of Agricultural Sciences, Aligarh Muslim University, Aligarh, Uttar Pradesh, India

Department of Biosciences and Biotechnology, Baba Ghulam Shah Badshah University, Rajouri, Jammu and Kashmir, India
e-mail: musarratj1@yahoo.com

Q. Saquib et al. (eds.), *Cellular and Molecular Toxicology of Nanoparticles*, Advances in Experimental Medicine and Biology 1048, https://doi.org/10.1007/978-3-319-72041-8_9

9.1 Introduction

Nanoparticles (NPs) have been defined as materials having at least one dimension in the nanoscale (1–100 nm), bearing size-dependent physiochemical properties different than their bulk counterparts. NPs have high surface-to-volume ratio, which provide them high reactivity and physicochemical dynamicity. The advantageous properties of more complex NPs retain the potential to drawn enormous research from different fields for its application in medical diagnostics and treatments [1]. Several commercially available consumer products contains NPs, particularly in cosmetics and sunscreens [2]. NPs have also been utilized in different areas of biology and medicine including tissue engineering, drug and delivery formulations, for hyperthermia induced tumor destruction, DNA structure probes and biosensors [3–5]. Nonetheless, there are uncertainties that the unique properties of NPs may also pose potential health risks in the occupationally and non-occupationally exposed populations [2].

With the greater demand of NPs, there is huge ambiguity related to their potential hazards towards environment and its different trophic levels. Special concerns have been raised on NPs mediated release of metal ions, reactive oxygen species (ROS) or its direct reactivity towards biological membranes [6, 7]. A major key question persisting with NPs toxicity is whether it is linked with the general properties shared by varying NPs or is it specific to individual NPs. Considering these facts, if toxicological effects are related to shape, agglomeration or the size of NP, then similar toxicological effects can be expected for different NPs [8]. On the other hand, if NPs composition controls the interaction, then definite toxicity would be expected [9, 10]. An understanding of the toxicological effects of each NPs is critical for any prediction of their immediate and long-term risks for humans upon occupational and consumer goods exposure (Fig. 9.1).

Exposure to NPs may occur during its production via handling, aerosolization may also occur during energetic processes, such as vortexing, weighing, sonication, mixing and blending [11]. Hence, inhalation is considered a relevant route of exposure [11, 12]. Presence of the NPs in the body may induces pathophysiological changes that might contribute to the development of cancer [13], cardiovascular diseases [14], respiratory tract inflammation [15], neurodegenerative diseases [16], and many other pathologies [17]. Researchers have summarized the potential mechanisms of nanotoxicity via greater ROS level and induction of inflammatory responses such as Parkinson's disease [18]. NPs can also alter the permeability of blood brain barrier and re-translocate from the site of deposition to other parts of the body via circulatory or lymphatic system [19]. The actual prophecy of the adversity of NPs cellular exposure and uptake are still lacking [20]. Cells can rapidly alter its transcriptomic output (gene expression profile) in response to extracellular and intracellular environmental changes, and get adapted for their survival and function. Nonetheless, biological functions and normal physiological activities can get disturbed under excessive environmental changes. Consequently, profiling of gene expression has been proven helpful in recognizing the NPs toxicity and its relevant molecular mechanism [21–25]. Application of toxicogenomics in NPs research can significantly contribute to unravel the mechanistic action of toxicity, parallel to traditional approaches and other omic technologies. Quantitation of mRNA transcripts in NPs treated cells not only provide the mechanistic information, rather it also provides "genetic fingerprinting" from the pattern of gene expression that the NPs elicit in vitro and in vivo. In this chapter, we have discusssed the cellular toxicity of NPs and the importance of toxicogenomics approaches in nanotechnology and NPs induced alterations in transcriptomic profile reportedly studied based on the RNA-Seq, and microarray techniques.

9.2 Toxicogenomics in Nanoparticles Research

Several biomarkers including mRNA transcripts, protein and enzyme expressions have been widely applied as health indicators for the range

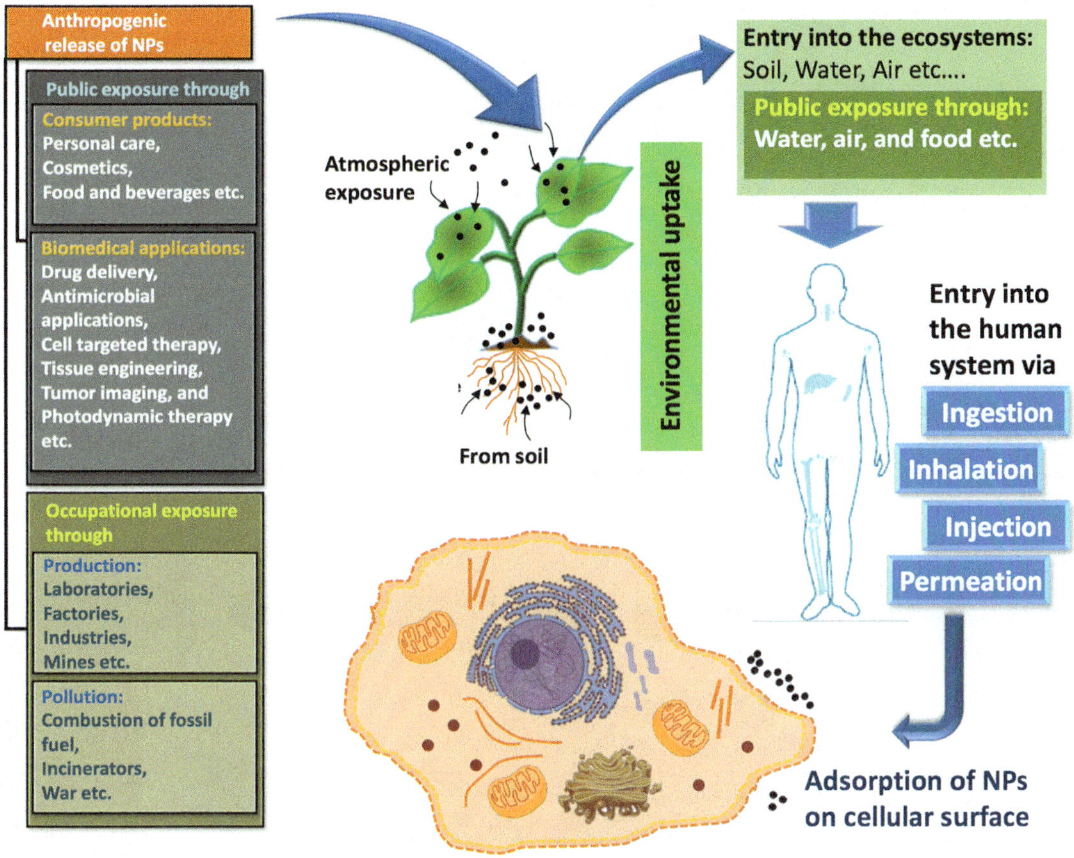

Fig. 9.1 Cellular interaction and uptake of natural and anthropogenic NPs

of human diseases, also it has been utilized to examine the xenobiotics induced changes in the health status of different model and non-model organisms [26–28]. The technology is based on the principle that the mRNA that codes for proteins, are expressed differently in an unexposed organism, vis-a-vis the xenobiotic exposed organisms. The quantification of genome-wide mRNA level by transcriptomic approaches involve the techniques like RNA sequencing (RNA-Seq) and oligonucleotide hybridization (microarray). Toxicogenomics, primarily comprising the hybridization technologies, has been a preferred choice in modern toxicological research. The vast data output and pathway based information represents toxicogenomics as a powerful approach, which has been used now for over decades for identifying novel mechanism of toxicity, disturbance in vital biological pathways, as well as biomarkers of toxicity [6]. In fact, the imperative benefit of toxicogenomics is the holistic approach, which provides a platform for discerning the genetic level changes from the perspective of altered pathways and networks revealing the novel mechanisms of toxicity and toxic responses (Fig. 9.2).

9.3 Nanoparticles Toxicity Analysis by RNA-seq

RNA-seq, being a novel and state-of-the art technique, allows the robust quantification of RNA transcripts on a genome-wide level [29]. Several properties of RNA-seq viz. accuracy and higher dynamic range allow detection of alternative splicing, and no preexisting knowledge of the genomic sequence makes it more advantages

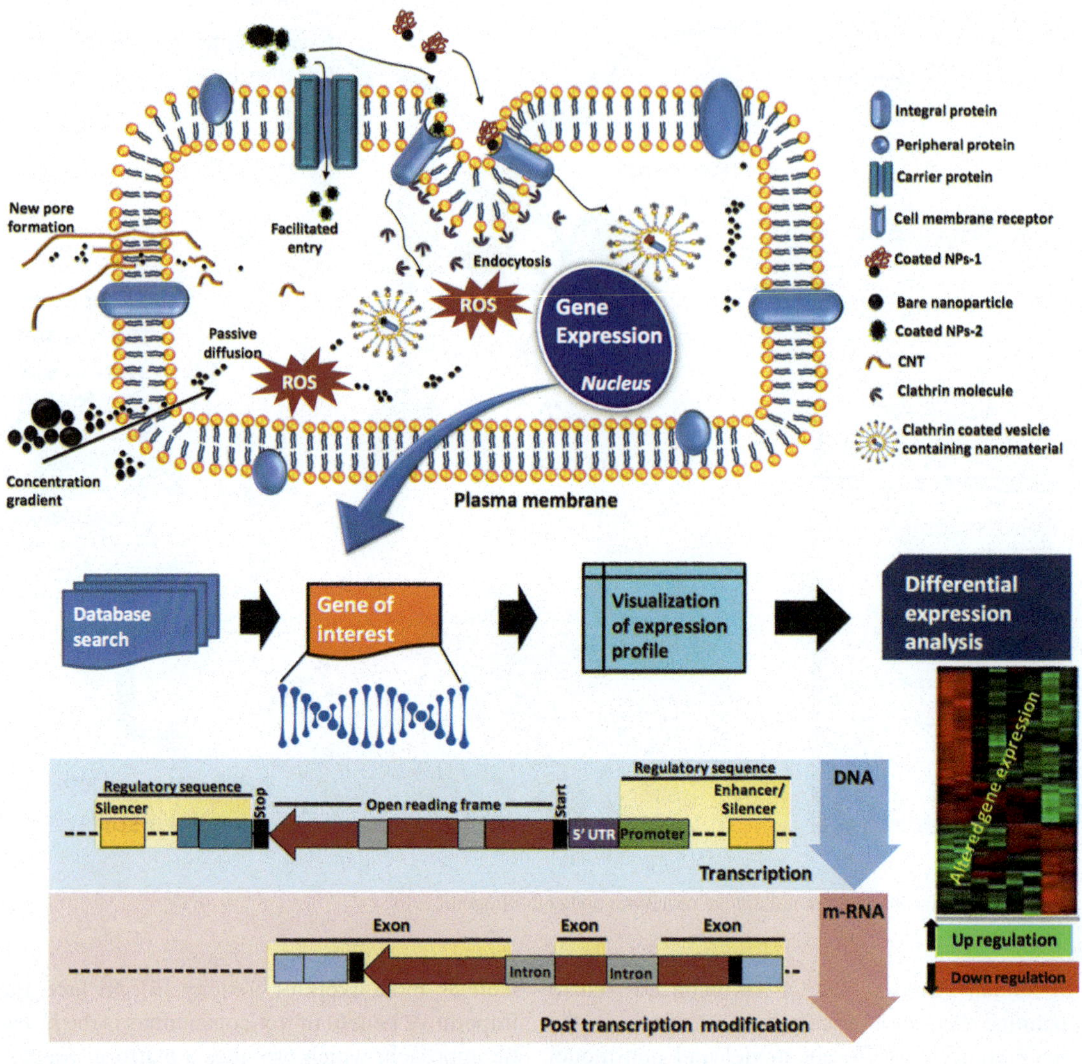

Fig. 9.2 Toxicogenomic approaches for evaluation of NPs toxicity

over microarray [29]. Consequently, the RNA-seq has recently gain entry into the field of nanotoxicology, bound to supersede microarrays in the toxicogenomics field [30]. RNA-seq can generate vast inventory of gene transcripts employing bioinformatics, DNA sequencing and sequence databases [29]. The low-abundance transcripts (approximately 30% of most transcriptomes) can easily be quantified by RNA-seq, also the technique can identify splice junctions and novel exons. RNA-seq accuracy and precision is equally comparable to quantitative real-time PCR [31, 32]. Data from various test models revealed that, about 40% changes in the protein level can be easily explained by knowing the mRNA abundance [33, 34]. Therefore, the application of RNA-seq based transcriptome profiling in NPs toxicity can provide detailed and substantial information, which can complement the results of other approaches.

The RNA-seq analysis of eukaryotic green alga *Chlamydomonas reinhardtii* exposed to metallic NPs (Ag, TiO_2, ZnO and CdTe/CdS quantum dots) revealed specific and different effects. Of the 1.2×10^8 total reads, 5.0×10^7 (42%) mapped uniquely with a maximum of two mismatches and no deletions or insertions. *Chlamydomonas reinhardtii* exposed to TiO_2, ZnO, Ag-NPs and quantum dots (QDs) exhibited 96, 156, 141 and 49 upregulated genes, whereas 80, 29, 86 and 55 genes were found downregulated. NPs exposure (TiO_2, ZnO, and QDs) increased the levels of transcripts encoding subunits of the proteasome, suggesting proteasome inhibition, which is regarded as inducer of several major diseases, including Alzheimer's disease, and used in chemotherapy against multiple myeloma [35]. The RNA-seq analysis of MCF-7 cells treated with fullerene derivative too exhibited strong negative effect on a number of fundamental and interconnected biological processes involved in cell growth and proliferation, mainly including mRNA transcription, protein synthesis/maturation, and cell cycle progression. Comparatively, a fullerene derivative-1 in the same study exhibited no cytotoxicity, although some pathway overlaps were observed with a another fullerene derivative-2 (Fig. 9.3).

Gene ontology (GO) analysis of fullerene derivatives reveled the repression of transcriptional activators (*DHX9, NR2F1, GATA4, AHR*), pre-mRNA complex components (*HNRNPM, WDR77, POP1*) and the transcription elongator factor elongin A (*TCEB3*). Ribosome biogenesis was negatively affected, also the three fundamental genes of pre-rRNA (*RRP9, BOP1, UTP20*) responsible for the maturation and two major subunits of RNA polymerase III (*POLR3B, POLR3G*) were found down regulated. The down regulation of mentioned genes also repressed the protein synthesis, particularly the maturation (*SEC11C, SRPRB*) involved in the recognition and processing of signal peptide. In addition, molecular chaperones responsible for the folding of newly synthetized proteins, including *HSP70* (*HSPA1A, HSPA8*) and *HSP90* (*HSP90AA1*) were down regulated [36].

RNA-seq of nano-hybrid (made of a gold nanoparticle core and a peptide coating; P12) treated human peripheral blood mononuclear cells (PBMNC) exhibited anti-inflammatory effects. Global gene expression of PBMNC exposed to P12 revealed suppression of 233 genes upregulated by LPS stimulation, and 29 genes downregulated by LPS. Overall, ca. 40% of genes that were upregulated by LPS in human PBMNC were suppressed by P12 (Fig. 9.4).

P12 exposure resulted in the activation of different signalling pathways including PKR, interferon, TLR, chemokine, JAK-STAT and TNF [37]. Deep sequencing-based RNA-seq analysis indicated 45 differentially expressed genes in living *Hydra vulgaris*, when exposed to SiO_2-NPs. Among these genes, 29 transcripts were upregulated (2-fold to 25-fold) and 16 were downregulated (2.2-fold to 5-fold). The authors have concluded that a sizeable number of genes remain unknown, providing a valid source of functional information to be further investigated [38]. Fifth instar of silkworm (*Bombyx mori*) when exposed to TiO_2-NPs exhibited differential expression of 11,268 genes in the silkworm fat body, out of which 341 genes showed significant differences, among which 138 were upregulated and 203 were downregulated (Fig. 9.5I). The GO map exhibited eleven biological processes accounting >10% of the annotated genes related to metabolic, cellular and single-organism process showing the highest percentages of annotated genes. Five cellular component subgroups >10% of annotated genes having cell, cell part and organelle showing the highest percentages of annotated genes. Three molecular functions related to binding, catalytic activity and structural molecular activity accounted for 10% or more of annotated genes have been observed in the silkworm (Fig. 9.5II). Further processing the RNA-seq data by KEGG (Kyoto Encyclopedia of Genes and Genomes) scatterplots indicated significant enrichments of all differentially expressed genes. The major affected pathways by TiO_2-NPs exposure in silkworm, includes insulin signaling pathway, which mediate primarily in insect growth and development, nutrient metabolism, lipids and carbohydrates homeostasis and protein synthesis (Fig. 9.5III) [39]. Transcriptome study on the effects of Ag^+ and Ag-NPs in earthworm *Eisenia fetida* employing RNA-seq reveled that Ag^+ verses control yielded 529 and 618 downregulated and upregulated transcripts. *Eisenia fetida*

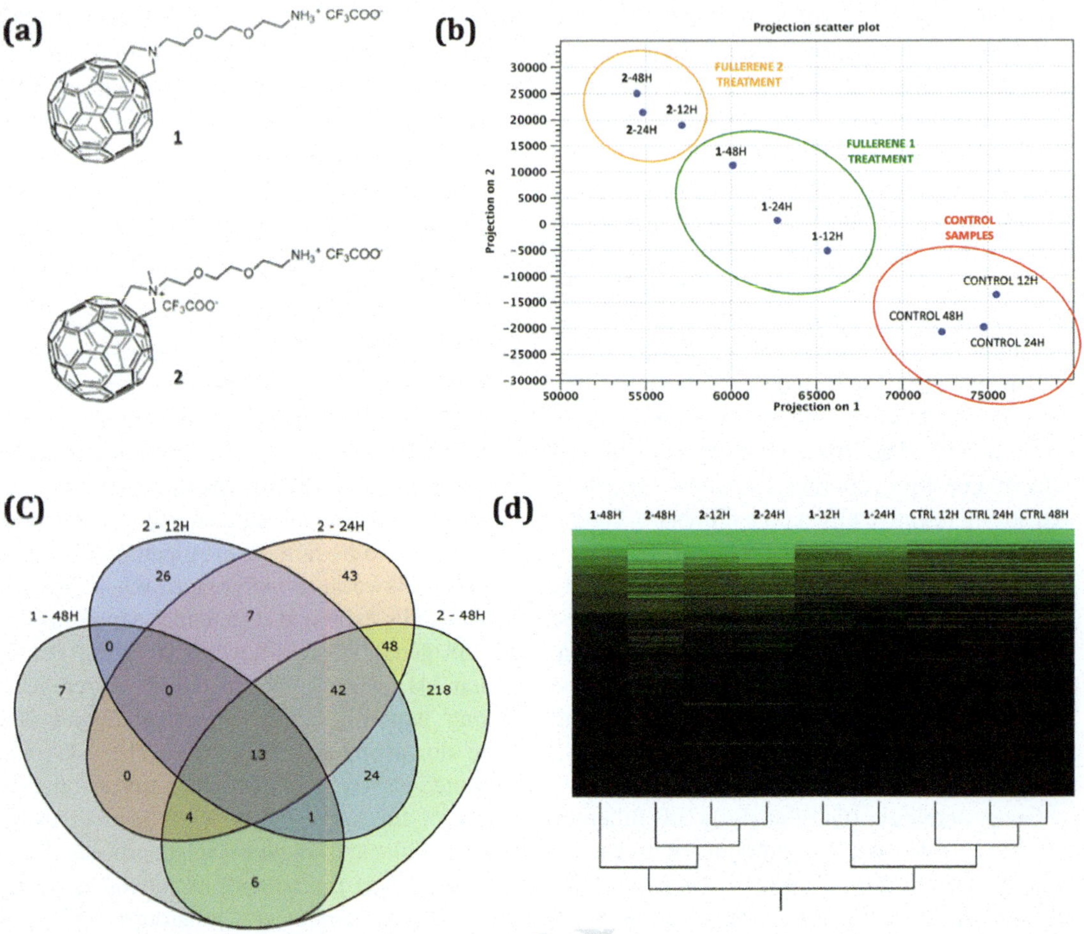

Fig. 9.3 (**a**) Structures of the two fullerenes *1* and *2*. (**b**) Principal component analysis showing the relationship between gene expression profiles of the six fullerene-treated and of the three control samples; a compound-dependent effect is visible, with fullerene 1 leading to minor effects compared to 2, as well as a time-dependent progressive deviation from the control samples. (**c**) Venn diagram depicting the differentially expressed genes identified by the Kal's Z-test on proportions common to the different experimental time points. (**d**) Hierarchical clustering of samples (Euclidean distance, complete linkage) based on the RNA-seq gene expression profiles (Reprinted from Toxicology, volume 314, Lucafò et al. [36], Profiling the molecular mechanism of fullerene cytotoxicity on tumor cells by RNA-seq, pages 183–192. Copyright (2013), with permission from Elsevier. See the reference list for full citation of proper credited)

exposed to Ag-NPs verses control exhibited 237 and 454 downregulated and upregulated transcripts, while Ag-NP verses Ag^+ showed 449 and 758 downregulated and upregulated transcripts. These alterations were related with the toxicity through pathways related to sugar and protein metabolism, disruption of energy production, ribosome function, molecular stress and histones gene alteration [40].

9.4 Microarray Analysis of Nanoparticles Toxicity

Microarrays analysis of genome changes is based on the probe, which is a pre-prescribed set of tens of thousands of genes at once. Comparative to RNA-seq, microarray is relatively cheaper, allowing multiple comparisons across treatments or individuals within an experiment. The transcrip-

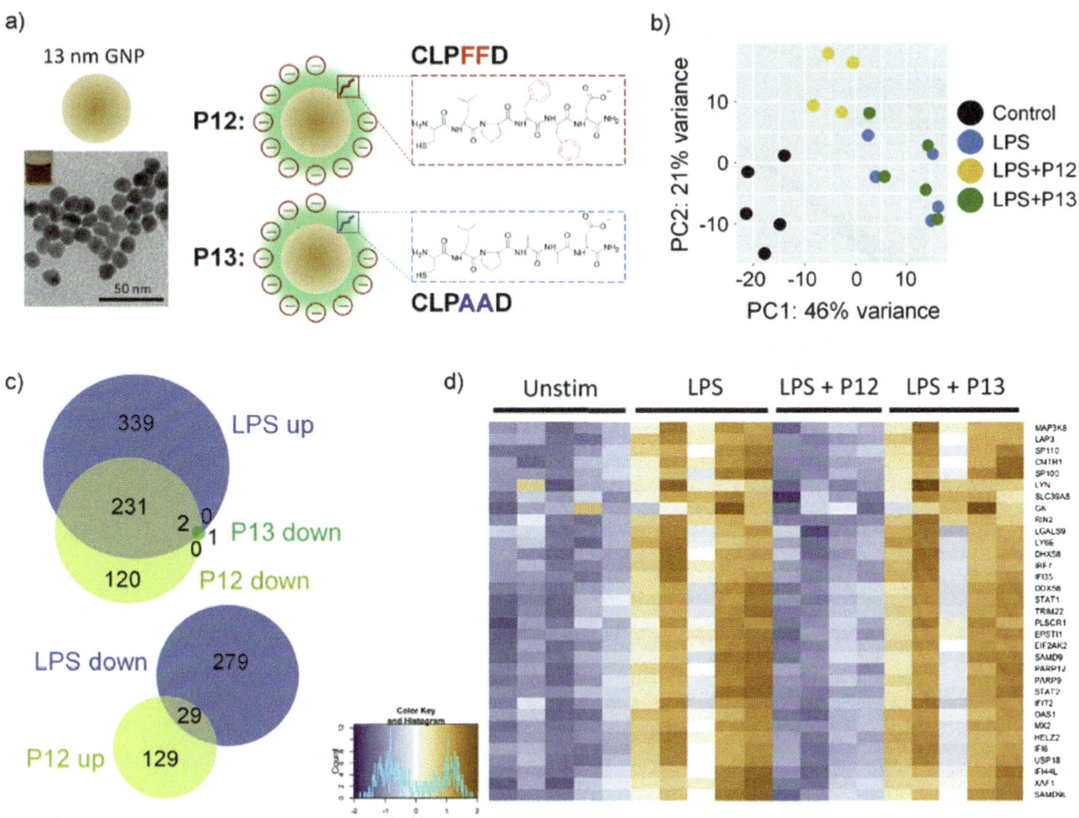

Fig. 9.4 Impact of anti-inflammatory nanoparticle P12 treatment on RNA-Seq transcriptome analysis of LPS-stimulated human PBMC. (**a**) Novel peptide-decorated nanoparticle hybrids with different surface chemistry (P12 vs. P13). (**b**) PCA plot. (**c**) Venn diagrams. (**d**) Differential expression profiles of top 33 genes ($p < 2.6 \times 10^{-15}$ and fold change >1.5) (Reprinted from Biomaterials, volume 111, Yang et al. [37], Endosomal pH modulation by peptide-gold nanoparticle hybrids enables potent anti-inflammatory activity in phagocytic immune cells, pages 90–102, Copyright (2016), with permission from Elsevier. See the reference list for full citation of proper credited)

tomic profiling using GeneChip or microarray technique allow the analysis of global gene expression in NPs exposed and unexposed test models. Exposure of retinoblastoma cell line (Y-79) with etoposide loaded NPs formulations modulated its gene activity. Microarray analysis exhibited differential expression of genes, showing upregulation of 171 genes, while 280 genes were found downregulated. The upregulated genes were mostly belonging to three groups including apoptosis, cell cycle and cell differentiation, and cell migration [41]. Time dependent (24 h and 48 h) exposure of zebrafish embryo with Ag-NPs, Ag-bulk, and Ag^+ exhibited differential expression of genes in the microarray analysis. After the 24 h post fertilization, embryos treated with Ag-NPs, Ag-bulk, and Ag^+ exhibited 35%, 71% and 89% of downregulated genes, while 65%, 29% and 11% genes were found upregulated. After 48 h of post fertilization 57%, 43%, 61% of genes were downregulated, and 43%, 57% and 39% genes were upregulated. The most significant over-represented in GO terms and KEGG pathways in all treatments at 24 h and 48 h were ribosome and oxidative phosphorylation. The prominent overlaps revealed that the toxicity of Ag-NP and Ag-bulk to zebrafish embryos has been predominantly associated with the toxicity of free Ag^+ [25]. The microarray analysis of human THP-1 derived macrophages exposed to single walled carbon nano tubes (SWCNT) reveled statistical significance of gene

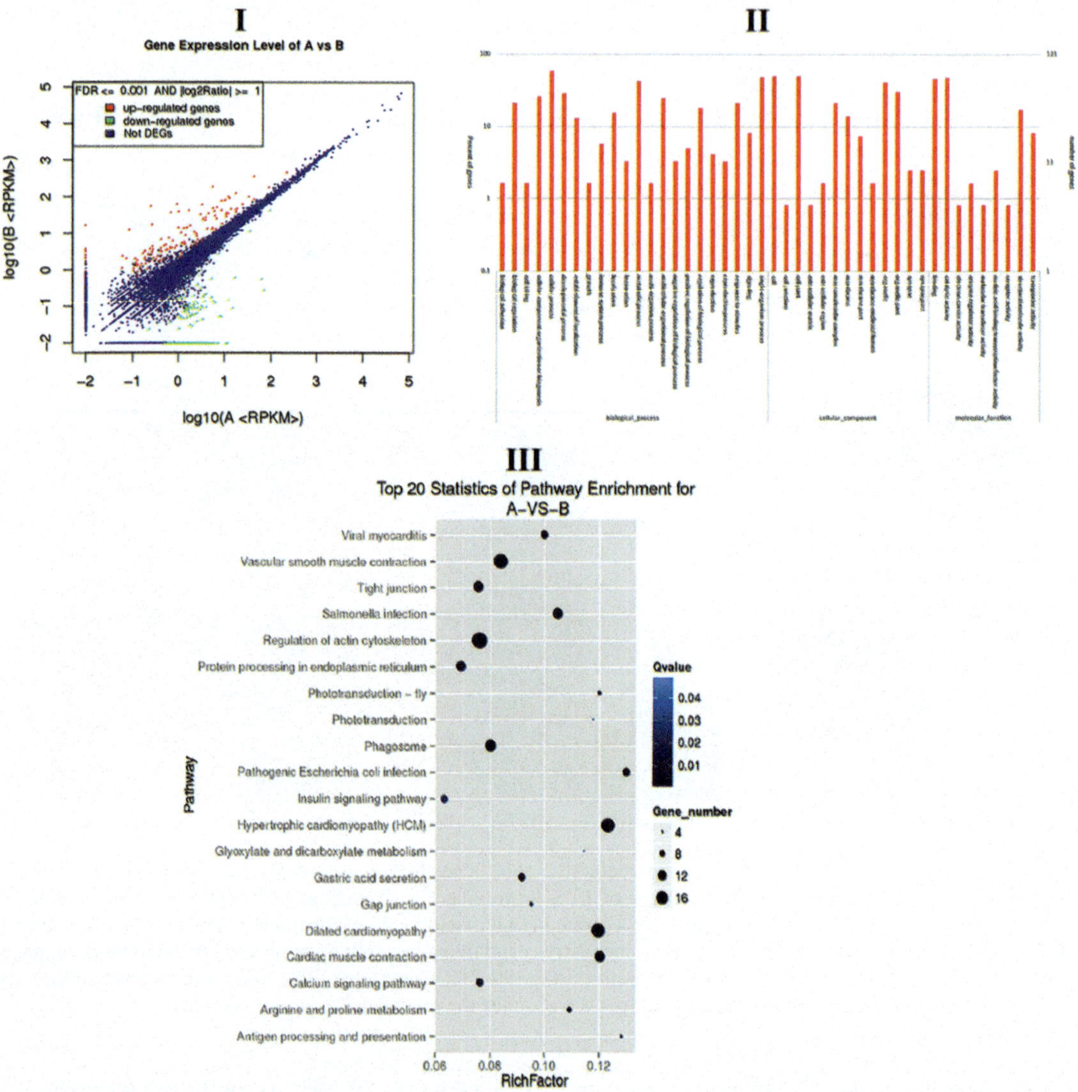

Fig. 9.5 (**a**) Statistical chart of significantly differentially expressed genes. A- represents the control group, while B- represents the experimental group. *RPKM* indicates the gene expression in samples. Red represents the upregulated genes in the figure; green represents downregulated genes; blue represents genes without significant differences. (**b**) Functional classification of significantly differentially expressed genes. A- represents the control group while B- represents the TiO_2-NP treatment group. The right ordinate represents the number of genes, with the maximum value of 121 indicating that a total of 121 genes underwent GO function classification. The left vertical axis represents the percentage of genes, indicating the percentage of functional genes to all annotated genes. (**c**) Scatter plot of KEGG pathway enrichment statistics. A- represents control group, and B- represents experimental group. Rich factor is the ratio of numbers of differentially expressed genes annotated in this pathway term to the numbers of all genes annotated in this pathway term. Greater rich factor means greater intensiveness. Q-value is corrected P-value ranging from 0 to 1, with a lower value means greater intensiveness. Top 20 pathway terms enriched are displayed in the figure (Adopted from Figs. 9.1, 9.2, and 9.3 of Tian et al. [39] (Copyright and Permission granted from the Biology Open Journal under the terms of the Creative Commons Attribution License (http://creativecommons.org/licenses/by/3.0), which permits unrestricted use, distribution and reproduction in any medium provided that the original work is properly attributed) (See the reference list for full citation of proper credited)

expression. Authors have analyzed the abundance changes in 406 pathways archived in KEGG and BioCarta databases. Microarray data exhibited significant alterations of at least 23 different vital pathways in SWCNT treated THP-1 derived macrophages. Considering the network data, authors have suggested that SWCNT uptake into macrophages activated the nuclear factor κB (NF-κB) and activator protein 1 (AP-1), leading to oxidative stress, proinflammatory cytokines activation, recruitment of leukocytes, induction of apoptotic genes and T cells [42].

In the same connection, human macrophage cell line exposed to 5 and 100 nm Ag-NPs were analyzed for approximately 28,000 cDNA profiles using GeneChip(R). Array profile revealed the expression of 45 genes between 5 nm Ag-NPs and the control, and 30 genes between 5 and 100 nm Ag-NPs. The stress genes (*HSP, HO, MT*) and one cytokine gene (*IL-8*) were the top expressed genes in 5 nm Ag-NPs treated cells. Also, the expression of *HO1, HSP*-70 and *MT-1E* RNA was significantly increased as the log2 ratio >2.0. *IL-8* was the only cytokine gene that was significantly induced among more than 70 cytokines [43]. Whole genome microarray analysis of the early gene expression changes induced by 10 and 500 nm amorphous silica NPs exhibited that the magnitude of change for the majority of genes affected were more tightly correlated with particle surface area, rather than either particle mass or number. The microarray data exhibited significant alteration of 503 and 502 genes by at least two-fold with either 10 or 500 nm silica NPs. The union of these gene sets included 753 genes, with 252 genes overlapping between the two particle size groups. Genes which were highly affected are known to play important role in lung inflammation, including *Cxcl2, Ccl4* (*MIP-1β*) and *Ccl3* (*MIP-1α*), along with the chemokine receptor *Cxcr4*. Nonetheless, gene set enrichment analysis revealed that among 1009 total biological processes, none were statistically enriched in one particle size group over the other [44]. Global gene expression of 11 nm dimercaptosuccinic acid (DMSA)-coated Fe-NPs on two cells (THP-1 and HepG2) revealed differential responses. Within ten top upregulated genes, Fe-NPs treated THP-1 cells induced maximum upregulation of *Cxcl13*, a humoral immune response gene. Other immune response genes upregulated were *Adamdec1, Ly96, Ifi44l, Ebi3, Clec7a*. While, in HepG2 cells treated with low and high doses of Fe-NPs induced maximum upregulation three genes (*Ifi27, Tagln and Ifi61*) (Fig. 9.6I). Four way Venn diagram revealed that two genes (*Ddx58* and *Ifi27*) were commonly induced by low and high doses of FeNPs in both cell lines (Fig. 9.6IIa). Eleven genes (*Ddx58, Ifi44, Fbxo16, Parp9, Ifit3, Serpini1, Ifi27, Nexn, Usp25, Rg9mtd2 Ccne2*) were commonly induced by low dose of FeNPs in both cell lines. On the other hand, ten genes (*Tmed2, Ifi27, Ddx58, Akap12, Nampt, Narg1, Usp16, Ifi6, Col9a2, and Zcchc2*) were commonly induced by high dose of Fe-NPs in THP-1 and HepG2 cells. However, no genes were commonly repressed at both doses of Fe-NPs in the cell lines (Fig. 9.6IIb). The authors also developed a heat map of top 55 genes commonly regulated in these two cell lines. The hierarchical clustering revealed that these genes were classified into four clusters. Some genes were steadily upregulated or downregulated in both types of cells, indicating cell-independent effects. Whereas some genes were inversely regulated in the two cell lines, suggesting cell-specific effects of the Fe-NPs (Fig. 9.7I). Interestingly, the authors have reported that Fe-NPs significantly enrich the hepatitis C pathway at both doses of treatments in the THP-1 cells (Fig. 9.7II) [45].

Kedziorek et al. [21] have examined the genome changes in *LacZ*-expressing mouse NSC cell line C17.2 exposed to superparamagnetic iron oxide nanoparticles (SPIONs). Microarray analysis of 2695 probe sets, representing 1399 genes, revealed that genes belonging to molecular functions of the cell describing activities such as catalytic reactions or binding that occur at the molecular level were significantly affected by the SPIONs exposure. A significant change in 970 genes of cellular function, 299 genes of gene expression, 312 genes of developmental, 431 genes of biologic regulation and 685 genes belonging to metabolic processes, have been reported. The treated cells also exhibited significant alterations in genes belonging to secretion,

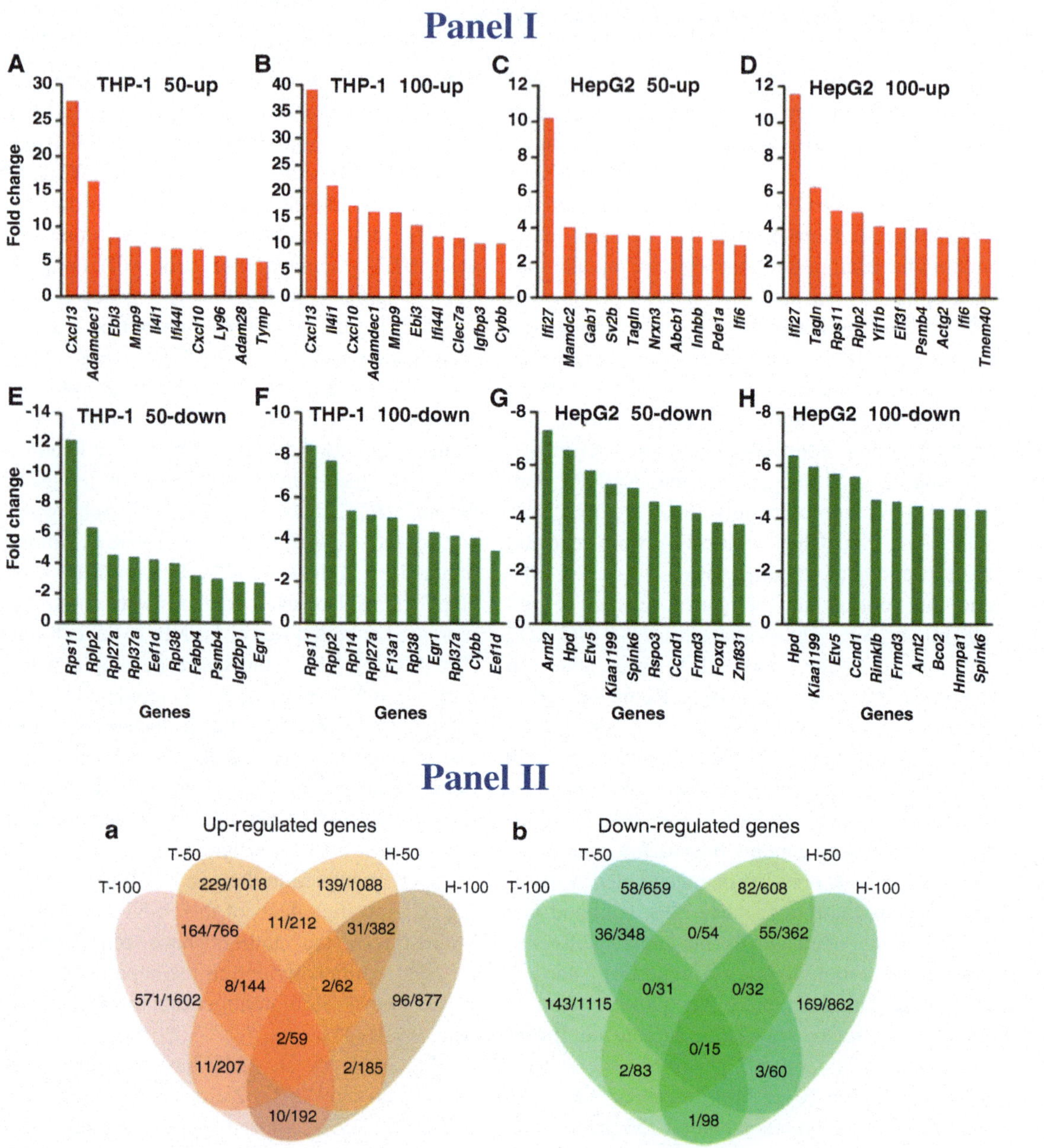

Fig. 9.6 *Panel I* showing the top 10 genes with the highest expression changes in the two cell types after treatment with Fe-NPs. (**a**, **b**) Induced genes in the THP-1 cells. (**c**, **d**) Induced genes in the HepG2 cells. (**e**, **f**) Repressed genes in the THP-1 cells. (**g**, **h**) Repressed genes in the HepG2 cells. 50-up and 100-up: induced genes in cells treated with 50 μg/mL and 100 μg/mL of FeNPs, respectively; 50-down and 100-down: repressed genes in cells treated with 50 μg/mL and 100 μg/mL of FeNPs, respectively. Some uncharacterized genes with fold changes greater than the lowest fold changes in these plots are not shown in this figure. *Panel II* Comparison of FeRGs in the THP-1 and HepG2 cells. **a** Comparison of induced genes in the two cell lines. **b** Comparison of repressed genes in the two cell lines. Each Venn diagram is divided into four areas labeled as T-50, T-100, H-50 and H-100. T-50 and T-100, THP-1 treated with 50 μg/mL and 100 μg/mL of FeNPs, respectively. H-50 and H-100, HepG2 cells treated with 50 μg/mL and 100 μg/mL of FeNPs, respectively. The number in overlapped area represents the overlapping genes. The numbers before and after the slash represent the genes with fold changes greater than 2 and 1.5, respectively (Reused from Figs. 9.3 and 9.4 of Zhang et al. [45] (Copyright and Permission granted from the Journal of Nanobiotechnology under the terms of the Creative Commons Attribution License (http://creativecommons.org/licenses/by/4.0), which permits unrestricted use, distribution, and reproduction in any medium, provided the original work is properly credited) (See the reference list for full citation of proper credited)

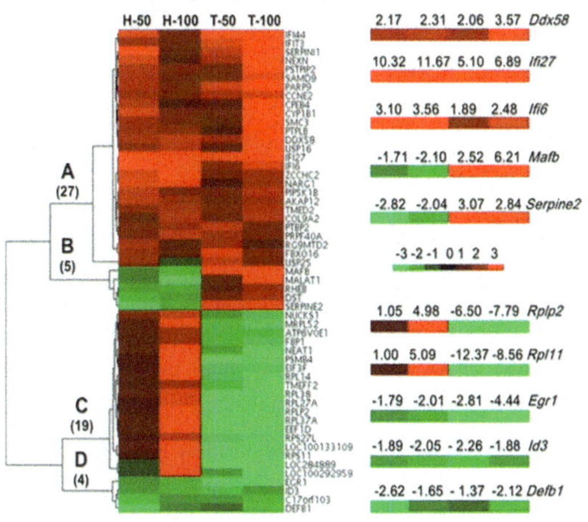

I

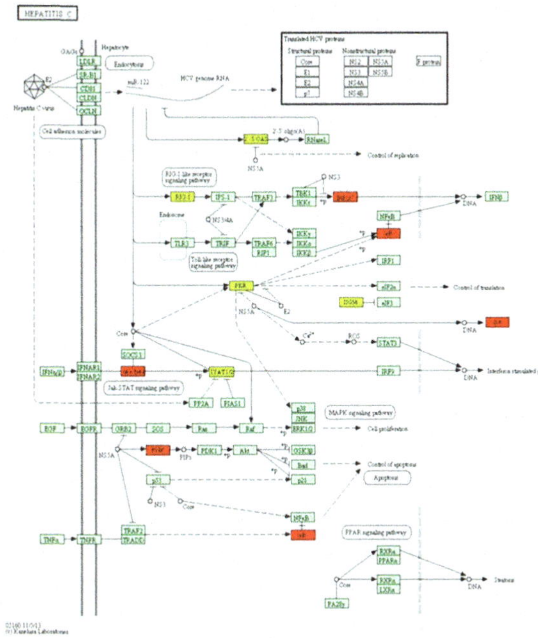

II

Fig. 9.7 (**I**) Cluster analysis of genes. Fifty-five FeRGs from the two cell lines were clustered according to their expression levels using a hierarchical clustering. The heatmap was drawn with Java TreeView. Red and green represent up- and down-regulation, respectively. The color depth reflects the expression level between −3 and +3 (marker). The numbers of genes in Clusters *A* to *D* are shown in parentheses. The fold changes of ten representative genes in four clusters are shown in the zoomed images. (**II**) KEGG pathway of hepatitis C in the FeNP-treated THP-1 cells. The genes in red refer to the FeRGs induced by 100 µg/mL of FeNPs. The genes in yellow refer to the FeRGs induced by both 50 µg/mL and 100 µg/mL of FeNPs. Abbreviations for the KEGG parameters can be found on the KEGG pathway webpage (Reused from Figs. 9.6 and 11 of Zhang et al. [45] (Copyright and Permission granted from the Journal of Nanobiotechnology under the terms of the Creative Commons Attribution License (http://creativecommons.org/licenses/by/4.0), which permits unrestricted use, distribution, and reproduction in any medium, provided the original work is properly credited) (See the reference list for full citation of proper credited)

transport, locomotion, reproductive processes, and establishment of localization. In addition, the expressions of heme oxygenase 1 (*Hmox1*) and transferrin receptor 1 (*Tfrc*) were repressed, while, genes responsible for detoxification (*Clu, Cp, Gstm2, and Mgst1*) and lysosomal function (*Sulf1*) were upregulated at later time points. Liu and Wang [24] have studied changes in the global gene expression of mouse macrophage (RAW264.7 cells) exposed to DMSA-Coated Fe_3O_4-NPs using GeneChips Mouse Genome 430 2.0 microarrays. The GO analysis revealed that several molecular functions and biological processes pertaining to metal ion transmembrane transporter activity, especially Fe ion transmembrane transporter activity and Fe ion binding, were significantly enriched in all DEGs. *Tfrc*, *Trf*, and *Lcn2* genes important to iron metabolism were frequently found in the GO terms. The microarray data of lung epithelial cell line (A549) exposed to 12.1 µg/ml Ag-NPs (EC20) for 24 and 48 h exhibited altered gene regulation of more than 1000 genes (>2-fold), while considerably fewer genes responded to Ag^+ (133 genes). The upregulated genes were belonging to the members of heat shock protein, metallothionein and histone families [23]. Human skin fibroblasts (HSF42) and human embryonic lung fibroblasts (IMR-90), both untransformed cells were exposed to multiwalled carbon nanotubes

(MWCNTs) and multiwalled carbon nano-onions (MWCNOs) exhibited profound gene expression changes, analyzed on new generation Affymetrix HTA GeneChip system. The low doses of both materials induced expressional changes in genes of secretory pathway, protein metabolism, golgi vesicle transport, fatty acid biosynthesis, and G_1/S transition of mitotic cell cycle. Also, an additional group of genes, involved in protein ubiquitination was found upregulated. Contrarily, high doses upregulated the genes in tRNA aminoacylation and amino acid metabolism pathways. Genes of inflammatory and immune response were also found upregulated [46].

BV2 microglia, which is an immortalized mouse cell line, when exposed to a mixture of anatase (70%) and rutile (30%) TiO_2-NPs (Degussia P25) exhibited global gene expressional changes in the microarray. The core canonical analysis reveled that BV2 exposed to P25 upregulated the signaling processes involved in B-cell death, ERK/MAPK receptors, apoptosis, calcium, and inflammation. P25 up-regulated the inflammatory (*NF-κB*), cell cycling and pro-apoptotic toxicity pathways. Core analysis of P25 induced downregulation of genes exhibited alteration of adaptive change and key energy production pathways, mainly associated with hypoxia, peroxisomes, and *Nrf2*-mediated oxidative stress [47]. In microarray analysis, male Sprague-Dawley rats fed with Synthetic Amorphous Silica (SAS) and NM-202 (a representative nanostructured silica for OECD testing) for 24 and 84 days exhibited non significant gene alteration in jejunal epithelial samples and liver homogenates. Although, fibrosis-related gene expression was significantly affected for NM-202 treated animals after 84-days of exposure, but not for SAS treated animals [48]. Osmond-McLeod et al. [49] demonstrated the transcriptome changes by the application of NPs based sunscreens. Mice treated with both the TiO_2-NPs sunscreen and UVR exposure (TiO_2-NPs + UVR) showed very low levels of differential regulation, as compared to untreated mice (Control-UVR) (14 genes). Pathway analysis exhibited that three of the top 5 canonical pathways in TiO_2-NPs + UVR were linked with metabolic functions (Heme Biosynthesis II; Tetrapyrrole Biosynthesis II; Mevalonate Pathway I). In addition, breast cancer regulation Stathmin1 and circadian rhythm signalling were also affected in the canonical pathways. While, ZnO-NPs sunscreen with UVR (ZnO-NPs + UVR) and ZnO-NPs sunscreen without UVR (ZnO-NPs-UVR) did not showed any transcriptome alterations [49].

HaCaT cells exposed to tungsten carbide (WC) and tungsten carbide cobalt (WC-Co) NPs exhibited whole genome transcription alterations. Fluorescence signal of microarray in all treatments revealed 1956 upregulated and 1146 downregulated differentially expressed genes, with more than two-fold expression level. HaCaT cells exposure to $CoCl_2$ salt, as metal source of Co induced strongest changes in the gene expression (373 and 826 genes for 3 h and 3 day) followed by WC-Co (37 and 248) and WC-NPs (28 and 49), respectively. Data analysis by enrichment method exhibited the fact that differentially expressed genes were related to hypoxia, endocrine pathways, carbohydrate metabolism, and targets of several transcription factors [50]. Human lung epithelial cells (A549) exposed to silica-NPs exhibited a dose dependent response, ranging from 5 to 2258 significantly differentially modulated transcripts compared with controls, with a fold change of at least 1.5 and p-value <0.05. Canonical analysis exhibited the coagulation system and intrinsic and extrinsic prothrombin activation pathways as most highly altered. Additionally, the acute phase response, xenobiotic metabolism, TREM1 signaling pathways and oxidative stress response were altered. The authors have extended the transcriptome into exproteome to understanding the NPs effect on proteins. Heat-shock proteins (*HSP70* and *HSP90*), detoxification enzymes such as glutathione reductase (*GSR*), glutathione S-transferase (*GSTP1*), lactate dehydrogenase (*LDH*A), peroxiredoxins (*PRDX*1, *PRDX6*), thioredoxin reductase (*TRXR1*), protein disulfide isomerase (*PDIA6*), and aldo-keto reductases (*AKR1B1*, *AKR1B10*, *AKR1C1/C2*, and *AKR1C3*) were found as affected proteins in the exproteome analysis [51]. Human intestinal epithelium model (Caco-2) cell line, when exposed to pristine (surface untreated) CeO_2-NPs exhibited 1643 modified genes. Comparatively, the manufactured CeO_2 Nanobyk™ NPs in the same study have not affected the gene regulation, while 344 and 428 modified genes were found for light (NB-DL) and acid (NB-DA) degraded CeO_2 Nanobyk™ NPs. Pristine CeO_2-NPs exhibited changes in the cellular growth and proliferation (274 genes) and cell death (265 genes) biological process. The canonical pathway analysis of pristine CeO_2-NPs revealed that it alters the mitochondrial function through the under expression of 27 genes of complexes I, III, IV and V [52]. Fisichella et al. [53] in their previous study on pangenomic oligomicroarrays (4 × 44,000 genes) demonstrated that TiO_2-STNPs have not altered the gene expression of Caco-2 cells when exposed to the highest concentration of 10 μg/ml.

Al_2O_3-NPs exposed human bronchial epithelial (HBE) cells significantly increased expression of 54 genes and decreased expression of 304 genes. GO analysis unravel the fact that total genes encoding proteins necessary for mitochondrial function were differentially expressed. KEGG pathway enrichment analysis of these 27 genes of mitochondrial function and neural system disease were significantly enriched. *NDUFA2, NDUFS, NDUFC2, NDUFA1, NDUFA4, UQCR11 (complex III), COX7B, COX17* (complex IV) and *ATP5H* (complex V, F0 unit) were among the most affected genes (Fig. 9.8) [54].

Global gene expression in the HepG2 cells upon 20 and 50 nm Ag-NPs treatment exhibited differential regulation of genes. After short exposure of 4 h, the 20 nm Ag-NPs induced alterations of 811 genes, out of which 649 were upregulated and 162 were downregulated. Comparatively, the 50 nm Ag-NPs treatment induced stark difference, only 21 genes were altered and all of them were upregulated. Extended exposure of 24 h did not made any massive alterations in the gene expression by both sizes of Ag-NPs. Overlapping of DEGs exhibited alterations of five common genes after 4 and 24 h of 20 nm Ag-NPs exposure, including members of the metallothionein (MT) family (*MT1B, MT1F, MT1G, MT1M* and *MT2A*). The 50 nm Ag-NPs exposure also showed four common genes of MT family, except an additional activation of SOX4 gene. Overall, *MT1B* and *MT1M* were shared by all treatment groups. Out of 108 GO terms derived from the 649 upregulated genes in 4 h 20 nm of Ag-NPs revealed 23 categories, on top listed the metabolism (47%), development (19%), protein metabolism (15%), cell differentiation (13%), biosynthesis (11%), death (9%) and cell communication (9%). The 162 downregulated genes exhibited 21 groups of GO terms enrichment, classified into 9 categories, mostly similar to upregulated GO terms, but with different percentage and orders. Within the pathways analysis, Ag-NPs affected the endocytosis, *MAPK*, *TGF-β*, *p53* signalling pathways, pathways in cancer and *NFR2*-mediated oxidative stress response [55].

9.5 Toxicological Potential of Nanoparticles

The gene expression profiling of TiO_2-NPs of varying sizes and surface properties has been reported to induce pulmonary inflammation. However, the different TiO_2-NPs vary in the magnitude of the inflammatory response induced in a property-dependent manner [56]. Our recent studies demonstrated that TiO_2-NPs preferentially bind in subdomains IB, IIA of HSA and minor groove of DNA [57]. TiO_2-NPs might be able to enter the human stratum corneum and interact with the immune system. Silica NPs are used in the synthesis of cosmetics, foods, drugs, and printing ink tonners, on a large industrial scale. The nanotoxicity of crystalline silica causes chronic obstructive pulmonary diseases such as silicosis [58]. Silica NPs exists in the nature in many diverse forms [59, 60]. Fumed silica showed dose dependent accumulation of alveolar macrophages [61–63]. Cerium oxides (CeO_2)-NPs are one of the most widely used types for UV protection in paints or as fuel additives [64, 65]. CeO_2-NPs can be used as a scavenger of superoxide anions [66, 67]. CeO_2-NPs were shown to exhibit superoxide dismutase (SOD) and catalase enzymes mimetic activities in a redox-state dependent manner. CeO_2-NPs has been shown to hold the neuroprotective

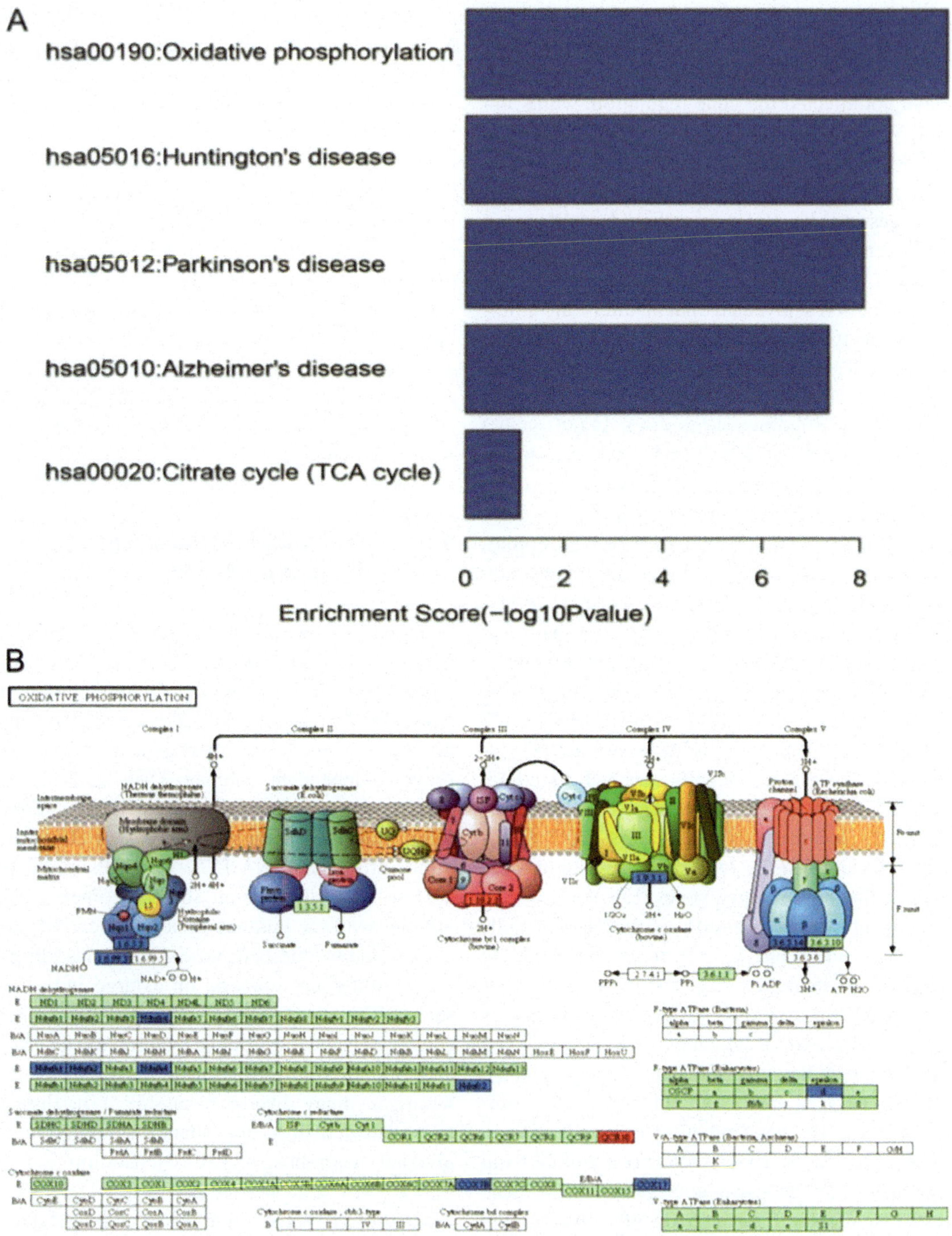

Fig. 9.8 (**a**) KEGG pathway enrichment analysis of mitochondria related genes. **a** A total of 27 mitochondria related genes were analyzed through DAVID functional annotation cluster tool. These genes mainly are involved in five KEGG pathways. (**b**) Oxidative phosphorylation is the most significant enrichment. **b** A schematic figure of the oxidative phosphorylation pathway by KEGG. mRNA microarray assay predicted up-regulated genes are stained red, and down-regulated genes are stained blue in this schematic figure (Reused from Fig. 9.2 of Li et al. [54] (Copyright and Permission granted from the Particle and Fibre Toxicology (http://creativecommons.org/licenses/by/4.0/), which permits unrestricted use, distribution, and reproduction in any medium, provided the original work is properly credited). (See the reference list for full citation of proper credited)

effects [68]. On the other hand, CeO_2-NPs has been reported to be cytotoxic to human hepatoma cells [69].

Ag-NPs have antimicrobial activity and are used in food packaging material, food supplements, odour-preventing textiles, cosmetics, kitchen utensils, toys, electronics, wound dressings, and room sprays [70]. Ag-NPs released Ag ions to exert antimicrobial properties by binding to sulphur- and phosphorous containing biomolecules and also causing damage to mammalian cells [71–75]. The in vivo inhalation data of Ag-NPs showing varied results from a minimal inflammatory response to the presence of inflammatory lesions in the lungs [76–79]. There are studies indicating that the dose-dependent increase of Ag-NPs might stimulate toxicity in the different organs [77, 78]. It has been hypothesized that small Ag-NPs will induce more prominent pulmonary toxicity compared to larger Ag particles because of the larger deposited dose in the alveoli and the higher dissolution rate. The dissolution of Ag-NPs depends on their particle size, the pH of the solution, the ions present in the solution [80–84]. The in vitro results show that the dissolution rate of the 15 nm Ag-NPs will probably be higher compared to the 410 nm Ag-NPs resulting in an increased ion release. None of the studies report complete dissolution of Ag-NPs, the effects observed after exposure to Ag-NPs can be induced by the released ions, the Ag-NPs itself or a combination of both [85]. Ag ions released from Ag-NPs caused more damage inside the nucleus as compared to Ag ions released from silver nitrate [86–88].

The harmful effects of carbon nanotubes (CNTs) to animals and cells appeared almost a decade ago [89]. Several mechanisms of toxicity, similar to the ones linked to asbestos-exposure, have been proposed for CNTs, such as (i) association of fibres with the cell membrane causing physical damage and cell membrane malfunction, (ii) protein-fibre interaction inhibiting protein function, and (iii) induction of ROS, either directly by the CNTs themselves or indirectly through mitochondrial dysfunctions or NADPH oxidase activation induced by so-called frustrated phagocytosis in e.g. macrophages [90–92]. It seems probable that a combination of different mechanisms could contribute to the toxicity of CNTs, as it has been considered to be the case with asbestos [93]. Multiwalled CNTs are efficient scavengers of $^{\bullet}OH$ and superoxide ($^{\bullet}O_2^-$) radicals in cell-free conditions [94]. The generation of free radicals by CNTs was suggested to be related to the amount and nature of defects in the CNTs, i.e. ruptures of the graphene framework [94]. In contrast, ROS formation by single walled CNTs was observed in cell media with and without FE1-Muta™ Mouse lung epithelial cells, at intermediate levels between that of Printex 90 and C60 fullerene and correlated with the order of genotoxicity [95]. Our toxicogenomic analysis on $ZnFe_2O_4$-NPs reveled its cytotoxicity and apoptosis through ROS generation and oxidative stress via *p53*, survivin, *bax/bcl-2* and caspase pathways in WISH cells [96]. We have also demonstrated that ZnO-NPs have the potential to induce DNA damage and alter the mitochondrial membrane potential of human lymphocytes [97]. Previously, we have reported that ZnO-QDs can induce dose dependent apoptosis induction in C2C12, HepG2 and MCF-7 cells via oxidative stress and alterations of apoptosis related genes [98, 99].

9.6 Conclusion and Future Perspective

The incessant use of NPs in various sectors and life domains may pose serious threats to ecosystem and adversely affect the living entities via interactions and accumulation of nanomaterials in the body of the organisms. Toxicogenomics approach notably contributes to our understanding of genetic changes at molecular levels. Significant change in the gene expression levels due to the NPs treatment provides information related to biochemical pathways and mechanism of action of nanotoxicants. Differential gene expression pattern may also yield molecular fingerprints of these nanotoxicants both with the in vitro and in vivo test model systems. Thus, the toxicogenomic methods have the power and potential to change nanotoxicology research landscape.

Acknowledgments SD gratefully acknowledge the financial support given by CSIR, New Delhi, India. The authors extend their appreciation to the International Scientific Partnership Program ISPP at King Saud University for funding this research work through ISPP# 0031.

References

1. Oberdorster G, Maynard A, Donaldson K et al (2005) A report from the ILSI Research Foundation/Risk Science Institute Nanomaterial Toxicity Screening Working Group. Principles for characterizing the potential human health effects from exposure: elements of screening strategy. Part Fibre Toxicol 6:2–8
2. Hoet PHM, Brüske-Hohlfeld I, Salata OV (2004) Nanoparticles: known and unknown health risks. J Nanobiotechnol 8:1–12
3. Salata O (2004) Applications of nanoparticles in biology and medicine. Nano Biotechnol 2:3
4. Wang K, Xu JJ, Chen HY (2005) A novel glucose biosensor based on the nanoscaled cobalt phthalocyanine-glucose oxidase biocomposite. Biosens Bioelectron 20:1388–1396
5. Yang MH, Jiang JH, Yang YH et al (2006) Carbon nanotube/cobalt hexacyanoferrate nanoparticle biopolymer system for the fabrication of biosensors. Biosens Bioelectron 21:1791–1797
6. Chen M, Zhang M, Borlak J et al (2012) A decade of toxicogenomic research and its contribution to toxicological science. Toxicol Sci 130:217–228
7. Ivask A, Bondarenko O, Jepihhina N et al (2010) Profiling of the reactive oxygen species-related ecotoxicity of CuO, ZnO, TiO_2, silver and fullerene nanoparticles using a set of recombinant luminescent Escherichia coli strains: differentiating the impact of particles and solubilised metals. Anal Bioanal Chem 398:701–716
8. Pujalté I, Passagne I, Brouillaud B et al (2011) Cytotoxicity and oxidative stress induced by different metallic nanoparticles on human kidney cells. Part Fibre Toxicol 8:10–26
9. Griffitt TJ, Weil R, Hyndman KA et al (2007) Exposure to copper nanoparticles causes injury and acute lethality in zebrafish (Danio rerio). Environ Sci Technol 41:8178–8186
10. Nair PMG, Choi J (2011) Characterization of a ribosomal protein L15 cDNA from Chironomus riparius (Diptera; Chironomidae): transcriptional regulation by cadmium and silver nanoparticles. Comp Biochem Physiol B Biochem Mol Biol 159:157–162
11. Oberdorster G, Oberdorster E, Oberdorster J (2005) Nanotoxicology: an emerging discipline evolving from studies of ultrafine particles. Environ Health Perspect 113(7):823–839
12. Donaldson K, Tran CL (2002) Inflammation caused by particles and fibers. Inhal Toxicol 14:5–27
13. Kumar A, Dhawan A (2013) Genotoxic and carcinogenic potential of engineered nanoparticles: an update. Arch Toxicol 87(11):1883–1900
14. Duan J, Yu Y, Li Y et al (2013) Cardiovascular toxicity evaluation of silica nanoparticles in endothelial cells and zebrafish model. Biomaterials 34(23):5853–5862
15. Khatri M, Bello D, Gaines P et al (2013) Nanoparticles from photocopiers induce oxidative stress and upper respiratory tract inflammation in healthy volunteers. Nanotoxicology 7(5):1014–1027
16. Shin JA, Lee EJ, Seo SM et al Nanosized titanium dioxide enhanced inflammatory responses in the septic brain of mouse. Neuroscience 165(2):445–454
17. Smita S, Gupta S, Bartonova A et al (2012) Nanoparticles in the environment: assessment using the causal diagram approach. Environ Health 11:S13
18. Chinta SJ, Andersen JK (2008) Redox imbalance in Parkinson's disease. Biochim Biophys Acta 1780(11):1362–1367
19. Husain M, Saber AT, Guo C et al (2013) Pulmonary instillation of low doses of titanium dioxide nanoparticles in mice leads to particle retention and gene expression changes in the absence of inflammation. Toxicol Appl Pharmacol 269:250–262
20. AshaRani P, Hande MP, Valiyaveettil S (2009) Antiproliferative activity of silver nanoparticles. BMC Cell Biol 10:65
21. Kedziorek DA, Muja N, Walczak P et al (2010) Gene expression profiling reveals early cellular responses to intracellular magnetic labeling with superparamagnetic iron oxide nanoparticles. Magn Reson Med 63:1031–1043
22. Li X, He Q, Shi J (2014) Global gene expression analysis of cellular death mechanisms induced by mesoporous silica nanoparticle-based drug delivery system. ACS Nano 8:1309–1320
23. Foldbjerg R, Irving ES, Hayashi Y et al (2012) Global gene expression profiling of human lung epithelial cells after exposure to nanosilver. Toxicol Sci 130:145–157
24. Liu Y, Wang J (2013) Effects of DMSA-coated Fe3O4 nanoparticles on the transcription of genes related to iron and osmosis homeostasis. Toxicol Sci 131:521–536
25. Van Aerle R, Lange A, Moorhouse A et al (2013) Molecular mechanisms of toxicity of silver nanoparticles in zebrafish embryos. Environ Sci Technol 47(14):8005–8014
26. Andersen ME, Krewski D (2009) Toxicity testing in the 21st century: bringing the vision to life. Toxicol Sci 107(2):324–330
27. Yang H, Liu C, Yang D et al (2009) Comparative study of cytotoxicity, oxidative stress and genotoxicity induced by four typical nanomaterials: the role of particle size, shape and composition. J Appl Toxicol 29(1):69–78
28. Van Hummelen P, Sasaki J (2010) State-of-the-art genomics approaches in toxicology. Mutat Res 705(3):165–171

29. Wang Z, Gerstein M, Snyder M (2009) RNA-seq: a revolutionary tool for transcriptomics. Nat Rev Genet 10:57–63
30. Costa PM, Fadeel B (2016) Emerging systems biology approaches in nanotoxicology: towards a mechanism-based understanding of nanomaterial hazard and risk. Toxicol Appl Pharmacol 299:101–111
31. González-Ballester D, Casero D, Cokus S et al (2010) RNA-seq analysis of sulfur-deprived Chlamydomonas cells reveals aspects of acclimation critical for cell survival. Plant Cell 22:2058–2084
32. Miller R, Wu G, Deshpande RR et al (2010) Changes in transcript abundance in Chlamydomonas reinhardtii following nitrogen deprivation predict diversion of metabolism. Plant Physiol 154:1737–1752
33. Maier T, Guell M, Serrano L (2009) Correlation of mRNA and protein in complex biological samples. FEBS Lett 583:3966–3973
34. De Sousa Abreu R, Penalva LO, Marcotte E et al (2009) Global signatures of protein and mRNA expression levels. Mol BioSyst 5:1512–1526
35. Simon DF, Domingos RF, Hauser C et al (2013) Transcriptome sequencing (RNA-seq) analysis of the effects of metal nanoparticle exposure on the transcriptome of Chlamydomonas reinhardtii. Appl Environ Microbiol 79(16):4774–4785
36. Lucafò M, Gerdol M, Pallavicini A et al (2013) Profiling the molecular mechanism of fullerene cytotoxicity on tumor cells by RNA-seq. Toxicology 314(1):183–192. https://doi.org/10.1016/j.tox.2013.10.001
37. Yang H, Kozicky L, Saferali A et al (2016) Endosomal pH modulation by peptide-gold nanoparticle hybrids enables potent anti-inflammatory activity in phagocytic immune cells. Biomaterials 111:90–102. https://doi.org/10.1016/j.biomaterials.2016.09.032
38. Ambrosone A, Scotto di Vettimo MR, Malvindi MA et al (2014) Impact of amorphous SiO2 nanoparticles on a living organism: morphological, behavioral, and molecular biology implications. Front Bioeng Biotechnol 2:37
39. Tian JH, Hu JS, Li FC et al (2016) Effects of TiO_2 nanoparticles on nutrition metabolism in silkworm fat body. Biol Open 5(6):764–769. https://doi.org/10.1242/bio.015610
40. Novo M, Lahive E, Díez-Ortiz M et al (2015) Different routes, same pathways: molecular mechanisms under silver ion and nanoparticle exposures in the soil sentinel *Eisenia fetida*. Environ Pollut 205:385–393
41. Mitra M, Dilnawaz F, Misra R et al (2011) Toxicogenomics of nanoparticulate delivery of etoposide: potential impact on nanotechnology in retinoblastoma therapy. Cancer Nanotechnol 2(1–6):21–36
42. Chou CC, Hsiao HY, Hong QS et al (2008) Single-walled carbon nanotubes can induce pulmonary injury in mouse model. Nano Lett 8(2):437–445
43. Lim DH, Jang J, Kim S et al (2012) The effects of sub-lethal concentrations of silver nanoparticles on inflammatory and stress genes in human macrophages using cDNA microarray analysis. Biomaterials 33(18):4690–4699
44. Waters KM, Masiello LM, Zangar RC et al (2009) Macrophage responses to silica nanoparticles are highly conserved across particle sizes. Toxicol Sci 107(2):553–569
45. Zhang L, Wang X, Zou J et al (2015) Effects of an 11-nm DMSA-coated iron nanoparticle on the gene expression profile of two human cell lines, THP-1 and HepG2. J Nanobiotechnol 13:3. https://doi.org/10.1186/s12951-014-0063-3
46. Ding L, Stilwell J, Zhang T et al (2005) Molecular characterization of the cytotoxic mechanism of multiwall carbon nanotubes and nano-onions on human skin fibroblast. Nano Lett 5(12):2448–2464
47. Long TC, Tajuba J, Sama P et al (2007) Nanosize titanium dioxide stimulates reactive oxygen species in brain microglia and damages neurons in vitro. Environ Health Perspect 115(11):1631–1637
48. Van der Zande M, Vandebriel RJ, Groot MJ et al (2014) Sub-chronic toxicity study in rats orally exposed to nanostructured silica. Part Fibre Toxicol 11:8
49. Osmond-McLeod MJ, Oytam Y, Rowe A et al (2016) Long-term exposure to commercially available sunscreens containing nanoparticles of TiO_2 and ZnO revealed no biological impact in a hairless mouse model. Part Fibre Toxicol 13(1):44
50. Busch W, Kühnel D, Schirmer K et al (2010) Tungsten carbide cobalt nanoparticles exert hypoxia-like effects on the gene expression level in human keratinocytes. BMC Genomics 11:65
51. Pisani C, Gaillard JC, Nouvel V et al (2015) High-throughput, quantitative assessment of the effects of low-dose silica nanoparticles on lung cells: grasping complex toxicity with a great depth of field. BMC Genomics 16:315
52. Fisichella M, Berenguer F, Steinmetz G et al (2014) Toxicity evaluation of manufactured CeO2 nanoparticles before and after alteration: combined physicochemical and whole-genome expression analysis in Caco-2 cells. BMC Genomics 15:700
53. Fisichella M, Berenguer F, Steinmetz G et al (2012) Intestinal toxicity evaluation of TiO2 degraded surface-treated nanoparticles: a combined physicochemical and toxicogenomics approach in caco-2 cells. Part Fibre Toxicol 9:18
54. Li X, Zhang C, Zhang X et al (2016) An acetyl-L-carnitine switch on mitochondrial dysfunction and rescue in the metabolomics study on aluminum oxide nanoparticles. Part Fibre Toxicol 13:4. https://doi.org/10.1186/s12989-016-0115-y
55. Sahu SC, Zheng J, Yourick JJ et al (2015) Toxicogenomic responses of human liver HepG2 cells to silver nanoparticles. J Appl Toxicol 35(10):1160–1168
56. Halappanavar S, Saber AT, Decan N et al (2015) Transcriptional profiling identifies physicochemical properties of nanomaterials that are determinants

of the in vivo pulmonary response. Environ Mol Mutagen 56:245–264
57. Ali K, Qais FA, Dwivedi S et al (2017) Titanium dioxide nanoparticles preferentially bind in subdomains IB, IIA of HSA and minor groove of DNA. J Biomol Struct Dyn. https://doi.org/10.1080/07391102.2017.1361339
58. Colvin VL (2003) The potential environmental impact of engineered nanomaterials. Nat Biotechnol 21(10):1166–1170
59. Mohamed BM, Verma NK, Prina-Mello A et al (2011) Activation of stress-related signalling pathway in human cells upon SiO_2 nanoparticles exposure as an early indicator of cytotoxicity. J Nanobiotechnol 9:29
60. Perkins TN, Shukla A, Peeters PM et al (2012) Differences in gene expression and cytokine production by crystalline vs. amorphous silica in human lung epithelial cells. Part Fibre Toxicol 9(1):6
61. Reuzel PG, Bruijntjes JP, Feron VJ et al (1991) Subchronic inhalation toxicity of amorphous silicas and quartz dust in rats. Food Chem Toxicol 29(5):341–354
62. Xu Z, Wang SL, Gao HW (2010) Effects of nano-sized silicon dioxide on the structures and activities of three functional proteins. J Hazard Mater 180(1–3):375–383
63. Rabolli V, Thomassen LC, Princen C et al (2010) Influence of size, surface area and microporosity on the in vitro cytotoxic activity of amorphous silica nanoparticles in different cell types. Nanotoxicology 4(3):307–318
64. Salata O (2004) Applications of nanoparticles in biology and medicine. J Nanobiotechnol 2:3
65. Pirmohamed T, Dowding JM, Singh S et al (n.d.) Nanoceria exhibit redox state dependent catalase mimetic activity. Chem Commun (Camb) 46:2736–2738
66. Heckert EG, Karakoti AS, Seal S et al (2008) The role of cerium redox state in the SOD mimetic activity of nanoceria. Biomaterials 29:2705–2709
67. Kong L, Cai X, Zhou X et al (2011) Nanoceria extend photoreceptor cell lifespan in tubby mice by modulation of apoptosis/survival signaling pathways. Neurobiol Dis 42:514–523
68. Schubert D, Dargusch R, Raitano J et al (2006) Cerium andyttrium oxide nanoparticles are neuroprotective. Biochem Biophys Res Commun 342:86–91
69. Cheng G, Guo W, Han L et al (2013) Cerium oxide nanoparticles induce cytotoxicity in human hepatoma SMMC-7721 cells via oxidative stress and the activation of MAPK signaling pathways. Toxicol in Vitro 27:1082–1088
70. Wijnhoven SWP, Peijnenburg WJGM, Herberts CA et al (2009) Nano-silver – a review of available data and knowledge gaps in human and environmental risk assessment. Nanotoxicology 3:109–138
71. Sotiriou GA, Pratsinis SE (2010) Antibacterial activity of nanosilver ions and particles. Environ Sci Technol 44:5649–5654
72. Sotiriou GA, Pratsinis SE (2011) Engineering nanosilver as an antibacterial, biosensor and bioimaging material. Curr Opin Chem Eng 1(1):3–10
73. Dwivedi S, Saquib Q, Al-Khedhairy AA et al (2015) Rhamnolipids functionalized AgNPs-induced oxidative stress and modulation of toxicity pathway genes in cultured MCF-7 cells. Colloids Surf B: Biointerfaces 132:290–298
74. Morones JR, Elechiguerra JL, Camacho A et al (2005) The bactericidal effect of silver nanoparticles. Nanotechnology 16:2346–2353
75. Xiu ZM, Zhang QB, Puppala HL et al (2012) Negligible particle specific antibacterial activity of silver nanoparticles. Nano Lett 12:4271–4275
76. Stebounova LV, Adamcakova-Dodd A, Kim JS et al (2011) Nanosilver induces minimal lung toxicity or inflammation in a subacute murine inhalation model. Part Fibre Toxicol 8:5
77. Song KS, Sung JH, Ji JH et al (2012) Recovery from silver-nanoparticle exposure- induced lung inflammation and lung function changes in Sprague Dawley rats. Nanotoxicology 7:169–180
78. Sung JH, Ji JH, Yoon JU et al (2008) Lung function changes in Sprague–Dawley rats after prolonged inhalation exposure to silver nanoparticles. Inhal Toxicol 20:567–574
79. Sung JH, Ji JH, Park JD et al (2009) Subchronic inhalation toxicity of silver nanoparticles. Toxicol Sci 108:452–461
80. Ma R, Levard C, Marinakos SM et al (2012) Size-controlled dissolution of organic-coated silver nanoparticles. Environ Sci Technol 46:752–759
81. Leo BF, Chen S, Kyo Y et al (2013) The stability of silver nanoparticles in a model of pulmonary surfactant. Environ Sci Technol 47:11232–11240
82. Stebounova LV, Guio E, Grassian VH (2011) Silver nanoparticles in simulated biological media: a study of aggregation, sedimentation, and dissolution. J Nanopart Res 13:12
83. Kent RD, Vikesland PJ (2012) Controlled evaluation of silver nanoparticles dissolution using atomic force microscopy. Environ Sci Technol 46:6977–6984
84. Zook JM, Long SE, Cleveland D et al (2011) Measuring silver nanoparticle dissolution in complex biological and environmental matrices using UV-visible absorbance. Anal Bioanal Chem 401:1993–2002
85. Pratsinis A, Hervella P, Leroux JC et al (2013) Toxicity of silver nanoparticles in macrophages. Small 9:2576–2584
86. Lubick N (2008) Nanosilver toxicity: ions, nanoparticles–or both? Environ Sci Technol 42:8617
87. Park EJ, Yi J, Kim Y et al (2010) Silver nanoparticles induce cytotoxicity by a Trojan-horse type mechanism. Toxicol in Vitro 24:872–878
88. Limbach LK, Wick P, Manser P et al (2007) Exposure of engineered nanoparticles to human lung epithelial cells: influence of chemical composition and catalytic activity on oxidative stress. Environ Sci Technol 41:4158–4163

89. Lam CW, James JT, McCluskey R et al (2004) Pulmonary toxicity of singlewall carbon nanotubes in mice 7 and 90 days after intratracheal instillation. Toxicol Sci 77:126–134
90. Sargent L, Reynolds S, Castranova V (2010) Potential pulmonary effects of engineered carbon nanotubes: in vitro genotoxic effects. Nanotoxicology 4:396–408
91. Shvedova AA, Pietroiusti A, Fadeel B et al (2012) Mechanisms of carbon nanotube-induced toxicity: focus on oxidative stress. Toxicol Appl Pharmacol 261:121–133
92. Sund J, Alenius H, Vippola M et al (2011) Proteomic characterization of engineered nanomaterial–protein interactions in relation to surface reactivity. ACS Nano 5:4300–4309
93. Jaurand MC, Renier A, Daubriac J (2009) Mesothelioma: do asbestos and carbon nanotubes pose the same health risk? Part Fibre Toxicol 6:1–14
94. Fenoglio I, Greco G, Tomatis M et al (2008) Structural defects play a major role in the acute lung toxicity of multiwall carbon nanotubes: physicochemical aspects. Chem Res Toxicol 21:1690–1697
95. Jacobsen NR, Pojana G, White P et al (2008) Genotoxicity, cytotoxicity, and reactive oxygen species induced by single-walled carbon nanotubes and C60 fullerenes in the FE1-Muta™Mouse lung epithelial cells. Environ Mol Mutagen 49:476–487
96. Saquib Q, Al-Khedhairy AA, Ahmad J et al (2013) Zinc ferrite nanoparticles activate IL-1b, NFKB1, CCL21 and NOS2 signaling to induce mitochondrial dependent intrinsic apoptotic pathway in WISH cells. Toxicol Appl Pharmacol 273(2):289–297
97. Musarrat J, Saquib Q, Azam A et al (2009) Zinc oxide nanoparticles-induced DNA damage in human lymphocytes. Int J Nanopart 2(1/2/3/4/5/6):402–415
98. Wahab R, Khan F, Yang YB et al (2016) Zinc oxide quantum dots: multifunctional candidates for arresting the C2C12 cancer cells and their role towards caspase 3 and 7 genes. RSC Adv 6:26111–26120
99. Wahab R, Siddiqui MA, Saquib Q et al (2014) ZnO nanoparticles induced oxidative stress and apoptosis in HepG2 and MCF-7 cancer cells and their antibacterial activity. Colloids Surf B: Biointerfaces 117:267–276

10

Nickel Oxide Nanoparticles Induced Transcriptomic Alterations in HEPG2 Cells

Quaiser Saquib, Maqsood A. Siddiqui, Javed Ahmad, Sabiha M. Ansari, Mohammad Faisal, Rizwan Wahab, Abdulrahman A. Alatar, Abdulaziz A. Al-Khedhairy, and Javed Musarrat

Abstract

Nickel oxide nanoparticles (NiO-NPs) are increasingly used and concerns have been raised on its toxicity. Although a few studies have reported the toxicity of NiO-NPs, a comprehensive understanding of NiO-NPs toxicity in human cells is still lagging. In this study, we integrated transcriptomic approach and genotoxic evidence to depict the mechanism of NiO-NPs toxicity in human hepatocellular carcinoma (HepG2) cells. DNA damage analysis was done using comet assay, which showed 26-fold greater tail moment in HepG2 cells at the highest concentration of 100 μg/ml. Flow cytometric analysis showed concentration dependent enhancement in intracellular reactive oxygen species (ROS). Real-time PCR analysis of apoptotic (p53, bax, bcl2) and oxidative stress (SOD1) genes showed transcriptional upregulation. Transcriptome analysis using qPCR array showed

Q. Saquib (✉)
Zoology Department, College of Sciences, King Saud University, P.O. Box 2455, Riyadh 11451, Saudi Arabia

A.R. Al-Jeraisy Chair for DNA Research, Zoology Department, College of Sciences, King Saud University, P.O. Box 2455, Riyadh, Saudi Arabia
e-mail: quaiser.saquib0@gmail.com

M. A. Siddiqui · J. Ahmad · R. Wahab
A.R. Al-Jeraisy Chair for DNA Research, Zoology Department, College of Sciences, King Saud University, P.O. Box 2455, Riyadh, Saudi Arabia

S. M. Ansari · M. Faisal · A. A. Alatar
Department of Botany and Microbiology, College of Sciences, King Saud University, P.O. Box 2455, Riyadh, Saudi Arabia

A. A. Al-Khedhairy
Zoology Department, College of Sciences, King Saud University, P.O. Box 2455, Riyadh 11451, Saudi Arabia

J. Musarrat
Department of Agricultural Microbiology, Faculty of Agricultural Sciences, Aligarh Muslim University, Aligarh, Uttar Pradesh, India

Department of Biosciences and Biotechnology, Baba Ghulam Shah Badshah University, Rajouri, Jammu and Kashmir, India

Q. Saquib et al. (eds.), *Cellular and Molecular Toxicology of Nanoparticles*, Advances in Experimental Medicine and Biology 1048, https://doi.org/10.1007/978-3-319-72041-8_10

over expression of mRNA transcripts related to six different cellular pathways. Our data unequivocally suggests that NiO-NPs induces oxidative stress, DNA damage, apoptosis and transcriptome alterations in HepG2 cells.

Keywords
Nanotoxicity · Apoptosis · DNA damage · Oxidative stress · Transcriptome · Nickel Oxide Nanoparticles · NiO

10.1 Introduction

In recent years' nanotechnology has exhibited exponential growth in various sectors to accomplish market commodities with higher prospective applications [1]. At least thousand consumer products are available which contains nanoparticles (NPs), ranging from everyday household items to medical diagnostic tools, imaging, drug delivery and aerospace engineering [2, 3]. Compared to the bulk counterpart, the small size and large specific surface area of NPs endow them with high chemical reactivity and intrinsic toxicity. Such unique physiochemical properties of NPs draw global attention of scientists and environmental watchdogs to keep concern about NPs potential risks and adverse effect on human health [4]. NPs find route to human body via skin penetration, ingestion, inhalation or injection and interact with cellular organelles for longer time period [5]. Consequently, NPs have been found to effortlessly interact with cells and organs by various mechanisms [6]. Since methodologies for exposure assessment are non-consistent, the toxicological research on NPs is still lagging. Therefore, in order to plug the gap between development and toxicity of NPs, a major effort is needed to study the effects of exposure to NPs.

In this study, we have selected nickel oxide nanoparticles (NiO-NPs) owing to its increasing application in ceramic material, catalysts, electronic component and biosensors [7–9]. Despite its wide use, NiO-NPs has raised concerns about its adverse effects on the environment and human health. NiO-NPs generated from welding fumes during the coastal region developments were considered as a potential nano-pollution source in coastal seawaters (IARC Monographs on the Evaluation of Carcinogenic Risks to Humans). Direct aerial emission of NiO-NPs has the tendency to pollute surface waters through leakages, spills and indirect storm-water runoff from land [10]. The metallic Ni-NPs has been recently used to catalyze the reversible hydration of CO_2 to carbonic acid, which is holding extreme importance in CO_2 capture technologies and mineralization processes. These advantages led to its utilization to point flue sources like air-conditions outlets on top of building or power plants [11, 12]. Being the 24th most abundant element in the Earth crust, nickel compounds ($NiSO_4$, NiO, nickel hydroxides and crystalline nickel) are well known as an environmental pollutant and classified as carcinogenic agents to humans (Group 1) by the International Agency for Research on Cancer (IARC) [13]. The in vivo studies on NiO-NPs have been mostly focused on pulmonary pathology. Female Wistar rats intratracheally instilled with NiO-NPs exhibited a significant increase in the bronchiolar alveolar lavage fluid (BLAF), activation of IL-1β, IFN-Y, MIP-2 and histological changes [14]. A short-term exposure of rats to 500 cm^2/ml of NiO-NPs induced polymorphonuclear neutrophils in the BALF [15]. Inhalation exposure of rats with NiO-NPs in nebulizer chamber exhibited biopersistance of NPs in lungs and inflammatory responses [16]. Long term intratracheally instillation of NiO-NPs in rats exhibited increased vacuolization in alveolar macrophages and CINC-1 concentrations [17]. Wistar rats instilled with NiO-NPs after 4 days of exposure showed eosinophilic and neutrophilic inflammation, along with release of eotaxin and cellular disintegration by the release of Ni ions [18]. Female Wistar rats exposed by pharyngeal route to NiO-NPs showed enhanced proinflammatory cytokines, LDH, lymphocytes, polymorphonuclear leukocytes in BALF [19]. DNA damage and low expression of HO-1 and Nrf2 proteins were observed in Male Sprague Dawley rats when intratracheally instilled with Ni-NPs for two weeks. In addition, the animals showed alterations in the normal morphology of lungs, liver and kidneys [20]. Ultrafine-size particles and NiO-NPs of nickel compounds have greater bioavailability and toxicity as compared to its fine-size nickel compounds [21]. We have recently reported that NiO-NPs induces liver toxicity, cytogenetic anomalies and apoptosis via p53, MAPK, caspase 8 and 3 signalling in rats [22]. Another recent study in the same line expressed NiO-NPs genotoxicity, chromosomal aberrations, DNA damage in lymphocytes, liver and kidney of female rats [23]. Zebrafish exposed to NiO-NPs for longer time showed higher bioaccumulation and toxicity [24]. It is well documented and established that solubilization of Ni^{2+} from NiO-NPs plays vital role in inducing toxicity in animal, invertebrate, cell line and plant [18, 25–27].

Concerning the in vitro studies, recent reports suggest NiO-NPs as neurotoxic in SH-SY5Y neuroblastoma cells and cytotoxic for human breast carcinoma cells (MCF7) [28, 29]. NiO-NPs induced *HIF-1α* transcription factor followed by upregulation of its target NRDG1 (Cap43) in human lung epithelial (H460) cells [21]. In the same line, NiO-NPs induced oxidative stress and cytotoxicity in human alveolar epithelial cells (A549) has also been reported [30]. HepG2 cells exposed to NiO-NPs resulted in cytotoxicity and apoptosis responses via reactive oxygen species (ROS) generation, which is likely to be mediated through bax/bcl-2 pathway [31].

Despite these facts, a systematic interpretation on the underlying mechanism of NiO-NPs induced hepatotoxicity is scarce. In this context, NiO-NPs has been reported to induce cytotoxicity and apoptotic cell death in HepG2 cells via bax/blc2 pathway [31]. However, authors did not explain the vital queries on HepG2 transcriptomic profile. To decipher these unattended queries, we have provided a concrete evidence on hepatotoxicity under in vitro condition. Primary human hepatocytes have been considered as gold standard model for xenobiotic metabolism and cytotoxicity studies [32]. However, the complexity in isolation procedures, short life-span, inter-individual variability, cost effectiveness and rare availability of fresh human liver samples, constitute serious limitations for the use of aforesaid in vitro systems in screening [33]. Such constrains were run-over by immortalized liver-derived cell lines, owing to their unlimited availability and phenotypic stability. A first alternative is the widely used HepG2 cells, as these cells are highly differentiated and display many of the genotypic features of normal liver cells [34]. HepG2 can be used to screen the cytotoxic potential of new chemical entities at the lead generation phase and imitate the normal metabolic pathway in vivo [35, 36]. In this study, we have selected HepG2 cells as a model system for studying the hepatotoxic effects of NiO-NPs.

Consequently, the current study was aimed to evaluate molecular mechanism of NiO-NPs in vitro toxicity in HepG2 cells by the measurement of (i) intracellular ROS generation (ii) DNA damage (iii) transcriptional activation of array of genes related to human stress and toxicity pathways.

10.2 Materials and Methods

10.2.1 NiO-NPs Characterization

NiO-NPs (Cat. No. 637130) was purchased from sigma chemical company (St. Louis, MO, USA). A stock of NiO-NPs (1 mg/ml) was prepared in MQ water and sonicated for 20 min at 40 W. TEM analysis of NiO-NPs were done by dropping the stock solution on copper grids and subjected to microscopic analysis at 200 KeV (JEM-2100 F, JEOL, Japan).

10.2.2 Cell Culture and NiO-NPs Exposure

Human liver hepatocellular carcinoma (HepG2) cells were cultured in DMEM with 10% FBS, antibiotic-antimycotic solution and incubated at 37 °C with 5% CO_2. HepG2 cells were seeded in 96 and 6-well plates and allowed to attach with the surface for 24 h prior to NiO-NPs treatment. Before each experiment, the ultra sonicated NiO-NPs (25, 50 and 100 µg/ml) solutions added to cell culture and grown for 24 h. Control groups were not added with NiO-NPs.

10.2.3 In Vitro DNA Damage Analysis by Comet Assay

The HepG2 cells exposed for 3 h with 25, 50 and 100 µg/ml of NiO-NPs were detached and centrifuged at 3000 rpm for 3 min to collect the pellets. The cells (4×10^4) from untreated and treated groups were suspended in 100 µl of Ca^{++} Mg^{++} free PBS and mixed with 100 µl of 1% low melting point agarose (LMA). The cell suspension (80 µl) was then layered on one-third frosted slides, pre-coated with normal melting agarose (NMA) (1% in PBS) and kept at 4 °C for 10 min. After gelling, a layer of 90 µl of LMA (0.5% in PBS) was added. After the solidification of agarose on slides, all of them were kept in lysis solution for overnight, followed by unwinding and electrophoresis at 24 V (300 mA) for 20 min. Cells were stained with ethidium bromide (20 µg/ml) and DNA damage were scored under fluorescence microscope.

10.2.4 ROS Measurements in HepG2 Cells

After the specified treatment, cells were trypsinized, pelleted and washed twice with cold PBS, followed by the resupension of cells in 500 µl PBS (Ca^{++} and Mg^{++} free) containing 5 µM of dichloro-dihydro-fluorescein diacetate (DCFH-DA) dye. All cells were incubated for 60 min at 37 °C in dark followed by washing and the fluorescence were recorded upon excitation at 488 nm at FL1 Log channel through 525 nm band-pass filter on Beckman Coulter flow cytometer (Coulter Epics XL/Xl-MCL, USA). Qualitative analysis of ROS in NiO-NPs treated cells were also done by staining the HepG2 cells with 5 µM of DCFH-DA for 60 min at 37 °C in CO_2 incubator. Fluorescence images were captured on microscope equipped with fluorescent lamp (Nikon Eclipse 80i, Japan).

10.2.5 RT² Profiler PCR Array Analysis

PCR array experiments were done with HepG2 cells exposed for 24 h with NiO-NPs (100 µg/ml). In brief, total RNA was isolated using the commercially available kit (RNeasy Mini Kit, Cat. No. 74106, Qiagen, USA), purification was done using iPrep™ PureLink™ kit (Invitrogen, USA) by Invitrogen® automated system. Purity of total RNA was verified by use of a Nanodrop 8000 spectrophotometer (Thermo Scientific, USA). The first-strand cDNA synthesis was performed with 1 µg of total RNA and 100 ng of oligo-p(dT)12-18 primer and MLV reverse transcriptase (GE Health Care, UK). Changes in the relative gene expression of 84 genes responsible for human stress and toxicity pathway were quantified using 96-well format of RT² Profiler™ PCR Array (Cat. No. PAHS-003 A, SABiosciences Corporation, Frederick, MD). cDNA equivalent to 1 µg of total RNA was used for each array. The arrays were run on Roche® LightCycler® 480 (Roche Diagnostics, Rotkreuz, Switzerland) following the recommended cycling programs. Online software from SABiosciences Corporation, Frederick, MD, was used to analyze the expression data. NiO-NPs expression results were normalized to the average Ct value of five housekeeping genes (*B2M*, *HPRT1*, *RPL13A*, *GAPDH* and *ACTB*) and expressed with respect to the untreated control. RT-PCR array data were evaluated from at least three independent experiments and the resultant $^{\Delta}$Ct values were combined to calculate the average fold regulation values. Genes that were significantly different for NiO-NPs versus control were determined by Students t-test ($p < 0.05$) by comparing the $^{\Delta}$Ct values for the triplicate trials for each test sample with the $^{\Delta}$Ct values for the control. Then PCR array data were validated by measuring the mRNA expression of some selected genes (*P53*, *BAX*, *BCL2*, *SOD1*) using real-time PCR analysis (Table 10.1).

Table 10.1 Primers of candidate genes for qPCR array-qPCR validation

Gene symbol	Sense primer	Antisense primer
P53	CCCAGCCAAAGAAGAAACCA	TTCCAAGGCCTCATTCAGCT
Bax	TGCTTCAGGGTTTCATCCAG	GGCGGCAATCATCCTCTG
Bcl2	AGGAAGTGAACATTTCGGTGAC	GCTCAGTTCCAGGACCAGGC
SOD1	AGGGCATCATCAATTTCGAG	TGCCTCTCTTCATCCTTTGG
GAPDH	CCACTCCTCCACC TTTGAC	ACCCTGTTGCTGTAGCCA

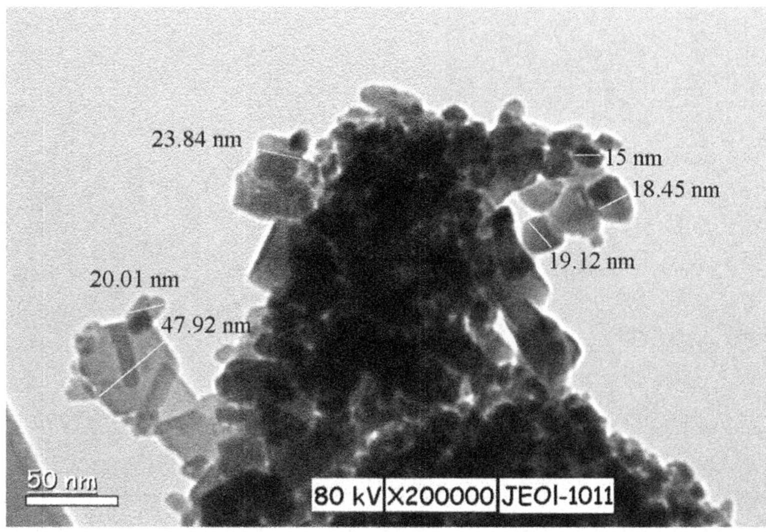

Fig. 10.1 Depicts the particle characterization of NiO-NPs by TEM at 200000× magnification

10.3 Results

10.3.1 NiO-NPs Characterization

The size and morphology of NiO-NPs were measured by transmission electron microscopy (TEM). In TEM analysis, NiO-NPs appeared as an aggregate showing crystallite's spheres. The particles size of NiO-NPs analyzed from six TEM images were determined to be 24.05 ± 2.9 nm (Fig. 10.1).

10.3.2 DNA Damage in HepG2 Cells

HepG2 cells exposed to NiO-NPs for 3 h resulted in DNA damage. The representative comet image from NiO-NPs (100 μg/ml) treatment clearly demonstrates the broken DNA liberated from the comet head (Fig. 10.2). NiO-NPs at 25, 50 and 100 μg/ml induced significant 25.1, 25.3 and 26.7-fold higher Olive tail moment (OTM) parameter of comet assay *vis-à-vis* the control showed a background of 0.28 ± 0.05 OTM (Table 10.2). The advantage of comet assay is that it is capable of analysing population of cells with various degrees of DNA damage. Nevertheless, the differences in distribution of DNA damage exist in the cell population. Variation in distribution of DNA damage by NiO-NPs exposure in-terms of frequency is shown in Fig. 10.2.

10.3.3 Quantitative and Qualitative Analysis of Intracellular ROS

A concentration dependent increase in the intracellular ROS generation in HepG2 cells, as evident by the shift of DCF peaks in treated groups (Fig. 10.3a). Compared to the 100% DCF fluorescence in control, cells treated with 25, 50 and 100 μg/ml of NiO-NPs showed significant 134%,

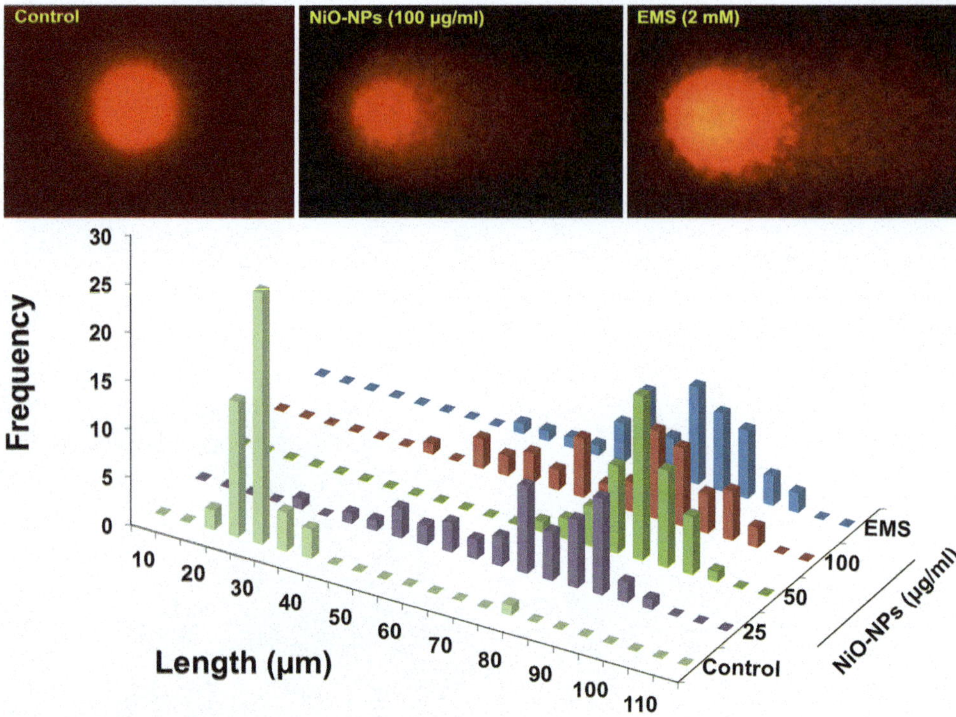

Fig. 10.2 Photomicrographs showing DNA strand breaks analysis by comet assay in NiO-NPs treated HepG2 cells. Histograms showing frequency distribution of DNA damage in HepG2 cells treated with varying concentrations of NiO-NPs for 3 h

Table 10.2 NiO-NPs induced DNA damage in HepG2 cells analyzed using different parameters of alkaline comet assay

Groups	Olive tail moment (arbitrary unit)	Tail length (µm)	Tail intensity (%)
Control	0.28 ± 0.05	27.65 ± 1.87	2.63 ± 0.04
EMS (1 mM)	6.42 ± 0.32**	81.45 ± 3.88**	30.53 ± 1.24**
NiO-NPs (µg/ml)			
25	7.05 ± 0.43**	72.71 ± 2.61**	33.12 ± 1.11**
50	7.11 ± 0.23**	77.43 ± 4.35**	43.13 ± 1.09**
100	7.48 ± 0.46**	76.88 ± 3.09**	42.64 ± 2.43**

Data represent the mean±S.D. of three independent experiments done in duplicate

EMS ethyl methanesulphonate

$^{**}p < 0.01$ vs. control

150% and 143% ($p < 0.01$) increase in ROS generation (Fig. 10.3a, Inset). Fluorescence images further validated the flow cytometric data by showing an enhanced level of DCF fluorescence in the NiO-NPs treated cells (Fig. 10.3b).

10.3.4 qPCR Array of HepG2 Cells

HepG2 cells treated with NiO-NPs (100 µg/ml) for 24 h exhibited differential expression of genes in the RT^2 profiler PCR array. The corresponding heat map suggested strong oxidative or metabolic stress, growth arrest and senescence, apoptosis signalling, proliferation and carcinogenesis, and activation of proinflammatory responses upon NiO-NPs exposure (Fig. 10.4). *CYP2E1* gene in oxidative or metabolic stress group has exhibited a maximum of 3.4-fold up-regulation. Considerable number of genes in this pathway was up-regulated, and 1.4, 1.5 and 1.1-fold of maximum up-regulation has been recorded for *HMOX1*, *SOD2* and *SOD1* genes. Among the set of seven genes responsible for growth arrest and senescence, *GDF15*, *DDIT3*, *GADD45A*, *MDM2* and *P53* genes have exhibited 4.6, 2.4, 1.6, 1.2 and 1.2-fold up-regulation. *TNFSF10*, *TNFRSF1A*, *CASP8* and *NFKB1A* genes in apoptosis signalling group showed maximum up-

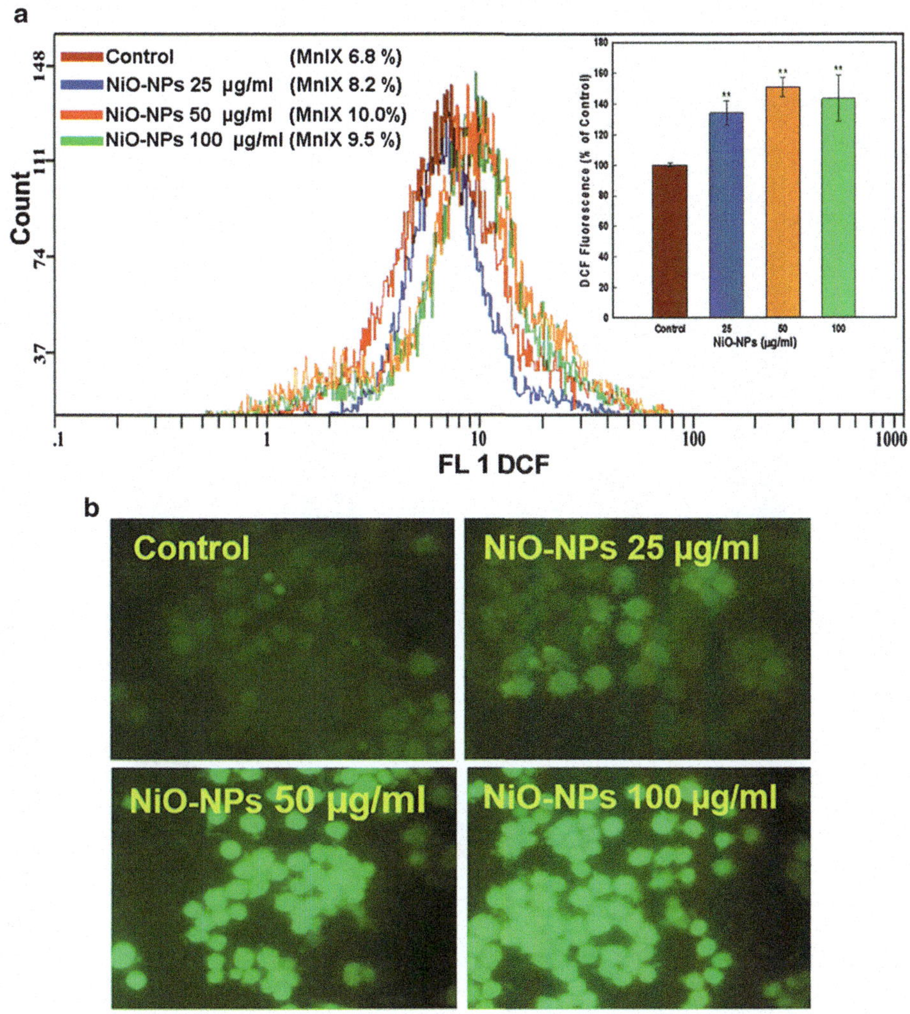

Fig. 10.3 (**a**) Fluorescence enhancement of DCF indicating ROS production with increasing NiO-NPs concentrations in HepG2 cells analyzed by flow cytometry. Each histogram in inset represents the values of mean±SD of three independent experiments done in triplicate wells (**$p < 0.01$ vs. control). (**b**) Fluorescence microscopic images of treated cells showing an enhancement in green fluorescence of DCF in treated cells

regulation of 1.8, 1.1, 1.4 and 1.0-fold, while the *BCL2L1* expression was down-regulated to 1.1-fold. Within the proliferation and carcinogenesis pathway *EGR1* showed 1.2-fold up-regulation. Among the proinflammatory genes, *NOS2* was maximally up-regulated to 2.3-fold. *HSPA6* gene in heat shock group, showed up-regulation of 2.5-fold. qPCR array data validation was done by measuring the expression of selective genes (*P53*, *BAX*, *BCL2* and *SOD1*) by real-time PCR. The expressional analysis also showed 1.0, 1.2, and 1.1-fold up-regulation of *P53*, *BAX* and *SOD1*. *BCL2* was found under expressed to 1.2-fold (Fig. 10.5).

10.4 Discussion

An integrated approach was used to identify toxicity mechanism induced by NiO-NPs. In this study, low, medium and high (25, 50 and 100 µg/ml) doses of NiO-NPs has been chosen to expose the human liver cells. Lowest concentration was chosen with the aim to imitate the potential

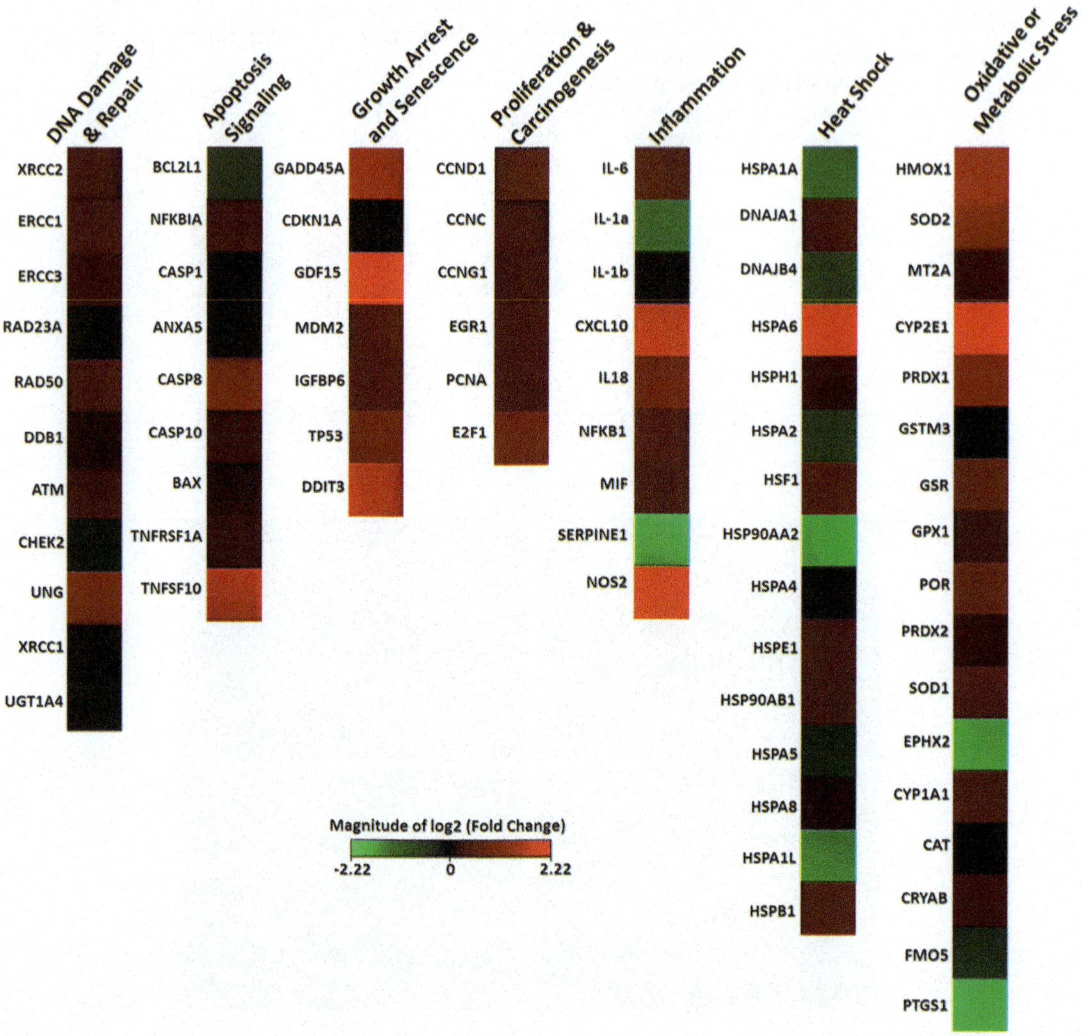

Fig. 10.4 Effect of NiO-NPs on oxidative stress and toxicity pathway genes in HepG2 cells. Heat map showing the relative gene expression of different genes responsible for human stress and toxicity pathway in NiO-NPs (100 µg/ml) treated HepG2 cells after 24 h of exposure

human exposure, on the other hand highest concentration was selected to reflect toxicological effects upon accidental exposure of NiO-NPs. In this line an enhanced level of ROS has been observed in NiO-NPs treated cells. These results corroborate with enhanced ROS level in NiO-NPs treated HepG2 and A549 cells [31, 37]. We suggest oxidative stress in HepG2 cells. The current study demonstrates that NiO-NPs can induce DNA damage after short exposure of 3 h, and corresponds with previous reports on DNA damage in HepG2 and WISH cells exposed to NiO, TiO_2 and $ZnFe_2O_3$-NPs [31, 38, 39]. The appearance of comet tail with NiO-NPs exposure unequivocally suggest the impairment of DNA repair machinery. The enhancement in intracellular ROS and DNA damage data are in agreement with our recent report on NiO-NPs induced liver toxicity in rats [22]. Ni^{2+} is involved in ROS generation and accounted for inducing high level of damage via direct oxidative damage by H_2O_2 production [40]. Hence, the elevated toxicity and damage in our study could also be an additive oxidative action of Ni^{2+} released from NiO-NPs.

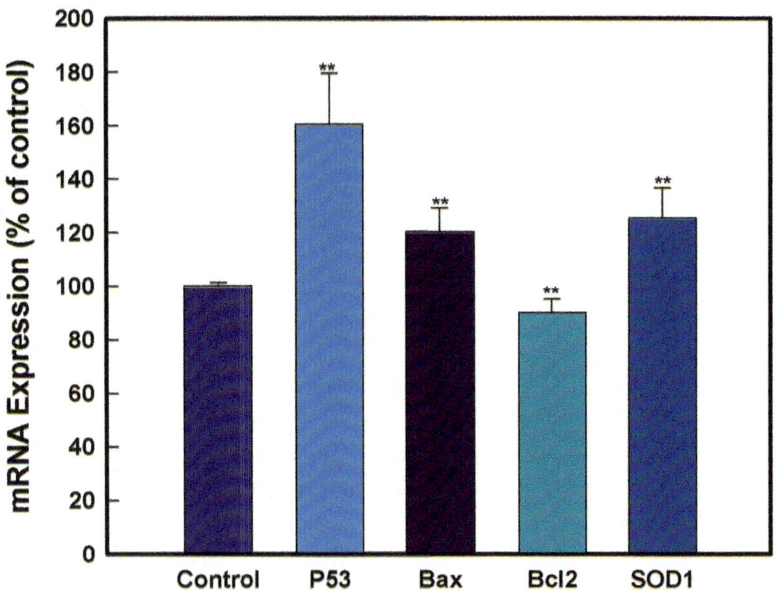

Fig. 10.5 Transcriptional activation of apoptotic and oxidative stress genes in NiO-NPs treated HepG2 cells. Transcript levels were determined by real-time quantitative PCR. $^{**}p < 0.01$ using one-way ANOVA (Dunnett's multiple comparison test) significantly different when compared to control

PCR array revealed that NiO-NPs treatment resulted in the up-regulation of genes related to different pathways. We found that *TNFSF10*, *TNFRSF1A*, *CASP8* and *NFKBIA* genes in apoptosis signalling pathways were up-regulated. *TNFSF10* and *TNFRSF1A* belong to the tumor necrosis factor receptor (*TNFR*) family and their up-regulation has been suggested to induce cell death [41]. Up-regulation of *CASP8* expression has been linked to execute the apoptotic signaling mainly through extrinsic pathway [42]. Activation of the above genes strongly suggests the participation of death receptor-mediated *TNFR* family members to induce apoptosis via intrinsic as well as extrinsic pathways. *TNFR* genes can act through an autocrine pathway to induce cell growth arrest and apoptosis through *NFKB* activation [43]. Therefore, the *NFKB* pathway and related genes could also be an important molecular mechanism by which NiO-NPs induces apoptosis in HepG2 cells.

NiO-NPs treatment resulted in the up-regulation of *EGR1*, *MDM2*, *GADD45A* and *DDIT3* genes. Although the induction of *EGR1*, a family of zinc finger transcription factors, is directly linked with oxidative stress per se, other condition like mitochondrial dysfunction contributes well in its up-regulation [44]. We have found in vivo mitochondrial dysfunction in rats exposed to NiO-NPs for 7 and 14 days [22]. Therefore, the observed mitochondrial dysfunction and oxidative stress data strongly substantiate the likelihood that NiO-NPs may function as an initiator to increase the expression of *EGR1* in treated HepG2 cells. Up-regulation of *DDIT3* and *GADD45A* transcripts can be correlated with the fact that under stressed condition *EGR1* is known to initiate *DDIT3* and *GADD45* family genes by binding to 5′-flanking regions [45]. The oxidative stress related genes (*SOD1*, *SOD2*, *GPX1* and *HMOX1*) were found up-regulated after NiO-NPs exposure. In view of the higher ROS generation by NiO-NPs, we suggest that cytoplasmic (*SOD*1), mitochondrial (*SOD2*), glutathione system (*GPX1*) and *HMOX1* might have involved in scavenging the free radicals and cytoprotection. However, the excessive oxidative stress was beyond the attenuation capacity of these enzymes to subtle the DNA damage in treated cells. Up-regulation of above genes corresponds with increased expression of *SOD*, *GPX*, and *HMOX1* in human cells, when treated with ZnO-NPs and polyphenolic compounds [46, 47]. *NOS2* is a hallmark of inflammatory response and its up-regulation is governed by oxidative stress, metals and lipopolysaccharides [48]. *NOS2* expression

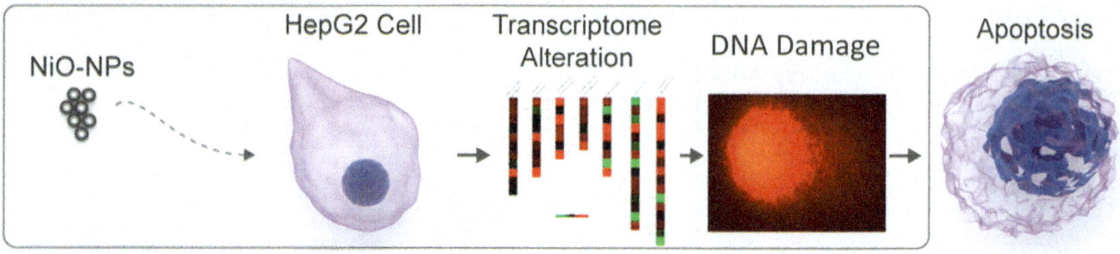

Fig. 10.6 Scheme showing NiO-NPs induced transcriptomic alterations and DNA damage leading to cell death in HepG2 cells

is in accordance with our previous work on $ZnFe_2O_4$-NPs, exhibiting its induction in WISH cells [38]. *GDF15* overexpression corresponds well with *p53-GDF15* link, and points towards its important role during inflammatory responses after NiO-NPs treatment [49]. Within the set of heat shock genes, *HSPA6* was found highly up-regulated in NiO-NPs treated cells. Heat shock proteins (HSP) are highly conserved class of stress response proteins, which work as molecular chaperons to correct the protein conformation under stress condition to maintain cellular homeostasis and protect the cells from apoptotic cell death [50]. Nonetheless, the DNA damage in NiO-NPs treated HepG2 cells supports the view that HSP fails to intervene the apoptotic process, as depicted in the image (Fig. 10.6).

10.5 Conclusion

We conclude that NiO-NPs have the potential to alter the transcriptome of HepG2 cells. We observed that NiO-NPs generated ROS and these free radicals induce heavy oxidative stress, which has affected the cell survival and promoted DNA. Transcriptional analysis of PCR array revealed overall up-regulation of different pathway genes, suggesting a pleiotropic effect of NiO-NPs to induced HepG2 cell death. The analysis of transcriptome was helpful to reveal potential molecular mechanism underlying NiO-NPs induced effects on HepG2 cells. The observed toxicity in HepG2, corresponds well with our recent study on rat's showing hepatotoxicity. Hence, NiO-NPs widespread application should be given meticulous attention for potential adverse biological effects.

Acknowledgements The authors are grateful to the Deanship of Scientific Research, King Saud University for funding through Vice Deanship of Scientific Research Chairs.

References

1. Singh SP, Chinde S, Kamal SSK et al (2016) Genotoxic effects of chromium oxide nanoparticles and microparticles in Wistar rats after 28 days of repeated oral exposure. Environ Sci Pollut Res 23:3914–3924(2)
2. Tomankova K, Horakova J, Harvanova M et al (2015) Cytotoxicity, cell uptake and microscopic analysis of titanium dioxide and silver nanoparticles in vitro. Food Chem Toxicol 82:106–115
3. Kim T, Hyeon T (2014) Applications of inorganic nanoparticles as therapeutic agents. Nanotechnology 25:012001
4. Beaudrie CE, Kandlikar M, Satterfield T (2013) From cradle-to-grave at the nanoscale: gaps in U.S. regulatory oversight along the nanomaterial life cycle. Environ Sci Technol 47(11):5524–5534
5. Nel A, Xia T, Mädler L et al (2006) Toxic potential of materials at the nano level. Science 311:622–627
6. Nel AE, Mädler L, Velegol D et al (2009) Understanding biophysicochemical interactions at the nano-bio interface. Nat Mater 8:543–557
7. Salimi A, Sharifi E, Noorbakhsh A et al (2007) Direct electrochemistry and electrocatalytic activity of catalase immobilized onto electrodeposited nano-scale islands of nickel oxide. Biophys Chem 125(2–3):540–548
8. Mu Y, Jia D, He Y (2011) Nano nickel oxide modified non-enzymatic glucose sensors with enhanced sensitivity through an electrochemical process strategy at high potential. Biosens Bioelectron 26(6):2948–2952
9. Rao KV, Sunandana CS (2008) Effect of fuel to oxidizer ratio on the structure, micro structure and EPR of combustion synthesized NiO nanoparticles. J Nanosci Nanotechnol 8(8):4247–4253
10. Wiesner MR, Lowry GV, Alvarez P et al (2006) Assessing the risks of manufactured nanomaterials. Environ Sci Technol 40(14):4336–4345
11. Bhaduri GA, Siller L (2012) Carbon capture. United Kingdom Patent Application GB1208511.4

12. Bhaduri GA, Siller L (2013) Nickel nanoparticles catalyse reversible hydration of carbon dioxide for mineralization carbon capture and storage. Cat Sci Technol 3:1234e1239
13. Cameron KS, Buchner V, Tchounwou PB (2011) Exploring the molecular mechanisms of nickel-induced genotoxicity and carcinogenicity: a literature review. Rev Environ Health 26(2):81–92
14. Cho WS, Duffin R, Poland CA et al (2010) Metal oxide nanoparticles induce unique inflammatory footprints in the lung: important implications for nanoparticle testing. Environ Health Perspect 118(12):1699–1706
15. Lu S, Duffin R, Poland C et al (2009) Efficacy of simple short-term in vitro assays for predicting the potential of metal oxide nanoparticles to cause pulmonary inflammation. Environ Health Perspect 117(2):241–247
16. Oyabu T, Ogami A, Morimoto Y et al (2007) Biopersistence of inhaled nickel oxide nanoparticles in rat lung. Inhal Toxicol 19(1):55–58
17. Nishi K, Morimoto Y, Ogami A et al (2009) Expression of cytokine-induced neutrophil chemoattractant in rat lungs by intratracheal instillation of nickel oxide nanoparticles. Inhal Toxicol 21(12):1030–1039
18. Lee S, Hwang SH, Jeong J et al (2016) Nickel oxide nanoparticles can recruit eosinophils in the lungs of rats by the direct release of intracellular eotaxin. Part Fibre Toxicol 13(1):30
19. Jeong J, Kim J, Seok SH et al (2016) Indium oxide (In2O3) nanoparticles induce progressive lung injury distinct from lung injuries by copper oxide (CuO) and nickel oxide (NiO) nanoparticles. Arch Toxicol 90(4):817–828
20. Magaye R, Gu Y, Wang Y et al (2016) In vitro and in vivo evaluation of the toxicities induced by metallic nickel nano and fine particles. J Mol Histol 47(3):273–286
21. Pietruska JR, Liu X, Smith A et al (2011) Bioavailability, intracellular mobilization of nickel, and HIF-1alpha activation in human lung epithelial cells exposed to metallic nickel and nickel oxide nanoparticles. Toxicol Sci 124(1):138–148
22. Saquib Q et al (2017) p53, MAPKAPK-2 and caspases regulate nickel oxide nanoparticles induce cell death and cytogenetic anomalies in rats. Int J Biol Macromol.: pii: S0141-8130(17)31498-8. https://doi.org/10.1016/j.ijbiomac.2017.07.032
23. Dumala N et al (2017) Genotoxicity study of nickel oxide nanoparticles in female Wistar rats after acute oral exposure. Mutagenesis. https://doi.org/10.1093/mutage/gex007
24. Kovriznych JA, Sotnikova R, Zeljenkova D et al (2014) Long-term (30 days) toxicity of NiO nanoparticles for adult zebrafish Danio rerio. Interdiscip Toxicol 7(1):23–26
25. Kanold JM, Wang J, Brümmer F et al (2016) Metallic nickel nanoparticles and their effect on the embryonic development of the sea urchin Paracentrotus lividus. Environ Pollut 212:224–229
26. Duan WX, He MD, Mao L et al (2015) NiO nanoparticles induce apoptosis through repressing SIRT1 in human bronchial epithelial cells. Toxicol Appl Pharmacol 286(2):80–91
27. Faisal M, Saquib Q, Alatar AA et al (2013) Phytotoxic hazards of NiO-nanoparticles in tomato: a study on mechanism of cell death. J Hazard Mater 250–251:318–332
28. Abudayyak M, Guzel E, Özhan G (2017) Nickel oxide nanoparticles are highly toxic to SH-SY5Y neuronal cells. Neurochem Int 108:7–14. https://doi.org/10.1016/j.neuint.2017.01.017
29. Umaralikhana L, Jaffar MJM (2016) Antibacterial and anticancer properties of NiO nanoparticles by co-precipitation method. J Adv App Sci Res 1(4):24–35
30. Lu S et al (2015) Comparison of cellular toxicity caused by ambient ultrafine particles and engineered metal oxide nanoparticles. Part Fibre Toxicol 12:5
31. Ahamed M, Ali D, Alhadlaq HA et al (2013) Nickel oxide nanoparticles exert cytotoxicity via oxidative stress and induce apoptotic response in human liver cells (HepG2). Chemosphere 93(10):2514–2522
32. Guillouzo A et al (2007) The human hepatoma HepaRG cells: a highly differentiated model for studies of liver metabolism and toxicity of xenobiotics. Chem Biol Interact 168:66–73
33. Madan A et al (2003) Effects of prototypical microsomal enzyme inducers on cytochrome P450 expression in cultured human hepatocytes. Drug Metab Dispos 31:421–431
34. Sassa S, Sugita O, Galbraith RA et al (1987) Drug metabolism by the human hepatoma cell, HepG2. Biochem Biophys Res Commun 143:52–57
35. Gerets HHJ et al (2009) Selection of cytotoxicity markers for the screening of new chemical entities in a pharmaceutical context: a preliminary study using a multiplexing approach. Toxicol In Vitro 23:319–332
36. Liu X et al (2014) The role of lysosomes in BDE 47-mediated activation of mitochondrial apoptotic pathway in HepG2 cells. Chemosphere 124:10–21
37. Capasso L, Camatini M, Gualtieri M (2014) Nickel oxide nanoparticles induce inflammation and genotoxic effect in lung epithelial cells. Toxicol Lett 226(1):28–34
38. Saquib Q, Al-Khedhairy AA, Ahmad J et al (2013) Zinc ferrite nanoparticles activate IL-1b, NFKB1, CCL21 and NOS2 signaling to induce mitochondrial dependent intrinsic apoptotic pathway in WISH cells. Toxicol Appl Pharmacol 273(2):289–297
39. Saquib Q, Al-Khedhairy AA, Siddiqui MA et al (2012) Titanium dioxide nanoparticles induced cytotoxicity, oxidative stress and DNA damage in human amnion epithelial (WISH) cells. Toxicol In Vitro 26(2):351–361
40. Kawanishi S, Oikawa S, Inoue S et al (2002) Distinct mechanisms of oxidative DNA damage induced by carcinogenic nickel subsulfide and nickel oxides. Environ Health Perspect 110:789–791
41. Liu ZG (2005) Molecular mechanism of TNF signaling and beyond. Cell Res 15(1):24–27

42. Luzio A, Monteiro SM, Fontainhas-Fernandes AA et al (2013) Copper induced upregulation of apoptosis related genes in zebrafish (Danio rerio) gill. Aquat Toxicol 128–129:183–189
43. Kang JX, Liu J, Wang J et al (2005) The extract of huanglian, a medicinal herb, induces cell growth arrest and apoptosis by upregulation of interferon-beta and TNF-alpha in human breast cancer cells. Carcinogenesis 26(11):1934–1939
44. Han MH, Kim GY, Yoo YH et al (2013) Sanguinarine induces apoptosis in human colorectal cancer HCT-116 cells through ROS-mediated Egr-1 activation and mitochondrial dysfunction. Toxicol Lett 220(2):157–166
45. Frank MB, Yang Q, Osban J et al (2009) Frankincense oil derived from Boswellia carteri induces tumor cell specific cytotoxicity. BMC Complement Altern Med 9:6
46. Setyawati MI, Tay CY, Leong DT (2013) Effect of zinc oxide nanomaterials-induced oxidative stress on the p53 pathway. Biomaterials 34(38):10133–10142
47. Wang X, Stavchansky S, Kerwin SM et al (2010) Structure-activity relationships in the cytoprotective effect of caffeic acid phenethyl ester (CAPE) and fluorinated derivatives: effects on heme oxygenase-1 induction and antioxidant activities. Eur J Pharmacol 635(1–3):16–22
48. Lee JK, Sayers BC, Chun KS et al (2012) Multi-walled carbon nanotubes induce COX-2 and iNOS expression via MAP kinase-dependent and-independent mechanisms in mouse RAW264.7 macrophages. Part Fibre Toxicol 9:14
49. Yang H, Filipovic Z, Brown D et al (2003) Macrophage inhibitory cytokine-1: a novel biomarker for p53 pathway activation. Mol Cancer Ther 2(10):1023–1029
50. Wang X, Bynum JA, Stavchansky S et al (2014) Cytoprotection of human endothelial cells against oxidative stress by 1-(2-cyano-3,12-dioxooleana-1,9(11)-dien-28-oyl)imidazole (CDDO-Im): application of systems biology to understand the mechanism of action. Eur J Pharmacol 734:122–131

11 Nanoparticle-Protein Interaction: The Significance and Role of Protein Corona

Saad Mohammad Ahsan, Chintalagiri Mohan Rao, and Md. Faiz Ahmad

Abstract

The physico-chemical properties of nanoparticles, as characterized under idealized laboratory conditions, have been suggested to differ significantly when studied under complex physiological environments. A major reason for this variation has been the adsorption of biomolecules (mainly proteins) on the nanoparticle surface, constituting the so-called "biomolecular corona". The formation of biomolecular corona on the nanoparticle surface has been reported to influence various nanoparticle properties *viz.* cellular targeting, cellular interaction, in vivo clearance, toxicity, etc. Understanding the interaction of nanoparticles with proteins upon administration in vivo thus becomes important for the development of effective nanotechnology-based platforms for biomedical applications. In this chapter, we describe the formation of protein corona on nanoparticles and the differences arising in its composition due to variations in nanoparticle properties. Also discussed is the influence of protein corona on various nanoparticle activities.

Keywords

Protein corona · Nanoparticles · Biomolecular corona · Nanoparticles-protein interaction

11.1 Introduction

After entering into a biological system, foreign particles interact with various biomolecules. As compared to bulk matter, such interactions are more prominent in case of nanomaterials due to their large surface area to volume ratio and a high surface free energy. With rapid developments in the field of nanomedicine and the realisation of

S.M. Ahsan (✉) · C.M. Rao
Centre for Cellular and Molecular Biology (CSIR-CCMB), Hyderabad, Telangana State, India
e-mail: saadmahsan@gmail.com

Md.F. Ahmad (✉)
Department of Pharmacology, School of Medicine, Case Western Reserve University, Cleveland, OH, USA
e-mail: fam37@case.edu

Q. Saquib et al. (eds.), *Cellular and Molecular Toxicology of Nanoparticles*, Advances in Experimental Medicine and Biology 1048, https://doi.org/10.1007/978-3-319-72041-8_11

regulatory/safety features in nanoparticle based therapeutics, it has become important to study the interaction of components from biological fluids with such nanoscale materials. The interaction of nanomaterials with the surrounding biological environment results in the formation of a layer of adsorbed biomolecules on the nanomaterial surface (Fig. 11.1a–c), leading to the modification of its properties and imparting it with a new identity [1]. This layer of biomolecules is termed as the "biomolecular corona" [2]. The lipid profile of the biomolecular corona has been mainly studied in context of the pulmonary environment and is shown to be relatively conserved across different nanoparticle types and depends on the relative abundance of the lipids [3, 4]. The protein composition of the biomolecular corona however, has been a subject of extensive study due to its variations and significance in dictating nanoparticle properties [3, 5].

11.2 Formation of Protein Corona

The process of continuous adsorption and desorption of plasma proteins on a surface with regards to time of incubation, surface properties, etc. is termed as the "Vroman effect" [6–8]. The adsorption of plasma proteins and the subsequent formation of protein corona on nanoparticle surface is a rapid event that occurs within a few seconds of nanoparticle exposure to a biological environment. The proteins adsorbed onto a surface stay in a continuous state of flux and the composition of these proteins may vary with time (Fig. 11.1d) [9]. On a nanoparticle surface, the initial corona consists of proteins with high association rates and high abundance in the plasma. With time, proteins with higher affinities and longer residence times replace proteins with lower affinities and short residence times.

Proteins which bind to the nanoparticle surface with high affinities and display high association rates constitute the "hard corona". The proteins constituting the hard corona directly interact with the nanoparticle surface. On the other hand, proteins that are loosely bound to the nanoparticle surface with low binding affinities and display low association rates constitute the "soft corona" (Fig. 11.1e).

The interaction of proteins present in the soft corona with the nanoparticle surface is indirect and meditated via the hard corona [1, 5].

The proteins constituting the soft corona rapidly exchange from the surrounding environment on short time scales. The hard corona is more stable and tightly associated with the nanoparticles and is considered to play an important role in nanoparticle interaction with its surrounding cells [11]. As a nanoparticle passes through a variety of biological environments, the soft corona is altered. However, the hard corona is altered to a lesser extent and retains the proteins adsorbed during initial exposures.

A variety of techniques are used to study the structure and composition of the protein corona. These techniques include dynamic light scattering (DLS), differential centrifugal sedimentation (DCS), size exclusion chromatography (SEC) and transmission electron microscopy (TEM) for determining the thickness of the corona layer; poly(acrylamide) gel electrophoresis (PAGE) and liquid chromatography tandem mass spectrometry (LC-MS/MS) for identification and

Fig. 11.1 (continued) adsorb with low affinity. These proteins are subsequently displaced by proteins with a higher affinity (blue). The adsorption of certain proteins (blue and green) may aid in the binding of other proteins (yellow) with an initial low affinity for the particle surface. Other proteins (red) do not adsorb at all (Reprinted by permission from Macmillan Publishers Ltd.: [Nature Nanotechnology] (Biomolecular coronas provide the biological identity of nanosized materials), Copyright (2012) [2]). (**e**) Schematic representation depicting the hard/soft corona and the dynamic nature of protein corona: Proteins bound to a nanoparticle surface via protein-nanoparticle interactions constitute the hard corona, while proteins adsorbed on the nanoparticles through protein-protein interactions constitute the soft corona. The dynamic nature of protein corona arises due to competitive particle-protein interactions, continuous particle-protein association/dissociation and a characteristic protein profile specific to a particular biological fluid. A continuous association and dissociation of proteins alters the protein corona composition prior to equilibrium after which it is stabilized (Reproduced from "Nanoparticle–protein corona complexes govern the biological fates and functions of nanoparticles" with permission of The Royal Society of Chemistry [10])

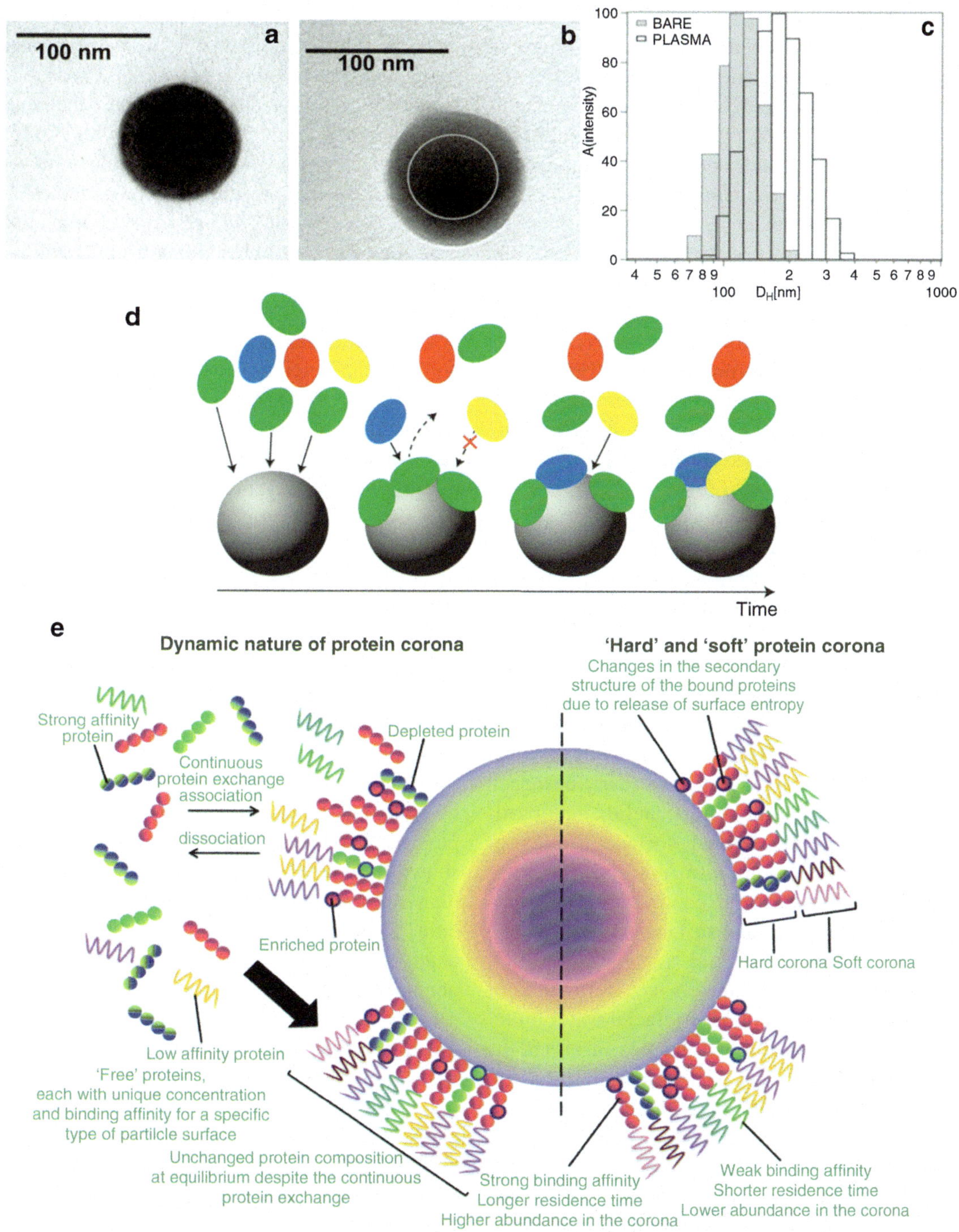

Fig. 11.1 TEM image of (**a**) bare 100 nm surface-carboxylated polystyrene (PSCOOH) nanoparticles and (**b**) protein-nanoparticle complex. (**c**) DLS (intensity-weighted size distribution) for 100 nm surface-carboxylated polystyrene (PSCOOH) nanoparticles (bare) (gray bars) and protein-PSCOOH nanoparticle complex (white bars). Protein-PSCOOH nanoparticles show an increase in size (Reprinted with permission from "What the Cell sees in Bionanoscience". Copyright (2010) American Chemical Society [1]) (**d**) Time based evolution of the biomolecular corona (left to right): Initial corona consists of highly abundant proteins (green) that

quantification of the proteins; circular dichroism spectroscopy (CD), fluorescence quenching and computational simulations for determining the conformation of the bound proteins; size exclusion chromatography (SEC), surface plasmon resonance (SPR) and isothermal titration calorimetry (ITC) for estimating the affinities of the bound proteins with the nanomaterial surface [5]. Various factors dictate the composition of the protein corona on the nanoparticle surface and are discussed in detail later.

An important parameter that determines the adsorption of proteins on a nanomaterial surface is the thermodynamic feasibility of the interaction. It can be defined in terms of the Gibbs free energy of adsorption (ΔG_{ADS}) [5, 12].

$$\Delta G_{ADS} = \Delta H_{ADS} - T\Delta S_{ADS} < 0$$

where, ΔH_{ADS} and ΔS_{ADS} are the enthalpy and entropy during adsorption and T is the temperature. The stability of a protein-nanoparticle interaction is determined by the net free energy of adsorption (ΔG_{ADS}). Proteins adsorbed with a large ΔG_{ADS} have a low probability of desorption and hence a prolonged association with the nanoparticles. On the other hand, proteins adsorbed with a low ΔG_{ADS} are easily desorbed and return to solution. In general, nanoparticles with either charged surface groups or hydrophobic surfaces form more stable interactions with proteins as compared to nanoparticles with neutral surface charges and hydrophilic surfaces.

A number of parameters *viz.* formation of bonds, rearrangement of interfacial water molecules, or conformational changes in either the protein or the nanomaterial surface contribute to a favourable change in enthalpy or entropy. Hydrogen bonding, solvation forces, Van der Waals interactions, etc. between the nanoparticle surface and the adsorbed proteins have been reported [13]. In the case of human serum albumin adsorption on polymeric nanoparticles, the interaction has been shown to be driven with a release of heat suggesting an enthalpy change favouring the interaction [14]. For other proteins such as fibrinogen, lysozyme, ovalbumin and human carbonic anhydrase II, no change in enthalpy is observed. The binding in these cases is suggested to be entropy driven and is accompanied by a release of bound water molecule from the nanoparticle surface. Such an entropy driven mechanism of protein binding with the release of water molecule from the nanoparticle surface does not alter the protein conformation [15]. In other examples, protein conformation may be altered if it is thermodynamically favourable and allows for hydrophobic patches or charged sequences within proteins to interact with a complementary region on the nanomaterial surface [16]. In some cases, specific domains within proteins may enhance interaction with nanoparticles. As an example, the interaction of high molecular weight kininogen (HMWK) to iron oxide nanoparticles occurs via the histidine-rich domain, with the imidazole side chains of histidines interacting with iron oxide nanoparticles [17]. Apart from directly interacting with a nanoparticle surface, proteins may indirectly interact with nanoparticles via protein-protein interactions. These interactions may be specific where a conformation dependent protein-protein interaction takes place. Such interactions and adsorption of proteins on to a nanoparticle surface is observed for complement factors and coagulation proteins, and is shown to play an important role in mediating a humoral response [18]. Whereas, in non-specific interactions an initial conformational change in a protein, post adsorption on a nanoparticle surface, may lead to an exposure of a hydrophobic or charged patch which subsequently mediates the secondary interactions. As an example, the interaction of α-chymotrypsin (ChT) and lysozyme (Lyz) with gold nanoparticles (AuNP) (either nanospheres;AuNS or nanorods;AuNR), has been shown to result in a loss of protein structure and enzyme activity. The effect of nanoparticle morphology is found to be primarily protein-specific in this case. The binding of proteins to gold nanoparticles follows three steps (Fig. 11.2) *viz.* (i) an initial protein-surface interactions (Region I), (ii) an intermediate stage characterized by protein-sur-

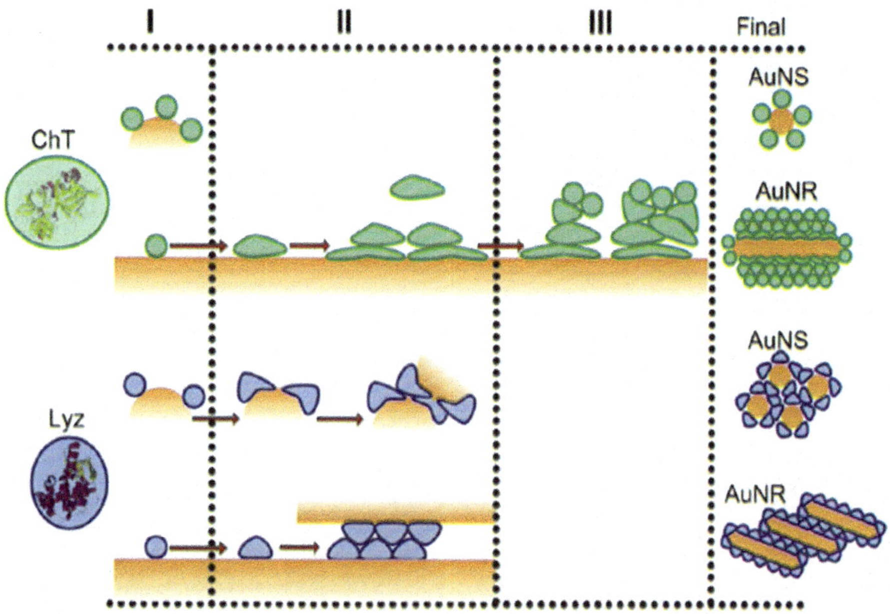

Fig. 11.2 Schematic representation of Chymotrypsin (ChT) and Lysozyme (Lyz) adsorption on gold nanoparticles (AuNP). Initially (region I) both proteins get adsorbed onto AuNP with minimal structural perturbation. Later, as depicted in region II, protein-surface interactions (as in the case of ChT-AuNR, Lyz-AuNS and Lyz-AuNR) lead to changes in protein conformation. This stage is characterized by protein-surface and protein-protein interactions. The final stage (region III) is dominated by protein-protein interactions resulting in the stacking of multiple layers of protein on the particle surface. ChT-AuNS stay in Region I as stable conjugates with ChT in a native conformation. ChT-AuNR form stable conjugates and proceed to region III with multiple layers of ChT adsorbed on the nanoparticle surface. Lyz-AuNP conjugates progress to region II, where a mixture of protein-surface and protein-protein interactions result in the aggregation of the conjugates (Reprinted from "Effect of gold nanoparticle morphology on adsorbed protein structure and function", 32(29), Gagner JE, Lopez MD, Dordick JS & Siegel RW, 7241–52, Copyright (2011), with permission from Elsevier [19])

face and protein-protein interactions (Region II), and (iii) a final stage dominated by protein-protein interactions (Region III). The interaction of ChT with gold nanospheres exists in Region I, and does not lead to structural alteration of ChT. ChT-AuNR conjugates, on the other hand progress to Region III, with alterations in ChT conformation and formation of a protein multilayer. The Lyz adsorption on AuNPs proceeds to Region II, characterized by Lyz adsorption and denaturation leading to the exposure of hydrophobic patches in the protein. This phenomenon is observed regardless of the AuNP (AuNS or AuNR) morphology. The Lyz-surface and Lyz-Lyz interactions thereby lead to aggregation of the conjugates (Fig. 11.2) [19].

11.3 Composition of Protein Corona

The biomolecular corona formed on nanoparticles is constituted by a variety of plasma proteins which fall into the following major categories *viz.* apolipoproteins, immunoglobulins, complement factors, coagulation factors, acute phase proteins, tissue leakage components and other miscellaneous proteins [20–22]. As an example, the adsorption of various plasma proteins on silica nanoparticles (SiNPs) of different sizes is provided in Fig. 11.3 [21]. The physiological function of the major components of the adsorbed proteins is mainly lipid transport, blood coagulation, complement activation, pathogen recognition and ion transport. The relative concentrations

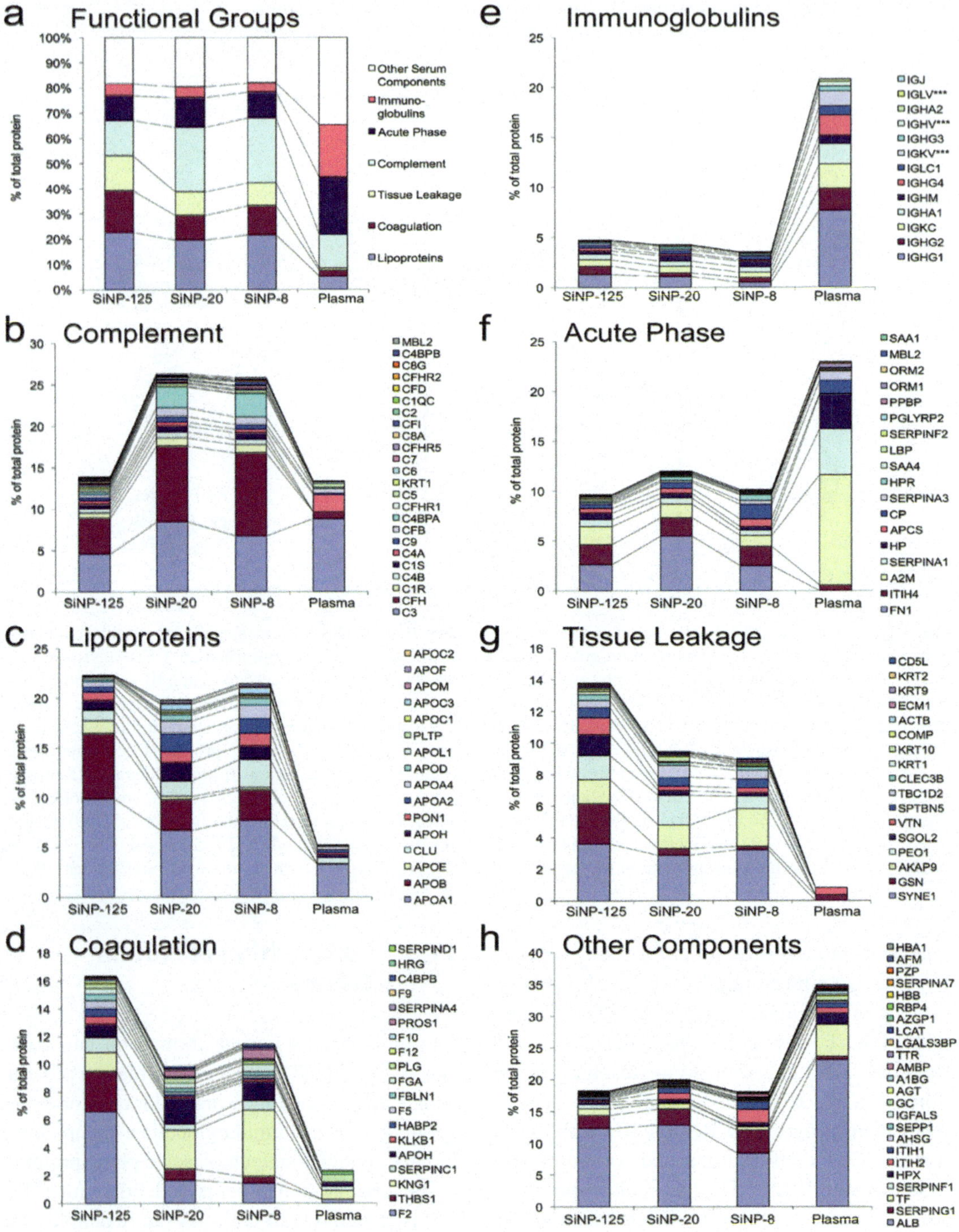

Fig. 11.3 Bioinformatic classification (according to their functions) of corona proteins identified on the surface of silica nanoparticles (SiNPs) of 125 nm (SiNP-125), 20 nm (SiNP-20) and 8 nm (SiNP-8). (**a**) Proteins identified in the respective SiNPs corona are classified according to their biological functions. (Relative percentages of the proteins compared to crude plasma are shown). Significant enrichment of plasma proteins involved in complement activation (**b**), lipoproteins (**c**), coagulation (**d**) and tissue leakage proteins (**g**) was evident in the corona. Other protein groups viz. immunoglobulins (**e**), acute phase response proteins (**f**), and other components (mainly serum albumin) (**h**) although present in high amounts in the plasma, displayed a lower affinity for the SiNPs (Reprinted with permission from "Nanoparticle Size Is a Critical Physicochemical Determinant of the Human Blood Plasma Corona: A Comprehensive Quantitative Proteomic Analysis". Copyright (2011) American Chemical Society [21])

of different proteins on the nanoparticle surface constituting the corona does not correspond to their plasma concentrations, suggesting a selective enrichment of some proteins. The selective adsorption of proteins on the surface of nanoparticles has been shown to be a complex interplay of a variety of nanoparticle features (discussed in detail later). A detailed account of various proteins adsorbed on different nanoparticles, termed as the "adsorbome", has been provided by Walkey and Chan [5].

Some of the abundant proteins identified on the corona formed on most nanoparticles are albumin, α-2-macroglobulin, apolipoprotein A-I, apolipoprotein A-III, apolipoprotein C-III, immunoglobulin kappa chain, different forms of light- and heavy-chains of immunoglobulins, complement C3, complement C4, haptoglobin, alpha-1-antitrypsin, kininogen, plasminogen, keratin, vitronectin, etc. [21, 22]. While the binding of albumin and apolipoproteins helps in the transport of nanoparticles, the binding of complement proteins and immunoglobulins leads to particle opsonization thereby promoting nanoparticle clearance from circulation [23].

Walkey and Chan have reported around two to six proteins to be adsorbed at high abundance on most nanomaterials, with the most abundant identified protein (across all nanomaterials) representing approximately 29% of the total proteins adsorbed. Collectively the three most abundant proteins represent an average 56% of the total adsorbed proteins. By defining a threshold of 10% of the total adsorbed protein mass as "high abundance", Walkey and Chan have identified 21 out of 125 proteins above this threshold on at least one type of nanomaterial. These proteins are albumin, transferrin, fibrinogen, haptoglobin, hemoglobin, histidine-rich glycoproteins, Ig gamma chain, Ig light chain, Ig mu chain, inter alpha trypsin inhibitor H1, mannose-binding protein C, paraoxonase-1, antithrombin-III, apolipoprotein A-1, apolipoprotein A-IV, apolipoprotein B-100, apolipoprotein C-II, apolipoprotein C-III, apolipoprotein E, clusterin and complement C3 [5]. The interaction of these proteins with the nanoparticle surface, as mentioned earlier, is not related to their physiological roles or relative plasma abundance. The interaction is mediated by various nanoparticle properties as discussed in the following section.

11.4 Factors Influencing Protein Corona

11.4.1 Nanoparticle Size

The size of nanomaterials determines its surface curvature and also influences its surface area. It is thus an important factor in determining the qualitative and quantitative composition of the adsorbed proteins on a nanoparticle surface. While some reports suggest variations only in the amount of bound protein with particle size and curvature [22], other reports have shown both quantitative and qualitative size dependent changes in the protein corona [20, 21, 24, 25]. A difference of as low as 10 nm has been found to significantly affect the nanoparticle corona composition. In a study on gold nanoparticles, an increase in size leads to an increase in the binding constant of proteins along with an increase in the degree of cooperativity of nanoparticle-protein binding [26]. In a separate study on adsorption of plasma proteins on silica nanoparticles of different sizes, it was revealed that the binding of approximately 37% of all corona proteins was significantly affected by particle size [21]. Lundqvist et al. have reported a comparison of protein corona formed on 50 and 100 nm polystyrene nanoparticles with different surface functionalities. While the plain polystyrene nanoparticles of the two sizes showed almost 80% similarity in the composition of protein corona, the size effect was more pronounced for the carboxyl modified nanoparticles of the two sizes which showed only 50% similarity [20].

11.4.2 Nanoparticle Surface Properties

Surface features of nanoparticles *viz.* charge and hydrophobicity/hydrophilicity are important factors in determining the protein corona composition. As a general concept, negatively charged nanoparticles attract positively charged proteins and vice versa. Proteins with pI >5.5 show a preference for negatively charged (nanoparticles with acidic functional groups) nanoparticles, while those with pI <5.5 show a preference for positively charged (nanoparticles with basic functional groups) nanoparticles. In such cases, the protein adsorption on nanoparticle surfaces is expected to be mainly driven by coulombic interactions [27]. However, as most plasma proteins possess a negative charge at physiological pH (most plasma proteins have a pI in the range of 6–8), charge based interaction of plasma proteins with negatively charged nanoparticles is a far more complex interaction. In such cases, coulombic interactions take place between specific surface domains of the proteins that possess a complementary charge to that of the nanoparticle surface. As the coulombic interactions are achieved between molecules that are in close vicinity of each other, such charged based interactions are expected to involve individual/specific domains on the protein and nanoparticle surface rather than the entire surface [28]. In a study on the adsorption of three human serum proteins *viz.* serum albumin, apolipoprotein A-I and apolipoprotein E4 on negatively charged (carboxylated) FePt nanoparticles, an increase in the hydrodynamic radius of the nanoparticles with protein concentration was observed suggesting corona formation. Equilibrium dissociation coefficient studies suggested electrostatic interactions between protein and nanoparticles. Structural studies revealed the presence of positively charged patches on the protein surface, through which the protein molecules were shown to interact electrostatically with the nanoparticle surface [29]. Studies on the effect of nanoparticle surface charge density have revealed a direct relation between the surface charge and protein adsorption profiles. A higher surface charge leads to an increased plasma protein adsorption, without much difference in the qualitative profile of the of the detected proteins [7, 30].

In general, charged nanoparticles adsorb more protein than nanoparticles with a neutral surface. This differential adsorption further leads to a difference in opsonization rates of the nanoparticles. Neutrally charged nanoparticles show slow opsonization rates as compared to charged nanoparticles, suggesting a direct correlation between the corona composition and nanoparticle uptake [23, 31]. As an example, a charged based cellular uptake for albumin nanoparticles has been reported by Roser et al. Albumin nanoparticles with neutral surface charge showed a reduced phagocytic uptake in comparison with albumin nanoparticles which were modified to carry a net positive or negative surface charge [31]. The serum proteins identified in the opsonization process were C3b-complement, IgG and fibronectin, whereas albumin was shown to reduce phagocytic uptake of the particles.

Apart from surface charge, the hydrophilicity/hydrophobicity of nanoparticles has also been reported to influence the protein corona formation both quantitatively and qualitatively [22, 32]. The major proteins affected by surface hydrophobicity are albumin, fibrinogen and apolipoproteins [33]. As an example, a comparison of proteins adsorbed on a less hydrophobic 85:15 NIPAM/BAM copolymer particle and a more hydrophobic 50:50 copolymer particle revealed clear differences in the amount and type of proteins bound. While the less hydrophobic 85:15 NIPAM/BAM copolymer particle bound virtually no protein (trace amounts of albumin), the more hydrophobic 50:50 copolymer particles preferentially bound apolipoproteins (AI, AII, AIV, and E), human serum albumin, fibrinogen and various other proteins [22]. As mentioned earlier, the differences in the amount and type of plasma proteins adsorbed on nanoparticles leads to variation in their opsonization rates. Nanoparticles with hydrophobic surfaces thus show high opsonization rates due to an enhanced adsorption of plasma proteins [23, 34].

11.4.3 Time of Exposure

The Vroman effect exhibited by protein adsorption on to a nanoparticle surface can be divided into "early" and "late" stages. The early stage is characterized by the adsorption of proteins with high mobility although low binding affinity. Thereafter, the low binding affinity proteins are progressively displaced by the ones with higher binding affinities. The initial interaction of low binding affinity proteins is an attempt on the part of the nanoparticles to lower its surface energy followed by exchange of proteins on shorter time scales from the available set of proteins that may diffuse to the surface [35]. Various studies have revealed the importance of plasma exposure time on the quantity and composition of the corona proteins. Although the corona deposition is dictated by the NP composition and surface functionalization, exposure time is also an important factor. As demonstrated by Tenzer et al. the corona on nanoparticles with different types of surface functionalization, evolves in a completely different manner [9]. While the protein quantity on commercial 30 nm silica nanoparticles (AmSil30) decreased with time, it was shown to increase for negatively and positively charged polystyrene nanoparticles (nPsNP and PPsNP) (Fig. 11.4a, b).

Apart from quantitative variations, time based differences in composition of the proteins constituting the corona were also observed for different nanoparticles (Fig. 11.4c). In a report on the characterization of adsorbed proteins on the surface of gold nanoparticles, it has been reported that at initial time points (1 h), high molecular weight (HMW) and low molecular weight (LMW) proteins are adsorbed while the medium molecular weight (MMW) proteins are down represented. However, after prolonged

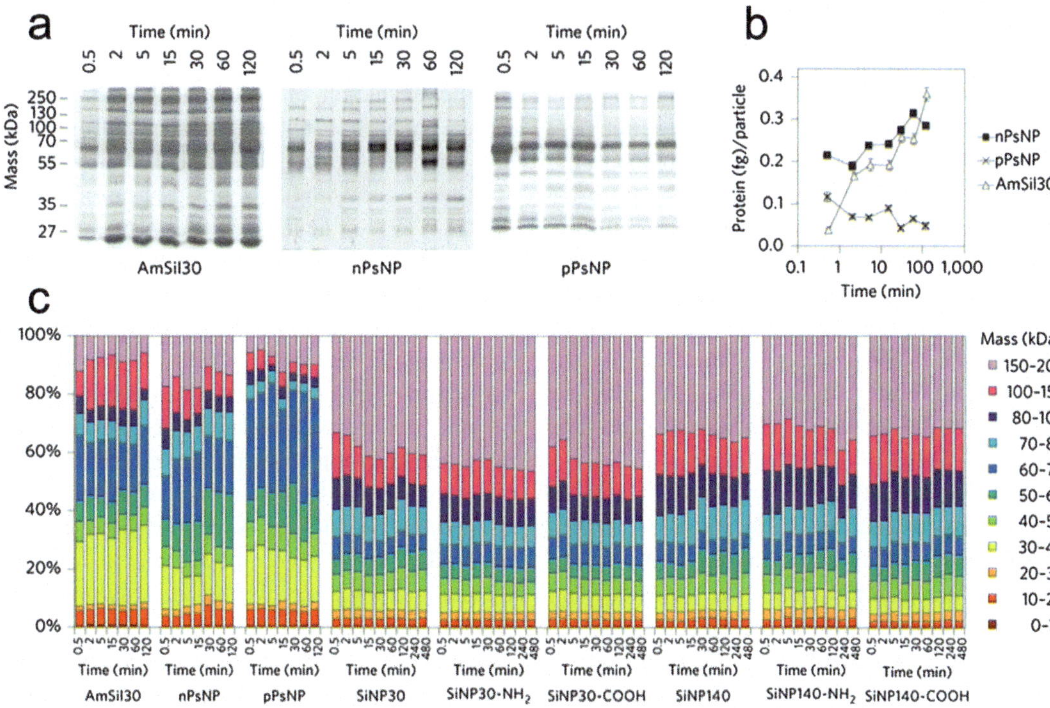

Fig. 11.4 (**a**) SDS–PAGE of plasma proteins bound to different nanoparticles. (**b**) Quantification of proteins in the corona (protein (fg) per nanoparticle) at different time points. (**c**) Identification and classification of corona proteins identified on different nanoparticles. (30 nm commercial silica nanoparticles (AmSil30); 30 nm laboratory synthesized silica nanoparticles (SiNP30); Amine functionalized SiNP30 (SiNP30-NH_2); carboxylate functionalized SiNP30 (SiNP30-COOH); 140 nm laboratory synthesized silica nanoparticles (SiNP140); Amine functionalized SiNP140 (SiNP140NH_2); Carboxylate functionalized SiNP140 (SiNP140-COOH); negatively or positively charged polystyrene nanoparticles (nPsNP and pPsNP) (Reprinted by permission from Macmillan Publishers Ltd.: [Nature Nanotechnology] (Rapid formation of plasma protein corona critically affects nanoparticle pathophysiology), Copyright (2013) [9])

incubation (48 h) the concentration of the MMW proteins was found to increase while that of LMW was found to decrease. The major proteins identified were MMW proteins and some LMW proteins, involved in important biological processes such as transport and trafficking (apolipoprotein A1, transferrin, vitamin D-binding protein, etc.), blood coagulation (protein C inhibitor, antithrombin III, coagulation factor V, etc.) and tissue development (fibulin 1, periostin, thrombospondin-1, galectin 3 binding protein, etc.). The concentration of the HMW proteins was however not effected significantly with time [36]. A sequential adsorption of plasma proteins has also been demonstrated on N-iso-propylacrylamide/N-tert-butylacrylamide (NIPAM/BAM) nanoparticles through mathematical modelling. For these nanoparticles, the initial phase is characterized by the deposition of high abundance and fast dissociation proteins *viz.* albumin and fibrinogen followed by higher affinity and slow-exchanging apolipoproteins AI, AII, AIV and E [22, 37]. The relative abundance of proteins in the serum also plays an important role in determining the initial binding partners. The first proteins to be adsorbed on nanoparticles are albumin, IgG and fibrinogen which may later be replaced by apolipoproteins and coagulation factors. In a surface plasmon resonance (SPR) based study on silver coated slides, it was observed that initially the smallest and most abundant protein *viz.* albumin adsorbs to the surface followed by IgG and fibrinogen [38].

As the quantitative and qualitative profile of protein corona changes with incubation time, it also influences the cellular uptake of nanoparticles. As an example, the hepatic uptake of lecithin-coated polystyrene nanospheres has been found to increase significantly with an increase in incubation time. This is mainly due to an increase in the opsonin concentration in the adsorbed layer and an opsonin-mediated uptake by Kupfer cells. The major opsonins identified were complement C3 (C3) and immunoglobulin G (IgG). Their concentrations were found to increase with time, thereby accounting for an increase in cellular uptake [39].

While the initial phase of corona formation may be a rapid event occurring on a time scale of few seconds to minutes, the late stage may take a few hours to days [5, 40]. However, depending on the evolution stage and the protein corona at that particular moment, the physiological response to the nanoparticle may vary.

11.4.4 Temperature

Physiological temperatures are known to fluctuate and vary between 35 and 41 °C under different conditions. Exploring the effect of these subtle temperature changes on the surface adsorption of proteins thus becomes essential. Mahmoudi et al. studied the effect of temperature on the adsorption of albumin and apo-transferrin (apo-Tf) on polymer coated iron-platinum (FePt) nanoparticles. At lower temperatures (13 and 23 °C) the deposition of both albumin and apo-Tf took place till the formation of a monolayer, after which increasing the concentration of the respective protein had no effect on the corona thickness. However, at a higher temperature (43 °C) a decrease in the thickness of the corona shell was observed for both HSA and apo-Tf, which was proposed due to (i) conformation changes in the protein molecules, (ii) decrease in the number of protein molecules adsorbed on the nanoparticle surface and (iii) an increase in the flexibility of the polymer coating on the nanoparticle surface leading to the insertion of the protein molecules in the polymer shell [41]. Mahmoudi et al. have also studied the protein corona composition on gold nanorods (AuNR) after plasmonic heating and thermal heating under two different serum conditions mimicking an in vitro (10% serum) environment and an in vivo (100% serum) environment. Subtle changes in zeta-potential of AuNRs were observed following both plasmonic and thermal heating suggesting some compositional differences arising due to the change in temperature. The decrease in surface charge was slightly more in AuNRs incubated in 100% serum as compared to those incubated in 10% serum. Analysis of the protein composition revealed an increase in the amount of several important proteins *viz.* serum albumin, α-2-HS-glycoprotein precursor, apolipoprotein A-II precursor and apolipoprotein C–III precursor in the plasmonic heating 10% serum model as compared to the sample incubated at 37 °C. The levels of apolipoprotein A-I precursor protein was found to be decreased. However, there were significant differences in the in vitro (10% serum) and in vivo (100% serum) models. In terms of the molecular weight of the proteins adsorbed, it was observed that for the 10% serum model a heat treatment whether thermal or plasmonic did not have a significant effect on the adsorption of the low molecular weight proteins (<50 kDa). However, the amount of high molecular weight proteins (50–100 kDa) were observed to increase. On the contrary, in the in vivo model the amount of low molecular weight proteins (<30 kDa) were observed to be significantly decreased upon plasmonic heating but not thermal heating. No major change in the high molecular weight fraction (>30 kDa) was observed in the 100% serum model with either plasmonic or thermal heating [42].

Elevated temperatures have been reported to bring variations in the composition of serum, most notable being aggregation and depletion of complement proteins and immunoglobulins. The binding of heat inactivated serum on polystyrene nanoparticles of 100 nm reveals an increase in nanoparticle size at physiological temperature suggesting a thicker corona layer. However, no such differences were observed for smaller nanoparticles (40 nm). These differences in the amount of proteins adsorbed on the nanoparticles effect their uptake in A549 cells, with nanoparticles of smaller sizes (20 and 40 nm) showing a higher cellular uptake in non-heat inactivated serum as compared to heat inactivated serum. On the other hand, nanoparticles of larger sizes (100 and 200 nm), showed a complex behaviour, with nanoparticle to serum ratios playing an important role. A higher cellular uptake was observed in heat inactivated serum at low nanoparticle concentrations and non-heat inactivated serum at high nanoparticle concentrations. Interestingly, in all the conditions tested in this study, formation of a protein rich corona layer was found to inhibit cellular uptake of nanoparticles [43]. This may be due to the heat induced depletion of complement factors and the unresponsiveness of the cell line (A549) to complement based particle uptake.

11.4.5 Biological Environment

The biomolecular corona formed on nanoparticles may also vary depending on the variations in the physiological environments. Differences have been reported in the corona composition of nanoparticles when incubated with normal serum and hyperlipidemic serum, used as a model for cardiovascular diseases and obesity. Incubation with hyperlipidemic serum results in an increased amount of cholesterol adsorption on Fe_3O_4 nanoparticles. In this study, apart from differences in lipid composition, differences in protein composition were also reported. A total of 29 unique proteins were found to associate with nanoparticles incubated in the lipid-rich media. The corona formed on nanoparticles upon incubation in a lipid rich media resulted in an upregulation of genes associated with inflammation and cell adhesion in rat aortic endothelial cells (RAECs) to a larger extent as compared to nanoparticles with biocorona formed after incubation with normal serum [44].

The corona formation and nanoparticle uptake studies in vitro are usually performed in the commonly used cell culture media *viz.* RPMI and DMEM. Interestingly, in a study with gold nanoparticles, it was found that the corona formation is dependent on the nanoparticle dimensions for RPMI media. However, in case of DMEM, no such correlation on nanoparticle dimensions was observed. The total amount of hard protein corona was found to be higher in nanoparticles incubated in DMEM as compared to those incubated in RPMI media resulting in an increased toxicity of the nanoparticles in RPMI media [36].

Another variation in media arises due to the use of plasma or serum as a protein source. The use of plasma or serum has also been shown to determine the corona composition and eventual fate of nanoparticles. In a study on silver and silica nanoparticles incubated with plasma and serum it was shown that a similar amount of protein with drastic variations in composition adsorbed on these nanoparticles. The variation in the corona composition led to a difference in the nanoparticle uptake and cell viability with nanoparticles incubated in plasma showing a higher viability and lower cellular uptake as compared to nanoparticles incubated in serum. This difference was attributed to the concentration of apolipoprotein J (clusterin) adsorbed on the nanoparticles upon incubation with the two fluids [45].

A comparative study on the effect of static (unconditioned) media *viz.* human plasma (HP), human serum (HS), free albumin, free fibrinogen, etc. and dynamic (conditioned) media *viz.* media obtained after exposure to cells, on the quantitative and qualitative profile of the protein corona formed on polymeric nanoparticles has also been studied (Fig. 11.5a). Since cells secrete out a variety of molecules, the composition of cell-conditioned media is drastically different from that obtained from a defined source [46]. Incubation in a protein rich static media results in the adsorption of higher amount of proteins on the nanoparticle surface and leads to a larger reduction in surface charge (neutralization) as compared to incubation in a dynamic in vitro cell-conditioned media (Fig. 11.5b, c). In general the hard protein corona composition, nanoparticle composition and the cell phenotype determines the interaction of nanoparticles with cells. As an example, cell conditioned media obtained from LoVo cell culture was found to inhibit the nanoparticle association with undifferentiated and differentiated THP-1 cells to a larger extent than conditioned media obtained from HeLa or THP-1 cell cultures, suggesting the importance of the conditioned media in dictating nanoparticle fate in vivo (Fig. 11.5d). Nanoparticles with a hard protein corona from different sources *viz.* human serum, human plasma, cell conditioned media from HeLa, LoVo and THP-1 cells also effect cytokine production from target cells. While the protein corona from FBS, HP and LoVo conditioned media increased the IL-1β production and decreased MCP-1 secretion, the presence of protein corona from HS and conditioned media from HeLa and THP-1 cells resulted in a decreased production of IL-1β and an increased secretion of MCP-1 (Fig. 11.5e). These results suggest the complexities of the corona composition and its consequences on nanoparticle activity [47].

11.4.6 Plasma Concentration

As a nanoparticle passes through a variety of microenvironments, it is expected to encounter plasma proteins in different concentrations. To mimic this situation, the effect of gradient plasma on the corona composition of nanoparticles has been studied and compared to the corona formed after incubation in plasma of defined concentrations. A clear difference in terms of the amount and composition of the corona was observed [48]. A major difference in the corona composition of nanoparticles exposed to a plasma gradient was the depletion of low molecular weight proteins including apolipoprotein precursor A-1, A-II, C-I, C-II, C-III, plasma retinol-binding protein precursor, transthyretin precursor, Isoform 1 of haptoglobin-related protein precursor, beta-2 microglobulin precursor and serum amyloid A protein precursor [48]. A similar observation of selective enrichment of low molecular weight proteins at low plasma concentrations has been made by Monopoli et al. for hydrophilic silica and hydrophobic sulphonated polystyrene ($PSOSO_3$) nanoparticles [35].

Fig. 11.5 (continued) by arrow). Numbers above the image indicate the source from which the hard protein coronas (hPCs) were derived. (**c**) Total intensity of each lane, indicating the total amount of protein in the gel. (**d**) Association of PMA CAPs (green bars) or CSPs (blue bars) coated with different hPCs with THP-1 cells, as measured by flow cytometry. (**e**) Heat maps showing the relative pro-inflammatory cytokine secretion levels by THP-1 cells upon treatment by CAPs (green) or CSPs (blue) coated with hPCs derived from various environments (as indicated by the numbers above the maps) (Reprinted with permission from "Cell Conditioned Protein Coronas on Engineered Particles Influence Immune Responses". Copyright (2017) American Chemical Society [47])

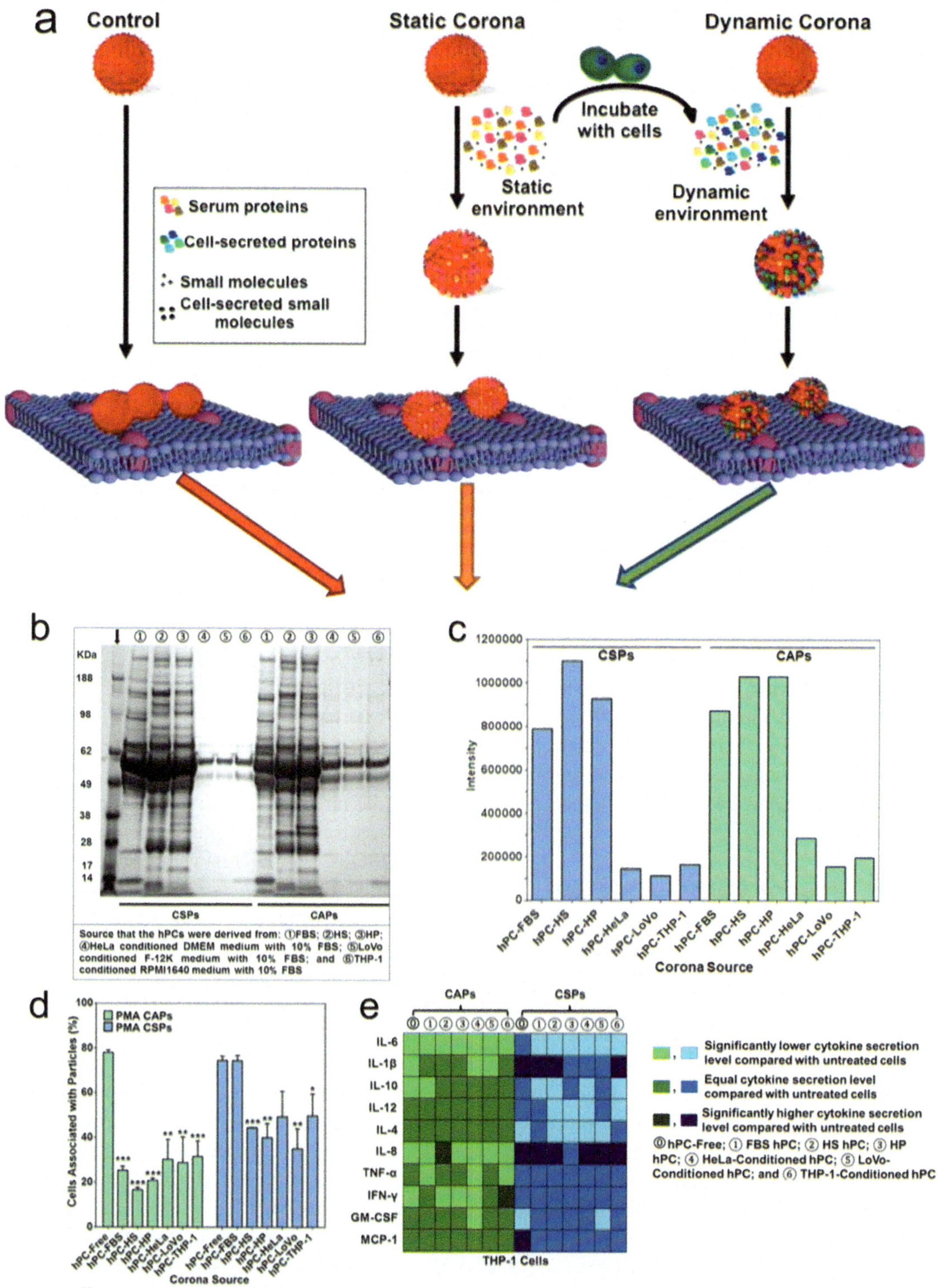

Fig. 11.5 (**a**) Schematic illustration of the formation of protein corona on poly (methacrylic acid) (PMA) particles in unconditioned and cell-conditioned environments. (**b**) SDS-PAGE image of corona proteins derived from various environments on core-shell particles (CSPs) and hollow capsules (CAPs). (Molecular weight marker is shown

In general, an increase in the plasma protein concentration leads to an increased adsorption of proteins on nanoparticles till the formation of a monolayer [41, 49]. While the amount of protein adsorbed is dependent on the plasma concentration, increasing the plasma concentration decreases the number of proteins present in the corona of the nanoparticles. As an example, for zeolite nanoparticles, an increased selective adsorption of apolipoprotein C-III (APOC-III) and fibrinogen (FIBA, FIBB and FIBG) was observed at 100% plasma concentration, while exposure to low plasma concentration (10%) resulted in the selective enrichment of immunoglobulin gamma (IGHG1, IGHG2, and IGHG4) [50].

11.5 Consequences of Protein Corona

The formation of protein corona has been reported to play an important role in defining the biological properties and physiological responses *viz.* toxicity and immunogenicity of nanoparticles [51–53]. The various effects and consequences of protein corona formation on the nanoparticle surface are discussed below.

11.5.1 Effect on Nanoparticle Size

The adsorption of plasma proteins results in the formation of a thick layer on the nanoparticle surface. Various studies have reported the layer to be around 20–40 nm for different types of nanoparticles. As the hydrodynamic diameters of most plasma proteins are in the range of 3–15 nm, the thickness of protein corona formed on nanoparticles suggests the existence of multiple layers of adsorbed proteins. The primary binders may be the ones interacting with the nanoparticle surface directly. The secondary binders then interact with the primary binders through protein-protein interactions. It could thus be assumed that the primary binders constitute the hard corona and the secondary binders comprise the soft corona. The contribution of the hard and soft corona to the overall thickness of the protein corona is not known. However, a study by Monopoli et al. [35], on sulphonated polystyrene nanoparticles ($PSOSO_3$) suggests that the soft corona is less thick in comparison to the hard corona.

11.5.2 Effect on Zeta-Potential

Apart from increasing the size of the nanoparticles, the adsorption of proteins on nanoparticle surface also leads to an alteration in its zeta-potential. As most plasma proteins exhibit a negative zeta-potential at neutral physiological pH (7.4), the formation of protein corona on nanoparticle surface imparts a zeta-potential in the range of −10 to −20 mV. This negative zeta-potential is observed irrespective of the nanoparticle physicochemical properties [5].

11.5.3 Cellular Targeting of Nanoparticles

The formation of corona has been reported to reduce the targeting abilities of nanoparticles. In a study on transferrin conjugated silica nanoparticles (SiO_2-PEG8-Tf), it was observed that the targeting ability of transferrin, when exposed to a biological environment (FBS), was compromised due to the shielding effect of the plasma proteins. Although, the nanoparticles were still found to enter cells, their specificity for transferrin receptors was found to be decreased (Fig. 11.6) [54].

Apart from a loss of targeting ability, the deposition of corona on nanoparticle surface may also lead to undesirable localization of nanoparticles resulting in unintended toxicities. Such a case has been reported for superparamagnetic iron oxide nanoparticles (SPIONs), where the deposition of protein corona leads to the crossing of the nanoparticles through the blood brain barrier (BBB) [55]. The BBB crossing ability was attributed to the presence of apoliprotein,

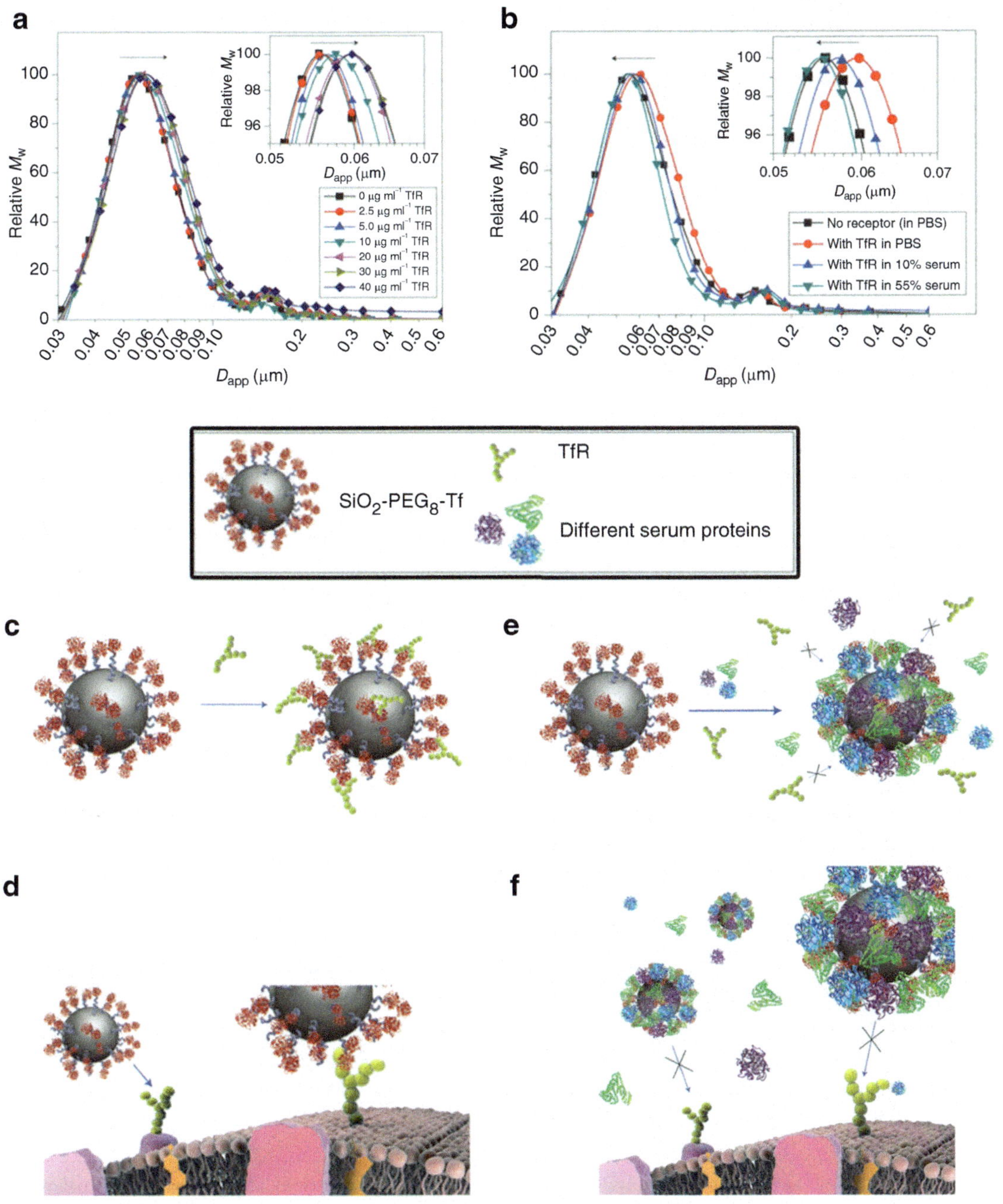

Fig. 11.6 (**a**) Apparent diameter (Dapp) as obtained by differential centrifugal sedimentation (DCS) for Tf-decorated silica nanoparticles with PEG8 spacers (SiO_2-PEG8-Tf) incubated with increasing amounts of transferrin receptor (TfR). A shift in the Dapp with increasing concentrations of TfR suggests interaction of SiO_2-PEG8-Tf nanoparticles with TfR. (**b**) DCS assessment of TfR binding to SiO_2-PEG8-Tf nanoparticles in the presence of fetal bovine serum (FBS) proteins. With increasing concentration of FBS a reduction in Dapp is observed, suggesting a loss of TfR binding due to the shielding effect of protein corona. Shown in the box is a cartoon representation of SiO_2-PEG8-Tf nanoparticles, soluble TfR and different serum proteins. (**c**) Schematic representation of nanoparticle binding to soluble TfR and (**d**) cell surface TfR. (**e**) Schematic representation of inhibition of nanoparticle binding to soluble TfR and (**f**) cell surface TfR in the presence of FBS proteins (Reprinted by permission from Macmillan Publishers Ltd.: [Nature Nanotechnology] (Transferrin-functionalized nanoparticles lose their targeting capabilities when a biomolecule corona adsorbs on the surface), Copyright (2013) [54])

ApoA-I. The nanoparticle corona of most nanoparticles has been shown to consist of apolipoproteins including ApoA-I. ApoA-I has been reported to cross the BBB and hence its association with nanoparticles may lead to an unintended accumulation of nanoparticles in the brain tissue.

11.5.4 Cellular Uptake of Nanoparticles

Generally a protein corona enriched with opsonins and coagulation proteins is believed to activate immune cells promoting phagocytosis thereby leading to the clearance of the NPs from the bloodstream. Conversely, enrichment of the protein corona with dysopsonins such as human serum albumin (HSA), apolipoproteins, and so forth promotes blood circulation [56]. The dependence of nanoparticle uptake by macrophages on the adsorbed complement factors has been reported for gold nanoparticles with different hydrophobicities. Immunoglobulins were shown to have a negative effect on macrophage uptake [57].

It is worth mentioning that different cell types interact with nanoparticles in completely different ways. As an example, in a study on the cellular uptake of silica nanoparticles by A549 cells, in the presence and absence of corona, the membrane association and internalization of silica nanoparticles was reported to be much higher in serum free media as compared to the ones with a preformed corona. Nanoparticles recovered from the cell surface also developed a corona layer consisting mainly of membrane and cytoskeletal proteins as well as membrane lipids. Also the number of proteins identified in the corona of these nanoparticles was higher (approximately 800) as compared to the total proteins identified in the corona of nanoparticles incubated in serum (approximately 300). This was attributed to cellular damage observed in the case of corona free nanoparticles. The cellular uptake for the corona free nanoparticles was found to be higher and the nanoparticles were observed to be localized both in the cytosol and lysosome of cells as compared to nanoparticles with a preformed corona, which localized exclusively in the lysosomal compartment of the cells. The results clearly demonstrated differences in the nanoparticle uptake efficiency and routes in the presence and absence of protein corona on nanoparticles [58].

The use of polyethylene glycol (PEG) as a stealth molecule for preventing nonspecific nanoparticle uptake has been well established. However, Schottler et al. reported this phenomenon to be not only dependent on the PEG polymer, but a consequence of the protein corona formed due to the PEGylation or polyphophoester (PEEP) conjugation. The corona on PEGylated and PEEP conjugated silica nanoparticles was found to be enriched in apolipoprotein J (clusterin). The binding of clusterin (Apo J) on PEG and PEEP was reported to be responsible for decreasing the macrophage uptake and providing a stealth effect (Fig. 11.7) [59].

The role of clusterin in preventing the macrophage uptake of non-PEGylated silver and silica nanoparticles has been demonstrated by Aoyama et al. [45]. Nanoparticles incubated with plasma and serum showed variations in their corona compositions and macrophage uptake. Clusterin,

Fig. 11.7 (continued) (**a**) Quantitative analysis of human plasma proteins adsorbed on different nanocarriers (Polystyrene nanoparticles ($PS-NH_2$) functionalized with polyethylene glycol with degree of polymerization 44 (PS-PEG44), and 110 (PS-PEG110), polystyrene nanoparticles functionalized with poly(ethyl ethylene phosphate) with degree of polymerization 49 (PS-PEEP49) and 92 (PS-PEEP92)). (**b**) Flow cytometry analysis of nanoparticle uptake by RAW264.7 cells. (**c**) Laser scanning microscopy images of nanoparticle uptake by RAW264.7 cells incubated in 100% human plasma or DMEM without plasma (Cell membrane is stained with CellMask Orange (red) and nanocarriers are shown in green; scale bars, 10 μm). (**d**) Classification of protein corona components according to their function. (**e**) Heat map of the most abundant proteins in the protein corona of different nanocarriers. (**f**) Flow cytometry analysis of cellular uptake of PS-PEEP92 by RAW264.7 cells after incubation in water, human plasma, clusterin or human serum albumin (Reprinted by permission from Macmillan Publishers Ltd.: [Nature Nanotechnology] (Protein adsorption is required for stealth effect of poly(ethylene glycol)- and poly(phosphoester)-coated nanocarriers), Copyright (2016) [59])

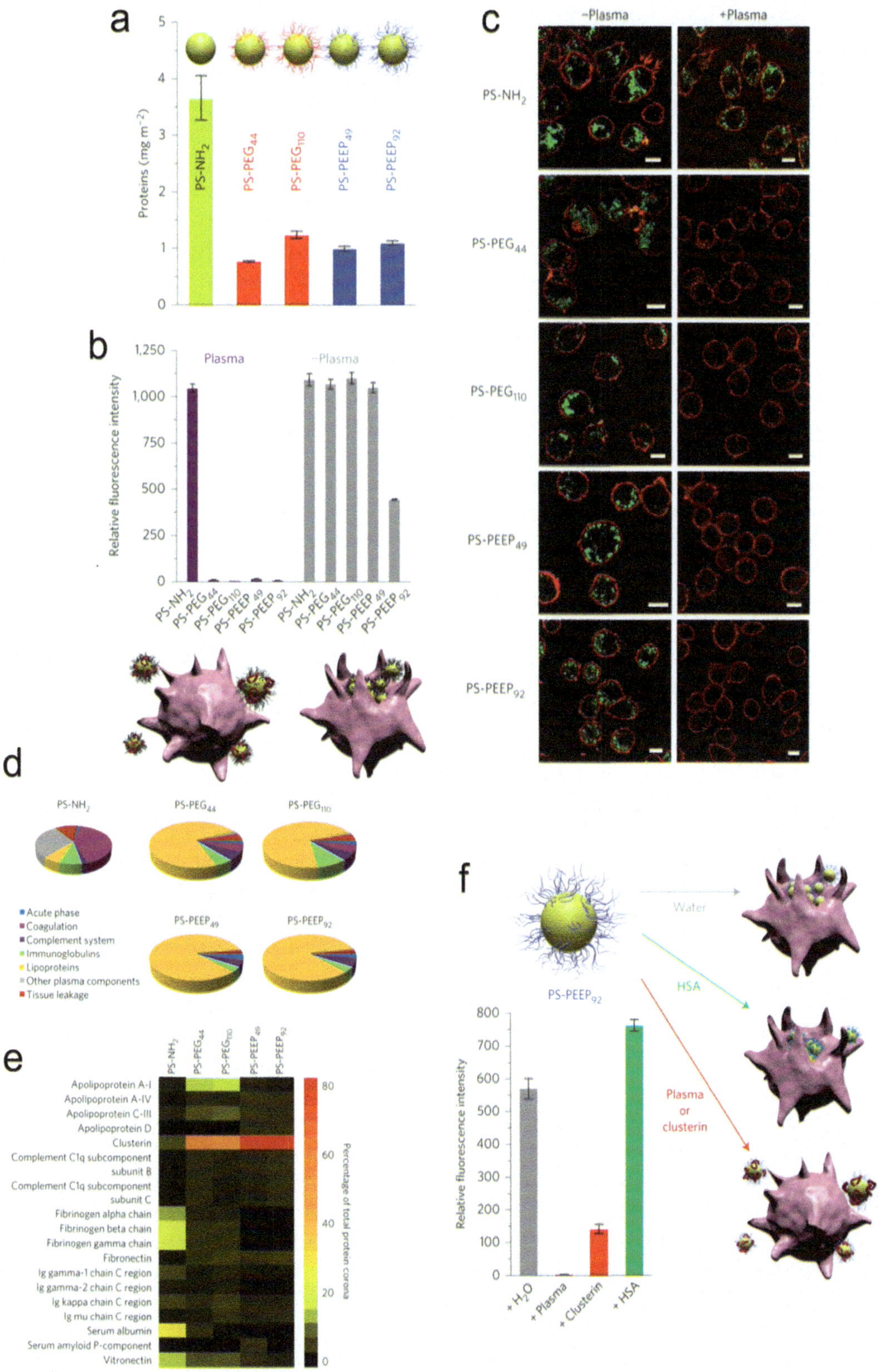
a
Proteins (mg m⁻²)
PS-NH2
PS-PEG44
PS-PEG110
PS-PEEP49
PS-PEEP92
b
Plasma
−Plasma
Relative fluorescence intensity
c
−Plasma
+Plasma
PS-NH2
PS-PEG44
PS-PEG110
PS-PEEP49
PS-PEEP92
d
Acute phase
Coagulation
Complement system
Immunoglobulins
Lipoproteins
Other plasma components
Tissue leakage
e
Apolipoprotein A-I
Apolipoprotein A-IV
Apolipoprotein C-III
Apolipoprotein D
Clusterin
Complement C1q subcomponent subunit B
Complement C1q subcomponent subunit C
Fibrinogen alpha chain
Fibrinogen beta chain
Fibrinogen gamma chain
Fibronectin
Ig gamma-1 chain C region
Ig gamma-2 chain C region
Ig kappa chain C region
Ig mu chain C region
Serum albumin
Serum amyloid P-component
Vitronectin
Percentage of total protein corona
f
Water
HSA
Plasma or clusterin
PS-PEEP92
Relative fluorescence intensity
+ H2O
+ Plasma
+ Clusterin
+ HSA

although present in the corona of nanoparticles incubated in serum and plasma, was found to be more in the nanoparticles incubated in plasma. Nanoparticles incubated in plasma therefore showed reduction in cellular uptake. Nanoparticle uptake based on PEG density has also been studied [60]. It was found that increasing the PEG density on gold nanoparticles led to decrease in the adsorption of a few proteins. These proteins were thus speculated to play an important role in inhibition of macrophage uptake. They included complement C3, transferrin, clusterin, alpha-2-macroglobulin, etc. [60].

The conformation of proteins bound to the nanoparticle surface is also important to mediate their interactions with cells. In a study, a prevention of macrophage uptake of protein-coated NPs was demonstrated for lipid and silica nanoparticles. While the corona formed on lipid nanoparticles was enriched in complement proteins, lipoproteins and immunoglobulins, the protein corona of silica NPs was found to be enriched in coagulation proteins and acute phase proteins. Although the proteins identified in this study are known to promote macrophage uptake through scavenger receptors, it was speculated that the functional motifs of these proteins may not be presented appropriately for recognition through these receptors. There is also a possibility that the opsonins present on these particles are shielded by the binding of the more abundant proteins such as albumin and lipoproteins, which are known to inhibit macrophage uptake [61].

11.5.5 Effect on Drug Release

As the formation of protein corona adds a layer on the nanoparticle surface it is expected to retard the drug release rates from nanoparticles. In case of nanocapsules, the thickness and the rigidity of the capsule shell is considered to be an important factor in determining the drug release rates rather than the protein corona, which is comparatively less thick than the capsule shell. However, in case of polymer coated SPIONs, the burst release of the drug from the polymer layer has been shown to be impeded by the corona layer. An increase of temperature was however found to increase the burst release suggesting that the protein layer is responsible for the impediment of the burst release. The release profile of paclitaxel from Abraxane also revealed a similar trend with both hard corona and soft corona playing an important role in decreasing the burst release from these nanoparticles [62].

11.5.6 Nanoparticle Biocompatibility and Toxicity

The interaction of plasma proteins with nanomaterials in general increases their biocompatibility [63]. The amount of protein corona deposited on nanomaterials directly governs their toxicity, with a thicker layer making the nanomaterial more biocompatible [36]. For instance the interaction of plasma proteins with graphene oxide (GO) nanosheets has been shown to mitigate its cytotoxic effects. The cytotoxicity of GO nanosheets arises due to its direct interaction and subsequent rupture of cellular membranes. However, the presence of protein corona reduced this cytotoxic effect drastically in a concentration dependent manner with a 10% serum having a more pronounced effect than 1% serum. Although GO nanosheets display a differential affinity towards serum proteins (in the order fibrinogen > Igs > Tf > BSA), the major role of the protein corona is to prevent a direct interaction between the cell membranes and GO nanosheets and the composition of the corona had no major role in imparting biocompatibility [64, 65]. Interestingly, the cytotoxicity of silica nanoparticles has been shown to depend on the time of exposure of nanoparticles to plasma, with nanoparticles subjected to a prolonged plasma exposure being relatively less toxic as compared to nanoparticles subjected to a brief exposure. The rapid corona formation on these nanoparticles also reduced thrombocyte activation, erythrocyte aggregation and prevented haemolysis [9].

Formation of a corona layer may also have an effect on the inflammatory response. As an example, the fibrinogen binding affinity of zeolite nanoparticles has been implicated in a potential pro-inflammatory response by these particles. Although, zeolite nanoparticles were shown to have no negative effect on major cellular processes *viz.* cell cycle, oxidative stress and cellular toxicity their high affinity towards fibrinogen (an acute phase protein) was speculated to induce a pro-inflammatory effect [66]. The immune activation and cytokine expression of nanoparticles has been reported to be dependent on the hydrophobicity of the nanoparticles with a more hydrophobic surface eliciting a higher immune response [67].

Apart from eliciting a toxic response, nanoparticles may have certain indirect toxic effects by inducing fibrillation, denaturation and conformational changes in the adsorbed proteins, leading to an exposure of new antigenic epitopes or a loss of function, etc. on the adsorbed proteins. It has been reported that various nanoparticles such as cerium oxide NPs, CNTs and copolymer NPs induce fibrillation in human beta-2 microglobulin. Although the mechanism of nucleation and fibril formation is not clear, it is speculated that the interaction with nanoparticle surface leads to appropriate conformational changes that may promote fibrillation and may also increase the local concentration of human beta-2 microglobulin monomers which may promote oligomer formation (Fig. 11.8a, b) [68]. Although the exact mechanism of protein unfolding at nanoparticle surface is not known, it is expected to involve contact forces such as the release of surface free energy through structural reorganization [69].

Conformational changes and unfolding of proteins upon interaction with nanoparticle surface has been a well reported phenomenon [16, 26, 70, 71]. It has been shown that nanomaterials with a high surface charge density or hydrophobicity alter the conformation of the adsorbed proteins to a larger extent than the nanomaterials with hydrophilic and neutral surfaces [72–74]. The effect of nanoparticles surface charge and hydrophobicity on protein structure has been studied for albumin, fibrinogen and cytochrome c with nanoparticles having a net positive or negative surface charge or high hydrophobicity showing an enhanced denaturation of proteins as compared to nanoparticles with a neutral surface charge or high hydrophilicity [75, 76]. The conformation changes in protein structures are usually irreversible even after the desorption of proteins from the nanoparticle surface [77]. As an example, the interaction of transferrin with SPIONs leads to an irreversible change in transferrin structure with a release of Fe. The transferrin protein does not regain its original structure upon desorption [78]. The degree of denaturation in proteins is dependent on the nanomaterial properties and the protein itself, with some proteins being more susceptible to conformation change than others [79]. Also, the surface functionalization on nanoparticles may have an effect on proteins, for example CdSe nanoparticles functionalized with oligoethylene glycol (OEG) terminated with hydroxyl groups do not interact with chymotrypsin, however carboxylate- terminated thioalkyl ligands and carboxylate-terminated OEG leads to denaturation and inhibition of enzyme activity [80]. A similar PEGylated gold nanoparticle mediated unfolding and aggregation has also been reported for lysozyme (Fig. 11.8c) [81]. In this case, the PEGylation density was shown to play an important role, with a low PEGylation density resulting in a protein aggregation and protein-Au nanoparticle assemblies. A high PEGylation density was shown to inhibit protein adsorption and aggregation even after prolonged exposures.

11.6 Attempts on the Fabrication of Corona Free Nanoparticles

The fabrication of corona free nanoparticles has been a long standing challenge for researchers. A classical approach has been the use of polyethylene glycol (PEG), that is known to prevent the adsorption of proteins on to surfaces [82]. However, as mentioned earlier, recent studies have revealed the adsorption of plasma proteins on PEG functionalized nanoparticles [59]. An alternative to PEG has been the utilization of zwitterionic molecules such as amino acids and poly(carboxybetadiene) [83–87]. Surface functionalization of silica nanoparticles with cysteine has been reported to increase their stability in protein (lysozyme and albumin) solutions [85]. Also, the functionalization of gold nanospheres with cysteine and lysine has been reported to reduce the adsorption of proteins upon incubation with

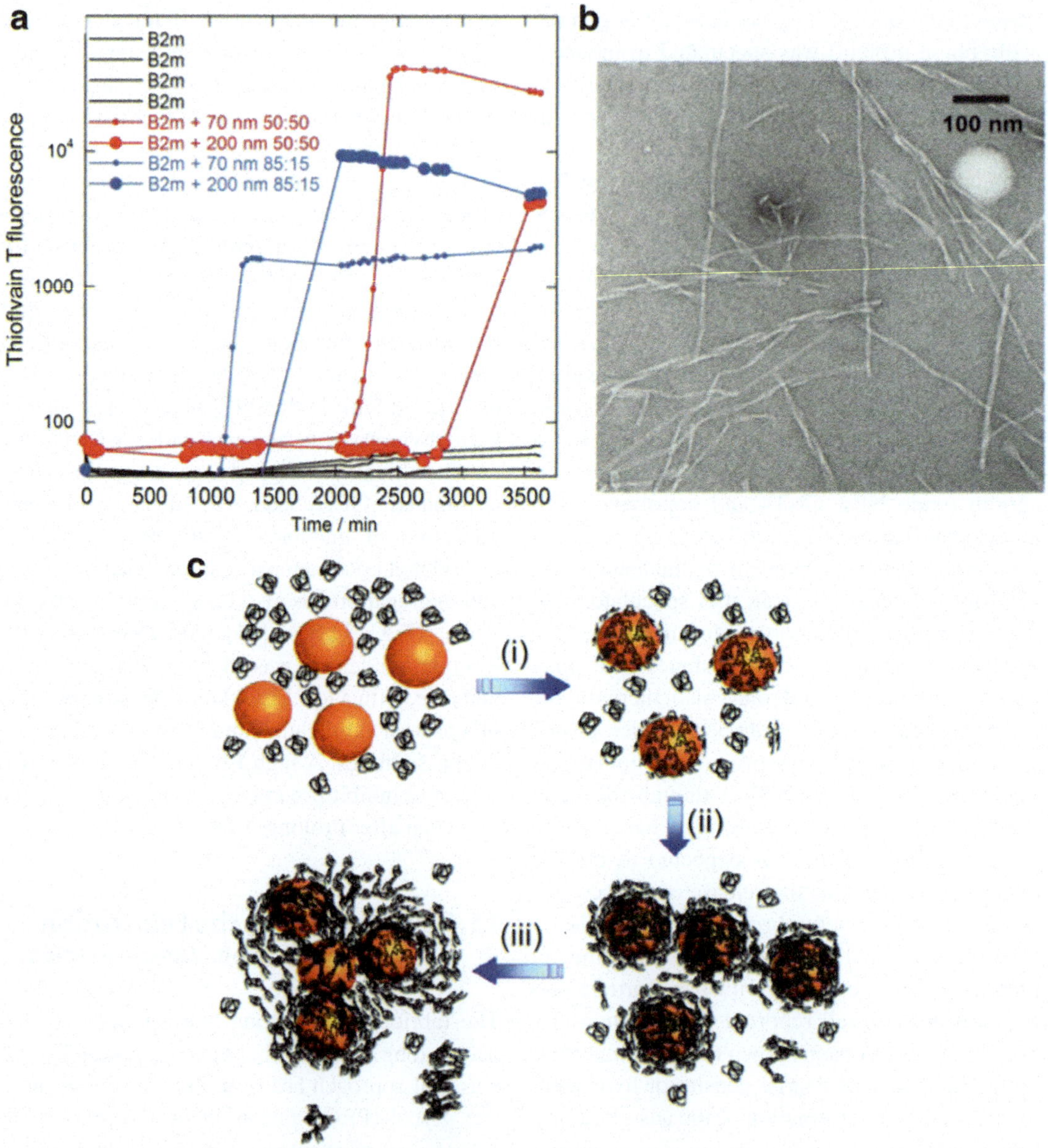

Fig. 11.8 (**a**) Thioflavin T fluorescence to study β2m fibrillation without (black) or with 85:15 NIPAM/BAM copolymer nanoparticles (blue) or 50:50 copolymer NIPAM/BAM nanoparticles (red). (Smaller symbols are used for 70 nm particles and larger symbols are used for 200 nm particles). (**b**) Transmission electron microscopy (TEM) image of β2m fibres grown in the presence of 70 nm 85:15 NIPAM/BAM copolymer nanoparticles (Reprinted with permission from "Nucleation of protein fibrillation by nanoparticles". Copyright (2007) National Academy of Sciences [68]). **c** Schematic representation describing the various steps of protein unfolding and aggregation in the presence of gold nanoparticle (AuNPs). The first step (*i*) is characterized by a conformational change in the protein molecules upon adsorption onto AuNPs. In the intermediate step (*ii*) partially unfolded proteins on the AuNP surfaces act as a seed for further protein aggregation. In the final step (*iii*), coalescence of protein-coated AuNPs takes place (Reprinted with permission from "Gold Nanoparticles Can Induce the Formation of Protein based Aggregates at Physiological pH". Copyright (2009) American Chemical Society [81])

fetal bovine serum. The low interaction observed for zwitterion-functionalized nanoparticles with proteins is due to the lack of a net charge and strong hydration of these molecules [86]. Functionalization with amino acids and other zwitterions provide an efficient mode of nanoparticle fabrication, as most of these zwitterions possess functional groups for surface modifications and prevent non specific adsorption and shielding effect from surrounding proteins.

11.7 Conclusion

It has become clear that the mechanism of nanomaterial interaction with biological fluids is significantly different from the way in which bulk matter interacts. Such interactions greatly influence nanoparticle properties and their physiological functions. With the rapid advancement in nanoparticle mediated drug delivery in pharmaceutical sciences, it becomes essential to study the interaction of such nanomaterials with their surrounding environment. A variety of nanomaterials have been studied in this context, and it could be stated that the interaction and influence of biomolecules in general and proteins in particular is unique to the nanoparticle type and cannot be generalized. However, with the large volume of data available, the biological identity and hence physiological response of nanomaterials can be predicted based on the synthetic compositions of the nanoparticles. Also, the design of novel nanomaterials may be done accordingly depending on the desired physiological response.

Acknowledgements SMA acknowledges Council for Scientific and Industrial Research (CSIR) for the research associate fellowship. CMR acknowledges the Department of Science and Technology, India for Sir J.C. Bose Fellowship. MFA acknowledges Dr. Chris Dealwis and NIH for supporting Postdoctoral/Senior Research associate fellowship. Also, we would like to thank Uzma Nisar for proofreading.

References

1. Walczyk D, Bombelli FB, Monopoli MP et al (2010) What the cell "sees" in bionanoscience. JACS 132:5761–5768
2. Monopoli MP, Åberg C, Salvati A et al (2012) Biomolecular coronas provide the biological identity of nanosized materials. Nat Nanotechnol 7:779–786
3. Raesch SS, Tenzer S, Storck W et al (2015) Proteomic and lipidomic analysis of nanoparticle corona upon contact with lung surfactant reveals differences in protein, but not lipid composition. ACS Nano 9:11872–11885
4. Kapralov AA, Feng WH, Amoscato AA et al (2012) Adsorption of surfactant lipids by single-walled carbon nanotubes in mouse lung upon pharyngeal aspiration. ACS Nano 6:4147–4156
5. Walkey CD, Chan WC (2012) Understanding and controlling the interaction of nanomaterials with proteins in a physiological environment. Chem Soc Rev 41:2780–2799
6. Vroman L, Adams A, Fischer G et al (1980) Interaction of high molecular weight kininogen, factor XII, and fibrinogen in plasma at interfaces. Blood 55:156–159
7. Aggarwal P, Hall JB, McLeland CB et al (2009) Nanoparticle interaction with plasma proteins as it relates to particle biodistribution, biocompatibility and therapeutic efficacy. Adv Drug Deliv Rev 61:428–437
8. Vroman L (1962) Effect of adsorbed proteins on the wettability of hydrophilic and hydrophobic solids. Nature 196:476–477
9. Tenzer S, Docter D, Kuharev J et al (2013) Rapid formation of plasma protein corona critically affects nanoparticle pathophysiology. Nat Nanotechnol 8:772–781
10. Gunawan C, Lim M, Marquis CP et al (2014) Nanoparticle–protein corona complexes govern the biological fates and functions of nanoparticles. J Mater Chem B 2:2060–2083
11. Maiolo D, Bergese P, Mahon E et al (2014) Surfactant titration of nanoparticle–protein corona. Anal Chem 86:12055–12063
12. Norde W (1994) Protein adsorption at solid surfaces: a thermodynamic approach. Pure Appl Chem 66:491–496
13. Saptarshi SR, Duschl A, Lopata AL (2013) Interaction of nanoparticles with proteins: relation to bio-reactivity of the nanoparticle. J Nanobiotechnol 11:26
14. Cedervall T, Lynch I, Lindman S et al (2007) Understanding the nanoparticle–protein corona using methods to quantify exchange rates and affinities of proteins for nanoparticles. Proc Natl Acad Sci U S A 104:2050–2055

15. Lynch I, Dawson KA (2008) Protein-nanoparticle interactions. Nano Today 3:40–47
16. Chakraborty S, Joshi P, Shanker V et al (2011) Contrasting effect of gold nanoparticles and nanorods with different surface modifications on the structure and activity of bovine serum albumin. Langmuir 27:7722–7731
17. Simberg D, Park J-H, Karmali PP et al (2009) Differential proteomics analysis of the surface heterogeneity of dextran iron oxide nanoparticles and the implications for their in vivo clearance. Biomaterials 30:3926–3933
18. Andersson J, Ekdahl KN, Larsson R et al (2002) C3 adsorbed to a polymer surface can form an initiating alternative pathway convertase. J Immunol 168:5786–5791
19. Gagner JE, Lopez MD, Dordick JS et al (2011) Effect of gold nanoparticle morphology on adsorbed protein structure and function. Biomaterials 32:7241–7252
20. Lundqvist M, Stigler J, Elia G et al (2008) Nanoparticle size and surface properties determine the protein corona with possible implications for biological impacts. Proc Natl Acad Sci U S A 105:14265–14270
21. Tenzer S, Docter D, Rosfa S et al (2011) Nanoparticle size is a critical physicochemical determinant of the human blood plasma corona: a comprehensive quantitative proteomic analysis. ACS Nano 5:7155–7167
22. Cedervall T, Lynch I, Foy M et al (2007) Detailed identification of plasma proteins adsorbed on copolymer nanoparticles. Angew Chem Int Ed 46:5754–5756
23. Owens DE, Peppas NA (2006) Opsonization, biodistribution, and pharmacokinetics of polymeric nanoparticles. Int J Pharm 307:93–102
24. Dobrovolskaia MA, Patri AK, Zheng J et al (2009) Interaction of colloidal gold nanoparticles with human blood: effects on particle size and analysis of plasma protein binding profiles. Nanomed Nanotechnol Biol Med 5:106–117
25. Deng ZJ, Mortimer G, Schiller T et al (2009) Differential plasma protein binding to metal oxide nanoparticles. Nanotechnology 20:455101
26. Lacerda SHDP, Park JJ, Meuse C et al (2009) Interaction of gold nanoparticles with common human blood proteins. ACS Nano 4:365–379
27. Gessner A, Lieske A, Paulke BR et al (2003) Functional groups on polystyrene model nanoparticles: influence on protein adsorption. J Biomed Mater Res A 65:319–326
28. Treuel L, Brandholt S, Maffre P et al (2014) Impact of protein modification on the protein corona on nanoparticles and nanoparticle–cell interactions. ACS Nano 8:503–513
29. Maffre P, Nienhaus K, Amin F et al (2011) Characterization of protein adsorption onto FePt nanoparticles using dual-focus fluorescence correlation spectroscopy. Beilstein J Nanotechnol 2:374–383
30. Gessner A, Lieske A, Paulke BR et al (2002) Influence of surface charge density on protein adsorption on polymeric nanoparticles: analysis by two-dimensional electrophoresis. Eur J Pharm Biopharm 54:165–170
31. Roser M, Fischer D, Kissel T (1998) Surface-modified biodegradable albumin nano-and microspheres. II: effect of surface charges on in vitro phagocytosis and biodistribution in rats. Eur J Pharm Biopharm 46:255–263
32. Lück M, Paulke BR, Schröder W et al (1998) Analysis of plasma protein adsorption on polymeric nanoparticles with different surface characteristics. J Biomed Mater Res A 39:478–485
33. Gessner A, Waicz R, Lieske A et al (2000) Nanoparticles with decreasing surface hydrophobicities: influence on plasma protein adsorption. Int J Pharm 196:245–249
34. Müller RH, Wallis KH, Troester SD et al (1992) In vitro characterization of poly (methyl-methaerylate) nanoparticles and correlation to their in vivo fate. J Control Release 20:237–246
35. Monopoli MP, Walczyk D, Campbell A et al (2011) Physical– chemical aspects of protein corona: relevance to in vitro and in vivo biological impacts of nanoparticles. JACS 133:2525–2534
36. Maiorano G, Sabella S, Sorce B et al (2010) Effects of cell culture media on the dynamic formation of protein– nanoparticle complexes and influence on the cellular response. ACS Nano 4:7481–7491
37. Dell'Orco D, Lundqvist M, Oslakovic C et al (2010) Modeling the time evolution of the nanoparticle-protein corona in a body fluid. PLoS One 5:e10949
38. Green R, Davies M, Roberts C et al (1999) Competitive protein adsorption as observed by surface plasmon resonance. Biomaterials 20:385–391
39. Nagayama S, Ogawara K, Fukuoka Y et al (2007) Time-dependent changes in opsonin amount associated on nanoparticles alter their hepatic uptake characteristics. Int J Pharm 342:215–221
40. Casals E, Pfaller T, Duschl A et al (2010) Time evolution of the nanoparticle protein corona. ACS Nano 4:3623–3632
41. Mahmoudi M, Abdelmonem AM, Behzadi S et al (2013) Temperature: the "ignored" factor at the nano-bio interface. ACS Nano 7:6555–6562
42. Mahmoudi M, Lohse SE, Murphy CJ et al (2013) Variation of protein corona composition of gold nanoparticles following plasmonic heating. Nano Lett 14:6–12
43. Lesniak A, Campbell A, Monopoli MP et al (2010) Serum heat inactivation affects protein corona composition and nanoparticle uptake. Biomaterials 31:9511–9518
44. Shannahan JH, Fritz KS, Raghavendra AJ et al (2016) From the cover: disease-induced disparities in formation of the nanoparticle-biocorona and the toxicological consequences. Toxicol Sci 152:406–416
45. Aoyama M, Hata K, Higashisaka K et al (2016) Clusterin in the protein corona plays a key role in the stealth effect of nanoparticles against phagocytes. Biochem Biophys Res Commun 480:690–695

46. Albanese A, Walkey CD, Olsen JB et al (2014) Secreted biomolecules alter the biological identity and cellular interactions of nanoparticles. ACS Nano 8:5515–5526
47. Dai Q, Guo J, Yan Y et al (2017) Cell-conditioned protein coronas on engineered particles influence immune responses. Biomacromolecules 18:431–439
48. Ghavami M, Saffar S, Emamy BA et al (2013) Plasma concentration gradient influences the protein corona decoration on nanoparticles. RSC Adv 3:1119–1126
49. Röcker C, Pötzl M, Zhang F et al (2009) A quantitative fluorescence study of protein monolayer formation on colloidal nanoparticles. Nat Nanotechnol 4:577–580
50. Rahimi M, Ng E, Bakhtiari K et al (2014) Zeolite nanoparticles for selective sorption of plasma proteins. Sci Rep 5:17259–17259
51. Corbo C, Molinaro R, Parodi A et al (2016) The impact of nanoparticle protein corona on cytotoxicity, immunotoxicity and target drug delivery. Nanomedicine 11:81–100
52. Miclăuş T, Beer C, Chevallier J et al (2016) Dynamic protein coronas revealed as a modulator of silver nanoparticle sulphidation in vitro. Nat Commun 7:11770
53. Lee YK, Choi E-J, Webster TJ et al (2015) Effect of the protein corona on nanoparticles for modulating cytotoxicity and immunotoxicity. Int J Nanomedicine 10:97–113
54. Salvati A, Pitek AS, Monopoli MP et al (2013) Transferrin-functionalized nanoparticles lose their targeting capabilities when a biomolecule corona adsorbs on the surface. Nat Nanotechnol 8:137–143
55. Mahmoudi M, Sheibani S, Milani AS et al (2015) Crucial role of the protein corona for the specific targeting of nanoparticles. Nanomedicine 10:215–226
56. Tabata Y, Ikada Y (1990) Phagocytosis of polymer microspheres by macrophages. In: New polymer materials, advances in polymer science, vol 94. Springer, Berlin, pp 107–141
57. Saha K, Rahimi M, Yazdani M et al (2016) Regulation of macrophage recognition through the interplay of nanoparticle surface functionality and protein corona. ACS Nano 10:4421–4430
58. Lesniak A, Fenaroli F, Monopoli MP et al (2012) Effects of the presence or absence of a protein corona on silica nanoparticle uptake and impact on cells. ACS Nano 6:5845–5857
59. Schöttler S, Becker G, Winzen S et al (2016) Protein adsorption is required for stealth effect of poly (ethylene glycol)-and poly (phosphoester)-coated nanocarriers. Nat Nanotechnol 11:372–377
60. Walkey CD, Olsen JB, Guo H et al (2012) Nanoparticle size and surface chemistry determine serum protein adsorption and macrophage uptake. JACS 134:2139–2147
61. Caracciolo G, Palchetti S, Colapicchioni V et al (2015) Stealth effect of biomolecular corona on nanoparticle uptake by immune cells. Langmuir 31:10764–10773
62. Behzadi S, Serpooshan V, Sakhtianchi R et al (2014) Protein corona change the drug release profile of nanocarriers: the "overlooked" factor at the nanobio interface. Colloids Surf B Biointerfaces 123:143–149
63. Landgraf L, Christner C, Storck W et al (2015) A plasma protein corona enhances the biocompatibility of Au@ Fe_3O_4 Janus particles. Biomaterials 68:77–88
64. Hu W, Peng C, Lv M et al (2011) Protein corona-mediated mitigation of cytotoxicity of graphene oxide. ACS Nano 5:3693–3700
65. Chong Y, Ge C, Yang Z et al (2015) Reduced cytotoxicity of graphene nanosheets mediated by blood-protein coating. ACS Nano 9:5713–5724
66. Laurent S, Ng E-P, Thirifays C et al (2013) Corona protein composition and cytotoxicity evaluation of ultra-small zeolites synthesized from template free precursor suspensions. Toxicol Res 2:270–279
67. Moyano DF, Goldsmith M, Solfiell DJ et al (2012) Nanoparticle hydrophobicity dictates immune response. JACS 134:3965–3967
68. Linse S, Cabaleiro-Lago C, Xue W-F et al (2007) Nucleation of protein fibrillation by nanoparticles. Proc Natl Acad Sci U S A 104:8691–8696
69. Nel AE, Mädler L, Velegol D et al (2009) Understanding biophysicochemical interactions at the nano–bio interface. Nat Mater 8:543–557
70. Smith JR, Cicerone MT, Meuse CW (2009) Tertiary structure changes in albumin upon surface adsorption observed via fourier transform infrared spectroscopy. Langmuir 25:4571–4578
71. Tsai DH, DelRio FW, Keene AM et al (2011) Adsorption and conformation of serum albumin protein on gold nanoparticles investigated using dimensional measurements and in situ spectroscopic methods. Langmuir 27:2464–2477
72. Buijs J, Vera CC, Ayala E et al (1999) Conformational stability of adsorbed insulin studied with mass spectrometry and hydrogen exchange. Anal Chem 71:3219–3225
73. Buijs J, Norde W, Lichtenbelt JWT (1996) Changes in the secondary structure of adsorbed IgG and F (ab ') 2 studied by FTIR spectroscopy. Langmuir 12:1605–1613
74. Buijs J, James WT, Norde W et al (1995) Adsorption of monoclonal IgGs and their F (ab′) 2 fragments onto polymeric surfaces. Colloids Surf B Biointerfaces 5:11–23
75. Roach P, Farrar D, Perry CC (2005) Interpretation of protein adsorption: surface-induced conformational changes. JACS 127:8168–8173
76. Aubin-Tam ME, Hamad-Schifferli K (2005) Gold nanoparticle− cytochrome C complexes: the effect of nanoparticle ligand charge on protein structure. Langmuir 21:12080–12084
77. Norde W, Giacomelli CE (2000) BSA structural changes during homomolecular exchange between the adsorbed and the dissolved states. J Biotechnol 79:259–268

78. Mahmoudi M, Shokrgozar MA, Sardari S et al (2011) Irreversible changes in protein conformation due to interaction with superparamagnetic iron oxide nanoparticles. Nanoscale 3:1127–1138
79. Kondo A, Murakami F, Higashitani K (1992) Circular dichroism studies on conformational changes in protein molecules upon adsorption on ultrafine polystyrene particles. Biotechnol Bioeng 40:889–894
80. Hong R, Fischer NO, Verma A et al (2004) Control of protein structure and function through surface recognition by tailored nanoparticle scaffolds. JACS 126:739–743
81. Zhang D, Neumann O, Wang H et al (2009) Gold nanoparticles can induce the formation of protein-based aggregates at physiological pH. Nano Lett 9:666–671
82. Prime KL, Whitesides GM (1991) Self-assembled organic monolayers: model systems for studying adsorption of proteins at surfaces. Science 252:1164
83. Cao Z, Jiang S (2012) Super-hydrophilic zwitterionic poly (carboxybetaine) and amphiphilic non-ionic poly (ethylene glycol) for stealth nanoparticles. Nano Today 7:404–413
84. Yang W, Zhang L, Wang S et al (2009) Functionalizable and ultra stable nanoparticles coated with zwitterionic poly (carboxybetaine) in undiluted blood serum. Biomaterials 30:5617–5621
85. Rosen JE, Gu FX (2011) Surface functionalization of silica nanoparticles with cysteine: a low-fouling zwitterionic surface. Langmuir 27:10507–10513
86. Murthy AK, Stover RJ, Hardin WG et al (2013) Charged gold nanoparticles with essentially zero serum protein adsorption in undiluted fetal bovine serum. JACS 135:7799–7802
87. Moyano DF, Saha K, Prakash G et al (2014) Fabrication of corona-free nanoparticles with tunable hydrophobicity. ACS Nano 8:6748–6755

Cellular and Molecular Toxicity of Iron Oxide Nanoparticles

12

Blanca Laffon, Natalia Fernández-Bertólez, Carla Costa, Fátima Brandão, João Paulo Teixeira, Eduardo Pásaro, and Vanessa Valdiglesias

Abstract

Iron oxide nanoparticles (ION) have attracted much attention because of their particular physico-chemical properties, including superparamagnetism. These features make them suitable for many purposes and several interesting biomedical applications, such as to increase contrast in magnetic resonance imaging (MRI), as drug delivery systems and as hyperthermia agents. However, they have also shown to be easily accumulated in diverse tissues and induce toxicity at different levels. This chapter reviews the different cellular and molecular effects induced by ION reported from in vitro studies with human and non-human cell lines. Those effects are mainly dependent on ION type and concentration, time of exposure, presence and nature of coating, and cell type evaluated. They include decreases in viability, plasmatic membrane disruption, oxidative damage, mitochondrial alterations, cell cycle impairments, cytoskeleton disruption, cell death, and alterations in cell motility, and in cell integrity. Despite these negative effects, the numerous advantages of ION together with their promising applications in biomedicine, make it necessary to clearly define their toxicity in order to discard potential health risks and to reach optimal benefits of their use.

B. Laffon (✉) · E. Pásaro
Department of Psychology, Area of Psychobiology, Universidade da Coruña, DICOMOSA Group, A Coruña, Spain
e-mail: blaffon@udc.es

N. Fernández-Bertólez
Department of Psychology, Area of Psychobiology, Universidade da Coruña, DICOMOSA Group, A Coruña, Spain

Department of Cell and Molecular Biology, Universidade da Coruña, A Coruña, Spain

C. Costa · F. Brandão · J.P. Teixeira
Department of Environmental Health, Portuguese National Institute of Health, Porto, Portugal

ISPUP-EPIUnit, Universidade do Porto, Porto, Portugal

V. Valdiglesias
Department of Psychology, Area of Psychobiology, Universidade da Coruña, DICOMOSA Group, A Coruña, Spain

ISPUP-EPIUnit, Universidade do Porto, Porto, Portugal

Q. Saquib et al. (eds.), *Cellular and Molecular Toxicology of Nanoparticles*, Advances in Experimental Medicine and Biology 1048, https://doi.org/10.1007/978-3-319-72041-8_12

Keywords

Iron oxide nanoparticles · Cell viability · Mitochondrial alterations · Oxidative damage · Cell membrane impairment · Genomic alterations

12.1 Introduction

Among the different types of magnetic nanoparticles, iron oxide nanoparticles (ION) arouse a particular interest due to their unique intrinsic magnetic properties, so called superparamagnetism. This feature, together with their high colloidal stability, makes them very attractive for a broad range of uses. From an industrial perspective, ION are frequently used in building materials, as pigments – which are low cost, colorfast, nontoxic and capable of imparting different colors – and as a food additive, which fortifies foods without altering their color or taste [1]. However, the most promising uses of ION are in the biomedicine field. Among others, they have applications in MRI, targeted drug delivery, tumor location and treatment, gene therapy, and tissue repair (reviewed in [2]). These biomedical uses, which require that nanoparticles are directly introduced in the human body, give rise to concerns regarding the potential toxic effects that may be associated with ION exposure.

12.1.1 Biomedical Applications

As mentioned before, most interesting applications of ION are in biomedicine and, particularly, their use as MRI contrast mediators, drug delivery systems and hyperthermia agents. ION are proved to be good contrast agents due to their superparamagnetic properties and high relaxivity, which provide significant advantages over traditional contrast agents [3, 4]. Since US Food and Drug Administration (FDA) approved the clinical use of ION for imaging (reviewed in [5]), these nanoparticles have been used in a wide number of MRI-based clinical applications, mainly for tumor detection and imaging of the gastrointestinal tract, and in central nervous system for neurovascular, neurooncological or neuroinflammatory processes (reviewed by [6]).

The functional properties of ION as imaging agents and their ability to be manipulated under a magnetic field, make them attractive delivery carriers in vitro and in vivo [7]. Because of these properties, magnetically-driven ION-based delivery systems have been already used in clinical applications for extracellular, intracellular, and site-specific targeted delivery of biotherapeutics to tumorous, inflammatory and infectious sites under the influence of an external magnetic field). Several biotherapeutic agents, including chemotherapeutics, radiotherapeutics, anti-inflammatory agents, antibodies, peptides, oligonucleotides as well as genes, can be integrated on ION in order to deliver those agents to specific locations [7, 9].

Hyperthermia has been used as a medical treatment for cancer in both traditional and modern medicine due to the fact that cancerous cells are more sensitive to temperature than normal cells). Magnetic nanoparticle-mediated hyperthermia is a promising therapy for the selective apoptosis of tumor cells through controlled heating of the altered tissue, which involves targeting ION to tumor tissue and then applying an external alternating magnetic field to generate heat). Compared to previous hyperthermia approaches, ION-mediated hyperthermia is more attractive since it offers a way to ensure that only specific target tissue is heated and eventually destroyed. Besides, it has been revealed that magnetic nanoparticles, either directly injected into tumors or delivered intravenously, can circulate through the blood stream and reach target organs by active targeting through surface ligands [12]. In this context, magnetic hyperthermia has the advantage that the heat source is directly in contact with the target cells [13].

12.1.2 Structure and Composition of Iron Oxide Nanoparticles

ION are usually made of a crystalline core and a surface coating for stabilizing the core properties and enhancing biocompatibility and bioavailability by preventing aggregation. Even though, technically speaking particles larger than 50 nm (size of core/shell) ION are classified as superparamagnetic iron oxides (SPIO), and particles smaller than 50 nm are ultra-small superparamagnetic iron oxides (USPIO) [14]; the term ION is usually employed in the literature to designate both of them. Likewise, in this chapter ION will be used to refer to both types of magnetic nanoparticles. ION may present multiple crystallographic structures that include: magnetite (Fe_3O_4), maghemite (γ-Fe_2O_3), hematite (α-Fe_2O_3), wüstite (FeO), ε-Fe_2O_3, and β-Fe_2O_3, among which hematite, maghemite and magnetite are the most common [1]. And, specifically, ION manufactured for biomedical purposes, both diagnostics and therapeutics, are typically formed by a core of magnetite or maghemite. This crystalline core is usually surface modified mainly because naked ION have high chemical reactivity, and are easily oxidized in air (especially magnetite), resulting in loss of magnetism and dispersibility. Surface modification prevents particle agglomeration, provides biocompatibility, and modifies cellular uptake efficiency of ION ([15–18]. The overall size of the nanoparticles is significantly increased with coating, which may be used to modify their toxicokinetic behavior (penetration, distribution, excretion) [19, 20]. Besides, use or administration of nanomaterials initially reported as toxic is possible by using stable surface coatings. Therefore, once synthesized, the iron oxide core is usually covered with a biocompatible coating. While the iron oxide core is responsible for the magnetic properties of ION, the ligand coat is essential to stabilize ION in physiological media. Different materials can be used to modify ION surface, including both polymeric and non-polymeric coatings (reviewed in [21, 22]). The choice of coating is mainly determined by the desired application concerning functionalization, stability or size, since every material has advantages and drawbacks [23]. Some of the most commonly used coatings for ION are polyethylene glycol, with good compatibility, favorable chemical properties, and solubility [24]; silica, largely used for bioimaging and biosensing purposes [25]; carboxydextran, used for cell labelling since it provides stability and increases intravascular retention time of nanoparticles [26]; and polyethylene imine, used as gene/drug delivery vehicle due to its high cellular uptake [27].

Together with this primary coating, targeting efficiency of ION can be further improved by employing conjugation biomarkers on their surface such as peptides, antibodies or small molecules [28, 29]. Thus, ION coating has frequently been modified with fluorescent dyes for imaging, targeting molecules [34, 35], drugs or nucleic acids [38–40]. Nevertheless, besides providing generally increased biocompatibility and enhancing ION properties to be used in biomedical applications, surface coating may also alter ION toxicity (reviewed in [41]). Therefore, it is important to carefully monitor the influence of surface modifications (chemical nature of coating, presence of functional groups, and net size) on ION toxicity. The purpose of this chapter was to review the ION-induced effects at cellular and molecular levels. In particular, studies addressing viability decrease, oxidative stress, mitochondrial alterations, cell membrane impairment and genetic damage on different cell systems by exposure to a wide variety of naked or coated ION were considered and described. Also, the most probable mechanism of toxicity of these nanoparticles, according the results obtained in all these studies, was discussed.

12.2 Cellular and Molecular Effects

As previously mentioned, ION have attracted much attention because of their particular physico-chemical properties, including superparamagnetism, which make them suitable for many purposes and several interesting biomedical applications. However, many studies indicate that ION can easily accumulate in different tissues and induce toxicity to different organs and systems. Thus, several toxic effects, including dullness, irritation, moribund condition, immunological alterations, biochemical and histopathological changes, inflammation in lung, or hepatic injuries, among others, have been reported in animals after ION exposure by various exposure routes [42–44]. Because of this, understanding the potential risks related to their exposure, not only on the whole organism but also at cellular and molecular levels, is imperative. A number of studies have been carried out in last decades to elucidate the toxicological profile of these nanoparticles as well as to define the specific conditions required to show this toxicity.

Generally speaking, most in vitro studies reported none or low toxicity of ION [45–47]. Consequently, these nanoparticles were initially classified as biocompatible and non-cytotoxic at low doses (<100 μg/ml) [48, 49]. However, there are also some studies reporting toxic effects of ION, even at such concentrations. Among them, several cellular and molecular effects were described after exposure to ION, including DNA damage, oxidative stress, mitochondrial membrane dysfunction and cell death [41]. Furthermore, exposure to ION might cause cellular stress, alterations in cell morphology and cell motility [50, 51]. Direct and indirect contact with ION can also affect the structure of DNA as well as its replication process [52]. These effects seem to be dependent on cell type, ION size and dose, exposure time and the presence and type of coating [53]. On this regard, several studies have compared the toxic effects induced by ION with different surface modifications. Thus, Soenen et al. [54] tested four different ION types (dextran-coated Endorem®, carboxydextran-coated Resovist®, lipid-coated magnetoliposomes, and citrate-coated very small iron oxide particles) on c17.2 neural progenitor cells. Different cytotoxic potential relating to the type of coating was observed in this case, resulting citrate-coated ION and lipid-coated ION the most and less toxic nanoparticles, respectively. Similarly, Rivet et al. [55] investigated the response of primary cortical neurons to aminosilane-, dextran- and polydimethylamine-coated ION (magnetite), and observed different effects depending on nanoparticle dose and coating. Other several studies have reported similar differences in toxic ION effects according to the presence and type of coating [56–58]. Schütz et al. [59] suggested that these effects were not only nanoparticle-specific but also cell-specific. And other authors pointed out that ION concentration may be an even more critical factor for cytotoxicity than surface modification or size [60, 61].

Although a variety of alterations has been reported after ION exposure, the ION-induced cellular and molecular effects most commonly reported in the literature include decreases in viability, oxidative damage, mitochondrial alterations, cell membrane disruptions, and DNA damage. Thus, these specific outcomes are addressed separately in the following subsections.

12.2.1 Viability

Most studies analyzing ION cytotoxicity are focused on evaluating decreases in viability of these nanoparticles on cell cultures with no conclusive results. Several studies reported no adverse cellular effects in different cell types treated with a variety of naked or differently coated ION. Among others, primary rat astrocytes [62, 63] and neurons [64], human T lymphocytes [65], monocytes [66], and amniotic fluid cells [67], murine microglial cells [68, 69], mouse fibroblasts [70], and both murine and human macrophages [71]. Nevertheless, other many in vitro studies have reported decreases in viability after ION exposure. Thus, ION (magnetite) was found to induce moderate time- and concentration-dependent decrease in viability of Vero cells [72]. Similarly, the same effect was described in other several cell systems treated with different ION. These cells include human alveolar epithelial A549 cells treated with Fe_2O_3 [47], human glioblastoma and urinary bladder carcinoma cells after treated with rhamnose-coated magnetite [73], cultured rat astrocytes exposed to aminosilane- or starch-coated magnetite [74], human hepatocarcinoma cells treated with cobalt ION [75], and human T lymphocytes treated with ION coated with carboxyl or amine groups [76].

As many of these studies pointed out, effects of ION on cellular viability depend on different factors including type of nanoparticle, dose, exposure time, cell system and culture medium. Together with these sources of variability, a recent study demonstrated that ION may indeed interfere with the main cytotoxicity assays employed to evaluate viability by significantly altering absorbance readings [77]. Standard protocols should be accordingly modified, in order to reduce these interferences and avoid false positive/negative results. The lack of interference evaluation in most of these previously mentioned studies might help to explain the high controversy found in the literature.

12.2.2 Oxidative Damage

Many studies have demonstrated that most of the toxicity of ION arises from the production of an excessive amount of reactive oxygen species (ROS), including free radicals such as the superoxide anion, hydroxyl radicals and hydrogen peroxide [78, 79] (Fig. 12.1). There are several known primary sources of oxidative stress in response to ION: direct generation of ROS from the surface of ION, production of ROS via leaching of iron ions, alteration of mitochondrial and other organelle functions, and induction of cell signaling pathways [80, 81].

Thus, an increased generation of ROS by ION exposure was previously observed in Chinese hamster ovary CHO-K1 cells [82], murine macrophage J774 cells [83], different vascular endothelial cells [84, 85], Chinese hamster lung cells [86], human lung A549 cells [87, 88], osteosarcoma cells [89], brain microglia cells [90], glial T98G and U251MG and bladder ECV304 cells [73], PC12 cells [91], mouse microglial Bv2 cells [92], and human brain-derived endothelial cells [93]. Not only increases in ROS levels but other oxidative damage was described after treatment with ION. Induction of ROS, depletion of glutathione and lower activity of superoxide dismutase enzyme were reported in HepG2 cells treated with cobalt ION [75]. Besides, intracellular ROS formation together with dose-dependent decline in GSH levels, and alterations in activity of antioxidant enzymes (lower superoxide dismutase and higher glutathione peroxidase) were reported in murine neural stem cells exposed to three different ION (γ-Fe_2O_3): uncoated, coated with d-mannose, and coated with poly-l-lysine [94].

However, other studies reported negative results on ROS production after ION treatment [95–99]. Several authors have suggested that oxidative stress and ROS generation induced by ION can be associated with the presence and type of surface coating [95, 96] but also with the cell type. Consequently, Petters et al. [90] found ROS generation in microglial cells, but not in astrocytes and neurons, treated with DMSA-coated ION (γ-Fe_2O_3).

12.2.3 Mitochondrial Alterations

ION can cause structural damage to mitochondria and alter this organelle functionality, significantly affecting electron flow and producing altered membrane potential (MMP) [100], cytochrome c release [101], and uncoupling of oxidative phosphorylation [102]. Mitochondrial dysfunction exacerbates nanoparticle-mediated toxicity, inducing multiple cell responses [103].

Baratli et al. [104] examined the effects of ION (Fe_3O_4) on mitochondrial respiratory chain complex activities and mitochondrial coupling in young (3 months) and middle-aged (18 months) rat livers, finding interesting differences depending on animal age. In young individuals, ION exposure did not alter mitochondrial function; however, nanoparticles dose-dependently impaired all complexes of the mitochondrial respiratory chain in middle-aged rat liver. A significant reduction of MMP was also found in ION-treated osteosarcoma [89] and breast cancer [105] cells, which indicated that cell death observed in these studies may originate from mitochondrial dysfunction. On the contrary, ION, irrespective of the type of surface coating, were

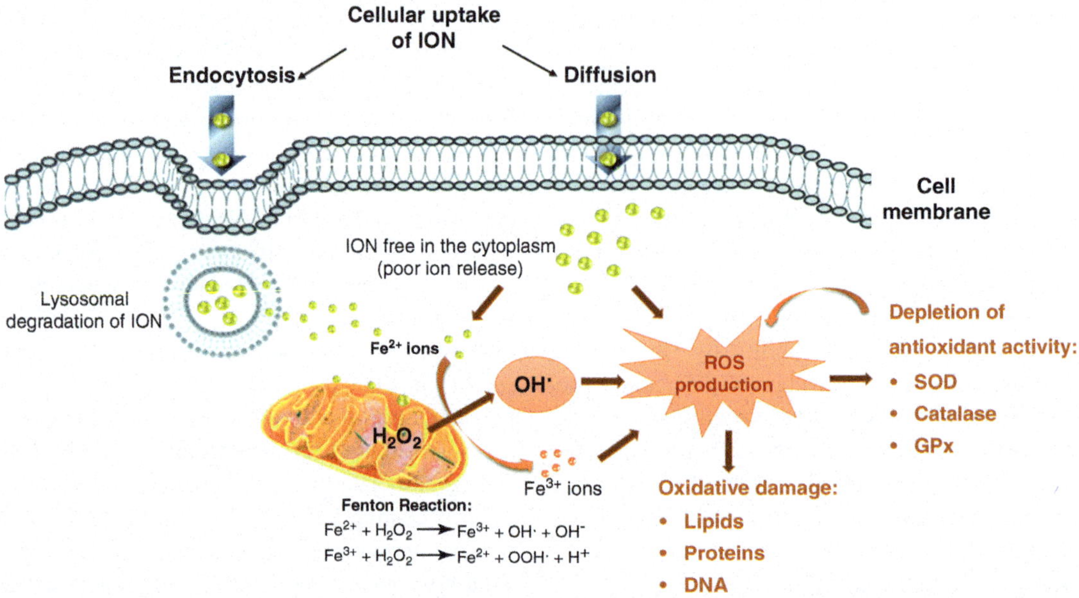

Fig. 12.1 Mechanisms of ION-mediated reactive oxygen species (*ROS*) generation. ION can be internalized into the cell mainly by endocytosis or passive diffusion. Upon internalization, iron ions may be presumably released by lysosome degradation or directly from the ION surface. This 'free iron' can potentially cross the nuclear or mitochondrial membrane. In the latter case, the free iron in the form of ferrous ions (*Fe^{2+}*) can react with hydrogen peroxide and oxygen produced by the mitochondria to produce highly reactive hydroxyl radicals (*OH·*) and ferric ions (*Fe^{3+}*) via the Fenton reaction. Hydroxyl radicals generated, among other ROS, impair the general redox status of the cell. ION are able to target mitochondria directly, which can lead to mitochondrial disruption and, in turn, to ROS production. Oxidative stress due to excess ROS generation induces over-expression of antioxidant enzymes in an attempt to control ROS levels. At high levels of oxidative stress, antioxidant defenses are overwhelmed, which leads to inflammatory and cytotoxic responses. Oxidative stress might induce collateral oxidative damage, such as lipid peroxidation, protein denaturation, DNA damage and immune reactivity. *ION* iron oxide nanoparticles, *CAT* catalase, *Cyto-c* cytochrome c, *GPx* glutathione peroxidase, *SOD* superoxide dismutase

found to increase MMP by 30–50% in murine neural stem cells compared to untreated cells [94].

12.2.4 Cell Membrane Disruptions

Several studies on different cells systems have reported different plasmatic membrane impairment after ION exposure [87, 89, 106]. Furthermore, a recent study confirmed that these effects may appear quickly, since several types of ION were found to induce significant and dose-dependent depolarization of the cell membrane in murine stem cells just after 30 s of exposure [94].

Lactate dehydrogenase (LDH) assay is a common approach to evaluate cell membrane impairment because LDH released into the culture medium is a sensitive indicator for cell membrane integrity disruption. Employing this technique, high levels of cytotoxicity in osteosarcoma cells exposed to magnetite [89], in human liver cells treated with cobalt-ION [75], in mouse macrophage cells treated with Fe_2O_3 nanoparticles [107], and in human periodontal ligament fibroblasts and mouse dermal fibroblasts exposed to magnetite [108] were reported. Similarly, Coccini et al. [106] also observed impairment in cell membrane integrity, evaluated by fluorescent calcein-acetoxymethyl/propidium iodide staining, in neuroblastoma and astrocytoma cells treated with magnetite nanoparticles. However, opposite to these findings, Kiliç et al. [109] did not find any significant alteration in the percentage of LDH activity at any medium, concentration or treatment time tested in the study.

As happens with other cellular effects, membrane impairment induced by ION seems to be related to the type and dose of nanoparticles. In this regard, a study investigating the effect of different surface coatings on cell behavior and morphology reported that dextran-coated magnetite nanoparticles induced more prominent membrane disruptions than uncoated nanoparticles; however, albumin-coated magnetite did not show cytotoxic effects [110]. But also cell type seems to be relevant in ION effects. Li et al. [111] carried out a comparative study measuring some cytotoxic effects, namely membrane disruption and intracellular enzymatic activity, in human cervical cancer (HeLa) cell line and immortalized normal human retinal pigment epithelial (RPE) cell line exposed to ION (magnetite). Results obtained showed that uncoated ION resulted toxic to both HeLa and RPE cells at high concentration (0.40 mg/ml); however, at low concentrations, cytotoxicity was cell-type specific, being RPE cells more susceptible than HeLa cells.

12.2.5 Genomic Alterations

A number of in vitro studies have evaluated the effects of ION exposure on the genetic material (reviewed in [1]) (Fig. 12.2). Different kinds of DNA damage, including strand breaks [60, 95, 112] and micronucleus (MN) formation [113], induced in cell systems after treatment with ION were reported. However, other several studies did not find any genotoxic alteration [47, 48, 86, 114]. Lack of consistence in the results of ION genotoxicity make it necessary to further investigate on this line. However, far from shedding light on this controversy, studies published these last years maintain the disagreement. The current section of this chapter is focused on the most recent studies carried out to further investigate the ION potential to induce genetic damage.

Rajiv et al. [115] observed DNA breaks and chromosome aberrations in human lymphocytes exposed to ION (Fe_2O_3). Similarly, Kiliç et al. [109] found increases in genetic damage (measured by comet assay) in neuronal SH-SY5Y

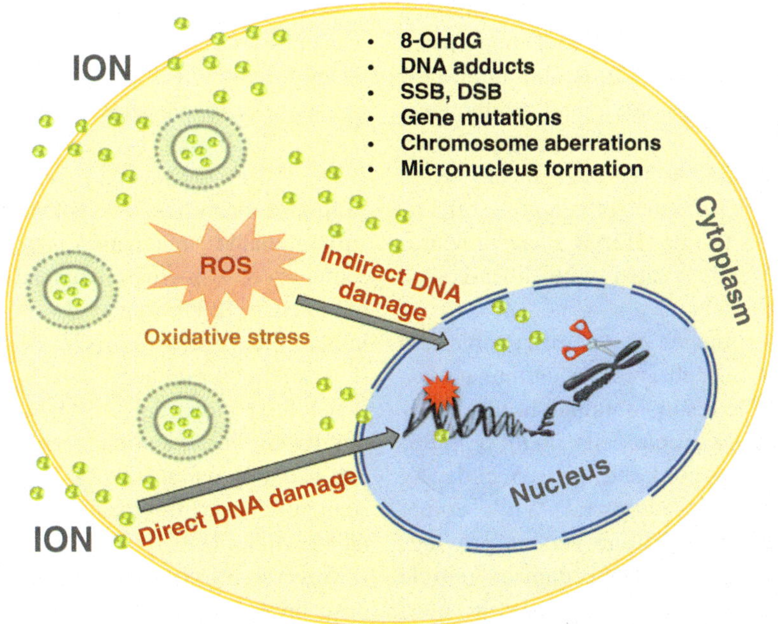

Fig. 12.2 ION-induced genomic alterations. ION may cause DNA damage through direct interaction with the DNA structure, or result in the generation of oxidative radicals that in turn have the potential to cause indirectly DNA damage, mainly through base oxidation (mostly 8-OHdG). Consequently, ION exposure may induce genotoxic clastogenic or aneugenic effects. *8-OHdG* 8 hydroxydeoxyguanosine, *DSB* double strand breaks, *ION* iron oxide nanoparticles, *ROS* reactive oxygen species, *SSB* single strand breaks

cells treated with silica-coated magnetite nanoparticles. And Pongrac et al. [95] observed that ION (γ-Fe_2O_3), uncoated or coated with d-mannose or poly-l-lysine, induced DNA damage (also evaluated by comet assay) in murine neural stem cells irrespective of the surface coating. In this case, lower dose of any ION induced heavier DNA damage, and the lack of genotoxic effects at higher doses was explained by the aggregation behavior of ION at such concentrations. In agreement with these studies, Cicha et al. [116] evaluated the levels of H2AX phosphorylated (γH2AX), as indicative of DNA double strand breaks, in human primary tubular epithelial cells exposed to lauric acid-coated ION functionalized with mitoxantrone (ION-MTO). They observed a significant increase in γH2AX foci upon treatment with ION-MTO; however, there was no correlation between particle content inside the cells and histone phosphorylation.

Opposite to these findings, Couto et al. [65] demonstrated absence of ION effects on the genetic material of human T-lymphocytes, reporting no chromosome aberrations in cells treated with polyacrylic acid-coated and non-coated nanomagnetite. In agreement, Paolini et al. [73] reported absence of genotoxic and carcinogenic effects of rhamnose-coated ION (magnetite) on mouse fibroblast Balb/c-3T3 cells. Finally, Kiliç et al. [109] reported absence of genotoxic effects, evaluated by MN test or γH2AX assay, in neuronal SH-SY5Y cells treated with silica-coated ION.

Apart from influence of type of nanoparticle and coating, dose, cell type, or culture media, controversy in the results obtained in the different studies on ION genotoxicity might be influenced by the assay used and, consequently, by the type of outcome addressed. While comet assay in its standard alkaline version provides information on primary DNA damage (single strand breaks, abasic sites, alkali-labile sites, and incomplete excision repair sites), γH2AX assay is an indicator of double strand breaks, and MN test reflects chromosome alterations (both chromosome loss and chromosome breakage) [109, 117].

12.2.6 Other Cellular Effects

Together with the cellular damage previously described, ION were also found to induce other different toxic effects at the cellular level, including cell cycle alterations, cell death, cytoskeleton disruption, and alterations in gene and protein expression (Fig. 12.3). Cell cycle alterations have been previously observed in different cells after ION exposure. Particularly, cell cycle arrest in G2/M phase was reported in PC12 cells treated with Fe_3O_4 nanoparticles [118], and in human colon HT29 cells treated with ION conjugated with doxorubicin [119]. Similarly, Fe_3O_4 nanoparticles also altered the cell-cycle progression of human mesenchymal stem cells through a decrease in the proportion of cells in the G0/G1 phase [120]. However, only slight alterations, at high doses (200 μg/ml) and long exposure time (24 h) were found in neuronal cells exposed to silica-coated magnetite nanoparticles [109]. And Lai et al. [121] also found that Fe_2O_3 nanoparticles did not induce cell cycle disruptions in lung epithelial cells.

Different types of cell death, including apoptosis, necrosis and autophagy, were also found to be induced by ION. Uncoated and dextran-coated magnetite nanoparticles induced apoptosis in human fibroblasts [122] and in hepatic HepG2 cells [75]. Fe_2O_3 nanoparticles induced necrosis in mouse macrophage cells [107], and graphitic carbon-coated magnetic nanoparticles were found to cause necrosis in human embryonic kidney cells [123]. Autophagy was reported in human blood cells [124], in osteosarcoma cells [89] and in breast cancer cells [105] treated with magnetite nanoparticles. Some studies suggested that ION-induced autophagy may indeed precede cell death by apoptosis [125], which would subsequently happen by excessive accumulation of nanoparticles [89].

Several morphological alterations have been described after ION treatment in different studies. Soenen et al. [54] demonstrated that several types of ION, namely dextran-coated Endorem®, carboxydextran-coated Resovist®, lipid-coated magnetoliposomes and citrate-coated ION, induced important cytoskeleton

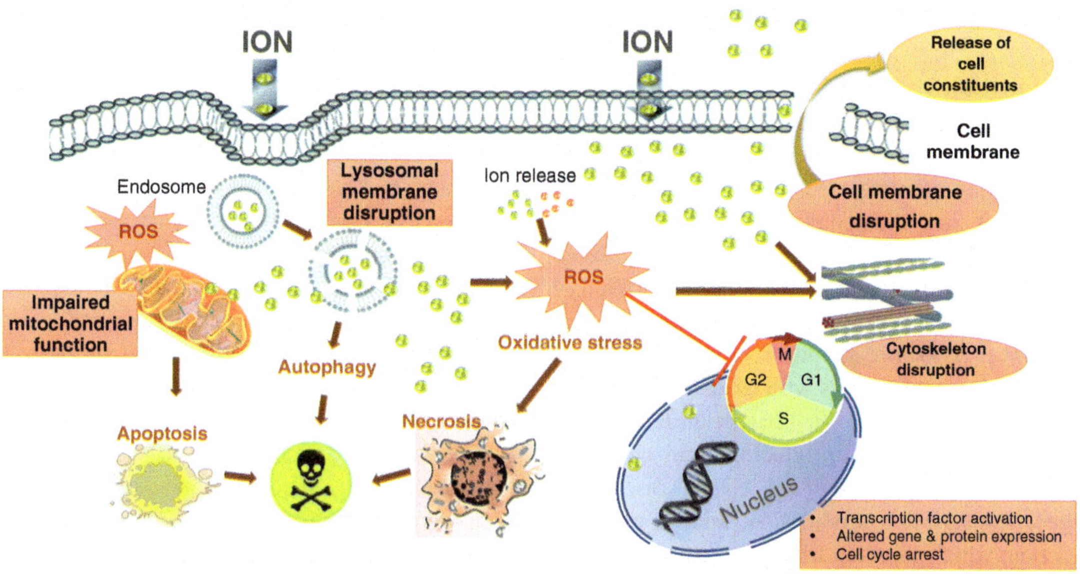

Fig. 12.3 Cellular toxicity induced by ION. ION exposure may lead to different cellular toxic effects including impaired mitochondrial function (and, consequently, apoptosis), lysosomal damage/dysfunction, cell membrane disruption, cytoskeleton disruption, DNA damage and cell cycle alterations. Besides, accumulation of high amounts of ION and iron in the cytoplasm leads (in fewer cases) to cell death by autophagy. All these effects may be produced by ION not only directly, but also indirectly through generation of reactive oxygen species (*ROS*) and iron ion release. Increased ROS levels would lead to enzyme depletion/inactivation, protein denaturation, genetic alterations or impacts on cell cycle or on cytoskeleton, among others; whereas ion release would cause genomic damage, iron imbalance and might eventually result in cell death. *ION* iron oxide nanoparticles, *ROS* reactive oxygen species

and morphology alterations in nervous C17.2 neural progenitor cells and PC12 cells, regardless the type of coating. Also, Buyukhatipoglu and Clyne [126] reported cytoskeleton and morphology alterations in porcine aortic endothelial cells after treatment with ION, and Gonnissen et al. [127] observed cytoskeleton disruption in primary monocytes and the leukemic monocyte MM6 cell line exposed to core-shell starch-coated ION.

Finally, other molecular alterations were also reported as a consequence of ION exposure. Wu and Sun [118] found activation of JNK- and p53-mediated pathways to regulate the cell cycle and apoptosis in PC12 cells treated with ION (Fe_3O_4). Similarly, the analysis of the cell cycling and proliferation regulatory molecules ERK, p53, and AKT, implied that cell cycle arrest was initiated and the cells were sensitized to necrosis [123]. Besides, Zhu et al. [85] found upregulation of intracellular cell adhesion molecule-1 (ICAM-1) and interleukin-8 (IL-8) expression in human aortic endothelial cells exposed to Fe_2O_3 and Fe_3O_4 nanoparticles. Other several genes, including CYP1A, TNF3, or E2F1 among others, were found overexpressed in response to Fe_3O_4 nanoparticles in human mesenchymal stem cells [120]. In this line, there is enough evidence that ION upregulate genes that are associated with endothelial layer integrity [128] and lysosomal function [129], caspase [130] and cytokine [65] activation, and produce changes in iron metabolism-related genes [131]. ION can also activate and upregulate plasma membrane proteins, such as nicotinamide adenine dinucleotide phosphate (NADPH) oxidase [132], which may also contribute to the underlying mechanisms associated with toxicity.

12.3 Toxicity Mechanism

Due to the high variability in the results from the ION cellular and molecular toxicity studies, the mechanisms by which these nanoparticles might impair the different processes involved in cellular functions and maintenance are not well understood. Still, one of the most commonly suggested mechanisms of ION toxicity is the generation of ROS [78, 79, 133, 134]. As previously described, a number of in vitro studies have associated cytotoxicity induced by ION with oxidative stress and ROS production [75, 89, 126]. High ROS levels can damage cells by producing lipid peroxidation, mitochondrial damage, DNA disruption, gene transcription modulation, and protein oxidation, which can then trigger a cascade of Ca^{+2}-dependent signaling mechanisms, resulting in decline of physiological functions and cell apoptosis/death [135]. Consequently, oxidative stress plays a very important role in various vital processes of proliferation and apoptosis [89]. Accordingly, presence of high intracellular ROS levels was reported to be associated with different ION-induced cytotoxic effects, including cytoskeleton disruption [126], apoptosis [75] and autophagy [89]. However, other studies reported no induction of oxidative stress-mediated toxic effects [95, 98, 99].

Together with ROS production, among the different mechanisms possibly involved in ION toxicity, it is also remarkable the ability of these nanoparticles to release iron ions. Ionic forms of iron, ferric (Fe^{3+}) and ferrous (Fe^{2+}), are essential in the cell since they play a critical role in important organic metabolic pathways, including cytochrome P450 function, mitochondrial oxidative phosphorylation, oxygen transport, DNA synthesis, and energy production [136]. Nevertheless, excess of this metal can be very toxic and, therefore, iron levels must be strictly controlled in the organism. Free iron released from ION can be incorporated to the normal cellular iron pool causing elevated intracellular iron concentrations [62, 69, 137]. This excess of iron has been found to cause elevated ROS generation through the Fenton reaction, resulting in oxidative stress that damages DNA, lipids and proteins [138, 139].

12.4 Conclusions

Initially ION seemed to be safe for biomedical use, since their potential cytotoxic effects, if any, were usually slight or limited to specific conditions (e.g., highest doses and/or longest exposure times). Moreover, ION cytotoxicity was initially considered only as a decrease in cell viability. Nevertheless, as many of the studies gathered in this chapter have demonstrated, these nanoparticles can exert other important effects on the cell dynamics and physiology. Indeed, different cellular and molecular effects – mainly dependent on ION type and concentration, time of exposure, presence and type of coating, and cell type evaluated – have been reported after ION exposure in in vitro studies. Those effects include decreases in viability, plasmatic membrane disruption, mitochondrial alterations, oxidative damage, cell cycle impairments, cytoskeleton disruption, cell death, and alterations in cell motility, and in cell integrity.

Hence, criteria to define ION toxicity must be clearly defined and, as previously suggested by other authors [41], terms such as 'biocompatibility' must be re-evaluated. Despite the reported negative effects, the numerous advantages of ION together with their promising applications in biomedicine make it necessary to define clearly their toxicity, in order to discard potential health risks and to reach optimal benefits of their use. Furthermore, the development of standardized nanotoxicity testing methods would help to generate results comparable across studies, thus assisting to define conditions to use ION under minimal reasonable risks for human health.

Acknowledgements This work was supported by Xunta de Galicia (grant number ED431B 2016/013) and the projects NanoToxClass (ERA-SIINN/001/2013) funded by FCT/MCTES (PIDDAC) and co-funded by the European Regional Development Fund (ERDF) through the COMPETE Programme, and CERASAFE (SIINN/0004/2014) funded by FCT (national funds of MES) and co-funded by the European Union (ERA-NET SIINN programme). V. Valdiglesias was supported by a Xunta de Galicia postdoctoral fellowship (reference ED481B 2016/190-0). N. Fernández-Bertólez was supported by an INDITEX-UDC fellowship. F. Brandão was supported by the grant SFRH/BD/101060/2014, funded by FCT (financing subsided by national fund of MES).

References

1. Dissanayake NM, Current KM, Obare SO (2015) Mutagenic effects of Iron oxide nanoparticles on biological cells. Int J Mol Sci 16:23482–23516
2. Revia RA, Zhang M (2016) Magnetite nanoparticles for cancer diagnosis, treatment, and treatment monitoring: recent advances. Mater Today 19:157–168
3. Lin MM, Kim DK, El Haj AJ et al (2008) Development of superparamagnetic iron oxide nanoparticles (SPIONS) for translation to clinical applications. IEEE Trans Nanobioscience 7:298–305
4. Rosen JE, Chan L, Shieh D-B et al (2012) Iron oxide nanoparticles for targeted cancer imaging and diagnostics. Nanomedicine 8:275–290
5. Anselmo AC, Mitragotri S (2016) Nanoparticles in the clinic. Bioeng Transl Med 1:10–29
6. Ittrich H, Peldschus K, Raabe N et al (2013) Superparamagnetic iron oxide nanoparticles in biomedicine: applications and developments in diagnostics and therapy. RöFo Fortschritte Auf Dem Gebiete Der Röntgenstrahlen Und Der Nukl 185:1149–1166
7. Kim JE, Shin JY, Cho MH (2012) Magnetic nanoparticles: an update of application for drug delivery and possible toxic effects. Arch Toxicol 86:685–700
8. Mok H, Zhang M (2013) Superparamagnetic iron oxide nanoparticle-based delivery systems for biotherapeutics. Expert Opin Drug Deliv 10:73–87
9. Wahajuddin AS (2012) Superparamagnetic iron oxide nanoparticles: magnetic nanoplatforms as drug carriers. Int J Nanomedicine 7:3445–3471
10. Laurent S, Dutz S, Häfeli UO et al (2011) Magnetic fluid hyperthermia: focus on superparamagnetic iron oxide nanoparticles. Adv Colloid Interf Sci 166:8–23
11. Wang YXJ, Xuan S, Port M et al (2013) Recent advances in superparamagnetic iron oxide nanoparticles for cellular imaging and targeted therapy research. Curr Pharm Des 19:6575–6593
12. Torres-Lugo M, Rinaldi C (2013) Thermal potentiation of chemotherapy by magnetic nanoparticles. Nanomedicine (London) 8:1689–1707
13. Grüttner C, Müller K, Teller J et al (2013) Synthesis and functionalisation of magnetic nanoparticles for hyperthermia applications. Int J Hyperth 29:777–789
14. Estelrich J, Escribano E, Queralt J et al (2015) Iron oxide nanoparticles for magnetically-guided and magnetically-responsive drug delivery. Int J Mol Sci 16:8070–8101
15. Zhu XM, Wang YXJ, Cham-Fai Leung K et al (2012) Enhanced cellular uptake of aminosilane-coated superparamagnetic iron oxide nanoparticles in mammalian cell lines. Int J Nanomedicine 7:953–964
16. Mahdavi M, Bin AM, Haron MJ et al (2013) Synthesis, surface modification and characterisation of biocompatible magnetic iron oxide nanoparticles for biomedical applications. Molecules 18:7533–7548
17. Mahmoudi M, Simchi A, Milani AS et al (2009) Cell toxicity of superparamagnetic iron oxide nanoparticles. J Colloid Interface Sci 336:510–518
18. Wu W, He Q, Jiang C (2008) Magnetic iron oxide nanoparticles: synthesis and surface functionalization strategies. Nanoscale Res Lett 3:397–415
19. Bjørnerud A, Johansson L (2004) The utility of superparamagnetic contrast agents in MRI: theoretical consideration and applications in the cardiovascular system. NMR Biomed 17:465–477
20. Wang YXJ, Hussain SM, Krestin GP (2001) Superparamagnetic iron oxide contrast agents: physicochemical characteristics and applications in MR imaging. Eur Radiol 11:2319–2331
21. Gupta AK, Gupta M (2005) Synthesis and surface engineering of iron oxide nanoparticles for biomedical applications. Biomaterials 26:3995–3402
22. Santhosh PB, Ulrih NP (2013) Multifunctional superparamagnetic iron oxide nanoparticles: promising tools in cancer theranostics. Cancer Lett 336:8–17
23. Petters C, Bulcke F, Thiel K et al (2014) Uptake of fluorescent iron oxide nanoparticles by oligodendroglial OLN-93 cells. Neurochem Res 39:372–383
24. Yu M, Huang S, Yu KJ et al (2012) Dextran and polymer polyethylene glycol (PEG) coating reduce both 5 and 30 nm iron oxide nanoparticle cytotoxicity in 2D and 3D cell culture. Int J Mol Sci 13:5554–5570
25. Alwi R, Telenkov S, Mandelis A et al (2012) Silica-coated super paramagnetic iron oxide nanoparticles (SPION) as biocompatible contrast agent in biomedical photoacoustics. Biomed Opt Express 3:2500–2509
26. Tong L, Zhao M, Zhu S et al (2011) Synthesis and application of superparamagnetic iron oxide nanoparticles in targeted therapy and imaging of cancer. Front Med 5:379–387
27. Xia T, Kovochich M, Liong M et al (2009) Polyethyleneimine coating enhances the cellular uptake of mesoporous silica nanoparticles and allows safe delivery of siRNA and DNA constructs. ACS Nano 3:3273–3286
28. Figuerola A, Di Corato R, Manna L et al (2010) From iron oxide nanoparticles towards advanced iron-based inorganic materials designed for biomedical applications. Pharmacol Res 62:126–143
29. Jin R, Lin B, Li D et al (2014) Superparamagnetic iron oxide nanoparticles for MR imaging and therapy: design considerations and clinical applications. Curr Opin Pharmacol 18:18–27
30. Bertorelle F, Wilhelm C, Roger J et al (2006) Fluorescence-modified superparamagnetic nanoparticles: intracellular uptake and use in cellular imaging. Langmuir ACS J Surf Colloids 22:5385–5391

31. Ge Y, Zhang Y, He S et al (2009) Fluorescence modified chitosan-coated magnetic nanoparticles for high-efficient cellular imaging. Nanoscale Res Lett 4:287–295
32. Petters C, Irrsack E, Koch M et al (2014) Uptake and metabolism of iron oxide nanoparticles in brain cells. Neurochem Res 39:1648–1660
33. Yan F, Wang Y, He S et al (2013) Transferrin-conjugated, fluorescein-loaded magnetic nanoparticles for targeted delivery across the blood-brain barrier. J Mater Sci Mater Med 24:2371–2379
34. Agemy L, Friedmann-Morvinski D, Kotamraju VR et al (2011) Targeted nanoparticle enhanced pro-apoptotic peptide as potential therapy for glioblastoma. Proc Natl Acad Sci U S A 108:17450–17455
35. Kumar M, Singh G, Arora V et al (2012) Cellular interaction of folic acid conjugated superparamagnetic iron oxide nanoparticles and its use as contrast agent for targeted magnetic imaging of tumor cells. Int J Nanomedicine 7:3503–3516
36. Hassan EE, Gallo JM (1993) Targeting anticancer drugs to the brain. I: enhanced brain delivery of oxantrazole following administration in magnetic cationic microspheres. J Drug Target 1:7–14
37. Jenkins SI, Pickard MR, Granger N et al (2011) Magnetic nanoparticle-mediated gene transfer to oligodendrocyte precursor cell transplant populations is enhanced by magnetofection strategies. ACS Nano 5:6527–6538
38. Choi JW, Park JW, Na Y et al (2015) Using a magnetic field to redirect an oncolytic adenovirus complexed with iron oxide augments gene therapy efficacy. Biomaterials 65:163–174
39. Krötz F, de Wit C, Sohn HY et al (2003) Magnetofection – a highly efficient tool for antisense oligonucleotide delivery in vitro and in vivo. Mol Ther J Am Soc Gene Ther 7:700–710
40. Kumar M, Yigit M, Dai G et al (2010) Image-guided breast tumor therapy using a small interfering RNA nanodrug. Cancer Res 70:7553–7561
41. Singh N, Jenkins GJS, Asadi R, Doak SH (2010) Potential toxicity of superparamagnetic iron oxide nanoparticles (SPION). Nano Rev 1:5358
42. Kumari M, Rajak S, Singh SP et al (2012) Biochemical alterations induced by acute oral doses of iron oxide nanoparticles in Wistar rats. Drug Chem Toxicol 36:1–10
43. Radu M, Din I, Hermenean A et al (2015) Exposure to iron oxide nanoparticles coated with phospholipid-based polymeric micelles induces biochemical and histopathological pulmonary changes in mice. Int J Mol Sci 16:29417–29435
44. Sadeghi L, Yousefi Babadi V, Espanani HR (2015) Toxic effects of the Fe2O3 nanoparticles on the liver and lung tissue. Bratisl Lek Listy 116:373–378
45. Hussain SM, Hess KL, Gearhart JM et al (2005) In vitro toxicity of nanoparticles in BRL 3A rat liver cells. Toxicol In Vitro 19:975–983
46. Jeng H, Swanson J (2006) Toxicity of metal oxide nanoparticles in mammalian cells. J Environ Sci Heal Part A Toxic/Hazard Subst Environ Eng 4112:2699–2711
47. Karlsson HL, Gustafsson J, Cronholm P et al (2009) Size-dependent toxicity of metal oxide particles-A comparison between nano- and micrometer size. Toxicol Lett 188:112–118
48. Karlsson HL, Cronholm P, Gustafsson J et al (2008) Copper oxide nanoparticles are highly toxic: a comparison between metal oxide nanoparticles and carbon nanotubes. Chem Res Toxicol 21:1726–1732
49. Kunzmann A, Andersson B, Vogt C et al (2011) Efficient internalization of silica-coated iron oxide nanoparticles of different sizes by primary human macrophages and dendritic cells. Toxicol Appl Pharmacol 253:81–93
50. Cromer Berman SM, Kshitiz CJ, Wang CJ et al (2013) Cell motility of neural stem cells is reduced after SPIO-labeling, which is mitigated after exocytosis. Magn Reson Med 69:255–262
51. Wu X, Tan Y, Mao H et al (2010) Toxic effects of iron oxide nanoparticles on human umbilical vein endothelial cells. Int J Nanomedicine 5:385–399
52. Patil US, Adireddy S, Jaiswal A et al (2015) In vitro/in vivo toxicity evaluation and quantification of iron oxide nanoparticles. Int J Mol Sci 16:24417–24450
53. Liu Y, Xia Q, Liu Y et al (2014) Genotoxicity assessment of magnetic iron oxide nanoparticles with different particle sizes and surface coatings. Nanotechnology 25:425101
54. Soenen SJH, Himmelreich U, Nuytten N et al (2011) Cytotoxic effects of iron oxide nanoparticles and implications for safety in cell labelling. Biomaterials 32:195–205
55. Rivet CJ, Yuan Y, Borca-Tasciuc DA et al (2012) Altering iron oxide nanoparticle surface properties induce cortical neuron cytotoxicity. Chem Res Toxicol 25:153–161
56. Magdolenova Z, Drlickova M, Henjum K et al (2013) Coating-dependent induction of cytotoxicity and genotoxicity of iron oxide nanoparticles. Nanotoxicology 5390:1–13
57. Sun Z, Yathindranath WM et al (2013) Characterization of cellular uptake and toxicity of aminosilane-coated iron oxide nanoparticles with different charges in central nervous system-relevant cell culture models. Int J Nanomedicine 8:961–970
58. Couto D, Freitas M, Vilas-Boas V et al (2014) Interaction of polyacrylic acid coated and non-coated iron oxide nanoparticles with human neutrophils. Toxicol Lett 225:57–65
59. Schütz CA, Staedler D, Crosbie-Staunton K et al (2014) Differential stress reaction of human colon cells to oleic-acid-stabilized and unstabilized ultrasmall iron oxide nanoparticles. Int J Nanomedicine 9:3481–3498
60. Hong SC, Lee JH, Lee J et al (2011) Subtle cytotoxicity and genotoxicity differences in superparamag-

netic iron oxide nanoparticles coated with various functional groups. Int J Nanomedicine 6:3219–3231
61. Cheng Z, Al Zaki A, Hui JZ et al (2012) Multifunctional nanoparticles: cost versus benefit of adding targeting and imaging capabilities. Science 338:903–910
62. Geppert M, Hohnholt MC, Thiel K et al (2011) Uptake of dimercaptosuccinate-coated magnetic iron oxide nanoparticles by cultured brain astrocytes. Nanotechnology 22:145101
63. Geppert M, Hohnholt MC, Nürnberger S et al (2012) Ferritin up-regulation and transient ROS production in cultured brain astrocytes after loading with iron oxide nanoparticles. Acta Biomater 8:3832–3839
64. Petters C, Dringen R (2015) Uptake, metabolism and toxicity of iron oxide nanoparticles in cultured microglia, astrocytes and neurons. Springerplus 4:L32
65. Couto D, Sousa R, Andrade L et al (2015) Polyacrylic acid coated and non-coated iron oxide nanoparticles are not genotoxic to human T lymphocytes. Toxicol Lett 234:67–73
66. Xu Y, Sherwood JA, Lackey KH et al (2016) The responses of immune cells to iron oxide nanoparticles. J Appl Toxicol 36:543–553
67. Bigini P, Diana V, Barbera S et al (2012) Longitudinal tracking of human fetal cells labeled with super paramagnetic iron oxide nanoparticles in the brain of mice with motor neuron disease. PLoS One 7:e32326
68. Bin NH, Palui G, Rosenberg JT et al (2012) Multidentate catechol-based polyethylene glycol oligomers provide enhanced stability and biocompatibility to iron oxide nanoparticles. ACS Nano 6:389–399
69. Rosenberg JT, Sachi-Kocher A, Davidson MW et al (2012) Intracellular SPIO labeling of microglia: high field considerations and limitations for MR microscopy. Contrast Media Mol Imaging 7:121–129
70. Jarockyte G, Daugelaite E, Stasys M et al (2016) Accumulation and toxicity of superparamagnetic iron oxide nanoparticles in cells and experimental animals. Int J Mol Sci 17:1193
71. Rojas JM, Sanz-Ortega L, Mulens-Arias V et al (2016) Superparamagnetic iron oxide nanoparticle uptake alters M2 macrophage phenotype, iron metabolism, migration and invasion. Nanomedicine Nanotechnol Biol Med 12:1127–1138
72. Szalay B, Tátrai E, Nyírő G et al (2012) Potential toxic effects of iron oxide nanoparticles in in vivo and in vitro experiments. J Appl Toxicol 32:446–453
73. Paolini A, Guarch CP, Ramos-López D et al (2016) Rhamnose-coated superparamagnetic iron-oxide nanoparticles: an evaluation of their in vitro cytotoxicity, genotoxicity and carcinogenicity. J Appl Toxicol 36:510–520
74. Schaub NJ, Rende D, Yuan Y et al (2014) Reduced astrocyte viability at physiological temperatures from magnetically activated iron oxide nanoparticles. Chem Res Toxicol 27:2023–2035
75. Ahamed M, Akhtar MJ, Khan MAM et al (2016) Cobalt iron oxide nanoparticles induce cytotoxicity and regulate the apoptotic genes through ROS in human liver cells (HepG2). Colloids Surf B: Biointerfaces 148:665–673
76. Ying E, Hwang H-M (2010) In vitro evaluation of the cytotoxicity of iron oxide nanoparticles with different coatings and different sizes in A3 human T lymphocytes. Sci Total Environ 408:4475–4481
77. Costa C, Brandão F, Bessa MJ et al (2016) In vitro cytotoxicity of superparamagnetic iron oxide nanoparticles on neuronal and glial cells. Evaluation of nanoparticle interference with viability tests. J Appl Toxicol 36:361–372
78. Nel A, Xia T, Mädler L et al (2006) Toxic potential of materials at the nanolevel. Science 311:622–627
79. Unfried K, Albrecht C, Klotz L-O et al (2007) Cellular responses to nanoparticles: target structures and mechanisms. Nanotoxicology 1:52–71
80. Liu G, Gao J, Ai H et al (2013) Applications and potential toxicity of magnetic iron oxide nanoparticles. Small Weinh Bergstr Ger 9:1533–1545
81. Pisanic TR, Jin S, Shubayev VI 2009 Iron oxide magnetic nanoparticle nanotoxicity: incidence and mechanisms. In: Sahu SC, Casciano DA (eds) Nanotoxicity. Wiley, New York, pp 397–425
82. Hanot CC, Choi YS, Anani TB et al (2015) Effects of iron-oxide nanoparticle surface chemistry on uptake kinetics and cytotoxicity in CHO-K1 cells. Int J Mol Sci 17:E54
83. Naqvi S, Samim M, Abdin MZ et al (2010) Concentration-dependent toxicity of iron oxide nanoparticles mediated by increased oxidative stress. Int J Nanomedicine 5:983–989
84. Hanini A, Schmitt A, Kacem K et al (2011) Evaluation of iron oxide nanoparticle biocompatibility. Int J Nanomedicine 6:787–794
85. Zhu M-T, Wang B, Wang Y et al (2011) Endothelial dysfunction and inflammation induced by iron oxide nanoparticle exposure: risk factors for early atherosclerosis. Toxicol Lett 203:162–171
86. Zhang T, Qian L, Tang M et al (2012) Evaluation on cytotoxicity and genotoxicity of the L-glutamic acid coated iron oxide nanoparticles. J Nanosci Nanotechnol 12:2866–2873
87. Watanabe M, Yoneda M, Morohashi A et al (2013) Effects of Fe3O4 magnetic nanoparticles on A549 cells. Int J Mol Sci 14:15546–15560
88. Könczöl M, Ebeling S, Goldenberg E et al (2011) Cytotoxicity and genotoxicity of size-fractionated iron oxide (magnetite) in A549 human lung epithelial cells: role of ROS, JNK, and NF-κB. Chem Res Toxicol 24:1460–1475
89. Du S, Li J, Du C et al (2017) Overendocytosis of superparamagnetic iron oxide particles increases apoptosis and triggers autophagic cell death in human osteosarcoma cell under a spinning magnetic field. Oncotarget 8:9410–9424

90. Petters C, Thiel K, Dringen R (2015) Lysosomal iron liberation is responsible for the vulnerability of brain microglial cells to iron oxide nanoparticles: comparison with neurons and astrocytes. Nanotoxicology 10:1743–5390
91. Wu J, Wang C, Sun J et al (2011) Neurotoxicity of silica nanoparticles: brain localization and dopaminergic neurons damage pathways. ACS Nano 5:4476–4489
92. Wang Y, Wang B, Zhu MT et al (2011) Microglial activation, recruitment and phagocytosis as linked phenomena in ferric oxide nanoparticle exposure. Toxicol Lett 205:26–37
93. Kenzaoui BH, Bernasconi CC, Hofmann H et al (2012) Evaluation of uptake and transport of ultrasmall superparamagnetic iron oxide nanoparticles by human brain-derived endothelial cells. Nanomedicine 7:39–53
94. Pongrac IM, Dobrivojevi M, Ahmed LB et al (2016) Improved biocompatibility and efficient labeling of neural stem cells with poly(L-lysine)-coated maghemite nanoparticles. Beilstein J Nanotechnol 7:926–936
95. Hildebrand H, Kühnel D, Potthoff A et al (2010) Evaluating the cytotoxicity of palladium/magnetite nano-catalysts intended for wastewater treatment. 158:65–73
96. Hohnholt M, Geppert M, Dringen R (2010) Effects of iron chelators, iron salts, and iron oxide nanoparticles on the proliferation and the iron content of oligodendroglial OLN-93 cells. Neurochem Res 35:1259–1268
97. Hohnholt MC, Geppert M, Dringen R (2011) Treatment with iron oxide nanoparticles induces ferritin synthesis but not oxidative stress in oligodendroglial cells. Acta Biomater 7:3946–3954
98. Lindemann A, Fraederich BM, Pries R et al (2014) Biological impact of superparamagnetic iron oxide nanoparticles for magnetic particle imaging of head and neck cancer cells. Int J Nanomedicine 9:5025–5040
99. Remya NS, Syama S, Sabareeswaran A et al (2016) Toxicity, toxicokinetics and biodistribution of dextran stabilized iron oxide nanoparticles for biomedical applications. Int J Pharm 511:586–598
100. Khan MI, Mohammad A, Patil G et al (2012) Induction of ROS, mitochondrial damage and autophagy in lung epithelial cancer cells by iron oxide nanoparticles. Biomaterials 33:1477–1488
101. Mukhopadhyay A, Joshi N, Chattopadhyay K et al (2012) A facile synthesis of PEG-coated magnetite (Fe 3O 4) nanoparticles and their prevention of the reduction of cytochrome C. ACS Appl Mater Interfaces 4:142–149
102. Sarkar A, Ghosh M, Sil PC (2014) Nanotoxicity: oxidative stress mediated toxicity of metal and metal oxide nanoparticles. J Nanosci Nanotechnol 14:730–743
103. Ding FA, Li YP, Liu J et al (2014) Overendocytosis of gold nanoparticles increases autophagy and apoptosis in hypoxic human renal proximal tubular cells. Int J Nanomedicine 9:4317–4330
104. Baratli Y, Charles AL, Wolff V et al (2014) Age modulates Fe3O4 nanoparticles liver toxicity: dose-dependent decrease in mitochondrial respiratory chain complexes activities and coupling in middle-aged as compared to young rats. Biomed Res Int 2014:474081
105. Zhang X, Zhang H, Liang X et al (2016) Iron oxide nanoparticles induce autophagosome accumulation through multiple mechanisms: lysosome impairment, mitochondrial damage, and ER stress. Mol Pharm 13:2578–2587
106. Coccini T, Caloni F, Ramírez Cando LJ et al (2017) Cytotoxicity and proliferative capacity impairment induced on human brain cell cultures after short- and long-term exposure to magnetite nanoparticles. J Appl Toxicol 37:361–373
107. Lee JH, Ju JE, Kim BI et al (2014) Rod-shaped iron oxide nanoparticles are more toxic than sphere-shaped nanoparticles to murine macrophage cells. Environ Toxicol Chem 33:2759–2766
108. Dönmez Güngüneş Ç, Şeker Ş, Elçin AE et al (2017) A comparative study on the in vitro cytotoxic responses of two mammalian cell types to fullerenes, carbon nanotubes and iron oxide nanoparticles. Drug Chem Toxicol:1–13
109. Kiliç G, Costa C, Fernández-Bertólez N et al (2016) In vitro toxicity evaluation of silica-coated iron oxide nanoparticles in human SHSY5Y neuronal cells. Toxicol Res 5:235–247
110. Berry CC, Wells S, Charles S et al (2003) Dextran and albumin derivatised iron oxide nanoparticles: influence on fibroblasts in vitro. Biomaterials 24:4551–4557
111. Li L, Mak KY, Shi J et al (2012) Comparative in vitro cytotoxicity study on uncoated magnetic nanoparticles: effects on cell viability, cell morphology, and cellular uptake. J Nanosci Nanotechnol 12:9010–9017
112. Bhattacharya K, Davoren M, Boertz J et al (2009) Titanium dioxide nanoparticles induce oxidative stress and DNA-adduct formation but not DNA-breakage in human lung cells. Part Fibre Toxicol 6:17
113. Singh N, Jenkins GJS, Nelson BC et al (2012) The role of iron redox state in the genotoxicity of ultrafine superparamagnetic iron oxide nanoparticles. Biomaterials 33:163–170
114. Guichard Y, Schmit J, Darne C et al (2012) Cytotoxicity and genotoxicity of nanosized and microsized titanium dioxide and iron oxide particles in Syrian hamster embryo cells. Ann Occup Hyg 56:631–644
115. Rajiv S, Jerobin J, Saranya V et al (2016) Comparative cytotoxicity and genotoxicity of cobalt (II, III) oxide, iron (III) oxide, silicon dioxide, and

aluminum oxide nanoparticles on human lymphocytes in vitro. Hum Exp Toxicol 35:170–183
116. Cicha I, Scheffler L, Ebenau A et al (2016) Mitoxantrone-loaded superparamagnetic iron oxide nanoparticles as drug carriers for cancer therapy: uptake and toxicity in primary human tubular epithelial cells. Nanotoxicology 5390:1–10
117. Valdiglesias V, Laffon B, Pásaro E et al (2011) Evaluation of okadaic acid-induced genotoxicity in human cells using the micronucleus test and γH2AX analysis. J Toxicol Environ Health A 74(15–16):980–992
118. Wu J, Sun J (2011) Investigation on mechanism of growth arrest induced by iron oxide nanoparticles in PC12 cells. J Nanosci Nanotechnol 11:11079–11083
119. Augustin E, Czubek B, Nowicka AM et al (2016) Improved cytotoxicity and preserved level of cell death induced in colon cancer cells by doxorubicin after its conjugation with iron-oxide magnetic nanoparticles. Toxicol In Vitro 33:45–53
120. Periasamy VS, Athinarayanan J, Alhazmi M et al (2016) Fe 3 O 4 nanoparticle redox system modulation via cell-cycle progression and gene expression in human mesenchymal stem cells. Environ Toxicol 31:901–912
121. Lai X, Wei Y, Zhao H et al (2015) The effect of Fe 2 O 3 and ZnO nanoparticles on cytotoxicity and glucose metabolism in lung epithelial cells. J Appl Toxicol 35:651–664
122. Berry CC, Wells S, Charles S et al (2004) Cell response to dextran-derivatised iron oxide nanoparticles post internalisation. Biomaterials 25:5405–5413
123. Kim JH, Sanetuntikul J, Shanmugam S et al (2015) Necrotic cell death caused by exposure to graphitic carbon-coated magnetic nanoparticles. J Biomed Mater Res – Part A 103:2875–2887
124. Shi M, Cheng L, Zhang Z et al (2015) Ferroferric oxide nanoparticles induce prosurvival autophagy in human blood cells by modulating the Beclin 1/Bcl-2/VPs34 complex. Int J Nanomedicine 10:207–216
125. Park EJ, Choi DH, Kim Y et al (2014) Magnetic iron oxide nanoparticles induce autophagy preceding apoptosis through mitochondrial damage and ER stress in RAW264.7 cells. Toxicol In Vitro 28:1402–1412
126. Buyukhatipoglu K, Clyne AM (2011) Superparamagnetic iron oxide nanoparticles change endothelial cell morphology and mechanics via reactive oxygen species formation. J Biomed Mater Res A 96:186–195
127. Gonnissen D, Qu Y, Langer K et al (2016) Comparison of cellular effects of starch-coated SPIONs and poly(lactic-co-glycolic acid) matrix nanoparticles on human monocytes. Int J Nanomedicine 11:5221–5236
128. Kalive M, Zhang W, Chen Y et al (2012) Human intestinal epithelial cells exhibit a cellular response indicating a potential toxicity upon exposure to hematite nanoparticles. Cell Biol Toxicol 28:343–368
129. Kedziorek DA, Muja N, Walczak P et al (2010) Gene expression profiling reveals early cellular responses to intracellular magnetic labeling with superparamagnetic iron oxide nanoparticles. Magn Reson Med 63:1031–1043
130. Alarifi S, Ali D, Alkahtani S et al (2013) Induction of oxidative stress, DNA damage, and apoptosis in a malignant human skin melanoma cell line after exposure to zinc oxide nanoparticles. Int J Nanomedicine 8:983–993
131. Liu Y, Wang J (2013) Effects of DMSA-Coated Fe3O4 nanoparticles on the transcription of genes related to iron and osmosis homeostasis. Toxicol Sci 131:521–536
132. Könczöl M, Weiss A, Stangenberg E et al (2013) Cell-cycle changes and oxidative stress response to magnetite in A549 human lung cells. Chem Res Toxicol 26:693–702
133. Shubayev VI, Pisanic TR, Jin S (2009) Magnetic nanoparticles for theragnostics. Adv Drug Deliv Rev 61:467–477
134. Soenen SJH, De Cuyper M (2009) Assessing cytotoxicity of (iron oxide-based) nanoparticles: an overview of different methods exemplified with cationic magnetoliposomes. Contrast Media Mol Imaging 4:207–219
135. Stroh A, Zimmer C, Gutzeit C et al (2004) Iron oxide particles for molecular magnetic resonance imaging cause transient oxidative stress in rat macrophages. Free Radic Biol Med 36:976–984
136. Shander A, Cappellini MD, Goodnough LT (2009) Iron overload and toxicity: the hidden risk of multiple blood transfusions. Vox Sang 97:185–197
137. Geppert M, Hohnholt M, Gaetjen L et al (2009) Accumulation of iron oxide nanoparticles by cultured brain astrocytes. J Biomed Nanotechnol 5:285–293
138. Toyokuni S (1996) Iron-induced carcinogenesis: the role of redox regulation. Free Radic Biol Med 20:553–566
139. Valko M, Rhodes CJ, Moncol J et al (2006) Free radicals, metals and antioxidants in oxidative stress-induced cancer. Chem Biol Interact 160:1–40

13 Detection of DNA Damage Induced by Cerium Dioxide Nanoparticles: From Models to Molecular Mechanism Activated

Tiago Alves Jorge de Souza, Thiago Lopes Rocha, and Leonardo Pereira Franchi

Abstract

This chapter will present an original effort to summarize the relevant data about the cyto-genotoxicity induced by cerium dioxide nanoparticles (nanoceria) in physiologically (in vivo and in vitro) relevant models. In this way, this chapter should be extremely useful to everyone who wants to plan their research and publishing their results. Massive application of nanoceria at different fields is increasing year after year, and it is urgent to address and discuss their use and its safety-related issues. Specifically, the nanoceria are being designed for nanomedicine, cosmetics, polishing materials and additives for automotive fuels. Their unique properties include the ability to absorb UV radiation, antioxidant potential and the rapid exchange of valence between Ce^{4+} and Ce^{3+} ions associated to oxygen storage. In this chapter, the state of the art regarding the physicochemical properties of nanoceria, nanogenotoxicity detected by in vitro and in vivo systems and the general aspects in the cyto-genotoxic mechanism of nanoceria are summarized. The cyto-genotoxicity will be discussed in terms of evaluations by Comet assay, Micronucleus test, DNA damage response and oxidative stress detected in cell culture systems and in vivo test. We also described the dose dependent cyto-genotoxic effects of nanoceria based on their physical-chemical nature. Paradoxically, these particles have been characterized as either pro-oxidant or anti-oxidant in dependence of microenvironment and physiological conditions such as pH. Finally, this chapter will contribute to point out aspects of the development of new in vitro and in vivo methodologies to detect cyto-genotoxic effects of the nanoceria.

T.A.J. de Souza
Department of Genetics, Faculty of Medicine of Ribeirão Preto; USP,
Ribeirão Preto, São Paulo, Brazil

T.L. Rocha
Institute of Tropical Pathology and Public Health,
Federal University of Goiás, Goiania, Goiás, Brazil

L.P. Franchi (✉)
Department of Chemistry, Faculty of Philosophy, Sciences and Letters of Ribeirão Preto; USP,
Ribeirão Preto, São Paulo, Brazil
e-mail: leonardofranchi@yahoo.com.br

Q. Saquib et al. (eds.), *Cellular and Molecular Toxicology of Nanoparticles*, Advances in Experimental Medicine and Biology 1048, https://doi.org/10.1007/978-3-319-72041-8_13

Keywords
Cerium oxide nanoparticles · Nanoceria · DNA damage · Nanotoxicity · Nanogenotoxicity

13.1 Introduction

The OECD (The Organization for Economic Co-operation and Development) presents a list of engineered nanoparticles (ENPs) that are representative of products on the market and therefore considers them as references for safety, risk and toxicology studies of ENPs. The list also includes ENPs that are currently produced on a large scale, such as fullerenes, carbon nanotubes (CNTs), and nanoparticles (NPs) from silver, iron, titanium dioxide, aluminum oxide, cerium dioxide, zinc oxide, silicon oxide NPs, dendrimers, gold and clays. According to the OECD, the materials presented are not prioritized and the list reflects only a momentary choice and other ENPs can be included as new needs appear (OECD 2010). From this list chapter will be focused on the studies of cyto- and genotoxicity of cerium dioxide NPs (nanoceria-CeO_2 NPs).

Among the properties of nanoceria which make them interesting include their ability to absorb UV radiation, their antioxidant properties and the rapid exchange of valence between Ce^{4+} and Ce^{3+} ions associated with oxygen storage [1]. Nanoceria are being designed for (bio)medical application, cosmetics, polishing materials and additives for automotive fuels [1, 2]. For example, the addition of 5 $mg{\cdot}L^{-1}$ of nanoceria (Envirox®) in diesel has been used to reduce fuel consumption by 5–8% and toxic gas emissions to the atmosphere by 15% [3]. Although these nanoceria products benefit their users and the global economy, they pose an increase risk to researchers, workers and consumers [4] and also to the environment [5–8].

The use of these particles as a fuel additive, for example, may increase the rate of nanoceria suspended in the air after disposal by cars and trucks. Actually, nanoceria have been detected by atmospheric monitoring in roadside airborne samples taken at London and Newcastle, United Kingdom [9, 10]. In addition, workers involved in the synthesis, transport or handling of ENPs are already being exposed to these materials which are mainly produced in the form of bulk powder [11]. Exposure to nanoceria may occur through inhalation, contact with the epidermis and ingestion. The inhalation has been the route most studied when compared to the others [12]. However, as they are also being designated for therapeutic and diagnostic applications, exposure to the digestive tract and blood vessels may also occur directly or indirectly [13] (Fig. 13.1).

The kinetics of nanoceria in biological systems is still poorly understood and the particokinetics was introduced [14] as an attempt to decipher the interaction and fate of NPs in biological systems which needs, urgently, too be addressed. In this context, the toxic effects of nanoceria can be discussed at molecular, cellular and organism levels. The difficulty of understanding the dynamics of nanoceria into organisms is due to the large number of variables involved, among them can be cited: [1] size, shape, and surface charge (zeta potential); [2] composition, density and stability of the material under biological/physiological conditions; [3] and properly the different physiological conditions such as the pH (mouth vs. stomach vs. intestine), the presence of the serum proteins (in the bloodstream). In particular, pH has an important influence on the cyto-genotoxicity of nanoceria [15].

At the molecular level an interaction between nanoceria and biomolecules (mainly to proteins from serum) known as corona effect changes the diffusion, agglomeration, sedimentation and interaction of these ENPs on cell surface, thus directly impacting the cell uptake and culminating in different toxic effects. The adsorptions of these proteins on ENPs are able to increase their dispersion and/or increase the cell uptake by facilitating the recognition by phagocytic cells.

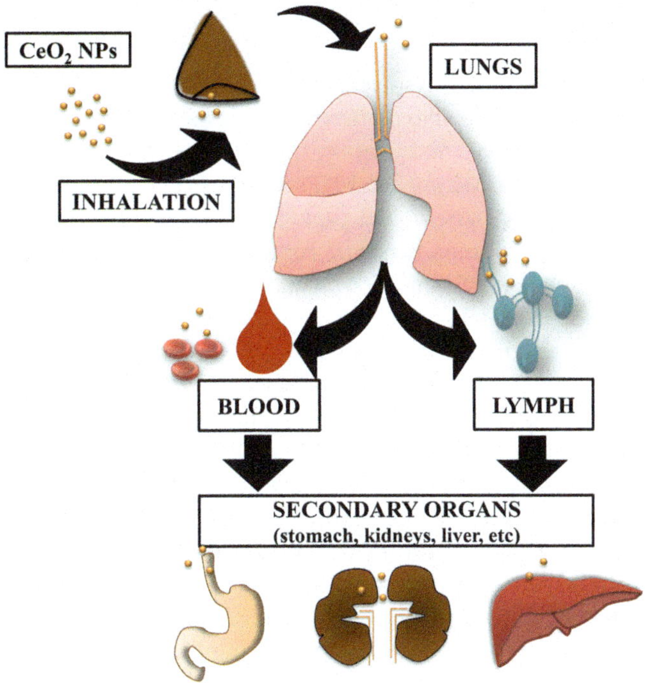

Fig. 13.1 Biodistribution of nanoceria into human organism after inhalation

Specially, nanoceria is involved in interactions with thiol-containing biomolecules [16]. At the cellular level, nanoceria kinetics may be affected by intracellular uptake pathways. Mammalian cells have five possible internalization pathways: phagocytosis (particles >0.5 μm), macropinocitosis (particles >1 μm), clathrin-mediated endocytosis (particles ~ 100–150 nn), caveolin-mediated endocytosis and caveolin- clathrin-independent endocytosis [17]. The studies performed until now detected nanoceria inside lysosomes (indicating an entry by phagocytosis or endocytosis), but nanoceria were also detected freely in cytoplasm showing that these NPs are also able pass through the membrane [18], another possibility is that nanoceria would disrupt the lysosome membrane and them escape directly to cytoplasm. Nanoceria can be internalized via clathrin-mediated endocytosis into cells. Moreover, the interaction between transferrin in the protein corona and the transferrin receptor in lung cells is involved in mediating the cellular entry of nanoceria. Interesting, under these conditions nanoceria does not affect cell growth, viability or metabolism. Otherwise in serum-free conditions these ENPs induce plasma membrane disruption and cause changes in cellular metabolism. Thus, this study provides significant insight into how different outcomes are dependent from microenvironment which nanoceria are found [19].

As we described before nanoceria effects are influenced by pH. Thus, once inside of cells these ENPs can entry in contact with different microenvironments and consequently to distinct pH. Into endosomes they will be submitted to an acidic pH (pH 6.2–6.5) and later in the lysosomes an even more acidic microenvironment (pH 4.5–5.5) [15, 18] which are distinct from pH in the cell cytoplasm (pH 7.2–7.8). Nanoceria was already co-localized with mitochondria, lysosomes and endoplasmic reticulum as well as being abundant in the cytoplasm and the nucleus [20]. Therefore, in all these microenvironments nanoceria will still be submitted to different pHs and moreover to several enzymatic reactions which are able to modify their properties and initial characteristics. Also in this context, exocytosis appears as a cellular excretion pathway for this ENPs [21]. However, until now few studies were dedicated to issues related to biotransformation, transport and secondary localization of nanoceria inside cells and their excretion or elimination. Thus, more efforts are urgently necessary to solve these questions.

Finally, the in vivo particokinetics is based on the absorption, distribution, metabolism and excretion (ADME) of the ENPs in the organism and also depends on the type of organism or model of exposure, exposure routes (inhalation, ingestion, dermal, injection), exposure time and nano-specific properties [15, 18, 22]. As previously described, nanoceria are able to move through biological systems, allowing a variety of interactions (cellular, biochemical and molecular), leading to toxicity to the organism. It is already known that ENPs are able to penetrate cells by different mechanisms, and subsequently enter the nucleus by diffusion through the nuclear membrane (if they are small enough); transport through the nuclear pore; or may enter the nucleus by chance when mitosis occurs (during the process of cell division the nuclear membrane dissolves and forms again later in daughter cells). In this sense, it is of greater concern the potential of nanoceria to induce damage to DNA molecule, and consequently a direct relation to the process of carcinogenesis [23]. Thus, in the next sections we are going to discuss genotoxic effects induced by nanoceria in cell culture systems and in the entire organism as well as physicochemical characteristics correlated to these effects.

13.2 Physical-Chemical Characteristics of Nanoceria Governing Biological Outcomes

Among the parameters that must be taken into consideration in nano-genotoxicology size is one of the most important, because it is a critical feature to determine the interaction of nanoceria with biological systems. A variety of methods can be applied for this purpose, such as Brunauer-Emmett-Teller (BET), transmission electron microscopy (TEM), scanning electron microscopy (SEM), atomic force microscopy (AFM) and dynamic light scattering (DLS – dynamic light scattering) [24, 25]. However, it is noteworthy that these different techniques may result in different sizes for the same particles. This is mainly due to the intrinsic differences of each methodology, among which we can mention: variations in the methods of preparation of the samples and procedures of operation of the instruments [24, 26].

Thus, with the idea of a practical explanation it is possible to divide the particle size into primary and secondary sizes. We can define as primary the post-synthesized size of the particles obtained, for example, with the TEM technique. The secondary one refers to the modifications undergone to this primary size, as that obtained in physiological solutions with the adsorption of proteins and nutrients on the particles – denominated corona effect; or by agglomeration due to non-covalent interaction such as van der Waals forces inducing formation of larger sized grouped ENPs which can be equally or heterogeneously distributed [27].

The DLS technique measures time-dependent fluctuations in the scattering intensity of the light produced by particles in Brownian motion, and the particle size is calculated by applying the Stokes-Einstein equation. The size obtained by DLS is generally larger than the values reported by other techniques, such as TEM, BET, etc. This can be attributed to the fact that the DLS measurement is performed on a set of particles in liquid solution and to that extent some solvent layers are included on the particles [28]. Furthermore, during DLS measurements, there is a tendency for the particles to form agglomerates in physiological solutions, so that the measured size is influenced by the agglomerated ENPs [24, 25]. Therefore, the DLS technique is an indispensable tool in toxicity studies of nanoceria. For it measures the hydrodynamic diameter under conditions very similar to in vitro or in vivo exposure conditions, providing data on the stability of the particle suspension relative to the time, type of culture medium or type of exposure condition [11, 29, 30].

The zeta potential (ζ) is a measure of the charge of repulsion or attraction between the ENPs, and is one of the known fundamental parameters that influence the stability of the suspensions and the formation of agglomerates.

Empirical data demonstrate that electrostatic stability requires a zeta potential of at least ± 30 mV, however stability with low zeta potential can occur by means of steric stabilization (by coating them superficially with a layer of polymer, protein or dendrimer). Patil et al. (2007) reported that nanoceria with positive zeta potential adsorb many proteins on their surface [31]. In addition, albumin present in fetal bovine serum (FBS) used to supplement culture medium to in vitro experiments, has a negative charge of −20 mV at physiological pH [32], and it is expected to influence the charge detected for nanoceria into in vitro conditions. Moreover, the use of this technique has demonstrated that ENPs suspended in culture media form agglomerates influencing responses in the cyto-genotoxicity assays [33].

Thus, it is important to note that the deposition of the ENPs on the cells has a direct impact on the concentration unit of NPs *per* cell. In this context, there is a definition of three different dose levels for toxicology of NPs in vitro: administered dose, delivered dose and cellular dose. The administered dose or concentration in culture medium is the most common measure in classical toxicology (e.g.: mass/volume); but unlike soluble chemicals, agglomerated nanoceria can be deposited on the cells; this effect could be determined by SEM analyzes which detected a diffuse deposition of these nanoceria agglomerates on cell surfaces of fibroblast cells [34]. Moreover, the cell density will influence the direct contact with those ENPs and their agglomerates (delivered dose) [35]. Finally, the cellular dose is influenced by the mechanism of uptake of the ENPs [endocytosis/phagocytosis or nanopenetration (diffusion through the plasma membrane)] by the cells [36, 37].

13.3 DNA Damage Response Pathways Integrated to Cellular Outcomes Detected by In Vitro Systems

Since ENPs can enter into nucleus, a direct interaction between them and DNA can lead to physical damage to the genetic material. Nanoceria (primary size of 2.5 nm and hydrodynamic size of 580 nm) was identified into nucleus of macrophage cells after 24 h of treatment, but no signs of necrosis, apoptosis or chromatic condensation were observed [38]. On the other hand, Benameur et al. (2014) observed that human fibroblasts treated with cells cerium oxide (CeO2) NPs presented a statistically significant ($p < 0.001$) increasing of micronuclei frequency when compared to untreated cells, which points to the clastogenic potential of these NPs [39].

Alternatively, DNA damage can occur through indirect mechanism which there is no physical interaction with DNA molecule, but with proteins involved in the process of replication and/or DNA damage repair. In addition, they may activate other cellular responses that induce genotoxicity, such as reactive oxygen species (ROS) production, oxidative stress, inflammation and erroneous cell signaling [4]. Currently, ROS generation is considered to be the major cause of ENP cytotoxicity [40]. As a consequence, exposure to ENPs can affect, through the formation of ROS, the cellular signaling cascade that controls cell proliferation, inflammatory processes and cell death. The induction of oxidative DNA damage has been shown to be one of the effects that govern the genotoxicity of ENPs [41, 42].

The literature shows a conflicting data about the cyto-genotoxic effects induced by nanoceria [43–45] Paradoxically, nanoceria have also been characterized as either pro-oxidant or antioxidant agent [46]. A genotoxic effect was observed in human fibroblast exposed to nanoceria (primary size of 25 nm and hydrodynamic size of 580 nm) by a period of 24 h detected by Comet Assay using hOGG1-modified protocol to identify oxidative DNA damage [34]. Mittal and Pandey (2014) also reported the induction of oxidative DNA damage (Comet Assay with FPG) in lung (A549) cells after exposure to 25 μg/mL, 50 μg/mL, and 100 μg nanoceria for 6 h [47]. Auffan et al. (2009) also described the induction of DNA damage (Comet Assay) detected in primary fibroblasts exposed to nanoceria (6×10^{-6} – 1.2 g/L) for 24 h and with the addition of an antioxidant to the treatment a reduction of the damage was observed [48].

Pierscionek et al. (2010) reported the absence of genotoxic effect in human lens epithelial cells exposed to 10, 20 and 100 µg/mL during 48 h; however increasing the exposure time (72 h) there was a significant difference in DNA damage between all nanoceria dosages and the positive control [49], indicating that the genotoxicity of these ENPs after in vitro exposure is time-dependent and related to dissolution rate.

DNA fragmentation is also revealed by flow cytometry using iodide bromide as DNA marker and characterized as subG1 cell population. The increase of cells in subG1, after treatment with nanoceria was reported to human bronchial epithelial cells after exposure to 45 nm nanoceria (40 µg/mL) for 24 h [50]. An increase in subG1 cell population was also observed to human fibroblast after exposure to these ENPs for 24 h; however, this effect was not observed at later times of incubation (48 and 72 h), demonstrating a cell recovery of these transient effect [34]. Other toxicological studies have shown a decrease in the viability of mammalian cells as a result of ROS formation induced by nanoceria exposure [3, 48, 51]. The antioxidant response pathway mediated by p38-NRF2 proteins was also activated after exposure to these ENPs [50]. The mitogen-activated protein kinase (MAPK) pathway was also triggered in response to the oxidative stress (detected by fluorometric assay using intracellular oxidation of 2,7-dichlorofluoroscein diacetate -DCFH-DA as marker) in liver cell carcinoma (SMMC-7721). Specially, phosphorylation of the ERK, JNK and p38 proteins (the three main components of the MAPK pathway) were up-regulated by nanoceria treatment. The phosphorylation of these proteins occurs under stress conditions and regulates survival, death and cellular adaptation to stressful stimuli [44].

As consequence of excessive DNA damage cell undergoes cell death mainly by apoptosis. Nanoceria (5 and 10 µg/mL) was able to induce apoptosis characterized by activation of Bax, loss of mitochondrial membrane potential and DNA fragmentation in human monocytes after 40 h of the exposure, while caspase activation or ROS production were not observed [52]. Autophagy can also be triggered as resistance mechanism to the stress stimuli generated by these ENPs. Moreover, pharmacological inhibition of TP53 resulting in an increased autophagic response with apoptosis levels remaining unchanged after nanoceria treatment [45]. All these results together demonstrated the role of the pro-oxidant effect triggered by nanoceria resulting in DNA damage and consequently leading to cell death mainly by apoptosis.

Although several deleterious effects of nanoceria have been attributed to their oxidative potential, paradoxically, there are also reports of antioxidant and anti-inflammatory properties for these ENPs [53, 54]. A radioprotective activity was described [55, 56], which suggested that these ENPs act as ROS scavenger and also induce increased expression of the superoxide dismutase 2 (SOD2) enzyme, counterbalancing the radiation-induced oxidative stress. In this sense, nanoceria was described as a tool to control oxidative stress in cardiac progenitor cells (CPCs) [43]. It was demonstrated that treatment with nanoceria did not affect the growth and function of CPCs, nor promoted ROS generation. In fact, these ENPs were able to protect CPCs from H_2O_2-induced cytotoxicity. Based on these results, some authors suggest the use of nanoceria in therapies, since many pathologies are linked to oxidative stress and inflammatory processes, and these ENPs could have benefits on these effects [57–62]. Therefore, the dual role of nanoceria which present both oxidant and antioxidant properties makes these ENPs important for biological applications. This double role is influenced by the presence of oxygen vacancies and, mainly, by the redox cycle (coexistence of Ce^{3+} and Ce^{4+} ions) occurring on the ENP surface. For example, the concentration of Ce^{3+} ions is described as responsible for the anti-apoptotic effect to nanoceria observed in two lymphoid cell lines (U937 and Jurkat) treated with etoposide [57].

Considering that oxidative stress is one of the main toxicity mechanisms of nanoceria, particular attention should be given for the potential of Ce^{3+} and Ce^{4+} ions to mimic the action of superoxide dismutase (SOD) and catalase enzymes, respectively [46]. When there are a higher proportion of Ce^{3+} ions, the particles exhibit properties similar to that of the SOD inducing H_2O_2 production. However, when Ce^{4+} ions are found in a higher proportion into nanoceria they may mimics properties of the catalase enzyme, which is capable of degrading H_2O_2. Therefore, the ratio of Ce^{3+}/Ce^{4+} ions defines the specific "enzymatic" properties of these ENPs [46]. Furthermore, taking into account the great cytotoxicity of H_2O_2, the imbalance of redox cycle is intrinsically related with nanoceria toxicity. On the other hand, when the two nanoceria activities are coordinated, that is, when the degradation rate of H_2O_2 is greater than or equal to its production, these NPs present an antioxidant activity [46].

As mentioned above, the nanoceria dissolution kinetics is intrinsically related to ROS production and consequently to the toxicity of these NPs. In this context, for example, it has been shown that the increase of Ce^{3+} ions on the surface of CeO_2 NPs induce DNA damage and lipid peroxidation. In addition, it has been demonstrated that the administration of Ce^{3+} ions in cells and animals damaged several proteins, such as the enzyme superoxide dismutase, which is directly related to ROS production [39, 63–65]. According to reports in the literature, ENP sizes may influence the proportion of Ce^{3+}/Ce^{4+} ions [66] and, consequently, affect the properties of nanoceria which would reflect directly on biological responses after exposure to these ENPs. Moreover, it is very important to note that nanoceria NPs toxicity is also related with their dissolution kinetics in water. Thus, the employment of agents to maintain the stability of these NPs in aqueous dispersions is essential to ensure their safety use [67].

It is described enzyme mimetic properties of nanoceria that it is also related to the coexistence of Ce^{3+} and Ce^{4+} ions [46]. Moreover, the ratio of Ce^{3+}/Ce^{4+} ions defines the specific "enzymatic" properties and toxicity of these ENPs. When there are a higher proportion of Ce^{3+} ions, the particles exhibit properties similar to that of the enzyme SOD (catalyzing superoxide molecules). However, when Ce^{4+} ions are found in a higher proportion into nanoceria it may mimics properties of the catalase enzyme (capable of degrading H_2O_2). It is important to report that H_2O_2 production occurs when nanoceria act as SOD, and H_2O_2 has a potent cytotoxicity [46]. As the two activities (SOD and catalase) are present in the nanoceria, the H_2O_2 produced by the SOD function enters the catalase mimetic cycle, generating H_2O and O_2 as final product. However, the antioxidant property of nanoceria is only effective when the two activities are coordinated, that is, when the degradation rate of H_2O_2 is greater than or equal to its production. If this rule is not obeyed there is an imbalance and other effects can be observed [46]. According to reports in the literature, ENP sizes may influence the proportion of Ce^{3+}/Ce^{4+} ions [66] and, consequently, affect the properties of nanoceria which would reflect directly on biological responses and toxicity after exposure to these ENPs. In this sense, the revised data indicated that the antioxidant property and toxicity of nanoceria is related to size and to dissolution kinetics of nanoceria in the aqueous medium.

The properties of nanoceria may also be influenced by pH including a key role in the cytotoxic profile of nanoceria [68]. Asati et al. (2010) reported the localization of nanoceria (1.0 mM) in the cytoplasm and vesicles of the lysosomes type upon entry into the cell after exposure by 3 h. Minimal effects are detected when the ENPs are lodged in the cytoplasm. In contrast, nanoceria exhibit a strong cytotoxic effect when located on lysosomes of cancer cells. This effect is due to a pH ranging from acidic (4.5–5.0), in the interior of the lysosomes compared to the slightly basic cytosol (pH 7.2) that stimulates another mimetic property of nanoceria. Therefore, the microenvironment plays a fundamental role in the toxic aspects triggered by these ENPs [68].

13.4 Nanogenotoxicity Detected by In Vivo Systems

The genotoxicity induced by in vivo exposure to nanoceria may arise through direct and indirect mechanisms and are dependent of nano-specific properties, model system, concentration, time and route of exposure. Furthermore, the in vivo studies are more complex compared to in vitro one's due to distinct ENP behavior into organisms, such as formation of the nanoparticle-protein corona, as well as limitation to detect the exactly concentration reaching each tissue and cell and its correlation with toxic effects in a tissue/cell specific pattern [4].

Figure 13.2 describes the potential mechanism of genotoxicity of nanoceria after in vivo exposure. After uptake into the cell nucleus, ENPs can interact with the DNA and/or DNA-related proteins and induced direct damage to the DNA molecule. On the other hand, the indirect mechanisms involved in the nano-genotoxicity include toxic effects of ions released from soluble ENPs, ROS generation, oxidative stress, inflammatory response and abnormal cellular signalization. Furthermore, the long-term exposure to ENPs may lead to chronic inflammatory responses, which have the potential to induce oxidative DNA damage [69]. Although the epigenetic modifications induced by nanoceria after in vivo exposure remain unknown, recent studies indicated the potential of ENPs to induce epigenetic alterations, such as DNA methylation, histone modifications and microRNA mediated mechanisms [70, 71], indicating potential epigenetic effects of nanoceria. Garaud et al. also indicate that further studies about the co-exposures with pro-oxidant substances and nanoceria should also be conducted to confirm their potential antioxidant properties. In this sense, the revised data (Table 13.1) indicated that more studies about the in vivo genotoxicity of nanoceria are urgently needed [72].

13.5 General Aspects in the Cyto-Genotoxic Study of Nanoceria

Several different cell lines can be used in the cytotoxicity and genotoxicity studies of nanoceria, such as macrophage cells, which have a higher phagocytic activity than other cells lines [83, 84]. Due to this ability, these cells can phagocyte nanoceria more efficiently than other cells. Moreover, macrophages can cross biological barriers, determining the biopersistence of foreign particles and initiating inflammatory responses. Thus, macrophages [85] can be considered as a valuable cell model for assessing the nanoceria toxicity.

Moreover, the vast majority of in vitro studies used a two-dimensional cell layer approach. This type of approach turns out to be quite different from the natural in vivo three-dimensional environment of cells development, which is composed by different cell types. This is one of the reasons why some in vitro results are not equivalent to those obtained in vivo [86]. In this context, efforts have been made to combine the benefits derived from the use of in vitro methodologies with a microenvironment that more accurately represents the functioning of the organism [87]. Thus, multicellular 3D systems have been developed using different cell types and extracellular matrices in order to reproduce the complexity of an in vivo environment, these properties hardly affect the uptake and cell contact to nanoceria. In this context, some studies [88] have shown that the use of these methodologies effectively simulate the organic environment in which cells originally developed, providing more realistic morphological and physiological patterns than normal in vitro techniques. Thus, this kind of approach in an interdisciplinary methodology application is strongly required to a completely evaluation of the effects induced by nanoceria into biological systems.

13.6 Conclusion

In conclusion, the potential genotoxic effects of nanoceria after in vitro and in vivo exposure were revised in this study. The revised data indicates that the nanoceria genotoxicity is mediated by direct and indirect mechanisms, which are dependent on their nano-specific properties, experimental design, microenvironment and physiological conditions. Moreover, new models must be employed in order to elucidate the molecular mechanisms of nanoceria toxicity. In addition, efforts should be made to standardize

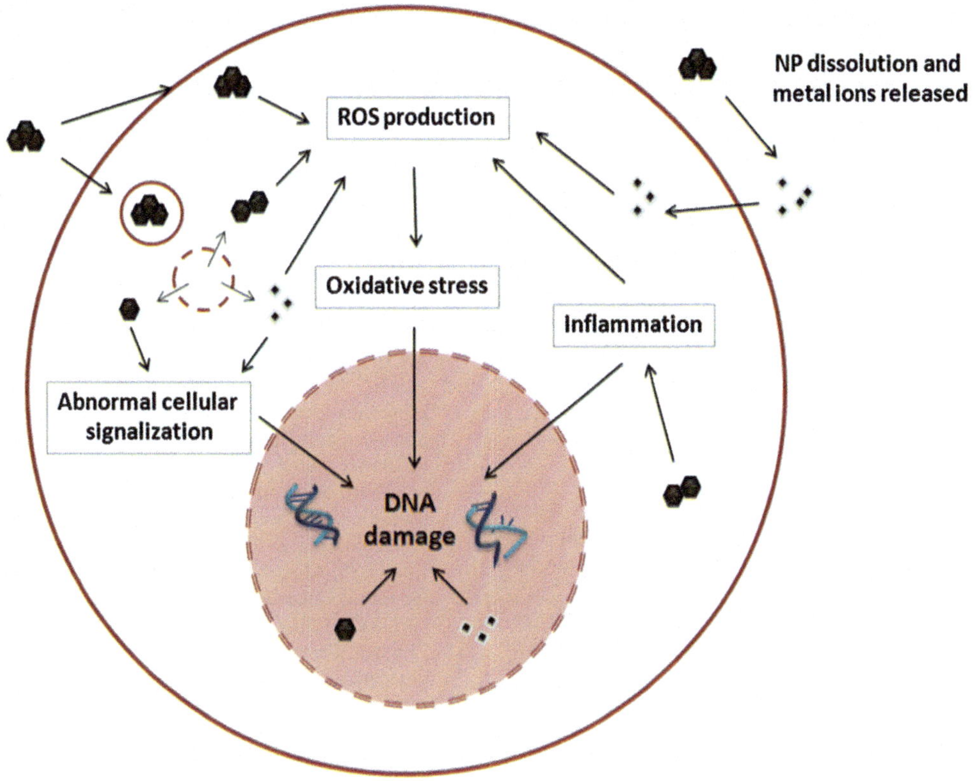

Fig. 13.2 General scheme illustrating the potential mechanism of genotoxicity (direct and indirect) of nanoceria after in vivo exposure. The epigenetic modifications induced by nanoceria deserve further studies and were not included in the scheme

Table 13.1 General data about nanoceria toxicity studies which employed in vivo systems

Species	Results	NP diameter	References
Caenorhabditis elegans	Induced ROS accumulation, oxidative damage and lead to a decreased lifespan	8.5 +/− 1.5 nm	[73]
Danio rerio	Caused no significant adverse effects	2.8–11.6 nm	[74]
Drosophila melanogaster	Caused no significant adverse effects	<25 nm	[75]
Lactuca sativa	Lettuce treated with 100 mg·kg-1 of CeO_2NPs grew significantly faster than other treatment groups	<25 nm	[76]
Mus musculus	Induced inflammation	16–22 nm	[77]
Mus musculus	Decreased fertilization rate	7 nm	[78]
Mus musculus	Increased the number of neutrophils and cytokines levels	15–30 nm	[79]
Paracentrotus lividus	Induced strictly packed membranes, reduced levels of GRP78	50–105 nm	[80]
Rattus norvegicus	Reduce apoptosis and lipid peroxidation	3–5 nm	[81]
Rattus norvegicus	Induced ROS generation and decreased NO production	4 ± 1 nm	[82]

the synthesis methods in order to control the size, shape, composition and surface loading of these ENPs as well as the determination of the nanoceria properties after entering in biological systems. These efforts will allow the determination of the ionic dissociation kinetics of nanoceria and a more accurate estimation of Ce^{3+}/Ce^{4+} ions ratio, allowing a better understanding of the cyto-genotoxicity of ENPs resulting in the safe application of nanoceria for therapeutic purposes as well as their use in industrial applications.

References

1. Ivanov VK, Shcherbakov AB, Usatenko AV (2009) Structure-sensitive properties and biomedical applications of nanodispersed cerium dioxide. Russ Chem Rev 78:855–871
2. Jung CR, Han J, Nam SW et al (2004) Selective oxidation of CO over CuO-CeO_2 catalyst: effect of calcination temperature. Catal Today 93–5:183–190
3. Park E-J, Choi J, Park Y-K et al (2008b) Oxidative stress induced by cerium oxide nanoparticles in cultured BEAS-2B cells. Toxicology 245:90–100
4. Singh N, Manshian B, Jenkins GJS et al (2009) NanoGenotoxicology: the DNA damaging potential of engineered nanomaterials. Biomaterials 30:3891–3914
5. Bour A, Mouchet F, Cadarsi S et al (2017) CeO_2 nanoparticle fate in environmental conditions and toxicity on a freshwater predator species: a microcosm study. Environ Sci Pollut Res 24:1–9
6. Gui X, Rui M, Song Y et al (2017) Phytotoxicity of CeO_2 nanoparticles on radish plant (Raphanus sativus). Environ Sci Pollut Res Int 24(15):13775–13781
7. Spielman-Sun E, Lombi E, Donner E et al (2017) Impact of surface charge on cerium oxide nanoparticle uptake and translocation by wheat (Triticum aestivum). Environ Sci Technol 51(13):7361–7368
8. Zhao X, Yu M, Xu D et al (2017) Distribution, bioaccumulation, trophic transfer, and influences of CeO_2 nanoparticles in a constructed aquatic food web. Environ Sci Technol 51(9):5205–5214
9. Park B, Martin P, Harris C et al (2007) Initial in vitro screening approach to investigate the potential health and environmental hazards of Enviroxtrade mark – a nanoparticulate cerium oxide diesel fuel additive. Part Fibre Toxicol 4:12
10. Park B, Donaldson K, Duffin R et al (2008a) Hazard and risk assessment of a nanoparticulate cerium oxide-based diesel fuel additive – a case study. Inhal Toxicol 20:547–566
11. Hussain SM, Braydich-Stolle LK, Schrand AM et al (2009) Toxicity evaluation for safe use of nanomaterials: recent achievements and technical challenges. Adv Mater 21:1549–1559
12. Bouillard JX, Vignes A (2014) Nano-Evaluris: an inhalation and explosion risk evaluation method for nanoparticle use. Part I: description of the methodology. J Nanopart Res 16:2149
13. Seaton A, Donaldson K (2005) Nanoscience, nanotoxicology, and the need to think small. Lancet 365:923–924
14. Zhu M-T, Feng W-Y, Wang Y et al (2009) Particokinetics and extrapulmonary translocation of intratracheally instilled ferric oxide nanoparticles in rats and the potential health risk assessment. Toxicol Sci 107:342–351
15. Zhu M, Nie G, Meng H et al (2013a) Physicochemical properties determine nanomaterial cellular uptake, transport, and fate. Acc Chem Res 46:622–631
16. Rollin-Genetet F, Seidel C, Artells E et al (2015) Redox reactivity of cerium oxide nanoparticles induces the formation of disulfide bridges in thiol-containing biomolecules. Chem Res Toxicol 28:2304–2312
17. Chou LYT, Ming K, Chan WCW (2011) Strategies for the intracellular delivery of nanoparticles. Chem Soc Rev 40:233–245
18. Zhu M, Perrett S, Nie G (2013b) Understanding the particokinetics of engineered nanomaterials for safe and effective therapeutic applications. Small 9:1619–1634
19. Mazzolini J, Weber RJM, Chen HS et al (2016) Protein corona modulates uptake and toxicity of nanoceria via clathrin-mediated endocytosis. Biol Bull 231:40–60
20. Singh S, Kumar A, Karakoti A et al (2010) Unveiling the mechanism of uptake and sub-cellular distribution of cerium oxide nanoparticles. Mol BioSyst 6:1813–1820
21. Sakhtianchi R, Minchin RF, Lee K-B et al (2013) Exocytosis of nanoparticles from cells: role in cellular retention and toxicity. Adv Colloid Interf Sci 201–202:18–29
22. Bai S, Zhao J, Wang L et al (2010) SO_2-promoted reduction of NO with NH_3 over vanadium molecularly anchored on the surface of carbon nanotubes. Catal Today 158:393–400
23. Kumari M, Kumari SI, Grover P (2014) Genotoxicity analysis of cerium oxide micro and nanoparticles in Wistar rats after 28 days of repeated oral administration. Mutagenesis 29:467–479
24. Dhawan A, Sharma V, Parmar D (2009) Nanomaterials: a challenge for toxicologists. Nanotoxicology 3:1–9
25. Dhawan A, Sharma V (2010) Toxicity assessment of nanomaterials: methods and challenges. Anal Bioanal Chem 398:589–605
26. Lin P-C, Lin S, Wang PC et al (2014) Techniques for physicochemical characterization of nanomaterials. Biotechnol Adv 32:711–726
27. Tenzer S, Docter D, Rosfa S et al (2011) Nanoparticle size is a critical physicochemical determinant of the human blood plasma corona: a comprehensive quantitative proteomic analysis. ACS Nano 5:7155–7167

28. Xu M, Li J, Iwai H et al (2012) Formation of nano-bio-complex as nanomaterials dispersed in a biological solution for understanding nanobiological interactions. Sci Rep 2:406. (2012). https://doi.org/10.1038/srep00406
29. Murdock RC, Braydich-Stolle L, Schrand AM et al (2008) Characterization of nanomaterial dispersion in solution prior to In vitro exposure using dynamic light scattering technique. Toxicol Sci 101:239–253
30. Montes-Burgos I, Walczyk D, Hole P et al (2010) Characterisation of nanoparticle size and state prior to nanotoxicological studies. J Nanopart Res 12:47–53
31. Patil S, Sandberg A, Heckert E et al (2007) Protein adsorption and cellular uptake of cerium oxide nanoparticles as a function of zeta potential. Biomaterials 28:4600–4607
32. Rezwan K, Meier LP, Rezwan M et al (2004) Bovine serum albumin adsorption onto colloidal Al_2O_3 particles: a new model based on zeta potential and UV-vis measurements. Langmuir 20:10055–10061
33. Prasad RY, Wallace K, Daniel KM et al (2013) Effect of treatment media on the agglomeration of titanium dioxide nanoparticles: impact on genotoxicity, cellular interaction, and cell cycle. ACS Nano 7:1929–1942
34. Franchi LP, Manshian BB, de Souza TAJ et al (2015) Cyto- and genotoxic effects of metallic nanoparticles in untransformed human fibroblast. Toxicol In Vitro 29:1319–1331
35. Teeguarden J, Hinderliter P, Orr G et al (2007) Particokinetics in vitro: dosimetry considerations for in vitro nanoparticle toxicity assessments. Toxicol Sci 95:300–312
36. Kuhnel D, Busch W, Meissner T et al (2009) Agglomeration of tungsten carbide nanoparticles in exposure medium does not prevent uptake and toxicity toward a rainbow trout gill cell line. Aquat Toxicol 93:91–99
37. Greulich C, Diendorf J, Simon T et al (2011) Uptake and intracellular distribution of silver nanoparticles in human mesenchymal stem cells. Acta Biomater 7:347–354
38. Szymanski CJ, Munusamy P, Mihai C et al (2015) Shifts in oxidation states of cerium oxide nanoparticles detected inside intact hydrated cells and organelles. Biomaterials 62:147–154
39. Benameur L, Auffan M, Cassien M et al (2014) DNA damage and oxidative stress induced by CeO_2 nanoparticles in human dermal fibroblasts: evidence of a clastogenic effect as a mechanism of genotoxicity. Nanotoxicology 9(6):696–705
40. Sarkar A, Ghosh M, Sil PC (2014) Nanotoxicity: oxidative stress mediated toxicity of metal and metal oxide nanoparticles. J Nanosci Nanotechnol 14:730–743
41. Kain J, Karlsson HL, Möller L (2012) DNA damage induced by micro- and nanoparticles-interaction with FPG influences the detection of DNA oxidation in the comet assay. Mutagenesis 27:491–500
42. Wan R, Mo Y, Feng L et al (2012) DNA damage caused by metal nanoparticles: involvement of oxidative stress and activation of ATM. Chem Res Toxicol 25:1402–1411
43. Pagliari F, Mandoli C, Forte G et al (2012) Cerium oxide nanoparticles protect cardiac progenitor cells from oxidative stress. ACS Nano 6:3767–3775
44. Cheng G, Guo W, Han L et al (2013) Cerium oxide nanoparticles induce cytotoxicity in human hepatoma SMMC-7721 cells via oxidative stress and the activation of MAPK signaling pathways. Toxicol In Vitro 27:1082–1088
45. Hussain S, Garantziotis S (2013) Interplay between apoptotic and autophagy pathways after exposure to cerium dioxide nanoparticles in human monocytes. Autophagy 9:101–103
46. Xu C, Qu X (2014) Cerium oxide nanoparticle: a remarkably versatile rare earth nanomaterial for biological applications. Npg Asia Mater 6:e90
47. Mittal S, Pandey AK (2014) Cerium oxide nanoparticles induced toxicity in human lung cells: role of ROS mediated DNA damage and apoptosis. Biomed Res Int 891934:14
48. Auffan M, Rose J, Orsiere T et al (2009) CeO_2 nanoparticles induce DNA damage towards human dermal fibroblasts in vitro. Nanotoxicology 3:161–171
49. Pierscionek BK, Li YB, Yasseen AA et al (2010) Nanoceria have no genotoxic effect on human lens epithelial cells. Nanotechnology 21:035102
50. Eom H-J, Choi J (2009) Oxidative stress of CeO_2 nanoparticles via p38-Nrf-2 signaling pathway in human bronchial epithelial cell, Beas-2B. Toxicol Lett 187:77–83
51. Brunner TJ, Wick P, Manser P et al (2006) In vitro cytotoxicity of oxide nanoparticles: comparison to asbestos, silica, and the effect of particle solubility. Environ Sci Technol 40:4374–4381
52. Hussain S, Al-Nsour F, Rice AB et al (2012) Cerium dioxide nanoparticles induce apoptosis and autophagy in human peripheral blood monocytes. ACS Nano 6:5820–5829
53. Tarnuzzer RW, Colon J, Patil S et al (2005) Vacancy engineered ceria nanostructures for protection from radiation-induced cellular damage. Nano Lett 5:2573–2577
54. Chen J, Patil S, Seal S et al (2006) Rare earth nanoparticles prevent retinal degeneration induced by intracellular peroxides. Nat Nanotechnol 1:142–150
55. Colon J, Herrera L, Smith J et al (2009) Protection from radiation-induced pneumonitis using cerium oxide nanoparticles. Nanomedicine 5:225–231
56. Colon J, Hsieh N, Ferguson A et al (2010) Cerium oxide nanoparticles protect gastrointestinal epithelium from radiation-induced damage by reduction of reactive oxygen species and upregulation of superoxide dismutase 2. Nanomedicine: NBM 6:698–705
57. Celardo I, De Nicola M, Mandoli C et al (2011a) Ce^{3}+ ions determine redox-dependent anti-apoptotic effect of cerium oxide nanoparticles. ACS Nano 5:4537–4549

58. Celardo I, Pedersen JZ, Traversa E et al (2011b) Pharmacological potential of cerium oxide nanoparticles. Nanoscale 3:1411–1420
59. Celardo I, Traversa E, Ghibelli L (2011c) Cerium oxide nanoparticles: a promise for applications in therapy. J Exp Ther Oncol 9:47–51
60. Alili L, Sack M, von Montfort C et al (2013) Downregulation of tumor growth and invasion by redox-active nanoparticles. Antioxid Redox Signal 19:765–778
61. Chaudhury K, Babu KN, Singh AK et al (2013) Mitigation of endometriosis using regenerative cerium oxide nanoparticles. Nanomedicine 9:439–448
62. Das S, Dowding JM, Klump KE et al (2013) Cerium oxide nanoparticles: applications and prospects in nanomedicine. Nanomedicine 8:1483–1508
63. Cheng Y, Li Y, Li R et al (2000) Orally administrated ceriumchloride induces the conformational changes of rat hemoglobin, the hydrolysis of 2,3-DPG and the oxidation of heme-Fe(II), leading to changes of oxygen affinity. Chem Biol Interact 125:191–208
64. Kuchma MH, Komanski CB, Colon J et al (2010) Phosphate ester hydrolysis of biologically relevant molecules by cerium oxide nanoparticules. Nanomedicine 6:738–744
65. Liu P, Ma L, Yin S et al (2008) Effect of Ce^{3+} on conformation and activity of superoxide dismutase. Biol Trace Elem Res 125:170–178
66. Deshpande S, Patil S, Kuchibhatla S et al (2005) Size dependency variation in lattice parameter and valency states in nanocrystalline cerium oxide. Appl Phys Lett 87:133113
67. Yokel RA, Hancock ML, Grulke EA et al (2017) Nanoceria dissolution and carboxylic acid stabilization in aqueous dispersions. FASEB J 31(1):lb624
68. Asati A, Santra S, Kaittanis C et al (2010) Surface-charge-dependent cell localization and cytotoxicity of cerium oxide nanoparticles. ACS Nano 4:5321–5331
69. Rim KT, Song SW, Kim HY (2013) Oxidative DNA damage from nanoparticle exposure and its application to workers' health: a literature review. Saf Health Work 4:177–186
70. Stoccoro A, Karlsson HL, Coppedè F et al (2013) Epigenetic effects of nano-sized materials. Toxicology 313:3–14
71. Smolkova B, El Yamani N, Collins AR et al (2015) Nanoparticles in food. Epigenetic changes induced by nanomaterials and possible impact on health. Food Chem Toxicol 77:64–73
72. Garaud M, Trapp J, Devin S et al (2015) Multibiomarker assessment of cerium dioxide nanoparticle ($nCeO_2$) sublethal effects on two freshwater invertebrates, *Dreissena polymorpha* and *Gammarus roeseli*. Aquat Toxicol 158:63–74
73. Zhang H, He X, Zhang Z et al (2011) Nano-CeO_2 exhibits adverse effects at environmental relevant concentrations. Environ Sci Technol 45(8):3725–3730
74. Wehmas LC, Anders C, Chess J et al (2015) Comparative metal oxide nanoparticle toxicity using embryonic zebrafish. Toxicol Rep 2:702–715
75. Alaraby M, Hernández A, Annangi B et al (2015) Antioxidant and antigenotoxic properties of CeO2 NPs and cerium sulphate: studies with drosophila melanogaster as a promising in vivo model. Nanotoxicology 9(6):749–759
76. Gui X, Zhang Z, Liu S et al (2015) Fate and phytotoxicity of CeO_2 nanoparticles on lettuce cultured in the potting soil environment. PLoS One 10(8):e0134261
77. Poma A, Ragnelli AM, de Lapuente J et al (2014) In vivo inflammatory effects of ceria nanoparticles on CD-1 mouse: evaluation by hematological, histological, and TEM analysis. J Immunol Res 2014:361419
78. Preaubert L, Courbiere B, Achard V et al (2016) Cerium dioxide nanoparticles affect in vitro fertilization in mice. Nanotoxicology 10(1):111–117
79. Kim YH, Boykin E, Stevens T et al (2014) Comparative lung toxicity of engineered nanomaterials utilizing in vitro, ex vivo and in vivo approaches. J Nanobiotechnol 12:47
80. Falugi C, Aluigi MG, Chiantore MC et al (2012) Toxicity of metal oxide nanoparticles in immune cells of the sea urchin. Mar Environ Res 76:114–121
81. Gao W, Wei X, Wang X et al (2016) A competitive coordination-based CeO_2 nanowire-DNA nanosensor: fast and selective detection of hydrogen peroxide in living cells and in vivo. Chem Commun (Camb) 52(18):3643–3646
82. Minarchick VC, Stapleton PA, Fix NR et al (2015) Intravenous and gastric cerium dioxide nanoparticle exposure disrupts microvascular smooth muscle signaling. Toxicol Sci 144(1):77–89
83. Lanone S, Rogerieux F, Geys J et al (2009) Comparative toxicity of 24 manufactured nanoparticles in human alveolar epithelial and macrophage cell lines. Part Fibre Toxicol 6:14
84. Eidi H, Joubert O, Attik G et al (2010) Cytotoxicity assessment of heparin nanoparticles in NR8383 macrophages. Int J Pharm 396:156–165
85. Nguea HD, de Reydellet A, Lehuede P et al (2008) A new in vitro cellular system for the analysis of mineral fiber biopersistence. Arch Toxicol 82:435–443
86. Edmondson R, Broglie JJ, Adcock AF et al (2014) Three-dimensional cell culture systems and their applications in drug discovery and cell-based biosensors. Assay Drug Dev Technol 12:207–218
87. Ravi M, Paramesh V, Kaviya SR et al (2015) 3D cell culture systems: advantages and applications. J Cell Physiol 230:16–26
88. Baharvand H, Hashemi SM, Ashtian SK et al (2006) Differentiation of human embryonic stem cells into hepatocytes in 2D and 3D culture systems in vitro. Int J Dev Biol 50:645–652

14 Mechanisms Underlying Neurotoxicity of Silver Nanoparticles

Lidia Strużyńska and Joanna Skalska

Abstract

The potent antimicrobial properties of nanoparticulate silver (AgNPs) have led to broad interest in using them in a wide range of commercial and medical applications. Although numerous in vivo and in vitro studies have provided evidence of toxic effects, rapid commercialization of AgNP-based nanomaterials has advanced without characterization of their potential environmental and health hazards. There is evidence that AgNPs can be translocated from the blood to the brain, regardless the route of exposure, and accumulate in the brain over time. As the brain is responsible for basic physiological functions and controls all human activities, it is important to assess the hazardous influence of AgNPs released from widely used nanoproducts and possible side effects of AgNP-based therapies. A number of studies have suggested that the size, shape and surface coating, as well as rates of silver ion release and interactions with proteins are the key factors determining the neurotoxicity of AgNPs. AgNPs target endothelial cells forming the blood-brain barrier, neurons and glial cells and leads finally to oxidative stress-related cell death. In this chapter, we review in detail current data on the impact of AgNPs on the central nervous system and discuss the possible mechanisms of their neurotoxic effects.

Keywords

Silver nanoparticles · AgNPs · Cell death · Oxidative stress · Neurotoxicity · Protein corona

L. Strużyńska (✉) · J. Skalska
Laboratory of Pathoneurochemistry, Department of Neurochemistry, Mossakowski Medical Research Centre, Polish Academy of Sciences, Warsaw, Poland
e-mail: lidkas@imdik.pan.pl; skalska_joanna@wp.pl

Q. Saquib et al. (eds.), *Cellular and Molecular Toxicology of Nanoparticles*, Advances in Experimental Medicine and Biology 1048, https://doi.org/10.1007/978-3-319-72041-8_14

14.1 Commercial and Medical Applications of AgNPs

In recent years, metallic nanoparticles in general and AgNPs in particular have attracted significant attention because of their unique properties and potential uses in a wide spectrum of applications. AgNPs have been subjected to numerous research investigations to gather information about their useful properties as well as their toxicity.

It has been shown that AgNPs are relatively easy to synthesize and possess a series of features such as stable morphology, high surface-to-volume ratio, useful chemical properties relating to surface and cell penetration capability which can be useful for many purposes, including therapy and diagnosis. The strong antimicrobial properties of AgNPs and a large plasmon field area have also proved their relevance in medical and pharmaceutical sciences. Incorporated into different materials, AgNPs can also be utilized in such areas as engineering, food and textiles [1, 2]. Currently, AgNPs are widely used in a vast array of commercial products [3]. They are present in toys, refrigerators, washing machines, cosmetic and hygienic products, and in food and drink containers (Table 14.1) [2, 4].

Table 14.1 Examples of medical and commercial applications of silver nanoparticles

AgNPs		
Medical applications		Consumer products
Based on antimicrobial properties:	Based on other properties:	Based on antimicrobial properties:
Wound dressings	Diagnostic agents	Cleansers to disinfect surfaces
Central venous and bladder catheters	Drug delivery carriers	Laundry and dishwashing detergents
Cardiovascular implants, stents	Cancer treatment	Cleansers to personal care, soaps and cosmetics
Endotracheal tubes	Biosensors	Water and air filters
Bone prostheses	Imaging tools	Clothing, underwear
Surgical instruments		Food storage containers
Endodontic cements		Nipples and nursing bottles
Dental implants		Children's toys

Although the antibacterial properties of silver have been known since ancient times [5], the development of nanotechnologies has contributed to a rediscovery of antimicrobial properties of silver for medical applications. Because of these properties, AgNPs can become promising alternative to conventional antimicrobial drugs and help us to reduce the serious public health problem of multidrug-resistant bacteria [6]. Moreover, AgNPs exert activity against many species of highly pathogenic fungi (ex. *Aspergillus sp.* and *Candida sp.*) [7, 8] as well as an antiviral effect against the herpes simplex virus [9], the hepatitis B virus [10] and the human immunodeficiency type 1 virus [11]. Because of this broad-spectrum antimicrobial action, we have witnessed increased development of AgNP-containing wound dressings and medical devices with AgNP-coated surfaces for use in surgical procedures (Table 14.1). Coating with AgNPs minimizes the risk of infections and prevents formation of biofilms, which are created by microorganisms sticking to each other and adhering to a surface [2, 8]. The efficiency of AgNPs against complex and invasive bacterial biofilms is very important in the context of antibiotic-resistant infections and host defenses [12]. Both the efficacy and safety of AgNP-based dressings have been confirmed in clinical studies in which such dressings were found to accelerate healing of wounds [13] and venous leg ulcers by inhibiting infection and the inflammatory response [14]. AgNPs are presently used in treatment of first and second degree burns, as well as venous and diabetic ulcers as commercially available nanosilver products. The Anticoat ™ Flex 3 Dressing is a typical example [8].

Due to their unique features, AgNPs may be also be employed in other fields of medicine such as bio-diagnostics, imaging and therapy (particularly methods of anticancer treatment), as well as in pharmacology as drug delivery systems. Their nanometer size enhances their biological activity and allows them to permeate cellular membranes and interact with biomolecules such as DNA and proteins. Moreover, they exhibit optical (mainly light scattering and absorption) and photothermal properties, and have surfaces which are easy to modify [15, 16].

The relative ease of modification of AgNPs with biomolecules, chemotherapeutic agents and biological stabilizing agents is one of their most advantageous properties [15]. The ability to functionalize the surfaces of AgNPs by conjugation of molecules capable of recognizing cellular or tissue-specific ligands, their nanometer-scale size and their ability to specifically accumulate within tumor tissue make AgNPs potential candidates for use in designing drug delivery systems [16, 17]. AgNPs can also be bound with a photosensitizer (such as protoporphyrin IX, for example) to improve photodynamic cancer therapy (PDT) based on light activation of singlet oxygen to induce cell death [18]. Additionally, the functionalization of AgNPs with DNA, peptides, proteins or antibodies can reduce their toxicity or allows to achieve specifically targeted AgNPs. Such modifications may be useful in biodiagnostic and biosensing applications. Attachment of doxorubicin to AgNPs has been achieved to create a DNA biosensor [19], and the capacity of AgNPs to interact with plasma proteins has been utilized in the development of a label-free, noninvasive tool for detection of the protein profile in gastric cancer cases [20].

The remarkable optical properties of AgNPs can be also readily utilized in biodiagnostic and imaging methods. Nanoparticles penetrate cellular and tumorigenic systems where they strongly scatter light and enhance the local electric field by confining photons within their structures. Such mechanisms allow creation of biological imaging or sensing agents which can be utilized for monitoring of disease progression or identification of diseased systems [21]. In this regard, the potential utility of AgNPs has been demonstrated for detection of Alzheimer's disease [22], apoptosis of cancer cells [16], as well as detection of neurotransmitters [23] and glucose [24].

AgNPs can also be applied as radiosensitizers in radiotherapy of cancer making tumor cells more sensitive to ionizing radiation. Radiosensitizers possess the ability to absorb X-ray radiation and then release the energy locally. This provides a means for minimizing damage of normal tissue during destruction of cancer cells [25]. Evidence exists that AgNPs may improve the outcome of radiotherapy of malignant glioma [26] and gastric cancer [27]. Silver nanostructures also have unique photothermal properties which enable them to convert photon energy into thermal energy [15]. The ability of metal NPs to heat the surrounding medium may be applied in plasmonic photothermal therapy. This is used to sensitize cancer cells to hyperthermia [28]. Additional studies suggest that a combined Au-Ag nanoparticle system may be a useful photothermal agent for cancer treatment [29, 30]. Another useful feature of AgNPs in cancer therapy is their anti-angiogenic activity. It has been reported that AgNPs inhibit cellular proliferation and migration in vascular endothelial growth factor (VEGF)-induced angiogenesis in bovine retinal epithelial cells [31].

A notable goal in modern medicine is to develop multifunctional therapeutic agents based on the above-mentioned NP properties to improve therapeutic efficacy, particularly of treatment-resistant cancers. Those nanoplatforms, conjugated with specific antibodies against cancer cell-surface targets, are capable of accumulation on the surfaces of cancer cells. Conjugation of AgNPs to agents which enhance the function of membrane transport systems provides new therapeutic agents for delivery inside the tumor cells. Apart from the effective targeted delivery of anticancer pharmaceuticals, such therapeutic systems provide controlled release of therapeutic substances bound to their surfaces at high loading capacity, as well as providing the ability to induce photothermal ablation of the cancer cells [16, 32, 33]. Another significant challenge for medicine is to identify efficient and safe strategies for treatment of brain disorders which have become a major health problem of a steadily aging global population. Conventional therapies are often ineffective because many compounds cannot translocate past the blood-brain barrier (BBB). Since different types of nanoparticles, including AgNPs, can cross BBB and accumulate in specific brain regions, they are expected to become the basis for development of tools suitable for therapy of brain pathologies such as neurodegenerative disorders. Furthermore, of AgNPs may provide the basis for development of efficient nanodrugs against neuroinfectious diseases such as AIDS and meningitis because of their strong antimicrobial activity [34].

Creation of multifunctional delivery nanoplatforms provides an attractive opportunity to develop theragnosis, a new concept in medicine, that combines treatment, diagnosis and monitoring of the therapeutic process. This new area of medicine may be particularly helpful for resolving the problems occurring during the process of early diagnosis and treatment of cancers and central nervous systems (CNS) disorders, because AgNPs may increase or quickly modify the therapeutic effects [34, 35]. The multifunctional biological activities of AgNPs, that could be simultaneously applied as antibacterial and anticancer agents, drug delivery systems, and imaging tools, makes them extremely attractive for development in these areas of medicine [33, 36].

14.2 Factors Influencing Toxicity of AgNPs

Physico-chemical characteristics of nanoparticles are relevant for interactions with biosystems and for possible toxic effects. The most important parameters determining the properties of AgNPs are: size, shape, chemical composition, porosity, crystallinity, and hydrophobicity/hydrophilicity of the nanomaterial. However, other characteristics such as surface charge/zeta potential, dispersion state, stability/biodegradability and hydration may also affect nano-bio interactions [37, 38].

14.2.1 Size and Shape

A number of reports have indicated that among the physico-chemical parameters of AgNPs, size is critically important for interactions between nanoparticles and biological systems. Particle size directly affects the surface-to-mass ratio and influences further the reactivity and the solubility of particulate systems. Smaller sizes provide increased surface area and increase the number of points for surface attachment of functional molecules, thereby enhancing the reactivity of NPs. Shannahan et al. [39] found that small (20 nm) AgNPs interact more strongly with hydrophobic proteins than larger AgNPs (110 nm) which is important for development of a protein corona (see: Sect. 14.2.4). However, it seems that the particle size does not determine the biodistribution of AgNPs, since the pattern of tissue deposition of two differently-sized citrate-coated AgNPs (10 and 25 nm) was found to be similar [40].

Higher reactivity is related to the smaller size of AgNPs, as well as higher efficiency of cellular uptake and toxicity towards biological systems. The size-dependency of the cytotoxic effect of AgNPs has been evaluated using in vitro systems such as fibroblasts [41], lung cells [42] and primary neuronal cultures [43]. Smaller AgNPs were shown to have significant effects on essentially all of the studied parameters including: cell morphology, cell viability and cellular membrane integrity, and were found to be even more potent in inducing oxidative stress [44]. Induction of oxidative stress was investigated using alveolar macrophages exposed to 15, 30 and 55 nm AgNPs [45]. In animal models, exposure to smaller AgNPs (20–40 nm) was found to induce severe myelin damage relative to larger AgNPs (50–60 and 130–150) [46].

It was reported that while large (100 nm) AgNPs, are not endocytosed by cells, they are capable of generating indirect cytotoxic effects, presumably via receptor-mediated signal transduction [47]. However, there are reports that have disagreed with the general conviction regarding the strong correlation between size and toxicity of AgNPs. Experiments performed by Gliga et al. [42] did not confirm cytotoxicity of larger (75 nm) citrate-coated AgNPs towards human lung cells. Similarly, the investigations of Espinosa-Cristobal et al. [48] revealed that there are no significant changes in the accumulation and toxicity of differently-sized AgNPs in the principal organs of exposed rats.

Shape is another factor relevant to the reactivity of AgNPs, particularly with respect to protein interactions. Ashkarran et al. [49] demonstrated that in the presence of differently-structured AgNPs (cube, sphere, wire, triangle) in fetal bovine serum (FBS), protein coronas are formed in different concentrations and compositions. This morphology-dependent corona effect is due to different surface energies caused by a variety of coordination complexes on the available surface specific for each shape [50]. In addition, activity towards bacteria was found to be higher for formulations of nanoplate-shaped AgNPs than for nanorod-shaped AgNPs [51].

14.2.2 Coating of the Surface

Nanoparticles are often coated with different compounds to enhance stability and prevent agglomeration. Among the most commonly used compounds are citrate, polyvinylpyrrolidone (PVP), chitosan, polysaccharides, carbon or peptides. Coating with PVP stabilizes AgNPs in aqueous solution for a longer time than coating with citrate [52]. In one comparison, uncoated AgNPs were found to agglomerate in solution faster than AgNPs capped with polysaccharide [53]. The chemical composition of the surface determines the behavior of AgNPs in physiological fluids. This is of particular importance in context of protein corona formation (see: Sect. 14.2.4) and further impact on nano-bio interactions. Citrate- and PVP-stabilized AgNPs exhibit different fates while interacting with bovine serum albumin (BSA). Coating of the surfaces of AgNPs with polymers such as PVP changes the surface area and this affects the behavior of proteins towards AgNPs [54]. There are strong correlations between the type of coating material, the bioavailability of AgNPs, and the profile of their tissue distribution. Also of note is the finding that toxicity of AgNPs depends on surface capping. It was demonstrated that carbon-coated AgNPs were less cytotoxic towards macrophages [55], whereas polysaccharide-coated AgNPs influence cell viability to a greater extent than uncoated AgNPs of similar size [53]. The issue of shaping the surface of AgNPs to modify their characteristics is still not fully understood and requires additional study, particularly for development of AgNP-based nanomedicines.

14.2.3 Liberation of Ag Ions

AgNPs continuously release relatively small amounts of silver ions in aqueous solution. The mechanism underlying the release of ions from particulate matter relies on cooperative oxidation with dissolved oxygen, protons and other constituents of the solution [56]. It has been previously stated, based on comparison of metal concentration in tissues from ionic silver- and AgNP-treated rats, that silver is mainly bioavailable in the ionic and not the particulate form [57]. However, a physiologically-based toxicokinetic (PBTK) model has been recently developed to predict the biodistribution of intravenously injected silver, both in ionic and nanoparticulate form [58]. PBTK modeling suggests that the majority of injected ionic silver is distributed systemically as de novo-formed secondary nanoparticles. Their presence in tissue of rats treated with ionic silver was confirmed in a TEM study [59]. It has also been suggested that de novo formation of small secondary AgNPs from silver ions is possible due to interactions with sulfhydryl groups of biomolecules inside cells [57, 59].

It is still not known whether the cellular response to AgNPs is driven by the particles themselves or by silver ions released from the particle surface inside the cell through the "Trojan horse effect" [60]. Some authors suggest that toxic effects are mainly caused by slowly liberated silver ions [42, 61], whereas others have advanced a contrary view which does not exclude nanoparticle-specific effects. According to Cronholm et al.

[62] AgNPs are taken up by the cell mostly in the particulate form (non-ionic form), and remain stable in this state for a long period of time inside the cell. Smaller AgNPs release Ag^+ more efficiently due to the larger surface area per unit mass, whereas particles larger than 10 nm release limited quantities of Ag^+, and thus their toxicity is attributed rather to the particulate state [42, 63]. The effect caused by AgNPs per se has also been confirmed in cultured cerebellar granule cells (CGCs), where they specifically interact with glutamate N-methyl-D-aspartate receptors (NMDARs) leading to excitotoxic cell death [64]. Sun et al. [65] also found that AgNPs, in contrast to silver ions, generate overproduction of ROS in a time- and concentration-dependent manner, as well as causing an increase in secretion of various cytokines in cultured astrocytic cells. Hence, multiple data provide solid proof for the distinctive toxicological effects caused by particulate and ionic silver. Furthermore, Beer et al. [66] concluded from an in vitro study using a human lung carcinoma epithelial-like cell line, that the cytotoxic effect of AgNPs depends on the fraction of silver ion liberated into the particle suspension. It was found that AgNPs have a measurable toxic effect under low ($\leq$2.6%) concentrations of silver ions in the suspension. This effect was not observed at high ($\geq$5.5%) concentrations of AgNPs. It can be presumed that the final effect observable in the cells is the result of a mixture of primary AgNPs, released silver ions, and secondary de novo-formed AgNPs.

14.2.4 Protein Corona

Upon entering a biological environment, AgNPs come into contact with various proteins which compete for adsorption on their surface. Adsorption is the final effect of protein-nanoparticle binding affinity and protein-protein interactions. Initially, low affinity proteins are loosely attached to the surfaces of the AgNPs, thereby forming a rapidly exchanging "soft corona." Subsequently, a tightly bound "hard corona" is formed over time by high affinity proteins. The affinity-dependent effect of substitution of certain proteins by others is known as Vroman's effect [67]. A typical composition of a human blood protein corona consists of albumins, immunoglobulins, fibrinogen, apolipoproteins, transferrin, complement proteins, and hemoglobin. Among these proteins, albumins and fibrinogen are the first to be adsorbed on the surface [50]. The composition and conformation of the protein corona will be unique for specific nanomaterials and will depend on parameters such as the nanoparticle size, shape, functional groups and surface charge [68]. The time and route of exposure and the nature of the biological environment (blood, interstitial fluid, cell cytoplasm) also have important effects on the protein composition and subsequent cellular response [69]. Different coatings of the nanoparticle surface (for example, citrate vs. PVP) will also determine interactions with proteins. The presence of a polymer was shown to strongly inhibit interactions between SH-groups and the metal surface. This effect may be useful in development of medical applications [54]. In turn, Shannahan et al. [39] have suggested that various surface components of AgNPs of the same size have different influences on the constituents of the protein corona by inducing different free energy requirements during the processes of protein folding and unfolding. Exposure to BSA leads to aggregation of PVP-coated AgNPs due to destabilization of electrostatic groups. Hence, electrostatic and hydrophobic interactions during protein corona formation appear to be important drivers of the process.

Current data indicate that the protein corona which forms on the surface of nanoparticles upon contact with biological fluids, creates an interface between the nanoparticles and the cellular system which determines biological response, including cytotoxicity and immunotoxicity, to a greater extent than the bulk material itself [70]. Likewise, the process of cellular internalization of NPs depends on the formation and composition of the corona [71]. In turn, the transfer of NPs between cellular compartments changes the composition of proteins. It has been found that each cellular compartment leaves a "fingerprint" inside the corona, as a result of the interplay of nanoparticles with specific organelles and macromolecules [69]. Changes in the protein corona can alter the characteristic of nanoparticles and provide them with new biological identities, thereby influencing the biological response to exposure. The parameters dictating the fate of NPs in a given organism such as transport kinetics, cellular uptake, organ accumulation and toxicity may be modified along with alterations of the corona structure. Silver ions, which are thought to be at least partially responsible for toxicity of AgNPs, are bound by protein components of the hard corona to form silver sulphide monocrystals as the effect of sulphidation. This process decreases the toxicity of AgNPs [72]. This may be of importance from a medical point of view. Recently, Duran et al. [50] found that protein-AgNPs interactions could lead to protein unfolding. Changes in the secondary structure of proteins may consequently result in unexpected cellular responses and/or reduction of antibacterial, antifungal or anticancer activity of AgNPs during biomedical applications. Thus, understanding the relationships between physicochemical parameters, the specificity of the protein corona, and biological systems are important issues for further development of AgNP-based nanotechnologies. This new subject has captured the attention of many researchers. Precise knowledge about NP-protein interactions and detailed characteristics of the corona will allow optimization of features of NPs to produce customized nanomedicines based on AgNPs. It is also expected that potential harmful effects of these applications will become better understood [50].

14.3 Cellular Targets for AgNPs in the Central Nervous System

AgNPs may gain access to the circulatory system via different routes, by oral or intravenous administration as well as via inhalation, intranasal application or through the skin. After entering the body, AgNPs are distributed via systemic circulation within organs such as liver, kidneys, lungs, spleen, heart, gonads and brain [40, 73, 74] regardless of the route of exposure. AgNPs are able to cross the intestinal barrier [41] being absorbed from the gastrointestinal tract in the nanoparticulate or ionic form. Transmission electron microscopy (TEM) studies enable visualization of particle-like structures in different tissues of exposed animals, including ileal macrophages [73], brain endothelial cells and neurons [75, 76], being localized mainly in such cellular compartments as endosomes, lysosomes and mitochondria [75]. Available toxicokinetic data indicate that AgNPs are primarily accumulated in the liver [73], as expected because of the metabolic and detoxifying roles of this organ. Highly effective antioxidant defense systems allow the liver to effectively maintain the oxidative/antioxidative balance during metabolism of toxic agents [77]. Absorbed AgNPs are in equilibrium with blood and are removed over the time from most of the tissues. However, in brain and testes unlike in other organs, AgNPs are deposited persistently [40, 57]. Thus, even under low-level exposure, elevated concentrations of nanoparticles may be expected in brain parenchyma. AgNPs have been shown to enter the rat brain when injected subcutaneously and accumulate in neuronal cells in the form of nanoparticles [78].

Administered intranasally to the rats, AgNPs induce impairment of hippocampal functions [79], and reach the brain through retrograde axonal transport via the olfactory nerve.

Nanoparticulate silver is taken up by the cell following the process of cellular membrane encapsulation and subsequent clathrin-mediated endocytosis, and macropinocytosis [80]. It is possible that endocytosis is not a cell type-specific process because it has been described in cultured astrocytes [65], lung fibroblasts and human glioblastoma cells [80], as well as in gut epithelial cells using an in vivo model of exposure [81]. These mechanisms may be different exclusively in endothelial cells of cerebral microvessels (see Sect. 14.4.1). It is also noteworthy that the pathway of entry of AgNPs determines the fate of the cell [82]. AgNPs internalized via endocytosis, unlike those bypassing this form of active transport, are identified exclusively within the endo-lysosomal compartment and exert a cytotoxic effect. Translocation of AgNPs into lysosomes leads to toxicity, first due to induction of ROS in these subcellular structures [83] and second, due to release of toxic Ag ions from the surfaces of the AgNPs which is favored by the acidic environment inside the lysosomes [84]. The nervous system is particularly sensitive to toxic insults. Thus, even low-level exposure to toxicants may lead to deleterious effects relative to other organs. For example, it has been reported that a low dose (0.2 mg/kg b.w.) of small (10 nm) citrate-stabilized AgNPs induces oxidative stress in the brain but not in the liver of exposed rats [85]. Moreover, as there is a tendency of AgNPs to accumulate in the nervous tissue over the time, neurotoxic effects are long-lasting after initial exposure.

Described in the next sections dysfunction of the brain microvasculature and astroglia, as well as degeneration of nerve endings and neurons, suggest the possible impairment or loss of nervous system functionality. Indeed, AgNPs altered spatial reference memory [86] and spatial cognition in exposed rats via their impact on hippocampal synaptic plasticity [79], likewise influenced locomotor activity when administered in neonatal period [87]. In turn, the study of Dąbrowska-Bouta et al. [88] designed specifically to investigate the behavioral effects of prolonged exposure of adult rats to low doses of small AgNPs (0.2 mg/kg b.w.) did not reveal significant behavioral alterations in the applied set of tests. It seems, that apart from the type/size of AgNPs, a dose, duration of exposure, and age of animal are important factors for neurobehavioral consequences. It should be stressed that extensive research are needed towards a comprehensive assessment of behavioral effects of AgNPs.

14.3.1 The Blood-Brain Barrier

Studies on the distribution of engineered AgNPs have revealed that they can enter the brain by moving along the olfactory nerve when administered by inhalation [86] or via the blood-brain barrier (BBB) during other routes of exposure [89]. Therefore, it is necessary to investigate the mechanisms of transvascular passage of AgNPs and their interactions with cellular components of the BBB.

The BBB is a unique anatomical and functional barrier between brain and systemic circulation which plays a key role in guarding a stable microenvironment in the brain parenchyma by protecting it against invading substances and microorganisms (Fig. 14.1).

It consists of highly specialized brain microvascular endothelial cells (BMECs) connected by tight junctions, pericytes, a capillary basement membrane composed of collagen, fibronectin and laminin, as well as astrocytic end-feet which enfold BMECs. BMECs and the basement membrane establish an anatomical barrier while being in close communication with other constituents. Pericytes enveloping cerebral microvessels fulfil many functions such as control of angiogenesis, cerebral blood flow and inhibition of barrier damage by infiltrating immune cells. Furthermore, astrocytes, which interact with pericytes and BMECs, maintain the integrity of tight junctions, regulate proliferation of endothelial cells and contribute to the homeostasis of metabolites, ions and water [90, 91].

Transport of blood-borne compounds across the BBB is strictly regulated by both physical and biochemical barriers. A physical barrier created by endothelial cell tight junctions, limits the cellular transport of hydrophilic molecules through the BBB. Tight junctions are dynamic and highly regulated structures established by transmembrane proteins (junction adhesion molecules, occludins and claudins) as well as supporting cytoplasmic proteins (zonula occludens and cingulin) which link transmembrane proteins to cytoskeletal proteins [91, 92]. A biochemical barrier is formed by receptors, ion channels and active influx/efflux transporter systems expressed by BMECs. There are a few different mechanisms involved in the process of passage of nanoparticles across the tight barrier including: passive diffusion, carrier-mediated active transport, transcytosis and endocytosis. The latter two mechanisms occur in membrane microdomains such as caveolae [90, 91] (Fig. 14.1).

Tight junctions between two endothelial cells create gaps of 4–6 nm, thus presumably only very small NPs fit this route of passage. As in vitro studies have revealed, different types of NPs sim-

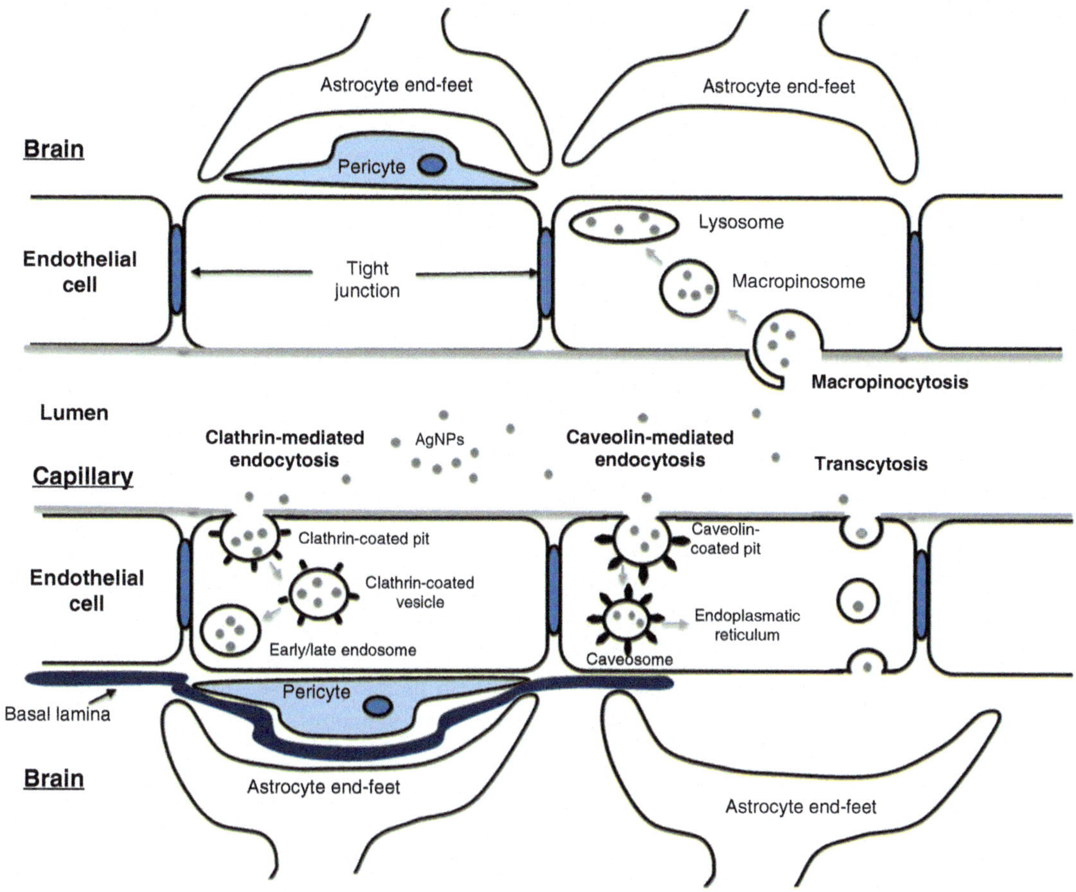

Fig. 14.1 Schematic representation of the structure of the blood-brain barrier and possible pathways for passage of AgNPs across endothelial cells

ply cross the endothelial cell membranes by caveolae- and clathrin-mediated endocytosis, and macropinocytosis [93, 94]. Moreover, there is evidence that the size (wherein smaller NPs are more effective) and the type of surface modification play relevant roles in this process [95].

AgNPs passing into capillary endothelial cells by transcytosis accumulate inside these cells [96], mainly in lysosomes [75], and hence interfere with the function of BMECs by increasing the permeability of microvessels. In vivo studies have revealed that single intravenous, intraperitoneal or intracerebroventricular administration of AgNPs causes increased BBB permeability and brain edema [89, 97]. Moreover, interactions of AgNPs with BMECs result in the release of pro-inflammatory mediators, mainly interleukin 1β (IL-1β), tumor necrosis factor α (TNF-α) and prostaglandin PGE_2, thus inducing inflammation which indirectly accelerates the existing leakage of microvessels [98, 99].

AgNPs may also induce a cascade of events leading to disruption of tight junctions. A study based on the BBB model (a triple co-culture of primary rat brain microvascular endothelial cells, pericytes and astrocytes) revealed that AgNPs enhance permeability of the BBB by decreasing the expression of ZO-1, one of the tight junction proteins [100]. Increased expression of cadherin-1 and claudin-1 were also observed after a 2-week oral exposure of rodents to AgNPs in a dose of 1 mg/kg b.w [101]. The influence of AgNPs on pericytes, which can be considered caretakers of BMECs, remains an unexplored issue. However, morphological changes (mainly vacuolization) of these cells under exposure to AgNPs have been identified [100].

Since astrocytes represent a functional part of the neurovascular unit, the impact of AgNPs on these cells is also important in this context (discussed in the Sect. 14.3.4), especially since astrocytes release pro-inflammatory cytokines under pathological conditions [102]. Local or systemic inflammation can alter functions of the BBB by triggering functional and morphological changes in brain microvasculature. Increased secretion of cytokines and cell adhesion molecules, the influx of immune cells to the CNS, and alterations in tight junction proteins finally increase the permeability of the BBB towards blood-derived chemical compounds, including nanoparticles [103]. The toxicity of AgNPs in the CNS under physiological and inflammatory conditions has been investigated using a double BBB model. Under normal conditions AgNPs trigger discontinuities in tight junction complexes by influencing protein constituents (claudin-5 and ZO-1), whereas in parallel, existing inflammation exacerbates the disruption of BBB and significantly increases its permeability rate [104].

A dysfunctional BBB is accompanied by a number of brain pathologies, including tumors, stroke, neuroinflammatory disorders such as multiple sclerosis (MS), and neurodegenerative diseases such as Parkinson's disease (PD), Alzheimer's disease (AD), and amyotrophic lateral sclerosis (ALS). It is believed that uncontrolled presence of nanoparticles in the environment may enhance the risk of neurodegeneration [105]. On the other hand, the presence of the BBB makes the diagnosis and treatment of CNS disorders difficult because it restricts access of drugs. Therefore, the strategy of using AgNPs as nanocarriers might potentially improve these processes. Presently, AgNPs are considered as drug nanocarriers in the AD [106].

In this light, identifying the mechanisms of interaction of AgNPs with components of the BBB is of crucial importance.

14.3.2 Neurons and Synapses

Toxic substances may exert neurotoxic effects by inducing alterations in neuronal structure and/or activity. It is generally accepted that neurotoxicants may damage neuronal cells by affecting the whole neuron or exclusively the axon by affecting its myelinated sheaths or the synaptic processes of neurotransmission [107]. However, neuronal death, either by necrosis or apoptosis, is often the final effect of these actions.

Current knowledge regarding the impact of AgNPs on the CNS is quite limited. However, there is a growing body of evidence suggesting that AgNPs may be considered as neurotoxic material. Toxicity towards neuronal cells has been proven without a doubt in in vitro experiments, wherein AgNPs were demonstrated to decrease viability of cultured neuronal cells [43, 64, 108, 109] and astroglial cells [43, 65]. In cortical neurons exposed to AgNPs in the concentration range of 1–50 μg/mL axonal outgrowth was found to be inhibited. Moreover, neurites were found to degenerate as a consequence of disrupted integrity of cytoskeletal and synaptic machinery (i.e. synaptic proteins such as synaptophysin and PSD-95) [108].

Regarding the harmful in vivo effects of AgNPs on the CNS, it appears that brain penetration and accumulation therein leads to formation of pyknotic and necrotic neurons [78]. Among the histopathological changes, destruction of the cerebellar granular layer with concomitant activation of glial cells [87] as well as cell degeneration and necrotic areas in the hippocampus [79] of AgNP-exposed rats were described. Furthermore, in the model of prolonged oral exposure to a low dose of AgNPs, ultrastructural and biochemical changes in synapses were revealed [75]. AgNPs induce considerable synaptic alterations, mainly in the hippocampus of exposed rats. Several of the observed ultrastructural features such as free synaptic vesicles located in the neuropil outside the discontinuous synaptic membrane and the presence of myelin-like bodies were found to be AgNP- but not Ag^+-specific effects. Other features, such as swollen synapses, enhanced density of synaptic vesicles and blurred structure of the synaptic cleft, were found to be common for both forms of silver.

In addition to the morphological indicators of synaptic destruction, decrease in the relative levels of pre- (synapsin I and synaptophysin) and post-synaptic (PSD-95) proteins is also observed. Reduced expression of these proteins, which are considered molecular markers of synaptic density, indicates synaptic loss under exposure to AgNPs. Moreover, pre-synaptic proteins are involved in the processes of neurotransmission, being responsible for movement of synaptic vesicles along the axonal actin filaments and exocytosis of neurotransmitters into the synaptic cleft [110]. Thus, reduction in expression of pre-synaptic proteins that regulate vesicle cycling, suggests the occurrence of damage to the secretory mechanism and further possible disturbances in neurotransmission. In turn, PSD-95 which is abundant in neuronal post-synaptic structures, plays a role in a synaptic plasticity phenomenon mediated by glutaminergic NMDA receptor signaling by modulating the clustering and function of these receptors [111]. The aberrant synapse ultrastructure together with decreased levels of PSD-95 may specifically suggest disrupted functioning of the excitatory synapses. Damaged and marker protein-deficient synapses have been observed in a variety of brain pathologies, including neurodegenerative diseases [112]. Skalska et al. [75] claimed that alterations in both synaptic structure and the level of synaptically-bound proteins, most frequently identified in hippocampus, indicate AgNP-induced synaptic degeneration and are predictors of changes in cognitive processes. However, no adverse effects either on hippocampal neurogenesis or the spatial learning and memory processes were observed in adult mice treated with AgNPs [113]. In contradiction are the results of behavioral tests estimating motor functions in AgNP-exposed animals. Dysfunction of motor coordination and impairment of locomotor activity has been reported in neonatal rats [87].

14.3.3 Myelin

Myelin is a spiral structure built by extensions of the plasma membrane of Oligodendrocytes. Myelin is characterized by a unique composition relative to other membranes in that it has a reverse lipid-to-protein ratio (70%: 30%) and a unique segmental structure (for a review, see [114]). The myelin sheath surrounding most of the axons in the CNS ensures proper transmission along with neurons enhancing efficiency of conduction of electrical impulses. Additionally, this membrane structure functions as a protective and isolating element. Disturbed myelin integrity may result in an inappropriate signal transfer and further behavioral abnormalities.

It has been reported that AgNPs administered to rodents may cause myelin damage. At the biochemical level, AgNPs influence the composition of myelin membrane by lowering the relative protein concentration. The decreased level of myelin basic protein (MBP) reflected by less intense immunostaining has been reported after exposure to a relatively high dose (50 mg/kg b.w) of AgNPs (50–60 nm) [46]. However, exposure to even a low dose (0.2 mg/kg b.w.) of AgNPs (10 nm) significantly decreases the level of several myelin-specific proteins: myelin/oligodendrocyte glycoprotein (MOG), myelin-associated glycoprotein (MAG) and 2′-3′-cyclic nucleotide 3′-phosphodiesterase (CNP) in a myelin fraction isolated from the exposed rats [88]. Since CNP, MAG and MOG play roles in stabilization of myelin membranes [114], deficiency of these proteins may influence formation and maintenance of proper myelin structure. Indeed, examination of the myelin ultrastructure under TEM revealed pathological changes of myelin sheaths in the form of focal detachments of the myelin lamellae and an increased volume of cytoplasm within the regions of compacted myelin [88]. Myelin disintegration may consequently lead to degeneration of myelinated fibers. Moreover, myelin glycoproteins such as MOG interact with the immune system and are potent inductors of anti-myelin antibodies [115]. This may be relevant to the pathological implications such as demyelinating diseases [114].

In low doses, AgNPs do not affect neurons directly. However, they cause both morphological and biochemical changes within synapses [75] and in the myelin sheaths surrounding the axons [88]. Synaptopathy and dysmyelination/demyelination may finally lead to axon degeneration. Hence, classification of AgNPs as axonopathy-inducing agents should be considered.

14.3.4 Glial Cells

Previous studies on toxicity of AgNPs have mainly focused on neurons and endothelial cells, whereas little is known about their effects on both resting and reactive glial cells. Lastly, astrocytes which are key players in maintaining homeostasis in the CNS (for a review see [116]), are of central interest. Together with neurons, glial cells form tripartite synapses which are the functional units wherein astrocytes interact closely with neurons to control neuronal activity and metabolism [117]. Glial cells also envelope microvessels with the end-feet of cellular processes to interact with pericytes and endothelial cells as a functionally-coupled structure known as the neurovascular unit. Being activated under pathological conditions, including exposure to metals, astrocytes may act in either a neuroprotective or a pro-neurodegenerative manner [118–121]. This depends on a several factors, such as the period and level of exposure. It is certain that an indirect neurotoxic effect will be the result of disruptions in glial-neuronal interactions. Moreover, the failure of astrocytic protective functions determines the degree of neuronal death or dysfunctionality.

The role of astroglia in AgNP-induced neurotoxicity and direct toxic effects of AgNPs towards astroglia are significant topics of current investigations. The existing results of in vitro experiments indicate that AgNPs are taken up by astrocytes and strongly affect cellular morphology [43]. However, AgNPs do not significantly disturb basal metabolism or the viability of astroglial cells [122]. TEM observations confirm that AgNPs enter astrocytes and are deposited inside the cytoplasmic phagocytotic vesicles [65, 100]. Ultrastructural studies indicate disturbances in cellular structure such as mitochondrial shrinkage, vacuolization, and expansion of the endoplasmic reticulum [108]. A pro-inflammatory response indicated by the secretion of multiple cytokines was also observed [65, 123].

Under cytotoxic levels of AgNPs (10 μg/mL) reactive oxygen species (ROS) were induced in a time and dose-dependent manner and caspase activity was found to be elevated, suggesting induction of apoptosis-like cell death in cultured astrocytes [65]. Apoptosis was also evidenced by the observation of AgNP-dependent modulation of the mitogen activated protein kinase (MAPK) pathway, Bcl-2 expression and mTOR activity. Moreover, AgNPs were found to significantly decrease expression of *Nr4a1* and *Dusp1* genes, whose role is to protect cells from oxidative stress, inflammation and apoptosis [101]. A comparative study of the affinity of 10-nm AgNPs for different types of cultured neuronal cells revealed that uptake of AgNPs is greater in astrocytic cells than in microglia or neurons. This correlates positively with the degree of cytotoxic activity towards respective cells [124]. This observation is consistent with that of Haase et al. [43] who found that astrocytes have a greater capability to accumulate 20 nm peptide-coated AgNPs relative to neurons and are more vulnerable. Recent results also suggest that affected astroglial cells may transform into neurotoxic partners due to disturbances in glial-neuronal interactions. ROS and NO production induced by AgNPs in astrocytes and microglia are indirect but key toxic factors for neurons [124].

Strong up-regulation of astroglial metallothioneins (MTs) by AgNPs has been reported [125]. MTs are proteins which have high capacity to bind heavy metals through their numerous thiol groups. Furthermore, this group of enzymes is involved in control of oxidative stress. By binding ROS such as superoxide and hydroxyl radicals, they contribute to a cellular mechanism of antioxidant defense [126]. This indicates that astroglia may play a protective role under conditions of AgNPs exposure.

14.4 Molecular Mechanisms of Neurotoxicity Exerted by AgNPs

The basic non-specific nano-neurotoxic mechanism relies on direct physical interactions between nanoparticles and cellular components at the nano/bio interface as the result of the unique size and surface properties of the nanoparticles [127]. The "non-oxidative stress effects" may be disruption of the membranes and disturbed transport processes, or even transmission of nanoparticles into the cell [128]. Further, neuronal cell death may be caused by of a variety of mechanisms specifically related to nanoparticles, such as an increase in transporter/receptor-related intracellular calcium levels, induction of oxidative stress and mitochondrial damage. A schematic overview of the possible mechanisms of AgNP-induced neurotoxicity is presented on Fig. 14.2.

14.4.1 Interactions with Receptors and Channels

Most physiological and pathological cellular reactions are controlled by calcium signaling systems. Calcium is a universal messenger which regulates life and death processes. The concentration of calcium ions (Ca^{2+}) inside the cell is under strict regulation by an array of transporters and channels which are responsible for proper distribution of calcium between intracellular compartments. Appropriate regulation is crucial, since overloading a cell with Ca^{2+}, activates pathways which trigger necrotic or apoptotic cell death. Thus, factors inducing or over-stimulating Ca^{2+}-dependent pathways, lead to cell damage [129]. There is evidence, that AgNPs influence cellular Ca^{2+} homeostasis. Increased levels of intracellular calcium have been identified in neuronal cells cultured in the presence of AgNPs [43, 64].

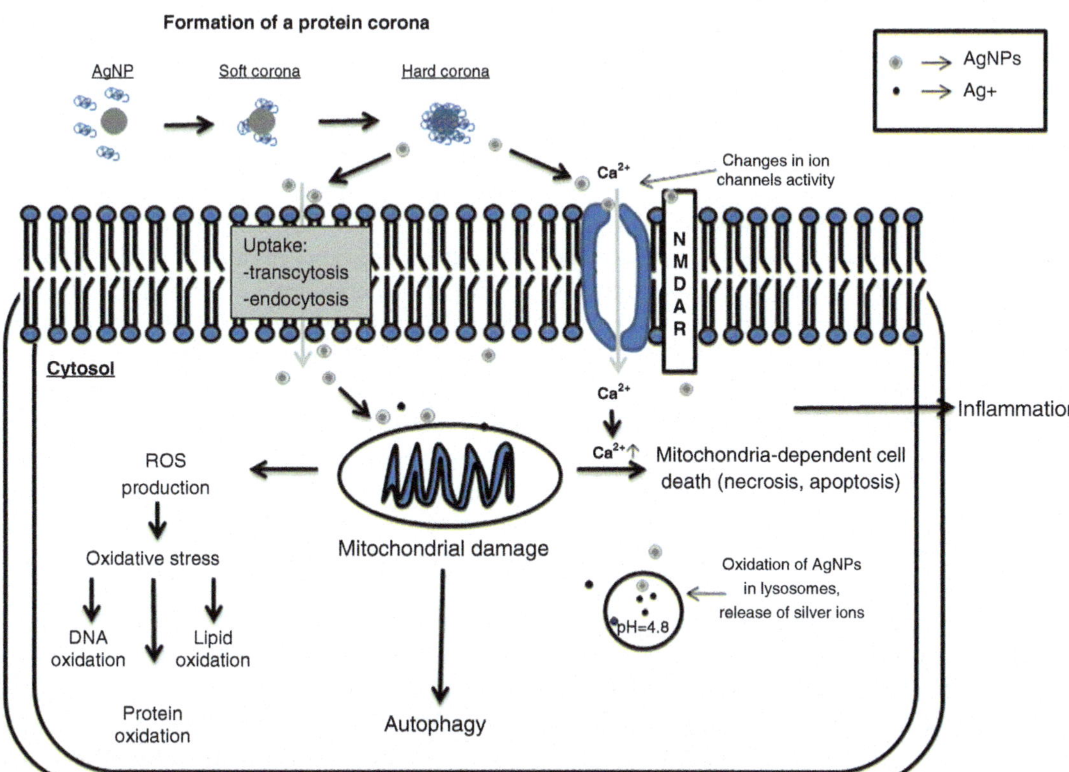

Fig. 14.2 Schematic diagram showing the possible cellular mechanisms of neurotoxic effects of nanoparticulate silver

A massive intracellular influx of Ca^{2+} from the extracellular space may occur via overstimulation of ion channels. The most important channels of this mechanism are ligand-operated Ca^{2+} channels, i.e. channels coupled with ionotropic glutamate receptors, mainly N-methyl-D-aspartate receptors (NMDARs) [130]. The mechanisms of cell death influenced by AgNPs have been extensively studied in CGCs. Yin et al. [87] demonstrated that AgNPs significantly reduce protein and mRNA levels of calcium channel protein (CACNA1A). Moreover, our group established a linkage between AgNP-induced overactivation of glutamate receptors, generation of ROS, and excitotoxic cell death [64]. Short-term incubation of CGCs with 0.2% PVP-coated AgNPs (mixture <100 nm) causes a dose-dependent increase in the uptake of $^{45}Ca^{2+}$ which is completely abolished by administration of MK-801, a noncompetitive inhibitor of glutaminergic NMDARs. Excessive entry of calcium ions through the channels associated with NMDA is then followed by an intracellular calcium imbalance leading finally to ROS generation, significant loss of mitochondrial potential and cell death. Again, addition of MK-801 was found to provide a protective effect. Simultaneously, 6-cyano-7-nitroquinoxaline-2,3-dione (CNQX), a potent antagonist of AMPA and kainate (AMPA/KA) glutamate receptors, did not improve the viability of AgNP-exposed CGCs, thereby excluding the contribution of receptors other than NMDA ionotropic glutamate receptors [131]. Thus, we concluded that AgNPs exert toxicity in CGCs by selective overactivation of NMDARs. This overactivation underlies the excitotoxicity phenomenon in which production of free radicals, mitochondrial dysfunction and neuronal damage are driven by overloading of neurons with calcium originating from the extracellular space [132]. Hence, excitotoxicity through the NMDAR-mediated dysregulation of intracellular calcium signaling pathways appears to be an important mechanism contributing to AgNP-evoked neurotoxicity. There is no direct evidence thus far that this mechanism operates in vivo as well. However, as discussed in Sect. 14.3.2 changes in the ultrastructure of synapses together with decreased levels of the protein PSD-95 [75] suggest that AgNPs may contribute to disturbances in NMDAR-mediated signaling. PSD-95 is a member of the membrane-associated guanylate kinase (MAGUK) family of proteins which interact with each other to form a scaffold for the clustering of NMDA receptors at the postsynaptic membrane [133].

14.4.2 Mitochondrial Damage

In addition to lysosomes, the mitochondrial compartment has also been identified as a potential subcellular target of AgNPs. Deleterious effects of AgNPs on mitochondrial functions have been observed in several in vitro and in vivo studies. AgNPs have been observed inside the mitochondria of liver cells, where they induce a collapse of the mitochondrial membrane potential, leading to impairment of mitochondrial function. A similar mechanism resulting in mitochondria-dependent cell death and degeneration of cytoskeletal components was observed in CGCs [64] and rat cortical neurons [108]. AgNPs with a range of different diameters between 5 and 45 nm were found to inactivate mitochondrial respiratory chain complexes I, II, III, and IV in brain and other tissues of exposed rats, uncoupling the oxidative phosphorylation system in the mitochondrial inner membrane [134]. Disruption of the mitochondrial respiratory chain decreases levels of cellular ATP and induces ROS production [135]. Presumably, interactions between protein thiol groups and nanoparticles are responsible for disrupting the structure and functions of proteins in mitochondrial membranes and/or mitochondrial enzymes [136].

14.4.3 Oxidative Stress

The results of thorough in vitro and in vivo studies on mechanisms of toxicity of AgNPs indicate that induction of ROS, which leads to oxidative stress in cells, is a predominant mechanism of their toxicity in general and neurotoxicity in particular. Levels of ROS generated by this mechanism depend on the unique physicochemical properties of AgNPs such as small size vs. large surface area and high reactivity, as well as the properties of the basic material. Cellular overproduction of ROS, when unbalanced by antioxidant defense mechanisms, causes oxidative stress. ROS are highly reactive with proteins, lipids and nucleic acids and thus, cells are unable to maintain physiological redox-regulated functions and become dysfunctional due to irreversible oxidative modifications. The modulation of gene expression through activation of redox-sensitive transcription factors may finally lead to genotoxic effects and cell death [137].

The unique characteristics of brain tissue underlie higher responsiveness to toxic insults relative to other tissues. A high rate of oxygen consumption, activity of the dopaminergic and glutaminergic systems or regionally high concentrations of redox-active transition metals all stimulate production of ROS. In addition, the combination of oxidation of polyunsaturated fatty acids and modestly active antioxidant defense systems, make the brain particularly vulnerable to the oxidative stress [138, 139]. Hence, it is not surprising that the brain is more responsive to even low concentrations of AgNPs than other organs, including the liver [85]. It has been observed in both in vitro and in vivo studies that oxidative stress contributes to the neurotoxic mechanisms exerted by AgNPs. Generation of ROS has been described in CGCs [64, 109] and mixed primary cortical neural cell culture [43]. In parallel, decreased levels of reduced glutathione were found in CGCs exposed to small (20–25 nm) AgNPs [109]. Exposure to AgNPs (24 nm) was also shown to elevate ROS in the cultured astroglial cells in the time-dependent manner. This finding was not replicated for silver ions [65]. Conversely, Luther et al. [125] found that incubation of astroglia-rich primary cultures with 75 nm AgNPs did not generate ROS. This discrepancy may indicate that the size of the AgNPs plays a crucial role in AgNP-induced ROS production.

Studies performed with animal models showed changes in expression of oxidative stress-related genes in the brain of AgNP-treated rodents [140, 141]. Overproduction of ROS has been demonstrated in rat brain after intranasal administration of AgNPs in a dose range of 3–30 mg/kg b.w./day [87] and oral administration of AgNPs at a dose of 0.2 mg/kg b.w./day [85]. One of the oxidative markers of excessive ROS is peroxidation of lipids in cellular membranes measured by increased levels of malondialdehyde (MDA). A significant increase of MDA in brain homogenates of rats exposed to a low dose of AgNPs was identified [85]. Cellular enzymatic and non-enzymatic antioxidant defense systems actively prevent oxidative stress and scavenge overproduced ROS. The first line of defense is provided by enzymes such as superoxide dismutase (SOD), catalase (CAT) and glutathione peroxidase (GPx). Current data indicate that in the brains of rats exposed to 20 nm AgNPs (5 mg/kg b.w.) SOD activity increases significantly whereas expression of *Sod 1* mRNA is lowered [140]. However, under conditions of exposure to a low dose of AgNPs (0.2 mg/kg b.w.) the activity of either SOD or CAT did not change, despite the presence of mild oxidative stress [85]. Generally, under these conditions, the enzymatic antioxidant system was not found to be significantly activated, with the exception of GPx. It was different in the case of the non-enzymatic glutathione system. A statistically significant decrease in the reduced-to-oxidized glutathione ratio (GSH/GSSG) was identified, predicting depletion of GSH. Similar lowering of the GSH concentration has been shown in rat cerebellar granule cells exposed to PVP-coated AgNPs [109] and in human liver cells [142]. In turn, administration of AgNPs to cultured astrocytes was found to evoke oxidative stress without inducing GSH depletion [65]. In brain, in contrast to the liver, treatment with a low dose of AgNPs induces oxidative stress as reflected by enhanced ROS production and membrane lipid peroxidation [85], indicating particular sensitivity of nervous tissue to AgNP-induced toxicity.

14.4.4 Inflammation

Overexpression of proinflammatory cytokines and induction of inflammatory response by AgNPs is the result of initial oxidative stress. Generation of ROS consequently leads to upregulation of transcription factor NF-κB and activation of the release of proinflammatory cytokines [143]. AgNPs have a higher rate of cellular uptake compared with ionic form of silver and induce oxidative stress and inflammatory response more potently, as shown by Prasad et al. [144] in liver HepG2 cells. However, few reports concerning AgNP-induced inflammatory responses in brain are available. An in vitro study revealed that AgNPs potently induce release of proinflammatory cytokines such as TNF-a, IL-1b and PGE2 in primary rat brain microvessels which further leads to increased permeability of microvessels [98]. Exposure of astroglial cultures to AgNPs significantly increases the secretion of multiple cytokines/chemokines. Moreover, unlike silver ions, AgNPs induce expression of CINC-2α/β and CINC-3 chemokines which are involved in recruitment of neutrophils, as well as fractalkine, interferon gamma-induced protein 10 (IP-10), L-selectin and thymus chemokine which are all pro-inflammatory mediators [65]. However, expression of interleukin-10 (IL-10) which has anti-inflammatory potential was also found to be increased. Similarly, Huang et al. [123] reported that AgNPs taken up in the process of endocytosis, induce pro-inflammatory IL-β cytokine secretion and gene expression of chemokine 13 (*CXCL13*) in murine ALT astrocytes, microglial BV-2 cells and neuronal N2a cells. These observations indicate that AgNPs induce pro-inflammatory activity in brain tissue of exposed animals.

14.4.5 Cell Death

As described above (see Sect. 14.4.1), cell death is accompanied by sustained elevation of intracellular Ca^{2+}. Dysregulation of homeostasis leads to pathological Ca^{2+}-mediated signaling which involves activation of intracellular pathways to trigger cell death by apoptosis or necrosis, depending on the type and status of the cell and the severity of pathological insult [129].

14.4.5.1 Apoptosis

Induction of apoptosis by AgNPs has been reported in a number of in vitro studies using different types of mammalian cells such as: human lung cancer cells [145], human liver cells [142], human colon cancer cells [146], fibroblast cells [147] and mouse embryonic fibroblasts [148]. Apoptosis is an energy-dependent process involving a complex cascade of events. A diverse array of stimuli may activate the intrinsic signaling pathways of apoptosis which are mitochondria-initiated processes. Changes in the inner mitochondrial membrane include opening of the mitochondrial permeability transition (MPT) pore, loss of the mitochondrial transmembrane potential and release of pro-apoptotic proteins (such as cytochrome c, for example) into the cytosol which further activate the caspase-dependent mitochondrial pathway (for a review see: [149]. AgNP-exposure has been reported to induce a drop of mitochondrial potential [64, 131] and release of cytochrome c from mitochondria [142, 147] (Fig. 14.3).

Caspases are enzymes with proteolytic activity which play important roles in the regulation of cellular pathways controlling apoptosis, which once activated, inevitably lead to cell death [149]. There is evidence that AgNPs activate caspase-3 in different cells [142, 146, 148, 150]. Moreover, AgNPs increase the activity of caspases and induce cell apoptosis of cultured astrocytes as opposed to silver ions which predominantly cause necrosis [65]. Exposure to AgNPs induces over-activation of caspase 3/7 while exposure to ionic silver fails to do so. This confirms that different mechanisms of cell death are triggered by different forms of silver [151]. Programmed cell death, which is presumed to be specifically evoked by nanoparticulate silver, has been shown to be induced by oxidative stress [109]. An increase in production of free radicals upon exposure to AgNPs is a further indication of activation of c-Jun N-terminal kinase (JNK) [147]. In the next step, members of the Bcl-2 family of proteins

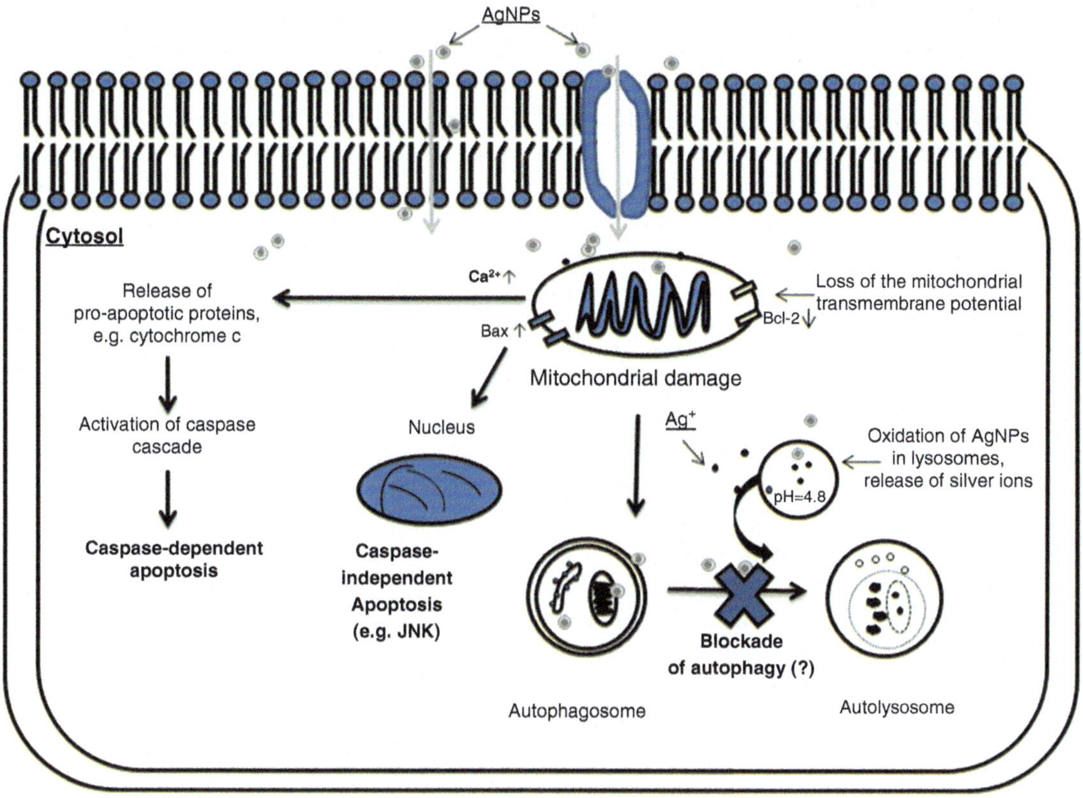

Fig. 14.3 Schematic diagram showing the pathways of cellular death induced by nanoparticulate silver

which control and regulate apoptotic mitochondrial events [152], have been shown to be activated upon exposure to AgNPs [147, 148, 150].

14.4.5.2 Autophagy

The potential of AgNPs to disrupt mitochondria leads to lower efficiency in ATP production and loss of cellular energy. Concomitantly, AgNP-induced oxidative stress may cause accumulation of damaged proteins, which in turn, may provide signals to activate the process of autophagy which promotes cell survival by using the misfolded proteins and damaged cellular structures as alternative energy sources. Autophagy is a major protein degradation pathway representing an evolutionarily conserved catabolic process which is crucial for cellular homeostasis. Autophagy is involved in proteolysis of distinct proteins and degradation of large protein complexes or damaged organelles [153]. At the ultrastructural level, autophagy starts with formation of a phagophore (isolation membrane), which subsequently elongates and closes into an autophagosome. The membrane used in construction of the autophagosome is presumed to originate from mitochondria or endoplasmic reticulum [154]. Inside this structure, cytoplasmic compartments containing damaged proteins or organelles are enclosed and subsequently, lysosomes are attached to form the autolysosome. Lysosomal enzymes hydrolyze protein substrates and recovering macromolecules to be reused by the cell. The physiological process of autophagy protects the cell against aggregated proteins and damaged organelles which, while agglomerating inside the cell, may lead to development of neurodegenerative diseases and cancer. However, prolonged autophagy may induce cell death through excessive autolysis or apoptosis [155]. In fact, this process has been described under both normal and pathological conditions, exhibiting pro-survival or pro-death effects and being involved in a variety of processes, such as differentiation, inflammation, cellular senescence or responses to metabolic stress [156].

As mentioned, after being internalized by the cell, AgNPs are stored within the endosomes and lysosomes. Hence, the suggestion has emerged that AgNPs may impair lysosomal functions thereby interfering with the autophagy process [157]. AgNPs have been shown to be an inductor of autophagy in the rat liver, where it may serve as a cellular defensive process against AgNP-induced energy reduction [148]. It has also been suggested that autophagy occurs in hippocampus of rats exposed to a low dose of AgNPs [75]. Although experiments directly confirming this process have not been conducted in the above-mentioned study, characteristic features of cell degeneration were observed such as myelin-like bodies containing synaptic vesicle clusters or mitochondrial remnants. The authors of this work hypothesize that these structures are indicative of the initial stages of autophagy and that the process occurs for efficient removal of oxidatively damaged proteins and/or organelles because of the response to AgNP-induced oxidative stress. Furthermore, the appearance of multi-vesicular bodies, a special kind of late endosomes crucial for efficient autophagic degradation [158], suggest the occurrence of autophagy in the brain tissue of AgNP-exposed rats.

Presumably, the induction of autophagy is a universal cellular effect evoked by nanomaterials. It has been claimed recently that the autophagic processes may be impaired after exposure to nanoparticulate materials such as AgNPs. The interruption of autophagy most frequently applies to the blockade of autophagosome-lysosome fusion. Accumulation of aberrantly enlarged autophagic vacuoles containing partially disintegrated material may be regarded as a phenotypic feature of defective autophagic flux [157] (Fig. 14.3).

In cultured mouse embryonic fibroblasts, AgNPs induce formation of cytoplasmic acidic vesicular organelles (i.e. autophagosomes and autolysosomes), several of which were found to contain cellular debris and/or particulate material. Furthermore, a dose-dependent increase in protein markers of autophagy was observed [148]. The authors suggest that increased conversion of LC3-I to LC3-II and accumulation of p62 proteins indicate that AgNP-induced autophagy occurs over time followed by blockade of later stages. The impairment of this crucial cellular process of adaptation may be an important aspect of the mechanism of toxicity exerted by AgNPs with far-reaching consequences of imbalanced homeostasis. More importantly, impairment of the autophagic process has been implicated in a variety of pathologies such as neurodegeneration, cancer, cardiovascular diseases, and immune system-related disorders [159]. This indicates a need for thorough investigations of this problem, particularly in the context of possible environmental exposure to AgNPs.

14.5 Summary

Increasing the number of AgNP-based products and medical tools creates potential hazards to human health. The high reactivity of AgNPs, and its ability to effectively cross cell membranes and accumulate in brain structures are of great concern in neuroscience. Most of our current knowledge on the neurotoxic impact of AgNPs has been collected from in vitro studies. However, it is known that transfer of nanoparticles throughout body compartments, and interactions with biological fluids and biomolecules (particularly proteins), causes their characteristics to change and influences their cellular fate and toxicity. Thus, the neurotoxic effects must be investigated thoroughly using animal models with realistic doses of exposure to assess the nano-bio interactions which influence structures and functions at the organism-level. Importantly, in attempting to relate the experimental doses and periods of exposure to realistic situations, we will provide answers which allow us to improve the biosafety of AgNP-based nanotools for medicine and to estimate the benefit-to-risk ratio of continued use of AgNP-containing products.

References

1. Marin S, Vlasceanu GM, Tiplea RE et al (2015) Applications and toxicity of silver nanoparticles: a recent review. Curr Top Med Chem 15:1596–1604
2. Skalska J, Strużyńska L (2015) Toxic effects of silver nanoparticles in mammals-does a risk of neurotoxicity exist? Folia Neuropathol 53:281–300
3. Rejeski D (2011) Project on emerging nanotechnologies. Woodrow Wilson International Center for Scholars, Washington, DC
4. Chen X, Schluesener HJ (2008) Nanosilver: a nanoproduct in medical application. Toxicol Lett 176:1–12
5. Clement JL, Jarrett PS (1994) Antibacterial silver. Met Based Drugs 1:467–482
6. Allahverdiyev AM, Abamor ES, Bagirova M et al (2011) Antimicrobial effects of TiO(2) and Ag(2) O nanoparticles against drug-resistant bacteria and leishmania parasites. Future Microbiol 6:933–940
7. Gajbhiye M, Kesharwani J, Ingle A et al (2009) Fungus-mediated synthesis of silver nanoparticles and their activity against pathogenic fungi in combination with fluconazole. Nanomedicine 5:382–386
8. Ge L, Li Q, Wang M et al (2014) Nanosilver particles in medical applications: synthesis, performance, and toxicity. Int J Nanomedicine 9:2399–2407
9. Baram-Pinto D, Shukla S, Perkas N et al (2009) Inhibition of herpes simplex virus type 1 infection by silver nanoparticles capped with mercaptoethane sulfonate. Bioconjug Chem 20:1497–1502
10. Lu L, Sun RW, Chen R et al (2008) Silver nanoparticles inhibit hepatitis B virus replication. Antivir Ther 13:253–262
11. Elechiguerra JL, Burt JL, Morones JR et al (2005) Interaction of silver nanoparticles with HIV-1. J Nanobiotechnol 3:6
12. Bryers JD (2008) Medical biofilms. Biotechnol Bioeng 100(1):1–18
13. Rigo C, Ferroni L, Tocco I et al (2013) Active silver nanoparticles for wound healing. Int J Mol Sci 14:4817–4840
14. Sibbald RG, Contreras-Ruiz J, Coutts P et al (2007) Bacteriology, inflammation, and healing: a study of nanocrystalline silver dressings in chronic venous leg ulcers. Adv Skin Wound Care 20:549–558
15. Jain PK, Huang XH, El-Sayed IH et al (2008) Noble metals on the nanoscale: optical and photothermal properties and some applications in imaging, sensing, biology, and medicine. Acc Chem Res 41:1578–1586
16. Austin LA, Mackey MA, Dreaden EC et al (2014) The optical, photothermal, and facile surface chemical properties of gold and silver nanoparticles in biodiagnostics, therapy, and drug delivery. Arch Toxicol 88:1391–1417
17. Parveen S, Misra R, Sahoo SK (2012) Nanoparticles: a boon to drug delivery, therapeutics, diagnostics and imaging. Nanomedicine 8:147–166
18. Hayden SC, Austin LA, Near RD et al (2013) Plasmonic enhancement of photodynamic cancer therapy. J Photochem Photobiol A 269:34–41
19. Ting BP, Zhang J, Gao ZQ et al (2009) A DNA biosensor based on the detection of doxorubicin-conjugated Ag nanoparticle labels using solid-state voltammetry. Biosens Bioelectron 25:282–287
20. Lin J, Chen R, Feng S et al (2011) A novel blood plasma analysis technique combining membrane electrophoresis with silver nanoparticle-based SERS spectroscopy for potential applications in noninvasive cancer detection. Nanomedicine 7:655–663
21. Petryayeva E, Krull UJ (2011) Localized surface plasmon resonance: nanostructures, bioassays and biosensing—a review. Anal Chim Acta 706:8–24
22. Haes AJ, Hall WP, Chang L et al (2004) A localized surface plasmon resonance biosensor: first steps toward an assay for Alzheimer's disease. Nano Lett 4:1029–1034
23. Cincotto FH, Canevari TC, Campos AM et al (2014) Simultaneous determination of epinephrine and dopamine by electrochemical reduction on the hybrid material SiO2/graphene oxide decorated with Ag nanoparticles. Analyst 139:4634–4640
24. Tashkhourian J, Hormozi-Nezhad MR, Khodaveisi J et al (2011) A novel photometric glucose biosensor based on decolorizing of silver nanoparticles. Sens Actuators B 158:185–189
25. Ceresa C, Bravin A, Cavaletti G et al (2014) The combined therapeutical effect of metal-based drugs and radiation therapy: the present status of research. Curr Med Chem 21:2237–2265
26. Liu P, Huang Z, Chen Z et al (2013) Silver nanoparticles: a novel radiation sensitizer for glioma? Nanoscale 5:11829–11836
27. Huang P, Yang DP, Zhang C et al (2011) Protein-directed one-pot synthesis of Ag microspheres with good biocompatibility and enhancement of radiation effects on gastric cancer cells. Nanoscale 3:3623–3626
28. El-Sayed MA (2001) Some interesting properties of metals confined in time and nanometer space of different shapes. Acc Chem Res 34:257–264
29. Jiang T, Song J, Zhang W et al (2015) Au-Ag-Au hollow nanostructure with enhanced chemical stability and improved photothermal transduction efficiency for cancer treatment. ACS Appl Mater Interfaces 7:21985–21994
30. Li J, Wang W, Zhao L et al (2015) Hydroquinone-assisted synthesis of branched au-ag nanoparticles with polydopamine coating as highly efficient photothermal agents. ACS Appl Mater Interfaces 7:11613–11623
31. Gurunathan S, Lee KJ, Kalishwaralal K et al (2009) Antiangiogenic properties of silver nanoparticles. Biomaterials 30:6341–6350
32. Noh MS, Lee S, Kang H et al (2015) Target-specific near-IR induced drug release and photothermal therapy with accumulated Au/Ag hollow nanoshells

on pulmonary cancer cell membranes. Biomaterials 45:81–92
33. Locatelli E, Naddaka M, Uboldi C et al (2014) Targeted delivery of silver nanoparticles and alisertib: in vitro and in vivo synergistic effect against glioblastoma. Nanomedicine (Lond) 9:839–849
34. Leite PE, Pereira MR, Granjeiro JM (2015) Hazard effects of nanoparticles in central nervous system: searching for biocompatible nanomaterials for drug delivery. Toxicol In Vitro 29:1653–1660
35. Kasyanenko N, Bakulev V, Perevyazko I et al (2016) Model system for multifunctional delivery nanoplatforms based on DNA-Polymer complexes containing silver nanoparticles and fluorescent dye. J Biotechnol 236:78–87
36. Mukherjee S, Chowdhury D, Kotcherlakota R et al (2014) Potential theranostics application of biosynthesized silver nanoparticles (4-in-1 system). Theranostics 4:316–335
37. Nel AE, Mädler L, Velegol D et al (2009) Understanding biophysicochemical interactions at the nano-bio interface. Nat Mater 8:543–557
38. Cheng LC, Jiang X, Wang J et al (2013) Nano-bio effects: interaction of nanomaterials with cells. Nanoscale 5:3547–3569
39. Shannahan JH, Lai X, Ke PC et al (2013) Silver nanoparticle protein corona composition in cell culture media. PLoS One 8:e74001
40. Lee JH, Kim YS, Song KS et al (2013) Biopersistence of silver nanoparticles in tissues from Sprague-Dawley rats. Part Fibre Toxicol 10:36
41. Park MV, Neigh AM, Vermeulen JP et al (2011) The effect of particle size on the cytotoxicity, inflammation, developmental toxicity and genotoxicity of silver nanoparticles. Biomaterials 32:9810–9817
42. Gliga AR, Skoglund S, Wallinder IO et al (2014) Size-dependent cytotoxicity of silver nanoparticles in human lung cells: the role of cellular uptake, agglomeration and Ag release. Part Fibre Toxicol 11:11
43. Haase A, Rott S, Mantion A et al (2012) Effects of silver nanoparticles on primary mixed neural cell cultures: uptake, oxidative stress and acute calcium responses. Toxicol Sci 126:457–468
44. Liu W, Wu Y, Wang C et al (2010) Impact of silver nanoparticles on human cells: effect of particle size. Nanotoxicology 4:319–330
45. Carlson C, Hussain SM, Schrand AM et al (2008) Unique cellular interaction of silver nanoparticles: size-dependent generation of reactive oxygen species. J Phys Chem B 112:13608–13619
46. Sharma A, Muresanu DF, Patnaik R et al (2013) Size- and age-dependent neurotoxicity of engineered metal nanoparticles in rats. Mol Neurobiol 48:386–396
47. Verano-Braga T, Miethling-Graff R, Wojdyla K et al (2014) Insights into the cellular response triggered by silver nanoparticles using quantitative proteomics. ACS Nano 8:2161–2175
48. Espinosa-Cristobal LF, Martinez-Castañon GA, Loyola-Rodriguez JP et al (2013) Toxicity, distribution, and accumulation of silver nanoparticles in Wistar rats. J Nanopart Res 15:1702
49. Ashkarran AA, Ghavami M, Aghaverdi H et al (2012) Bacterial effects and protein corona evaluations: crucial ignored factors. Chem Res Toxicol 25:1231–1242
50. Durán N, Silveira CP, Durán M et al (2015) Silver nanoparticle protein coronaand toxicity: a minireview. J Nanobiotechnol 13:55
51. Sadeghi B, Garmaroudi FS, Hashemi M et al (2012) Comparison of the anti-bacterial activity on the nanosilver shapes: nanoparticles, nanorods and nanoplates. Adv Powder Technol 23:22–26
52. Topuz E, Sigg L, Talinli I (2014) A systematic evaluation of agglomeration of Ag and TiO_2nanoparticles under freshwater relevant conditions. Environ Pollut 193:37–44
53. Ahamed M, Karns M, Goodson M et al (2008) DNA damage response to different surface chemistry of silver nanoparticles in mammalian cells. Toxicol Appl Pharmacol 233:404–410
54. Treuel L, Malissek M, Grass S et al (2012) Quantifying the influence of polymer coatings on the serum albumin corona formation around silver and gold nanoparticles. J Nanopart Res 14:1102
55. Nishanth RP, Jyotsna RG, Schlager JJ et al (2011) Inflammatory responses of RAW 264.7 macrophages upon exposure to nanoparticles: role of ROS-NFκB signaling pathway. Nanotoxicology 5:502–516
56. Liu J, Hurt RH (2010) Ion release kinetics and particle persistence in aqueous nano-silver colloids. Environ Sci Technol 44:2169–2175
57. van der Zande M, Vandebriel RJ, Van Doren E et al (2012) Distribution, elimination, and toxicity of silver nanoparticles and silver ions in rats after 28-day oral exposure. ACS Nano 6:7427–7442
58. Bachler G, von Goetz N, Hungerbuhler K (2013) A physiologically based pharmacokinetic model for ionic silver and silver nanoparticles. Int J Nanomed 8:3365–3382
59. Juling S, Bachler G, von Götz N et al (2016) In vivo distribution of nanosilver in the rat: the role of ions and de novo-formed secondary particles. Food Chem Toxicol 97:327–335
60. Hsiao IL, Hsieh YK, Wang CF et al (2015) Trojan-horse mechanism in the cellular uptake of silver nanoparticles verified by direct intra- and extracellular silver speciation analysis. Environ Sci Technol 49:3813–3821
61. Park EJ, Yi J, Kim Y et al (2010) Silver nanoparticles induce cytotoxicity by a Trojan-horse type mechanism. Toxicol In Vitro 24:872–878
62. Cronholm P, Karlsson HL, Hedberg J et al (2013) Intracellular uptake and toxicity of Ag and CuO nanoparticles: a comparison between nanoparticles and their corresponding metal ions. Small 9:970–982

63. Pratsinis A, Hervella P, Leroux JC et al (2013) Toxicity of silver nanoparticles in macrophages. Small 9:2576–2584
64. Ziemińska E, Stafiej A, Strużyńska L (2014) The role of the glutamatergic NMDA receptor in nanosilver-evoked neurotoxicity in primary cultures of cerebellar granule cells. Toxicology 315:38–48
65. Sun C, Yin N, Wen R et al (2016) Silver nanoparticles induced neurotoxicity through oxidative stress in rat cerebral astrocytes is distinct from the effects of silver ions. Neurotoxicology 52:210–221
66. Beer C, Foldbjerg R, Hayashi Y et al (2012) Toxicity of silver nanoparticles – nanoparticle or silver ion? Toxicol Lett 208:286–292
67. Vroman L, Adams AL, Fischer GC et al (1980) Interaction of high molecular weight kininogen, factor XII, and fibrinogen in plasma at interfaces. Blood 55(1):156–159
68. Li J, Mao H, Kawazoe N et al (2017) Insight into the interactions between nanoparticles and cells. Biomater Sci 5:173–189
69. Rahman M, Laurent S, Tawil N et al (2013) Protein-nanoparticle interactions. The bio-nano interface. In: Martinac B (ed) Springer series in biophysics, vol 15. Springer-Verlag, Berlin, pp 1–64
70. Lee YK, Choi EJ, Webster TJ et al (2015) Effect of the protein corona on nanoparticles for modulating cytotoxicity and immunotoxicity. Inter J Nanomed 10:97–113
71. Lesniak A, Fenaroli F, Monopoli MP et al (2012) Effects of the presence or absence of a protein corona on silica nanoparticle uptake and impact on cells. ACS Nano 6:5845–5857
72. Miclăuş T, Beer C, Chevallier J et al (2016) Dynamic protein coronas revealed as a modulator of silver nanoparticle sulphidation in vitro. Nat Commun 7:11770
73. Loeschner K, Hadrup N, Qvortrup K et al (2011) Distribution of silver in rats following 28 days of repeated oral exposure to silver nanoparticles or silver acetate. Part Fibre Toxicol 8:18
74. Sung JH, Ji JH, Park JD et al (2009) Subchronic inhalation toxicity of silver nanoparticles. Toxicol Sci 108:452–461
75. Skalska J, Frontczak-Baniewicz M, Strużyńska L (2015) Synaptic degeneration in rat brain after prolonged oral exposure to silver nanoparticles. Neurotoxicology 46:145–154
76. De Matteis V, Malvindi MA, Galeone A et al (2015) Negligible particle-specific toxicity mechanism of silver nanoparticles: the role of Ag+ ion release in the cytosol. Nanomedicine 11:731–739
77. Casas-Grajales S, Muriel P (2015) Antioxidants in liver health. World J Gastrointest Pharmacol Ther 6:59–72
78. Tang J, Xiong L, Wang S et al (2008) Influence of silver nanoparticles on neurons and blood-brain barrier via subcutaneous injection in rats. Appl Surf Sci 255:502–504
79. Liu Y, Guan W, Ren G et al (2012) The possible mechanism of silver nanoparticle impact on hippocampal synaptic plasticity and spatial cognition in rats. Toxicol Lett 209:227–231
80. Asharani PV, Low Kah Mun G, Hande MP et al (2009) Cytotoxicity and genotoxicity of silver nanoparticles in human cells. ACS Nano 3:279–290
81. Garcia-Alonso J, Khan FR, Misra SK et al (2011) Cellular internalization of silver nanoparticles in gut epithelia of the estuarine polychaete nereis diversicolor. Environ Sci Technol 45:4630–4636
82. Guarnieri D, Sabella S, Muscetti O et al (2014) Transport across the cell-membrane dictates nanoparticle fate and toxicity: a new paradigm in nanotoxicology. Nanoscale 6:10264–10273
83. Nohl H, Gille L (2005) Lysosomal ROS formation. Redox Rep 10:199–205
84. Sabella S, Carney RP, Brunetti V et al (2014) A general mechanism for intracellular toxicity of metal-containing nanoparticles. Nanoscale 6:7052–7061
85. Skalska J, Dąbrowska-Bouta B, Strużyńska L (2016) Oxidative stress in rat brain but not in liver following oral administration of a low dose of nanoparticulate silver. Food Chem Toxicol 97:307–315
86. Davenport LL, Hsieh H, Eppert BL et al (2015) Systemic and behavioral effects of intranasal administration of silver nanoparticles. Neurotoxicol Teratol 51:68–76
87. Yin N, Zhang Y, Yun Z et al (2015) Silver nanoparticle exposure induces rat motor dysfunction through decrease in expression of calcium channel protein in cerebellum. Toxicol Lett 237:112–120
88. Dąbrowska-Bouta B, Zięba M, Orzelska-Górka J et al (2016) Influence of a low dose of silver nanoparticles on cerebral myelin and behavior of adult rats. Toxicology 363–364:29–36
89. Sharma HS, Ali SF, Hussain SM et al (2009) Influence of engineered nanoparticles from metals on the blood-brain barrier permeability, cerebral blood flow, brain edema and neurotoxicity. An experimental study in the rat and mice using biochemical and morphological approaches. J Nanosci Nanotechnol 9:5055–5072
90. Serlin Y, Shelef I, Knyazer B et al (2015) Anatomy and physiology of the blood-brain barrier. Semin Cell Dev Biol 38:2–6
91. Ballabh P, Braun A, Nedergaard M (2004) The blood-brain barrier: an overview: structure, regulation, and clinical implications. Neurobiol Dis 16:1–13
92. Persidsky Y, Ramirez SH, Haorah J et al (2006) Blood-brain barrier: structural components and function under physiologic and pathologic conditions. J NeuroImmune Pharmacol 1:223–236
93. Landgraf L, Müller I, Ernst P et al (2015) Comparative evaluation of the impact on endothelial cells induced by different nanoparticle structures and functionalization. Beilstein J Nanotechnol 6:300–312

94. Ye D, Raghnaill MN, Bramini M et al (2013) Nanoparticle accumulation and transcytosis in brain endothelial cell layers. Nanoscale 5:11153–11165
95. Hanada S, Fujioka K, Inoue Y et al (2014) Cell-based in vitro blood-brain barrier model can rapidly evaluate nanoparticles' brain permeability in association with particle size and surface modification. Int J Mol Sci 15:1812–1825
96. Tang J, Xiong L, Zhou G et al (2010) Silver nanoparticles crossing through and distribution in the blood-brain barrier in vitro. J Nanosci Nanotechnol 10:6313–6317
97. Sharma HS, Hussain S, Schlager J et al (2010) Influence of nanoparticles on blood-brain barrier permeability and brain edema formation in rats. Acta Neurochir Suppl 106:359–364
98. Trickler WJ, Lantz SM, Murdock RC et al (2010) Silver nanoparticle induced blood-brain barrier inflammation and increased permeability in primary rat brain microvessel endothelial cells. Toxicol Sci 118:160–170
99. Trickler WJ, Lantz-McPeak SM, Robinson BL et al (2014) Porcine brain microvessel endothelial cells show pro-inflammatory response to the size and composition of metallic nanoparticles. Drug Metab Rev 46:224–231
100. Xu L, Dan M, Shao A et al (2015a) Silver nanoparticles induce tight junction disruption and astrocyte neurotoxicity in a rat blood-brain barrier primary triple coculture model. Int J Nanomedicine 10:6105–6118
101. Xu L, Shao A, Zhao Y et al (2015b) Neurotoxicity of silver nanoparticles in rat brain after intragastric exposure. J Nanosci Nanotechnol 15:4215–4223
102. Didier N, Romero IA, Creminon C et al (2003) Secretion of interleukin-1beta by astrocytes mediates endothelin-1 and tumour necrosis factor-alpha effects on human brain microvascular endothelial cell permeability. J Neurochem 86:246–254
103. Erickson MA, Dohi K, Banks WA (2012) Neuroinflammation: a common pathway in CNS diseases as mediated at the blood-brain barrier. Neuroimmunomodulation 19:121–130
104. Chen IC, Hsiao IL, Lin HC et al (2016) Influence of silver and titanium dioxide nanoparticles on in vitro blood-brain barrier permeability. Environ Toxicol Pharmacol 47:108–118
105. Heusinkveld HJ, Wahle T, Campbell A et al (2016) Neurodegenerative and neurological disorders by small inhaled particles. Neurotoxicology 56:94–106
106. Aliev G, Daza J, Herrera AS et al (2015) Nanoparticles as alternative strategies for drug delivery to the Alzheimer brain: electron microscopy ultrastructural analysis. CNS Neurol Disord Drug Targets 14:1235–1242
107. Costa LG, Giordano G, Guizzetti M et al (2008) Neurotoxicity of pesticides: a brief review. Front Biosci 13:1240–1249
108. Xu F, Piett C, Farkas S et al (2013) Silver nanoparticles (AgNPs) cause degeneration of cytoskeleton and disrupt synaptic machinery of cultured cortical neurons. Mol Brain 6:29
109. Yin N, Liu Q, Liu J et al (2013) Silver nanoparticle exposure attenuates the viability of rat cerebellum granule cells through apoptosis coupled to oxidative stress. Small 9:1831–1841
110. Masliah E, Terry R (1993) The role of synaptic proteins in the pathogenesis of disorders of the central nervous system. Brain Pathol 3:77–85
111. Keith D, El-Husseini A (2008) Excitation control: balancing PSD-95 function at the synapse. Front Mol Neurosci 1:4
112. Masliah E (2001) Recent advances in the understanding of the role of synaptic proteins in Alzheimer's disease and other neurodegenerative disorders. J Alzheimers Dis 3:121–129
113. Liu P, Huang Z, Gu N (2013) Exposure to silver nanoparticles does not affect cognitive outcome or hippocampal neurogenesis in adult mice. Ecotoxicol Environ Saf 87:124–130
114. Baumann N, Pham-Dinh D (2001) Biology of oligodendrocyte and myelin in the mammalian central nervous system. Physiol Rev 81:871–927
115. Ben-Nun A, Mendel I, Bakimer R et al (1996) The autoimmune reactivity to myelin oligodendrocyte glycoprotein (MOG) in multiple sclerosis is potentially pathogenic: effect of copolymer 1 on MOG-induced disease. J Neurol 243(4 Suppl 1):S14–S22
116. Oberheim NA, Goldman SA, Nedergaard M (2012) Heterogeneity of astrocytic form and function. Methods Mol Biol 814:23–45
117. Verkhratsky A, Nedergaard M, Hertz L (2015) Why are astrocytes important? Neurochem Res 40:389–401
118. Freitas-Andrade M, Naus CC (2016) Astrocytes in neuroprotection and neurodegeneration: the role of connexin43 and pannexin1. Neuroscience 323:207–221
119. Strużyńska L (2000) The protective role of astroglia in the early period of experimental lead toxicity in the rat. Acta Neurobiol Exp (Wars) 60:167–173
120. Strużyńska L, Dąbrowska-Bouta B, Koza K et al (2007) Inflammation-like glial response in lead-exposed immature rat brain. Toxicol Sci 95:156–162
121. Strużyńska L (2009) A glutamatergic component of lead toxicity in adult brain: the role of astrocytic glutamate transporters. Neurochem Int 55:151–156
122. Luther EM, Koehler Y, Diendorf J et al (2011) Accumulation of silver nanoparticles by cultured primary brain astrocytes. Nanotechnology 22:375101
123. Huang CL, Hsiao IL, Lin HC et al (2015) Silver nanoparticles affect on gene expression of inflammatory and neurodegenerative responses in mouse brain neural cells. Environ Res 136:253–263
124. Hsiao IL, Hsieh YK, Chuang CY et al (2017) Effects of silver nanoparticles on the interactions of neuron- and glia-like cells: toxicity, uptake mechanisms, and

lysosomal tracking. Environ Toxicol. https://doi.org/10.1002/tox.22397
125. Luther EM, Schmidt MM, Diendorf J et al (2012) Upregulation of metallothioneins after exposure of cultured primary astrocytes to silver nanoparticles. Neurochem Res 37:1639–1648
126. Ruttkay-Nedecky B, Nejdl L, Gumulec J et al (2013) The role of metallothionein in oxidative stress. Int J Mol Sci 14:6044–6066
127. Elsaesser A, Howard CV (2012) Toxicology of nanoparticles. Adv Drug Deliv Rev 64:129–137
128. Shvedova AA, Pietroiusti A, Fadeel B et al (2012) Mechanisms of carbon nanotube-induced toxicity: focus on oxidative stress. Toxicol Appl Pharmacol 261:121–133
129. Verkhratsky A (2007) Calcium and cell death. In: Carafoli E, Brini M (eds) Calcium signalling and disease, Handbook of subcellular biochemistry, vol 45. Springer, Heidelberg, pp 465–480
130. Kristian T, Siesjo BK (1996) Calcium-related damage in ischemia. Life Sci 59:357–367
131. Ziemińska E, Strużyńska L (2016) Zinc modulates nanosilver-induced toxicity in primary neuronal cultures. Neurotox Res 29:325–343
132. Zhivotovsky B, Orrenius S (2011) Calcium and cell death mechanisms: a perspective from the cell death community. Cell Calcium 50:211–221
133. Meyer D, Bonhoeffer T, Scheuss V (2014) Balance and stability of synaptic structures during synaptic plasticity. Neuron 82:430–443
134. Costa CS, Ronconi JV, Daufenbach JF et al (2010) In vitro effects of silver nanoparticles on the mitochondrial respiratory chain. Mol Cell Biochem 342:51–56
135. Asharani PV, Hande MP, Valiyaveettil S (2009) Antiproliferative activity of silver nanoparticles. BMC Cell Biol 10:65
136. Almofti MR, Ichikawa T, Yamashita K et al (2003) Silver ion induces a cyclosporine a-insensitive permeability transition in rat liver mitochondria and release of apoptogenic cytochrome C. J Biochem 134:43–49
137. Roberts RA, Laskin DL, Smith CV et al (2009) Nitrative and oxidative stress in toxicology and disease. Toxicol Sci 112:4–16
138. Friedman J (2011) Why is the nervous system vulnerable to oxidative stress? In: Gadoth N, Gobel HH (eds) Oxidative stress and free radical damage in neurology. Handbook of oxidative stress in applied basic research and clinical practice. Humana Press, Totowa, pp 19–27
139. Sayre LM, Perry G, Smith MA (2008) Oxidative stress and neurotoxicity. Chem Res Toxicol 21:172–188
140. Krawczyńska A, Dziendzikowska K, Gromadzka-Ostrowska J et al (2015) Silver and titanium dioxide nanoparticles alter oxidative/inflammatory response and renin-angiotensin system in brain. Food Chem Toxicol 85:96–105
141. Rahman MF, Wang J, Patterson TA et al (2009) Expression of genes related to oxidative stress in the mouse brain after exposure to silver-25 nanoparticles. Toxicol Lett 187:15–21
142. Piao MJ, Kang KA, Lee IK et al (2011) Silver nanoparticles induce oxidative cell damage in human liver cells through inhibition of reduced glutathione and induction of mitochondria-involved apoptosis. Toxicol Lett 201:92–100
143. Morgan MJ, Liu ZG (2011) Crosstalk of reactive oxygen species and NF-κB signaling. Cell Res 21:103–115
144. Prasad RY, McGee JK, Killius MG et al (2013) Investigating oxidative stress and inflammatory responses elicited by silver nanoparticles using high-throughput reporter genes in HepG2 cells: effect of size, surface coating, and intracellular uptake. Toxicol In Vitro 27:2013–2021
145. Foldbjerg R, Dang DA, Autrup H (2011) Cytotoxicity and genotoxicity of silver nanoparticles in the human lung cancer cell line, A549. Arch Toxicol 85:743–750
146. Sanpui P, Chattopadhyay A, Ghosh SS (2011) Induction of apoptosis in cancer cells at low silver nanoparticle concentrations using chitosan nanocarrier. ACS Appl Mater Interfaces 3:218–228
147. Hsin YH, Chen CF, Huang S et al (2008) The apoptotic effect of nanosilver is mediated by a ROS- and JNK-dependent mechanism involving the mitochondrial pathway in NIH3T3 cells. Toxicol Lett 179:130–139
148. Lee YH, Cheng FY, Chiu HW, Tsai JC, Fang CY, Chen CW, Wang YJ (2014) Cytotoxicity, oxidative stress, apoptosis and the autophagic effects of silver nanoparticles in mouse embryonic fibroblasts. Biomaterials 35:4706–4715
149. Elmore S (2007) Apoptosis: a review of programmed cell death. Toxicol Pathol 35:495–516
150. Gopinath P, Gogoi SK, Sanpui P et al (2010) Signaling gene cascade in silver nanoparticle induced apoptosis. Colloids Surf B: Biointerfaces 77:240–245
151. Sambale F, Wagner S, Stahl F et al (2015) Investigations of the toxic effect of silver nanoparticles on mammalian cell lines. J Nanomater 16(1):article No. 6
152. Cory S, Adams JM (2002) The Bcl2 family: regulators of the cellular life-or-death switch. Nat Rev Cancer 2:647–656
153. Kraft C, Peter M, Hofmann K (2010) Selective autophagy: ubiquitinmediated recognition and beyond. Nat Cell Biol 12:836–841
154. Hansen TE, Johansen T (2011) Following autophagy step by step. BMC Biol 9:39
155. Glick D, Barth S, Macleod KF (2010) Autophagy: cellular and molecular mechanisms. J Pathol 221:3–12

156. Shen HM, Codogno P (2012) Autophagy is a survival force via suppression of necrotic cell death. Exp Cell Res 318:1304–1308
157. Mao BH, Tsai JC, Chen CW et al (2016) Mechanisms of silver nanoparticle-induced toxicity and important role of autophagy. Nanotoxicology 10:1021–1040
158. Fader CM, Colombo MI (2009) Autophagy and multivesicular bodies: two closely related partners. Cell Death Differ 16:70–78
159. Zhang XJ, Chen S, Huang KX et al (2013) Why should autophagic flux be assessed? Acta Pharmacol Sin 34:595–599

15 Toxic and Beneficial Potential of Silver Nanoparticles: The Two Sides of the Same Coin

Lilian Rodrigues Rosa Souza,
Veronica Santana da Silva,
Leonardo Pereira Franchi,
and Tiago Alves Jorge de Souza

Abstract

Nanotechnology has allowed great changes in chemical, biological and physical properties of metals when compared to their bulk counterparts. Within this context, silver nanoparticles (AgNPs) play a major role due to their unique properties, being widely used in daily products such as fabrics, washing machines, water filters, food and medicine. However, AgNPs can enter cells inducing a "Trojan-horse" type mechanism which potentially leads to cellular autophagy, apoptosis or necrosis. On the other hand, this cytotoxicity mechanism can be optimized to develop drug nanocarriers and anticancer therapies. The increasing use of these NPs entails their release into the environment, damaging ecosystems balance and representing a threat to human health. In this context, the possible deleterious effects that these NPs may represent for the biotic and abiotic ecosystems components represent an obstacle that must be overcome in order to guarantee the safety use of their unique properties.

Keywords

Silver nanoparticles · AgNPs · Nanotoxicity · Aquatic toxicity · Cytotoxicity

L.R.R. Souza
Department of Chemistry, FFCLRP-USP, University of São Paulo – USP, Ribeirão Preto, SP, Brazil

V.S. da Silva
Department of Biology, ICB-UFMG, Federal University of Minas Gerais – UFMG, Belo Horizonte, MG, Brazil

L.P. Franchi
Department of Chemistry, Faculty of Philosophy, Sciences and Letters of Ribeirão Preto; USP, Ribeirão Preto, São Paulo, Brazil

T.A.J. de Souza (✉)
Department of Genetics, Faculty of Medicine of Ribeirão Preto; USP, Ribeirão Preto, São Paulo, Brazil
e-mail: tiagoajs@hotmail.com

Q. Saquib et al. (eds.), *Cellular and Molecular Toxicology of Nanoparticles*, Advances in Experimental Medicine and Biology 1048, https://doi.org/10.1007/978-3-319-72041-8_15

15.1 Introduction

The use silver nanoparticles (AgNPs) has become common in daily products employed in modern society. These nanoparticles (NPs) can be founded in wound dressings, food pack, medical devices and even in textiles industry [1–3]. Surprisingly, in a list of 1015 products present on the markets containing NPs, 259 of them contain AgNPs [4]. Therefore, due to their unique properties, these NPs have been entered in our houses without ask license exposing us to an unknown threat. Moreover, their physical, chemical and optical properties are being highly studied and exploited by researchers across the world for different purposes [5, 6]. However, the utilization of antibacterial potential of silver is not recent but it goes back to Hippocrates, the father of medicine, who used the silver to treat ulcers and spread the biocidal potential of this metal to the west civilization [7].

The AgNPs are formed by agglomerates of silver atoms ranging from 1 to 100 nm which are metallically bonded. Owing to their nano-size, these NPs present large surface area ratio and high reactivity being sensitive to oxygen [5]. The study of the relationship between silver nanotechnology's and its possible toxic effects to man health is relatively new and some studies have shown that these NPs can be toxic to mammalian cells [8–10].

In the sixteenth century, Paracelsus, known as the father of toxicology stated that all substances are potentially harmful, what makes something into a poison is just the dose. In this context, the elucidation of dose response of cytotoxic effect induced by AgNPs will allow their safe use for a multitude of industrial applications, as well as their employment for therapeutic purposes [11]. In addition to the concentration, it is important to determine the relationship between size, shape and toxicity, so that the NPs are synthesized with the desired properties.

Therefore, this chapter will explore the aspects related to the toxicity of these NPs discussing their use in the treatment of water and their consequent release into aquatic ecosystems, the influence of interaction with organic matter on their toxicity and the mechanisms of action of these NPs in cells and various organisms. The question of AgNPs safe use can be compared with decisions made from coin flipping. Initially used by Roman soldiers and known as "navia aut caput", the coin flipping offers only two different possibilities, heads or tails, which represent totally different results obtained from the same coin.

15.2 AgNP Toxicity in Aquatic Environments

Water is an essential need for life and the access to potable water is considered one of the most basic humanitarian goals. In view of this fact, the use of technologies including filtration, ultraviolet radiation, chemical treatment and desalination has been well established since the ancient civilizations [12–14]. As previously mentioned, NPs and especially AgNPs are present in a multitude of daily products. Thus, cleaning or disposal of these objects may result in the release of these NPs into the environment. Moreover, nanomaterials have become very useful in water treatment because of their different properties like high reactivity, high surface area, and high adsorption when compared to materials in macroscopic scale. Interestingly, one of the nanomaterials most used in water treatment are AgNPs. Among the various biocide mechanisms of AgNPs, attention is directed to their ability to attach to cell membrane and penetrate bacterias compromising their respiratory chain and cell division [15]. Another effect of AgNPs on bacteria is the Ag^{+} ion release. These ions interact with thiol groups resulting in enzymatic damage and preventing DNA replication [13]. Due to this bactericidal effect of AgNPs, there is a large variety of materials which employs these NPs for water disinfection (Table 15.1).

Membranes provide a physical barrier for undesirable matter based on their size. They provide a high level of automation, require less land and chemical use, and allows flexible design, besides the possibility of addition of components that improve the removal of pathogenic microor-

Table 15.1 Different AgNP coated materials used in water treatment

Material coated with AgNPs	Effectiveness	References
Membranes	Effective elimination of *E. coli, P. aeruginosa, B. subtilis, and S. aureus*	[14]
Foam	Effective elimination of *Escherichia coli*	[15]
Filter	Completely effective against *Escherichia coli*	[16]
Porous ceramic	Effective elimination of *Escherichia coli*, output count was zero	[17]
Woven fabric	100% efficient in elimination of *Escherichia coli*	[18]
Paper sheet	Significant biocidal action against *Escherichia coli* and *Enterococcus faecalis*	[19]

ganisms [13]. The addition of AgNPs in membranes is very common; these NPs can be anchored on a polymer (usually methacrylic acid copolymer due to its unique characteristics such as insolubility, mechanical strength, and macroporous nature). This method was effective to eliminate *E. coli*, *P. aeruginosa*, *B. subtilis*, and *S. aureus* [14] (Fig. 15.1).

AgNPs can be coated on foams and be used as antibacterial water filters through a non-toxic and cheap process. The polyurethane coated with AgNPs resists to storage, washing, and drying without AgNPs release and no bacterium was detected in the output water when the input water had a bacterial load of 1×10^5–1×10^6 CFU mL^{-1} (colony-forming units' mL^{-1}). Results from standard test such as "inhibition zone" and "test tube" are in agreement with WHO requirements for drinking water [15].

Another material used in water treatment is ceramics due to their long lifetime and resistance to high temperature, pressure and corrosive solutions. The addition of AgNPs makes them very useful to kill bacteria. At a flow rate of 0.01 L min^{-1}, the output count of *E. coli* was zero when the input water had a bacterial load of $\sim 10^5$ CFU mL^{-1}, proving the high efficient of this material in the water treatment [17]. The large utilization of different materials for water disinfection calls attention about the lifetime of them. Haider and colleagues [20] have studied aminated polyethersulfone-silver nanoparticles (AgNPs-APES) composite membranes and reported the release of ionic silver after 12 days [20]. Therefore, in order to ensure the safety use of these membranes and not exceed the allowed concentration for silver in drinking waters, it is important to find mechanisms to prevent the release of silver during these disinfection processes [21].

The release of silver (as NPs or ions) can affect the human health and the environment. AgNPs impregnated in coal of water filters used in home treatment devices can represent risks to human health due to the Ag^+ release in the purified water [14]. Furthermore, the sewage sludge resulting from water treatment is often used as fertilizer for agricultural soils, thus, AgNPs can be leached to aquatic systems and enter in food web by the primary producers. The first toxic effects on primary producers like algae are the decrease in chlorophyll content, damage in photosynthesis, increase of ROS (Reactive Species of Oxygen) and lipid peroxidation [22–24]. Secondary organisms like crustaceans (*Daphnia magna*) can be affected by AgNPs in water or by ingestion of primary producers and among these effects can be cited the abnormal swimming and decrease of reproduction [25, 26]. Malformations in embryos of zebrafish due to exposure to AgNPs have also been reported [10]. In this context, some studies have demonstrated that these NPs can be toxic to algae [27], fish [28], snails [29] and plants [30]. On the other hand, other analyzes have shown that the prolonged exposure to these NPs cannot be very harmful to the aquatic ecosystem. Jiang and coworkers [31], for example, demonstrated that the chronic exposure to AgNPs or $AgNO_3$ during 90 days does not significantly affect the phytoplankton biomass and the diversity of aquatic plants and animals [31].

AgNP toxicity is related to their size, shape and load, so the synthesis methods used in their manufacture must be controlled to obtain the desired properties. Moreover, the subsequent NP characterization analyzes by means of dynamic

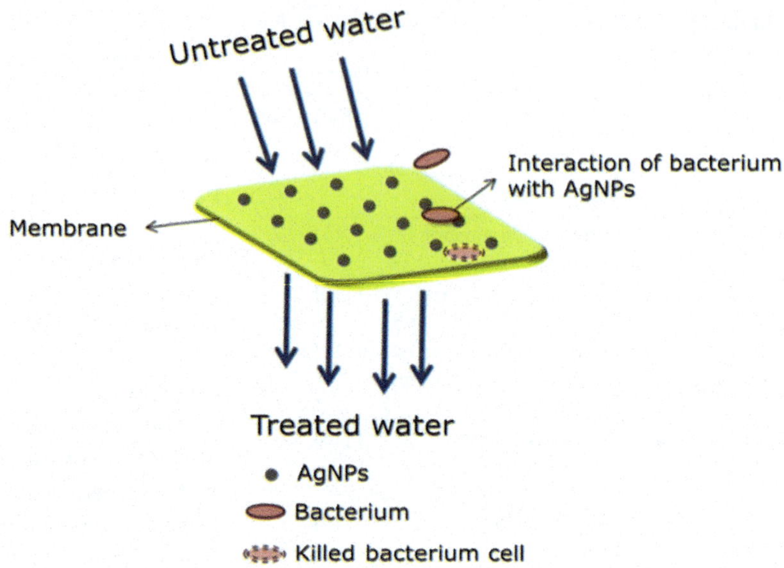

Fig. 15.1 Schematic representation of a membrane coated with AgNPs and its effects against bacteria

light scattering (DLS), scanning and transmission electron microscopy allow the comparison among different researches, thus elucidating the AgNP toxicity and helping to establish safe standards for their utilization. In this context, strategies comparing different sizes of AgNPs are extremely relevant because they elucidate the relation between size and toxicity. However, most of studies exploring AgNP toxicity disregard a very important step in determining the real risk offered by these NPs, which is the simulation of interactions between organic compounds with AgNPs after their releasing into the environment. These compounds may modulate or even inactivate the toxicity of AgNPs, demonstrating that the real damage caused by these NPs may be overestimated in in vitro and in vivo studies which do not consider these interactions.

15.3 AgNP Interaction with Natural Organic Matter (NOM)

The aquatic ecosystem is the most easily to be contaminated with NPs and their interactions and behavior with water, NPs and natural organic matter (NOM) are extremely important in studies of their toxic behavior in this environment. The NOM consists of a large variety of organic molecules that is referred as humic substances and is rich in humic and fulvic acids. These substances have great importance in binding metals, and because of this, can affect the transport and stability of metal NPs [32, 33]. When NOM is in contact with NPs, it can modify NPs proprieties by adsorption onto surface forming a coat [34].

When AgNPs enter into aquatic environment, they can interact with NOM. The adsorption of NOM by AgNPs depends on two factors: (i) the composition of NOM and (ii) the capping agent of AgNPs. If the amount of sulfur and nitrogen is higher in NOM, the adsorption increases. On the other hand, if the bonds between the capping agent and AgNPs were lost the NOM can bind to AgNPs, stabilizing them [35]. Low concentrations of NOM increase NP stability; however, high concentrations of NOM can stimulate the agglomeration of these NPs [36, 37]. The stabilization of AgNPs by NOM occurs by its adsorption on AgNPs´ surface preventing their agglomeration.

In the dissolution of AgNPs, Ag^+ species can be released; these Ag^+ ions are well known to be toxic to the environment because they induce oxidative stress [38]. However, the NOM adsorption on AgNPs affects their dissolution proprieties reducing the Ag^+ release in a dose-dependent manner. This ion releasing decreases by different mechanisms like the adsorption of NOM blocks,

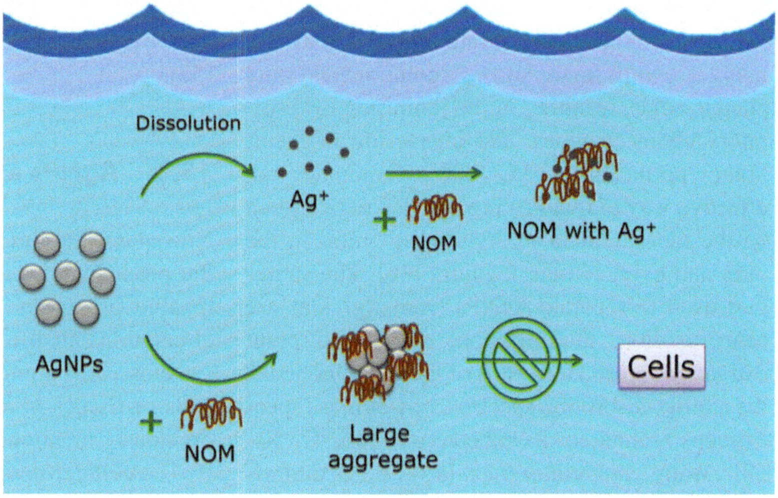

Fig. 15.2 Schematic representation of the interactions among AgNPs, NOM, and Ag^+ ions

the oxidation of AgNPs sites and the reversible reaction of Ag^+ formation from Ag^0 due to the humic/fulvic acids acting as reducing agents [39]. In addition, even when release of Ag^+ ions occurs, NOM can bind to these ions decreasing its toxicity [37, 40]. Interestingly, NOM binding to Ag^+ ions can reduce, coat, and stabilize these ions forming AgNPs naturally [41].

Due to the binding proprieties of NOM, the concentration of free Ag^+ ions can decrease, thus their toxicity also decrease [34, 35, 40]. More specifically, when NOM is in high concentrations, it stimulates the formation of AgNPs agglomeration; these agglomerates are large and cannot enter in cell membrane, thus decreasing the AgNPs toxicity, as represented in Fig. 15.2.

The decrease of toxicity of AgNPs coated with NOM is also related to their lower bioavailability. The formation of larger agglomerates with high molecular weight favors their removal into sediments, decreasing their bioavailability [41]. Besides all these factors, the composition of NOM is also an important factor in the toxicity of AgNPs: a higher amount of sulphur reduces the dissolution of AgNPs, and consequently the concentration of Ag^+ ions decrease and also their toxicity [42].

The effects caused by AgNPs in the environment are closely related to the interactions occurring in the ecosystem; and the organic matter is important in this context. These associations have impacts on the NP toxicity, decreasing the bioavailability and concentration of Ag^+ ions. Nevertheless, more studies exploring the interaction between AgNPs and organic matter should be done to unveil the real risk offered by the exposure to these NPs.

Some studies have also shown that the interaction between NPs and mammalian cells can cause lesions in the genetic material [9, 43, 44]. Our group used the micronucleus test and comet assay to demonstrate that AgNPs can induce chromosomal breaks and genotoxic damage [9]. Furthermore, high NP concentrations can be cytotoxic, causing cell death by various mechanisms such as apoptosis, necrosis and autophagy. Although numerous studies [9, 43–45] have confirmed the direct and indirect cyto- and genotoxic potential of AgNPs in vitro and in vivo, the mechanism of action of these NPs is still uncertain. However, increasing evidence has corroborated the Trojan-horse mechanism as responsible for AgNP toxicity [46, 47].

15.4 AgNPs and the Trojan Horse Mechanism

The antimicrobial potential of silver materials is related to Ag^+ ion release after the interaction with oxygen. AgNPs in aqueous solution release Ag^+ ions, which are biologically active and can mediate the bactericidal effect [6] as well as lead to significant cytotoxicity in mammalian cells [48, 48]. Studies demonstrate that Ag^+ ions can interact with cytoplasmic components and nucleic acids, resulting in the inhibition of respiratory chain enzymes, and interfering in membrane permeability [5, 49]. Thereunto, an effective way of quantifying AgNP's toxicity can be the measurement of AgNPs/Ag^+ ratio in the intra and extra cellular medium [46]. The strategies used to separate AgNPs from Ag^+ ions are centrifugation, ultrafiltration, and cloud point extraction. To quantify each of these components, the atomic adsorption spectroscopy or mass spectroscopy techniques have been used [46, 47, 50–52]. Wang and colleagues [47], for example, separated AgNPs and Ag^+ ions into erythroid cells (MEL) from mice by means of a cloud-point extraction and found AgNPs (82.1%) and Ag^+ ions (17.9%) together inside cells, which suggest the occurrence of a Trojan-horse type mechanism [47]. However, it is still necessary to compare the internalization rate of AgNPs and Ag^+ in order to determine if ionization is really occurring in the intracellular environment [46].

However, one study comparing AgNPs with others silver solutions revealed that these NPs have a greater antibacterial potential than free Ag^+ independent of elution [52]. Recent evidences show that AgNPs can produce many reactive oxygen species (ROS) including superoxide-radical (O_2^-), hydroxyl radical (OH^-), hydrogen peroxide (H_2O_2), and singlet molecular oxygen ($^1O_2$) [53]. Furthermore, H_2O_2 can react with internalized AgNPs to form more Ag^+ ions, thus, the Trojan-horse mechanism is a chain reaction which results in an increasing release of Ag^+ [46]. Inside cells, Ag^+ can react with other ions such as Cl^- and S^{2-}, forming AgCl and AgS_2, respectively [54, 55]. Moreover, other compounds such as Ag-cysteine and Ag_2O can be formed (Fig. 15.3). Oxidative stress can induce DNA and protein damage and lipid peroxidation [56], which partially explains how AgNPs can present antibacterial effects and are potential toxic to humans [57].

Moreover, Park and colleagues [58] reported that AgNPs can promote various genetic and physiological modifications, such as increased expression of matrix metalloproteinases and decreased intracellular glutathione expression [58].

15.5 AgNPs as Nanocarriers (NC)

Nowadays, the flexibility in NP synthesis allows the production of nanocarriers (NC) with some tunable properties like solubility, particle size, circulation half-life and degradation. These nanosystems can be engineered to target organelles, cells or tissues in a specific way reducing these previously mentioned side effects [11].

The cellular uptake of NCs can occur through a passive translocation across the bilayer membrane or by physicochemical techniques such as electroporation. Due to NP large surface area and curvature relative to their volume, cells activate the endocytosis process for their uptake [11]. This cellular uptake process can be used as a strategic pathway to deliver drugs in specific cells. Thus, drugs can be loaded with NPs and target moieties on the surface which will act against receptors without affect normal cells. Many receptors have been discovered for cancer drug targeting, the most commonly used is the folic acid [59]. NPs can spread over different tumors regions by blood vessels and then to interstitial space until arriving cancer cells, eradicating them [60, 61].

AgNPs are potential anticancer agents and some studies using biologically synthesis methods prove this potential [62]. Cytotoxicity studies of AgNPs using *Melia dubia* extracts against human breast cancer cells showed that low concentrations of these NPs were able eliminate 50% of cancer cells [63]. Moreover, AgNPs synthesized using *Malus domestica* (apple) extract showed significant cytotoxic effects against MCF-7 breast cancer cells [64]. Another study that used AgNPs engineered by *Nonotus obliquus* (Chaga mushroom) extract showed effective antiproliferative activity toward A549 human lung cancer and human breast cancer cells (MCF-7). These studies, demonstrate that AgNPs produced by green synthesis methods possess high cytotoxic activity against cancer cells which suggests

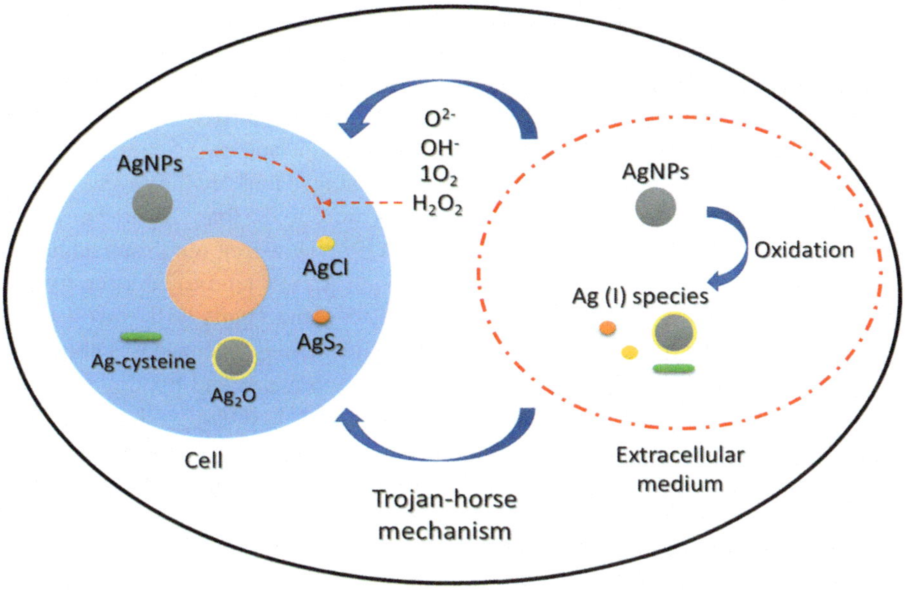

Fig. 15.3 Illustrative scheme of the "Trojan-horse" mechanism. AgNPs are internalized by cells and then are oxidized forming Ag^+ which can react with free ions such as Cl^-, S^{2-} and O^{2-} or cysteine in the intracellular medium. Ag^+ can also induce the formation of ROS inside cells (Modified from [46])

the potential therapeutic use of these NPs as a "Trojan Horse" strategy against cancer [65].

Recently, Pang and coworkers [66] used macrophages cells as "Trojan-horses" to carry drug-loading NPs which could pass through cellular barriers and offload them into brain tumor sites. Free anticancer drugs encapsulated into cells could cause damage to the carrier itself before arriving at tumor sites suppressing the functions of cells as transporters. Therefore, the researchers encapsulated anticancer drugs into NPs to reduce the damage of the drug to cell carriers. In this investigation, nanodrugs were encapsulated into patient macrophages, then these macrophages loaded with drugs-NPs were transferred back into the patient to achieve improved efficacy and to reduce immune responses [66].

Wang and colleagues [47] showed that AgNPs reduced the efficiency of cell transcription due to the direct binding of silver to RNA polymerase. The drugs that inhibit microbial or viral RNA polymerase activity have been used against invading pathogens [47]. Therefore, this work is an interesting example about how Trojan-horse mechanism can be used for medicine research. Evidence for AgNP toxicity through this mechanism was also found in mouse macrophages (RAW264.7) and in human bronchial epithelial cells (BEAS-2B) [58, 67, 68].

On the other hand, NPs can be captured by central nervous system through microglia and astrocytes cells, representing a threat to neuronal cells [69]. In vivo studies have been shown that AgNPs can accumulate on the developing brain, leading to developmental dysmorphologies [70]. The potential neurotoxicity of AgNPs is also related to ROS induced by NPs which may be associated with neurodegenerative disorders [46]. Nevertheless, studies evaluating the implications and applications of AgNPs in biological systems are still recent and how this NPs influence people health remains unanswered.

15.6 AgNP Cytotoxicity: Apoptosis, Necrosis and Autophagy

Cells which are unnecessary for the organism commit suicide by activating an intracellular programmed death known as apoptosis. This process is morphologically characterized by pyknosis (deep staining of nuclear mass), nuclear fragmentation, and formation of condensed cell bodies (apoptotic bodies). On the other hand, necrosis process occurs when cell suffer an unexpected and accidental damage. Therefore, toxic chemical or physical events like toxins and radiation exposure can result in an electron-lucent cytoplasm [71, 72].

In vitro studies have shown that AgNPs can be cytotoxic to human cells. Some researchers reported that AgNPs decrease the viability and proliferation of keratinocytes and human liver cells [73, 74]. The researchers discovered that AgNPs interfere in cell cycle and lead to increase of apoptosis both in brain tumor cells and normal fibroblasts [75]. However, the mechanisms related with AgNP cytotoxicity are still not totally clear.

As described in the last session, the oxidative stress mechanism mediated by AgNPs in cells is caused by an imbalance between oxidants and antioxidants and resulted in damage to cells organelles such as mitochondria and endoplasmic reticulum, [76–78] which activates apoptosis in mammalian cells [79–82].

ROS are mainly generated in the mitochondria [83], but some studies show that other cell organelles like endoplasmic reticulum (ER) also respond to oxidative stress playing an important role in the outcomes activated by AgNPs [84, 85]. The endoplasmic reticulum is involved in protein folding and assembly, lipid biosynthesis, vesicular traffic, and cellular calcium storage. This organelle is sensitive to alterations in homeostasis; thus, any change in cell metabolism can compromise its function, inducing cellular damage and apoptosis [86].

ER related changes such as inhibition of protein glycosylation, reduction in disulfide bond formation, calcium depletion from the ER lumen, impairment of protein transport from the ER to Golgi apparatus and expression of misfolded proteins may causes proteotoxicity in this organelle causing an endoplasmic reticulum stress [87–89]. Considering that this organelle is essential for cell survival, changes in ER function interfere in cell apoptosis and some studies already have reported the importance of ER in apoptotic process [90].

It is known that the toxic effect of AgNPs depends on their size and the coating material [91–93]. Liu and colleagues [92], for example, demonstrated that small AgNPs (3–4 and 5–7 nm) were more toxic than 10–40 nm NPs to mouse cells [92]. However, our group demonstrated that 100 nm AgNPs tend to be more toxic than their smaller counterparts (10 nm) [9]. Therefore, although there is no consensus about the relationship between size and toxicity in these NPs; the NP diameter is directly related with the biocide potential of AgNPs. On the other hand, Gliga and co-workers [91] demonstrated that citrate coating affect NP toxicity with the exception of 10 nm AgNPs [91]. Furthermore, the cytotoxic effects of AgNPs may cause different responses depending on the cell type. Kim and co-workers [94], for example, reported that AgNP cytotoxicity stimulated apoptosis in osteoblastic cells; otherwise it induced necrosis in adrenal medulla cells in mice [94]. These opposite effects can be activated in dependence of molecular mechanism differentially expressed in cells from diverse origins which can affect proactive pathways in those cells.

Asare and colleagues [95], for example, reported on in vitro study that AgNPs can caused DNA damage, apoptosis, necrosis and proliferation decrease in murine primary testicular cells as well as tumor cells [43, 95]. In this context, Li and coworkers [96] evaluated the cytotoxicity of AgNPs decorated by polyethylenimine (PEI) and paclitaxel (PTX) (Ag@ PEI™ PTX) in HepG2 cancer cells. Induction of apoptosis in these cells after exposure to Ag@PEI & gt; PTX was verified due to DNA fragmentation, depletion of mitochondrial membrane potential, activation of caspase 3 and increase in cell population in sub-G1 phase of cell cycle [96]. Assays using 7-AAD and Annexin-V dyes demonstrated that AgNPs have the potential to induce cell necrosis or accidental cell death. This induction has been shown to be related with size and time of exposure to AgNPs [97].

Autophagy is a degradation process of toxic proteins and damaged organelles in which portions of the cytoplasm are stocked in autophagosomes and then are fused with lysosomes forming autolysosomes. Posteriorly, the autolysosomes content is degraded by lysosomal hydrolases and recycled for energy utilization [98]. The autophagy process is mainly characterized as a survival mechanism from different environmental stresses such as AgNP exposure. It has been proposed that AgNPs can induce this type of cell response by interfering in the ubiquitination process. These NPs would be able to promote the increase of enzyme levels that participate in the ubiquitination and avoid the biological reactivity of ubiquitin [99].

NP-induced autophagy has been considered as a potential molecular target for NPs based chemotherapy [100–102]. The NP-activated autophagy has been associated with inflammation, oxidative stress, and induction of apoptosis [103]. Until now, studies involving autophagy and apoptosis after NPs exposure suggest that these NPs can cause irreversible damage to cell [104].

Considering the increasing use of NPs in many manufactured products, some researchers are also evaluating the possible effects that the interaction between AgNPs and others nanometals may cause to cells. In recently studies, the associations of AgNPs and metal ions of cadmium and mercury, which are found in various environmental contaminants (e.g. battery fluid, fertilizers, paints, plastic stabilizers, coal combustion, and seed treatment), were more toxic than AgNPs [105, 106]. The association caused a decrease in cell viability and changes the cell death type from apoptosis to necrosis [107]. Thus, these results suggest that there is still much to understand about NP interactions with other nanomaterials and how this may influence our lives.

15.7 Conclusion

Nowadays, AgNPs are the most widely used NPs in the industry because of their peculiar biocide features. The applications of these features to industrial and therapeutic purposes have been brought enormous benefits to our society. However, the employment of these NPs still runs into limitations mainly because of lack of standardization of size and shape and the absence of dose-dependent toxicity elucidation. Added to these obstacles is the scarcity of studies measuring the toxicity of these NPs after their interaction with organic matter and their intracellular mechanisms of action. Thus, further studies should explore these issues to potentiate the applications of the unique AgNPs properties.

References

1. Emam HE, Saleh NH, Nagy KS et al (2015) Functionalization of medical cotton by direct incorporation of silver nanoparticles. Int J Biol Macromol 7:249–256
2. Hebeish A, El-Bisi MK, El-Shafei A (2015) Green synthesis of silver nanoparticles and their application to cotton fabrics. Int J Biol Macromol 72:1384–1390
3. Ghavaminejad A, Park CH, Kim CS (2016) In situ synthesis of antimicrobial silver nanoparticles within antifouling zwitterionic hydrogels by catecholic redox chemistry for wound healing application. Biomacromolecules 17(3):1213–1223
4. Fabrega J, Luoma SN, Tyler CR et al (2011) Silver nanoparticles: behaviour and effects in the aquatic environment. Environ Int 37(2):517–531
5. Lok CN, Ho CM, Chen R et al (2007) Silver nanoparticles: partial oxidation and antibacterial activities. J Biol Inorg Chem 12(4):527–534
6. Chaloupka K, Malam Y, Seifalian AM (2010) Nanosilver as a new generation of nanoproduct in biomedical applications. Trends Biotechnol 28:580–588
7. Alexander JW (2009) History of the medical use of silver. Surg Infect 10:289–292
8. He W, Liu X, Kienzle A et al (2016) In vitro uptake of silver nanoparticles and their toxicity in human mesenchymal stem cells derived from bone marrow. J Nanosci Nanotechnol 6:219–228
9. Souza TA, Franchi LP, Rosa LR et al (2016) Cytotoxicity and genotoxicity of silver nanoparticles of different sizes in CHO-K1 and CHO-XRS5 cell lines. Mutat Res 795:70–83
10. Stenberg MC, Wei Q, McLamore ES et al (2011) Toxicological studies on silver nanoparticles: challenges and opportunities in assessment, monitoring and imaging. Nanomedicine 6:879–898
11. Zappavigna S, Misso G, Falanga AR et al (2016) Nanocarriers conjugated with cell penetrating peptides: new Trojan horses by modern ulysses. Curr Pharm Biotechnol 17:700–722

12. Hillie T, Hlophe M (2007) Nanotechnology and the challenge of clean water. Nat Nanotechnol 2:663–664
13. Qu X, Alvarez PJJ, Li Q (2013) Applications of nanotechnology in water and wastewater treatment. Water Res 47:3931–3946
14. Rosa LR, Rosa RD, Da Veiga MAMS (2016) Colloidal silver and silver nanoparticles bioaccessibility in drinking water filters. J Environ Chem Eng 4:3451–3458
15. Jain P, Pradeep T (2005) Potential of silver nanoparticle-coated polyurethane foam as an antibacterial water filter. Biotechnol Bioeng 90:59–63
16. Ahamed T, Imdad S, Yaldram K et al (2014) Emerging nanotechnology-based methods for water purification: a review. Desalin Water Treat 52:4089–4101
17. Lv Y, Liu H, Wang Z et al (2009) Silver nanoparticle-decorated porous ceramic composite for water treatment. J Membr Sci 331:50–56
18. Mecha CA, Pillay VL (2014) Development and evaluation of woven fabric microfiltration membranes impregnated with silver nanoparticles for potable water treatment. J Membr Sci 458:149–156
19. Dankovich TA, Gray DG (2011) Bactericidal paper impregnated with silver nanoparticles for point-of-use water treatment. Environ Sci Technol 45:1992–1998
20. Haider MS, Shao GN, Imran SM et al (2016) Aminated polyethersulfone-silver nanoparticles (AgNPs-APES) composite membranes with controlled silver ion release for antibacterial and water treatment applications. Mater Sci Eng C 62:732–745
21. Pradeep T, Anshup (2009) Noble metal nanoparticles for water purification: a critical review. Thin Solid Films 517:6441–6478
22. Li X, Schirmer K, Bernard L et al (2015) Silver nanoparticle toxicity and association with the alga Euglena gracilis. Environ Sci Nano 2:594–602
23. McGillicuddy E, Murray I, Kavanagh S et al (2017) Silver nanoparticles in the environment: sources, detection and ecotoxicology. Sci Total Environ 575:231–246
24. Oukarroum A, Bras S, Perreault F et al (2012) Inhibitory effects of silver nanoparticles in two green algae, Chlorella vulgaris and Dunaliella tertiolecta. Ecotoxicol Environ Saf 78:80–85
25. Asghari S, Johari SA, Lee JH et al (2012) Toxicity of various silver nanoparticles compared to silver ions in *Daphnia magna*. J Nanobiotechnol 10:14
26. Ribeiro F, Gallego-Urrea JA, Jurkschat K et al (2014) Silver nanoparticles and silver nitrate induce high toxicity to Pseudokirchneriella subcapitata, Daphnia magna and Danio rerio. Sci Total Environ 466–467:232–241
27. Huang J, Cheng J, Yi J (2016) Impact of silver nanoparticles on marine diatom Skeletonema costatum. J Appl Toxicol 36:1343–1354
28. Valerio-Garcia RC, Carbajal-Hernandez AL, Martinez-Ruiz EB et al (2017) Exposure to silver nanoparticles produces oxidative stress and affects macromolecular and metabolic biomarkers in the goodeid fish Chapalichthys pardalis. Sci Total Environ 583:308–318
29. Gonçalves FS, Pavlaki DM, Lopes R et al (2017) Effects of silver nanoparticles on the freshwater snail Physa acuta: the role of test media and snails' life cycle stage. Environ Toxicol Chem 36:243–253
30. Gorka DE, Liu J (2016) Effect of direct contact on the phytotoxicity of silver nanomaterials. Environ Sci Technol 50:10370–10376
31. Jiang HS, Yin L, Ren NN et al (2017) The effect of chronic silver nanoparticles on aquatic system in microcosms. Environ Pollut 223:395–402
32. De Paolis F, Kukkonen J (1997) Binding of organic pollutants to humic and fulvic acids: influence of pH and the structure of humic material. Chemosphere 34:1693–1704
33. Delay M, Dolt T, Woellhaf A et al (2011) Interactions and stability of silver nanoparticles in the aqueous phase: influence of natural organic matter (NOM) and ionic strength. J Chromatogr A 1218:4206–4212
34. Gao J, Powers K, Wang Y et al (2012) Influence of Suwannee River humic acid on particle properties and toxicity of silver nanoparticles. Chemosphere 89:96–101
35. Gunsolus IL, Mousavi MP, Hussein K et al (2015) Effects of humic and fulvic acids on silver nanoparticle stability, dissolution, and toxicity. Environ Sci Technol 49:8078–8086
36. Stankus DP, Lohse SE, Hutchison JE et al (2011) Interactions between natural organic matter and gold nanoparticles stabilized with different organic capping agents. Environ Sci Technol 45:3238–3244
37. Wirth SM, Lowry GV, Tilton RD (2012) Natural organic matter alters biofilm tolerance to silver nanoparticles and dissolved silver. Environ Sci Technol 46:12687–12696
38. Oukarroum A, Gaudreault MH, Pirastru L et al (2013) Alleviation of silver toxicity by calcium chloride (CaCl2) in Lemna gibba L. Plant Physiol Biochem 71:235–239
39. Liu JY, Hurt RH (2010) Ion release kinetics and particle persistance in aqueous nano-silver colloids. Environ Sci Technol 44:2169–2175
40. Kim JY, Kim KT, Lee BG et al (2013) Developmental toxicity of Japanese medaka embryos by silver nanoparticles and released ions in the presence of humic acid. Ecotoxicol Environ Safe 92:57–63
41. Navarro E, Baun A, Behra R et al (2008) Environmental behavior and ecotoxicity of engineered nanoparticles to algae, plants, and fungi. Ecotoxicology 17:372–386
42. Lee SW, Park SY, Kim Y et al (2016) Effect of sulfidation and dissolved organic matters on toxicity of silver nanoparticles in sediment dwelling organism, Chironomus riparius. Sci Total Environ 553:565–573

43. Asare N, Instanes C, Sandberg WJ et al (2012) Cytotoxic and genotoxic effects of silver nanoparticles in testicular cells. Toxicology 291:65–72
44. Franchi LP, Manshian BB, De Souza TA et al (2015) Cyto- and genotoxic effects of metallic nanoparticles in untransformed human fibroblast. Toxicol In Vitro 29:1319–1331
45. Samberg M, Monteiro-Riviere N, Oldenburg S (2010) Evaluation of silver nanoparticle toxicity in skin in vivo and keratinocytes in vitro. Environ Health Perspect 118:407–413
46. Hsiao IL, Hsieh YK, Wang CF et al (2015) Trojan-horse mechanism in the cellular uptake of silver nanoparticles verified by direct intra- and extracellular silver speciation analysis. Environ Sci Technol 49:3813–3821
47. Wang Z, Liu S, Ma J et al (2013) Silver nanoparticles induced RNA polymerase-silver binding and RNA transcription inhibition in erythroid progenitor cells. ACS Nano 7:4171–4186
48. Herzog F, Clift MJ, Piccapietra F et al (2013) Exposure of silver-nanoparticles and silver-ions to lung cells in vitro at the air-liquid interface. Part Fibre Toxicol 4:10–11
49. Russell AD, Hugo WB (1994) Antimicrobial activity and action of silver. Prog Med Chem 31:351–370
50. Gondrikas AP, Morris A, Reinsch BC et al (2012) Cysteine-induced modifications of zero-valent silver nanomaterials: implications for particle surface chemistry, aggregation, dissolution, and silver speciation. Environ Sci Technol 46:7037–7045
51. Liu J, Wang Z, Liu FD et al (2012) Chemical transformations of nanosilver in biological environments. ACS Nano 6:9887–9899
52. Yu SJ, Chao JB, Sun J et al (2013) Quantification of the uptake of silver nanoparticles and ions to HepG2 cells. Environ Sci Technol 47:3268–3274
53. Hee-Jin P, Jee Yeon K, Jaeeun K et al (2009) Silver-ion-mediated reactive oxygen species generation affecting bactericidal activity. Water Res 43:1027–1032
54. Liu J, Sonshine DA, Shervani S et al (2010) Controlled release of biologically active silver from nanosilver surfaces. ACS Nano 4(11):6903–6913
55. Lu M, Hu LF, Hu G et al (2008) Hydrogen sulfide protectsastrocytes against H_2O_2 -induced neural injury via enhancing glutamateuptake. Free Radic Biol Med 45:1705–1713
56. Sies H (1991) Oxidative stress: from basic research to clinical application. Am J Med 91:31S–38S
57. Carlson C, Hussain SM, Schrand AM et al (2008) Unique cellular interaction of silver nanoparticles: size-dependent generation of reactive oxygen species. J Phys Chem B 112:13608–13619
58. Park EJ, Yi J, Kim Y et al (2010) Silver nanoparticles induce cytotoxicity by a Trojan-horse type mechanism. Toxicol in Vitro 24:872–878
59. Kim YL (2006) Folate: a magic bullet or a double edged sword for colorectal cancer prevention? Gut 55:1387–1389
60. Firdhouse JM, Lalitha P (2015) Apoptotic efficacy of biogenic silver nanoparticles on human breast cancer MCF-7 cell lines. Prog Biomater 4:113–121
61. Wang J, Liu L, Zhu W et al (2014) Cancer cell uptake of polymer hydrogel nanotubes. J Biomed Nanotechnol 10:3329–3336
62. Raghunandan D, Ravishankar B, Sharanbasava G et al (2011) Anti-cancer studies of noble metal nanoparticles synthesized using different plant extracts. Cancer Nanotechnol 2:57–65
63. Kathiravan V, Ravi S, Ashokkumar S (2014) Synthesis of silver nanoparticles from melia dubia leaf extract and their in vitro anticancer activity. Spectrochim Acta A Mol Biomol Spectrosc 130:116–121
64. Lokina S, Stephen A, Kaviyarasan V et al (2014) Cytotoxicity and antimicrobial activities of green synthesized silver nanoparticles. Eur J Med Chem 76:256–263
65. Chung IM, Inmyoung P, Kim S et al (2016) Plant-mediated synthesis of silver nanoparticles: their characteristic properties and therapeutic applications. Nanoscale Res Lett 11:40
66. Pang L, Qin J, Han L et al (2016) Exploiting macrophages as targeted carrier to guide nanoparticles into glioma. Oncotarget 7:37081–37091
67. Lubick N (2008) Nanosilver toxicity: ions, nanoparticles or both? Environ Sci Technol 42:8617
68. Singh RP, Ramarao P (2012) Cellular uptake, intracellular trafficking and cytotoxicity of silver nanoparticles. Toxicol Lett 213:249–259
69. Haase A, Rott S, Mantion A et al (2012) Effects of silver nanoparticles on primary mixed neural cell cultures: uptake, oxidative stress and acute calcium responses. Toxicol Sci 126:457–468
70. Liu P, Huang Z, Gu N (2013) Exposure to silver nanoparticles does not affect cognitive outcome or hippocampal neurogenesis in adult mice. Ecotoxicol Environ Saf 87:124–130
71. Zong WX, Thompson CB (2006) Necrotic death as a cell fate. Genes Dev 20:1–15
72. Alberts B, Johnson A, Lewis J et al (2002) Molecular biology of the cell, 4th edn. Garland Science, New York
73. Yen HJ, Hsu SH, Tsal CL (2009) Cytotoxicity and immunological response of gold and silver nanoparticles of different sizes. Small 5:1553–1561
74. Zanette C, Pelin M, Crosera M et al (2011) Silver nanoparticles exert a long-lasting antiproliferative effect on human keratinocyte HaCaT cell line. Toxicol in Vitro 25:1053–1060
75. Asharrani PV, Low KMG, Hande MP et al (2009) Cytotoxicity and genotoxicity of silver nanoparticles in human cells. ACS Nano 24:279–290
76. Toyokuni S (1999) Reactive oxygen species-induced molecular damage and its application in pathology. Pathol Int 49:91–102
77. Zimmerman JJ (1998) Redox/radical repertoire rapport: pathophysiology and therapeutics. Acta Aaesthesiol Scand 42:1–3

78. Ercal N, Gurer-Orhan H, Aykin-Burns N (2001) Toxic metals and oxidative stress part I: mechanisms involved in metalinduced oxidative damage. Curr Top Med Che 1:529–539
79. Ahamed M, Posgai R, Gorey TJ et al (2010) Silver nanoparticles induced heat shock protein 70, oxidative stress and apoptosis in Drosophila melanogaster. Toxicol Appl Pharmacol 242:263–269
80. Foldbjerg R, Olesen P, Hougaard M et al (2009) PVP-coated silver nanoparticles and silver ions induce reactive oxygen species, apoptosis and necrosis in THP-1 monocytes. Toxicol Lett 190:156–162
81. Piao MJ, Kang KA, Lee IK et al (2011) Silver nanoparticles induce oxidative cell damage in human liver cells through inhibition of reduced glutathione and induction of mitochondria-involved apoptosis. Toxicol Lett 201:92–100
82. Xia T, Kovochich M, Nel A (2006) The role of reactive oxygen species and oxidative stress in mediating particulate matter injury. Clin Occup Environ Med 5:817–836
83. King BA, Oh DH (2004) Spatial control of reactive oxygen species formation in fibroblasts using two-photon excitation. Photochem Photobiol 80:1–6
84. Sanges D, Marigo V (2006) Cross-talk between two apoptotic pathways activated by endoplasmic reticulum stress: differential contribution of caspase-12 and AIF. Apoptosis 11:1629–1641
85. Walter L, Hajnóczky G (2005) Mitochondria and endoplasmic reticulum: the lethal interorganelle cross-talk. J Bioenerg Biomembr 37:191–206
86. Zhang Y, Dong L, Yang X et al (2011) α-Linolenic acid prevents endoplasmic reticulum stress-mediated apoptosis of stearic acid lipotoxicity on primary rat hepatocytes. Lipids Health Dis 10:81
87. Kaufman RJ (2002) Orchestrating the unfolded protein response in health and disease. J Clin Invest 110:1389–1398
88. Mori K (2000) Tripartite management of unfolded proteins in the endoplasmic reticulum. Cell 101:451–454
89. Oyadomari S, Araki E, Mori M (2002) Endoplasmic reticulum stress-mediated apoptosis in pancreatic beta-cells. Apoptosis 7:335–345
90. Kim CS, Park WH, Park JY et al (2004) Capsaicin, a spicy component of hot pepper, induces apoptosis by activation of the peroxisome proliferator-activated receptor gamma in HT-29 human colon cancer cells. J Med Food 7:267–273
91. Gliga AR, Skoglund S, Wallinder IO et al (2014) Size-dependent cytotoxicity of silver nanoparticles in human lung cells: the role of cellular uptake, agglomeration and Ag release. Part Fibre Toxicol 11:11
92. Liu HL, Daí SA, Fu KY et al (2010) Antibacterial properties of silver nanoparticles in three different sizes and their nanocomposites with a new waterborne polyurethane. Int J Nanomedicine 19:1017–1028
93. Park K, Park EJ, Chun IK et al (2011) Bioavailability and toxicokinetics of citrate-coated silver nanoparticles in rat. Arch Pharm Res 34(1):153–158
94. Kim TH, Kim M, Park HS et al (2012) Size-dependent cellular toxicity of silver nanoparticles. J Biomed Mater Res A 100:1033–1043
95. Asare N, Duale N, Slagsvold HH et al (2016) Genotoxicity and gene expression modulation of silver and titanium dioxide nanoparticles in mice. Nanotoxicology 10:312–321
96. Li Y, Guo M, Lin Z et al (2016) Polyethylenimine-functionalized silver nanoparticle-based co-delivery of paclitaxel to induce HepG2 cell apoptosis. Int J Nanomedicine 11:6693–6702
97. Kumar G, Degheidy H, Casey BJ et al (2015) Flow cytometry evaluation of in vitro cellular necrosis and apoptosis induced by silver nanoparticles. Food Chem Toxicol 85:45–51
98. Xie Z, Klionsky DJ (2007) Autophagosome formation: core machinery and adaptations. Nat Cell Biol 9:102–109
99. Mao BH, Tsai JC, Chen CW et al (2016) Mechanisms of silver nanoparticle-induced toxicity and important role of autophagy. Nanotoxicology 10:1021–1040
100. Adiseshaiah PP, Clogston JD, McLeland CB et al (2013) Synergistic combination therapy with nanoliposomal C6-ceramide and vinblastine is associated with autophagy dysfunction in hepatocarcinoma and colorectal cancer models. Cancer Lett 337:254–265
101. Stern JM, Cadeddu JA (2008) Emerging use of nanoparticles for the therapeutic ablation of urologic malignancies. Urol Oncol 26:93–96
102. Mishra AR, Zheng J, Tang X et al (2016) Silver nanoparticle-induced autophagic-lysosomal disruption and NLRP3-inflammasome activation in HepG2 cells is size-dependent. Toxicological Sciences 150:473–487
103. Johnson-Lyles DN, Peifley K, Lockett S et al (2010) Fullerenol cytotoxicity in kidney cells is associated with cytoskeleton disruption, autophagic vacuole ccumulation, and mitochondrial dysfunction. Toxicol Appl Pharmacol 248:249–258
104. Roy R, Singh SK, Chauhan LK et al (2014) Zinc oxide nanoparticles induce apoptosis by enhancement of autophagy via PI3K/Akt/mTOR inhibition. Toxicol Lett 227:29–40
105. Capaldo A, Gay F, Scudiero R et al (2016) Histological changes, apoptosis and metallothionein levels in Triturus carnifex (Amphibia, Urodela) exposed to environmental cadmium concentrations. Aquat Toxicol 173:63–73
106. Syversen T, Kaur P (2012) The toxicology of mercury and its compounds. J Trace Elem Med Biol 26:215–226
107. Miranda RR, Bezerra AGJR, Oliveira RCA et al (2017) Toxicological interactions of silver nanoparticles and non-essential metals in human hepatocarcinoma cell line. Toxicol in Vitro 5:134–143

16 Molecular and Cellular Toxicology of Nanomaterials with Related to Aquatic Organisms

Mohd Ashraf Rather, Irfan Ahmad Bhat, Niti Sharma, and Rupam Sharma

Abstract

The increasing application of nanomaterials both in commercial and industrial products has led their accumulation in the aquatic ecosystem. The rapid development and large scale production of nanomaterials in the last few decades have stimulated concerns regarding their potential environmental health risks on aquatic biota. Inorganic nanoparticles, due to their unique properties and associated material characteristics resulted in toxicity of these nanomaterials in aquatic organisms. Understanding their novel properties at nanoscale has established to be a significant aspect of their toxicity. Unique properties such as size, surface area, surface coating, surface charge, aggregation of particles and dissolution may affect cellular uptake, molecular response, in vivo reactivity and delivery across tissues of living organism. Already lot of research in the past three or four decades within the nano-ecotoxicology field had been carried out. However, there is not any standard technique yet to assess toxicity of nanoparticles (NPs) on different biological systems such as reproductive, respiratory, nervous, gastrointestinal systems, and development stages of aquatic organisms. Specific toxicological techniques and quantification of nanoparticles are vital to establish regulations to control their impact on the aquatic organism and their release in the aquatic environment. The main aim of this chapter is to critically evaluate the current literature on the toxicity of nanomaterials on aquatic organism.

M.A. Rather (✉)
Department of Fisheries Biology, College of Fisheries
Shirgaon, Rantagiri, Maharashtra, India
e-mail: mashraf38@gmail.com

I.A. Bhat • R. Sharma
Division of Fish Genetics and Biotechnology, Central Institute of Fisheries Education, Mumbai, India

N. Sharma
Regional Centre, Central Inland Fisheries Research Institute, Guwahati, Assam, India

Q. Saquib et al. (eds.), *Cellular and Molecular Toxicology of Nanoparticles*, Advances in Experimental Medicine and Biology 1048, https://doi.org/10.1007/978-3-319-72041-8_16

Keywords
Nanomaterial · Toxicity · Environment · Organism · Health risks

16.1 Introduction

Nanotechnology is an emerging technology for the twenty-first century and its application has been observed in all fields such as medical science, space science, semiconductor technology, information technology, cellular and molecular biology, agriculture and animal science, etc. It also has tremendous potential applications in fisheries and aquaculture. This is due to smaller size and targeted effects of the nanoparticles [1, 2]. United States National Nanotechnology Initiative (NNI) defined Nanotechnology as "understanding and assure of matter at proportions of roughly 1 to 100 nm where unique phenomena enable novel applications". However, Nanoforum, defined it as "Nanotechnology is the exploitation of the ability to assure matter at sizes between 0.1 and 100 nm, resulting in unique functionalities [3]. It includes the production and application of physical, chemical, and biological organizations at scales ranging from individual atoms or molecules to submicron dimensions, as well as the integration of the resulting nanostructures into larger systems. The prefix 'nano' is deduced from the Greek word for dwarf. One nanometre (nm) is equal to one-billionth of a metre, 10^{-9} m [4]. The conceptual underpinnings of nanotechnologies were first laid out in 1959 by Nobel Prize winner, the physicist Richard Feynman, in his lecture 'There's plenty of room at the bottom' [5]. The term 'nanotechnology' was not used until 1974, when Norio Taniguchi, a researcher at the University of Tokyo, Japan used it to mention, 'to the ability to engineer materials precisely at the nanometre level' [6]. It has two prime meanings: (1) Novel science and technology that takes benefit of properties functioning at the nanoscale and (2) building with atomic precision using molecular machine systems [7]. Nanoparticles (NPs) manufacturing is an essential component of nanotechnology because the specific properties are realized at the nanoscale, nanocrystal or nanolayer level.

NPs have been developed in many different inorganic and organic forms, the main types including nanometals (e.g. silver nanoparticles, gold nanoparticles), metal oxides (e.g. Iron, Copper, Nickel, Zinc etc.) carbon-based materials such as carbon nanotubes (CNTs) and carbon spheres (e.g. C60 fullerenes, often called 'bucky balls' by the media) as well as composites made of several substances such as nanoceramics and quantum dots [8, 9] and polymeric nanomaterials (Chitosan, sodium alginate, Poly lactic-co-glycolic acid). There are several other nanomaterials which are being used in medical science, biosensors, biomaterials, tissue engineering, DNA modification, etc. [10–12]. Quantum mechanism plays a significant role in reducing the size of the materials to nanoscale and also changes the electrical, mechanical and optical properties of the nanoparticles. Because of the change in their properties, nanoparticles gained its popularity to be used in various fields such as textiles and fabrics, drug delivery, food processing, electronics, innovative building materials, water treatment technology, remediation of contaminated land, etc. Presently, nanomaterials are utilized to solve various health related applications such as detection, diagnosis and treatment of diseases. The nanoparticles show a huge surface area/mass ratio to conjugate or encapsulate biomolecules. The small size of the nanoparticles facilitates their uptake by cells and the passage across epithelial or endothelial barriers into the blood or the lymph circulation. This in turn results in easy absorption of the nanoparticle or nanoconjugated biomolecules and thus acts more efficiently.

With the expansion in industrialization and the use in of NPs in new technology, there is an increase in the use of nanomaterials for the last few decades and this resulted in the surge in their release into the aquatic environment. There are innumerable sources of entry of NPs into the aquatic environment. NPs enter the environment mainly through waste water effluents, volcanic eruptions, accidental spillages, industrial run-offs, agricultural drainage water etc. However, there is a growing concern about the toxicity of the nanoparticle in general which needs to be addressed. If the nanoparticle used is not biodegradable, they will likely remain there for a long time, potentially accumulating to toxic levels. There are very limited studies on the long-term effects of these nanomaterials on the aquatic environment. There is scanty information about the negative effects of the nanoparticles on growth, reproduction, development, accumulation or about the elimination from the body. The methods to test for these materials in the environment are in their infancy. Therefore, an urgent need to know about the effect of the particles on aquatic organism is required. Because of increasing prevalence of nanoparticles in consumer products, international efforts are undertaken to verify nanoparticle safety and to understand the mechanism of their toxicity. The present chapter reviewed potential toxicity of NPs in fisheries and aquaculture providing a vital summary of recent scientific literature on potential hazardous effects of NPs on aquatic organisms.

16.2 Types of Nanomaterials

A wide range of NPs has been available in nowadays in commercial market as well as in natural environment. They can be classified based on their origin (natural) chemical composition (organic, inorganic), formation (biogenic, anthropogenic, and atmospheric), their size, shape and characteristics, their applications in research and industry, etc. As nanomaterials are small, they have a much better surface area to volume ratio than the conventional forms. They can be produced in one dimension (surface film), two dimension (strand or fibres) or three dimension (particles) and in different irregular and regular shape such as sphere, rod tubes wires etc. Due to their unique electrical, catalytic, magnetic and thermal features, these materials have received much attention among researchers in many fields

Table 16.1 Classification of nanomaterials based on nature of material

Nature of material	Examples	Applications
Inorganic or metal based	Nanosilver, Nanoglod, Quantum dots, Metal oxides etc.	Imaging, diagnostic, Antimicrobial etc.
Organic materials	Chitosan, Alginates, Gelatin.Starch, Liposomes, Dextran	Drug and gene delivery
Carbon based	Fullerenes, Dendrimers, Nanotube	Drug and gene delivery
Polymeric carrier	Poly lactic-co-glycolic acid (PLGA), Poly-alkylcyanoacrylates, Polyethyleneamine	Drug, vaccine, hormones and gene delivery

of biological science including fisheries and aquaculture. Nanomaterials can be broadly classified based on nature or material used for their manufacture as given in Table 16.1.

16.3 Nanotechnology in Aquaculture and Fisheries

In aquaculture industry, nowadays NPs are used for various purposes like rapid disease detection, drug delivery like vaccines, hormones, and nutrients, nanosensors etc., [13–15]. The areas associated with aquaculture and fisheries where nanotechnology can be applied are given in Fig. 16.1.

16.4 Toxicological Profiling of Different Types of NPs

Nanotechnology is a fast-growing technology. With the advantage of small size nanoparticles have applications in all fields. Its application has been observed in medical science, space science, semiconductor technology, information technology, cellular and molecular biology, agriculture and animal science including fisheries and aquaculture etc. Nanoparticle manufacturing is an indispensable component of nanotechnology because the properties are realized at the

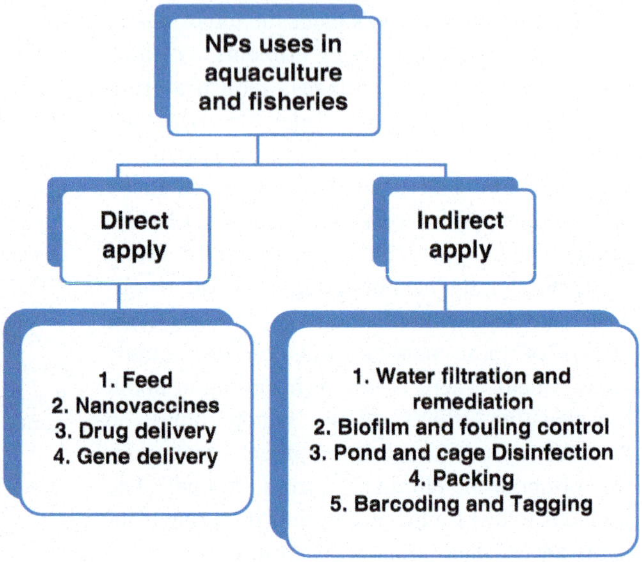

Fig. 16.1 Uses of nanomaterial's for different application in fisheries and aquaculture

nanoscale, nanocrystal or nanolayer level. These materials have been produced in many different organic, inorganic or chemical forms, the main types including nanometals like silver nanoparticle (AgNPs) gold nanoparticle (AuNPs) as well as composites made of several substances such as nanoceramics and quantum dots. However, with the tremendous use of these particles, there is a growing concern about the toxicity of the nanoparticle which needs to be addressed (Table 16.2). Most of the nanomaterials get accumulated in the aquatic environment through all the drainages and run off. If the nanoparticle used is not biodegradable, they will likely remain there for a long time, potentially accumulating to toxic levels. Exposure of these nanoparticles to aquatic organisms has been found to cause genotoxic, cytotoxic effects and get accumulated on aquatic organisms which may enter to human body through food chain.

16.5 Toxicity Mechanisms of Nanoparticles

The mechanism of nanoparticles to cause toxicity in a biological system is quite uncomplicated. In general, the nanoparticles upon any physico-chemical reactivity generate the highly reactive oxygen species (ROS) or free radicals directly or indirectly through oxidative enzymatic pathways that ultimately results in the oxidative stress. The free radicals may be superoxide radical anions and hydroxyl radicals [40–42]. The flowchart showing the behaviour of nanoparticle inside fish body to cause toxicity is presented in Fig. 16.2.

The toxic effect of different nanoparticles has been studied in several aquatic organisms and detailed of same is given below,

16.6 Silver Nanoparticles (AgNPs) and Aquatic Organism Toxicity

Metallic nanoparticles such as silver, gold, silica, etc. exhibit size and shape dependent properties. These materials are used for various applications like, catalysts, sensor, optics, antibacterial activity, data storage, drug delivery, etc. [43–45]. Exposure to AgNPs has been found to cause genotoxic and cytotoxic effects in various organisms. The silver particulates primarily accumulate in the liver and brain in mammals. Since these nanoparticles are used in commercial products, there is a possibility of reaching them in aquatic environment and cause nanotoxicity to the aquatic organisms which may enter to human

Table 16.2 Evaluation of nanoparticles toxicity in different fish species

Nature of nanoparticles	Nanoparticles	Species tested	Route of administration	Cell and tissue target	Major outcome	References
Polymer	Chitosan	*Oncorhynchus mykiss*	Water suspension	Whole physiological response	Mortality, lifting of lamellar epithelium, hypertrophy and hyperplasia of lamellar epithelial cells in gills	Bullock et al. [16]
		Danio rerio	Egg media	Embryos	Cellular death, mortality and hatchability rate reduction	Hu et al. [17]
	Alginates	–	Cell plate	hek 293 cells	Non-significant changes were found	Rafiee et al. [18]
Metallic	Silver	*Danio rerio*	Added to fish media and tank	Liver, gill and intestine	Oxidative stress, stress and immune response related genes upregulated, gill and intestine structure, effected with effect on $Na^+/K^+/$ atpase activity	Krishnaraj et al. [19] and Osborne et al. [20]
		Perca fluviatilis	Added to fish media	Metabolism and respiratory system	Less effect on metabolism (basal metabolic rate) but reduction in oxygen uptake.	Bilberg et al. [21]
		Labeo rohita	Water suspension, through live feed	Effect on gill, muscle, liver and immune responses	Alterations in histological structures of organs tested with effect on immune response	Rajkumar et al. [22] and Sharma et al. [23]
		Cyprinus carpio	Added to fish media	Gill, liver, brain, gastrointestinal tract and connective tissues	Nanoparticles got accumulated in different tissues	Jang et al. [24]
		D. rerio embryos and adults	Exposure in small dishes and dispersion in tank	Synaptic and behavioural changes in development	Developmental neurotoxicant in adults and dopamine and serotonin elevation in embryos	Powers et al. [25]
	Silver	*Pimephales promelas*	Exposure of embryos	Whole effect on embryos	Larval abnormalities and edema	Laban et al. [26]
	Gold	*Danio rerio*	From sediments into water tank	Overall impact on body	Overexpression of genes involved in oxidative stress, mitochondrial metabolism, altered neurotransmission	Dedeh et al. [27]
		Danio rerio embryos	In egg media	Accumulation and survival; development	No signs of toxicity	Asharani et al. [28]

(continued)

Table 16.2 (continued)

Nature of nanoparticles	Nanoparticles	Species tested	Route of administration	Cell and tissue target	Major outcome	References
Metallic	Zinc oxide	*Cyprinus carpio*	In water suspension	Brain, gill, kidney, muscle	Histopathological alterations in analysed tissues, superoxide dismutase activity decreased	Hao et al. [29]
		Danio rerio adult and embryos	In water suspension and mixed with egg media	Zinc accumulation and effect on development of hatching	Non-significant effect on adults while in embryos hatchability rate decreased followed by high motality	Johnston et al. [30] and Bai et al. [31]
	Iron	*Oryzias latipes*	In water suspension	Effect on respiratory system	Hypoxia and reactive oxygen species production	Chen et al. [32]
		Danio rerio	In media	Embryonic development	Malformation in embryo, decrease in hatching rate and survivability	Zhu et al. [33]
		Oreochromis mossambicus	In water suspension	Effect on haematological parameters	Changes in blood cell counts	Karthikeyeni et al. [34]
	Titanium dioxide (TiO_2)	*Cyprinus carpio*	In water suspension and dietary exposure.	Accumulation in tissues	Increase in cadmium accumulaton in presence of more tio_2	Zhang et al. [35]
		D. rerio	In media	Development of larva	Decreased hatching rate	Clemente et al. [36]
Non metallic	Single walled carbon nanotubes	*Oncorhynchus mykiss*	In water suspension	Osmoregulation and effect on respiratory system	Increases in Na^+K^+-Atpase activity in the gills and intestine, increase in respiratory rate and histopathological changes gills	Smith et al. [37]
	Quantum dots	*Oncorhynchus mykiss*	Water suspension and cell media	Stress and effect on gene expression	Induction of metallothioneins; increase hsp70 gene expression	Louis et al. [38] and Gagne et al. [39]

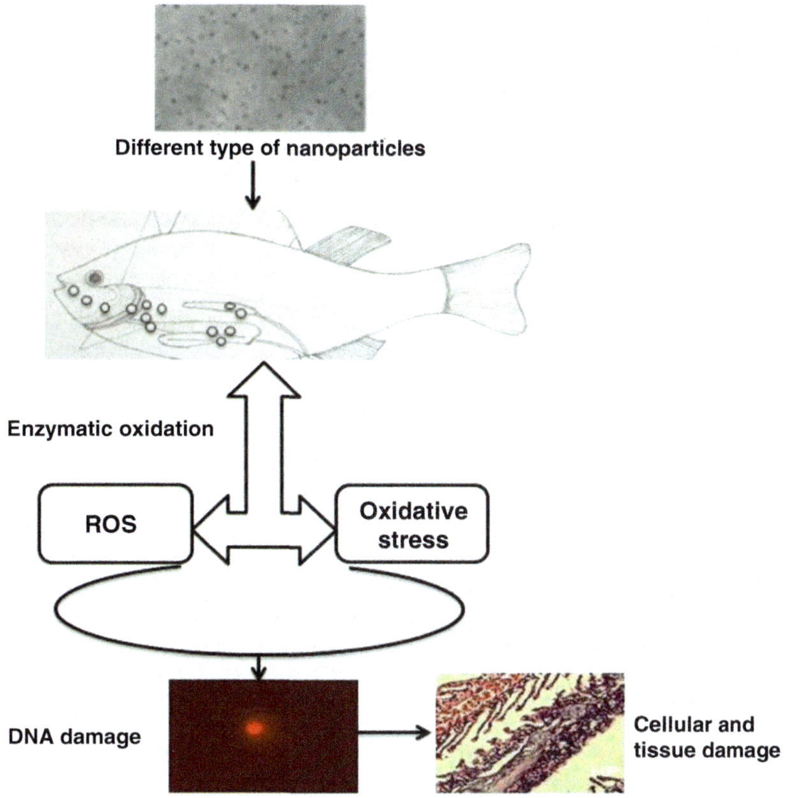

Fig. 16.2 Flowchart showing mechanism of toxicity of nanoparticles

body through food chain. These AgNPs can cross the egg membrane and move into fish embryo in less than a day [26, 46]. Therefore, toxicity studies should be carried out for the benefit of human health and aquatic species.

Nanoparticles of noble metals such as silver exhibited considerably different physical, chemical and biological properties from their bulk counterparts [47]. AgNPs have distinctive physico-chemical properties, including a high electrical and thermal conductivity, surface-enhanced Raman scattering, chemical stability, catalytic activity and non-linear optical behaviour [48, 49]. AgNPs are one of the fastest growing products in the nanotechnology industry due to its antimicrobial activity [50]. This increased the number of medical applications of AgNPs. Some of the products which are already accessible in the commercial market include wound dressings, contraceptive devices, surgical instruments and bone prostheses [35, 51, 52]. Apart from these, AgNPs are being used for water purification and indoor air quality management [53–55]. Silver is normally found in surface waters originated from natural leaching, mining and the photographic industry [56]. It is present in water at nanogram level [57]. Silver is considered relatively harmless to humans. A report published by World Health Organization (WHO) mentioned that the estimated acute lethal dose of silver nitrate is at least 10 g (WHO). In fact; bactericidal properties of silver have been utilized by certain groups to commercialize colloidal silver suspensions as 'health supplements'.

Use of AgNPs-containing products are likely to increase the amount of dissolved silver in water and its releases into the environment [58–60] which are likely to persist and bioaccumulate. This may be a matter of concern and needs to address by assessing its toxicity on the exposed organisms as well as on the ecosystems. Prior to the application of nanoparticles, the silver ion (Ag^+ (aq)) was considered as the most toxic form of silver in water [61]. The chemistry of the neighbouring environment affects the connotation of silver ions with various ligands, in turn influencing bioavailability and toxicity [62–64]. Fabrega et al. [46] reviewed the existing knowledge on AgNPs as a potential problem for environmental health, considering the characteristics, behaviour, bioavailability and biological effects of AgNPs in aqueous suspensions. Levard et al. [65] revealed the major transformation processes of AgNPs in aqueous environments. Transformations of the metallic Ag cores produced by reactions with organic ligands, and the effects of such transformations on physical, chemical stability and toxicity of AgNPs. Modelling approaches helps in predicting the life cycle of AgNP-containing products in the environment. It will also help in assessing release and accumulation of Ag compounds in the environment at different concentrations. One model has been developed by Dale et al. [66] to understand the complex chemical transformations of AgNPs in surface waters or sediments. The model assumes that AgNPs mix with the sediment at approximately the equal rate as the sediment particles themselves. As AgNPs are mostly expected to heteroaggregate with sediments in natural environments Quik, et al., [67], the model's assumption holds well. The AgNPs may be discharged to the environment at different stages of handling the nanoparticle which include, during synthesis, manufacturing and incorporation of the NPs into goods, usage of goods containing NPs and while recycling or during disposal of goods [68]. When risk assessment is conducted detailed information about the exposure and subsequent bioaccumulation of the particle need to be taken care of, since these are the precursors of toxicity of xenobiotics [69]. Bioaccumulation is an important process to understand when evaluating hazards and risks from AgNPs. Factors which govern the bioavailability and bioaccumulation of AgNPs as given by Luoma and Rainbow [70] are -

1. Concentration of the AgNPs
2. The nature of the nanoparticle
3. The nature of the environment
4. The route of exposure, and
5. The biology and functional ecology of the organism involved

Silver ions are mostly responsible for causing toxicity in the environment. Several researchers have worked on Ag ion toxicity in vivo in freshwater fish species [71, 72]. The LC_{10} value was reported as low as 0.8 μg L^{-1} for certain freshwater fish species [73]. At higher concentrations (μm), physiological changes such as blood acidosis which ultimately led to circulatory collapse and death were reported in trout and fathead minnows [74, 75] utilized silver as model metal to understand the effect of temperature and complexing agents such as thiosulphate on metal uptake of fish (rainbow trout, *Oncorhynchus mykiss*). Wood et al. [71] found an accumulation of silver in gills, liver and kidney when applied in higher concentration, in rainbow trout (*Oncorhynchus mykiss*).

AgNPs of 10–80 nm affect early life stage development that includes spinal cord deformities, cardiac arrhythmia and survival in zebrafish [76, 77]. AgNPs also accumulate in the gills and liver tissue affecting the ability of fish to cope with low oxygen levels and inducing oxidative stress [21]. However, the threshold at which such consequences occur is variable among these experiments, even for the same species. Such variability may reflect differences in choices of experimental conditions or alterations in particle behaviour or character that were vague. In general, juvenile zebrafish and Japanese medaka have shown to be more susceptible to AgNPs than to equal mass concentrations of $AgNO_3$, due to increase in free ion Ag concentrations for the latter [46].

In Eurasian perch (*Perca fluviatilis*) AgNPs and silver nitrate at different concentration caused respiratory problem by affecting the gills externally, reducing the diffusion conductance which then led to internal hypoxia during low water oxygen tensions [21]. Laban et al. [26] studied the effect of silver nanoparticles at different level in fathead minnows (*Pimephales promelas*) embryos. They noticed that uptake of AgNPs into the embryos was observed after 24 h causing larval abnormalities like edema. Toxicity study of silver nanoparticles on embryonic development of oysters, *Crassostrea virginica* revealed adverse effects on embryonic development on both adult and embryos [78]. AgNPs induced a variety of morphological malformations like edema, spinal abnormalities, finfold abnormalities, heart malformations and eye defect in Japanese medaka (*Oryzias latipes*) at early life stages [79]. Silver nanoparticles showed reduction in membrane integrity and cellular metabolic activity on rainbow trout (*Onchorhyncus mykiss*) [80]. They also reported that in this species effect of AgNPs at different concentration was mostly on gills. Cytotoxic effects were seen and silver were taken into the cells and found in epithelial layer [80]. Uptake of silver particles from the water was found to be more in gills and liver in rainbow trout [77]. AgNPs at different concentration induced neurotoxicant in larvae of zebrafish (*Danio rerio*) that causes persistent neurobehavioral effects [25]. Again, Choi et al. [81] observed the hepatotoxicity of AgNPs in zebrafish it was also responsible to cause oxidative stress and apoptosis. Increased mortality was concomitantly observed in zebrafish with the exposure to AgNPs [82]. The increased rate of operculum movement and surface respiration in zebrafish during the exposure was found in the above study suggesting respiratory toxicity. The study demonstrated that AgNPs are lethal to zebrafish.

A study conducted at Purdue University unveiled AgNPs suspended in solution is toxic and even lethal to the fathead minnows- an organism often used to test the effects of toxicity on aquatic life. These AgNPs are so small they can cross the egg membranes and move into the fish embryos in less than a day [26]. Rayner et al. [83] examined the effect of silver ion and silver nanoparticles in a worm (*Nereis virens*), in which overall deformities and accumulation of nano and ionic Ag in organelles and the enzymes were noticed. It was found that AgNPs cause endoplasmic reticulum stress signalling and induce apoptosis in human cell lines. The works showed that, AgNPs induce apoptosis, mediated by the ER stress-signaling pathway [84]. Rajkumar et al. [22] exposed to 5, 10, 25, 50 and 100 mg/kg of AgNPs in rohu (*Labeo rohita*) and the results unveiled after 7d of exposure to AgNPs, hematological parameters significantly reduced in comparison to control group. Antioxidant enzymes significantly upregulated in the AgNPs treated fish tissues like gill, liver and muscle when compared with control tissues. Histopathological examination of these tissues exhibited that different lesions were arosed in different concentration of AgNPs. Recently Sharma et al. [23] revealed AgNPs of different concentration in the liver of *Labeo rohita* led vacuolar degeneration and hepatocytic degeneration (Fig. 16.3). Also, the high accumulation of Ag-NPs were detected that showed both time and dose-dependent relationships.

The general mechanism of AgNPs to cause toxicity in animals depends on its transformation process in biological and environmental media, the release of silver ions, the surface oxidative power and its interaction with biological macromolecules like Nucleic acids, proteins, lipids etc. [85, 86]. Nanosilver can cause toxicity by activation of signalling pathways through interaction with membrane proteins, by entering into the cell directly through diffusion or endocytosis to damage the mitochondrial and generate ROS that damages the DNA, protein etc. and in the cell and bring the cell to death by necrosis and apoptosis [87, 88]. In zebrafish the toxic effects of AgNPs were explained with the same mechanism [89].

16.7 Gold Nanoparticles (AuNPs) and Aquatic Organism Toxicity

Metal nanoparticles such as gold are excellent candidates for bioconjugation with biological substances [90]. Many researchers have shown that biologically active substance with amine function can bind strongly with gold nanoparticles [91]. The tenable shape and size dependent optical properties of AuNPs have been exploited in various surface coating and biomedical application [92]. They are biocompatible and bind readily to a large range of biomolecules such as amino acids, protein/enzymes and DNA and expose large surface areas for immobilization of

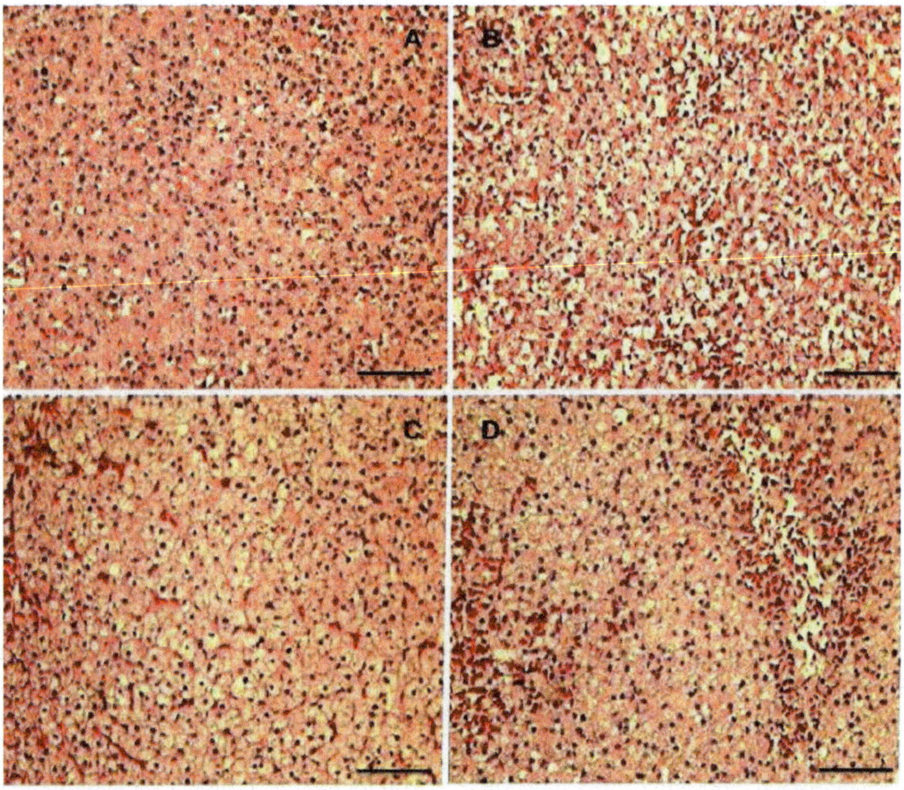

Fig. 16.3 Histopathology examination of fish liver treated with different concentration of silver nanoparticles. Section **a** (100 μg/l): Normal looking hepatic cells (No-Observed-Effect Level). Section **b** (200 μg/l): Diffuse degeneration and necrotic changes in hepatocytes (No-Observed-Adverse-Effect Level). Section **c** (400 μg/l): Swelling of hepatocytes with engorgement of sinusoids (No-Observed-Adverse-Effect Level). Section **d** (800 μg/l): Centrilobular and periportal congestion of sinusoids, and diffuse hepatic necrosis (Observed-Adverse-Effect Level). Scale bars = 100 μm (Permission have been taken from Sharma et al. [23] for publishing above figure)

such biomolecules [93]. The ability to modulate the surface chemistry of gold nanoparticles by binding suitable ligand has important application in many areas such as novel organic reaction, sensors (both inorganic and biological entities), drug/DNA delivery. However, various scientific literature has revealed AuNPs are toxic on aquatic biota and may induce cellular toxicity via different mechanisms such as oxidative stress, the disruption of cell membranes, inflammation, and DNA damage [94–96]. Asharani et al. [28] evaluated and compare the effect of silver, gold and platinum NPs on developing zebrafish embryos. Their results unveiled AgNPs in concentration dependent manner caused an increase in mortality rate and phenotypic changes as well as both AgNPs and platinum NPs induced hatching delay. In contrast, AuNPs did not show any sign of toxicity. Blue mussel (*Mytilus edulis*) were exposed to 750 ppm of Au-citrate NPs, after 24 h exposure, heamolymph surge and catalase activity, ubiquitination and caronylation reduced in digestive gland, gill and mantle cavity [97]. Garcia-Negrete et al. [98] on bivalve (*Ruditapes philippinarum*) concluded that AuNPs (21.5 ± 2.9 nm citrate-capped; 6–30 mg/L) accumulate more readily in digestive gland heterolysosomes (plateauing after 12 h), whilst accumulation of ionic Au was more in the gills. Truong et al. [96] revealed the toxic effects of AuNPs coated with three different functional groups (positively charged, negatively charged and neutral) using zebrafish. They found AuNPs functionalized with positively or negatively charged coatings cause toxic effects, however AuNPs functionalized with a neutral coating did not reveal any effect, verifying that coating chemistry plays an important role in AuNP toxicity. Bar-Ilan et al. [99] verified that zebrafish embryo toxicity depends more on its surface chemical composition instead than on particle size. This suggests that surface functionalization or coatings of AuNPs has a vast impact on the toxicity of nanomaterials on aquatic organism. Together, the previous studies confirm that size and coating chemistry of AuNPs are vital factors to understand AuNP toxicity, but on the other hand particle characteristics like aggregation or particles shape size, may also play an important role in toxicity. Generally, AuNPs toxicity is lower than that of AgNPs for aquatic animals [100].

16.8 Quantum Dots (QDS) and Aquatic Organism Toxicity

Quantum dots are very tiny particles of a semiconducting material with only several nanometres in size. These particles show unique electronic properties, due to high surface-to-volume ratio of these particles [101, 102]. Generally, QDS are made of binary compounds such as cadmium selenide, lead sulfide, lead selenide, cadmium sulfide, indium arsenide and indium phosphide. QDS, have been used in electronic bio-imaging, bio-sensing, water quality monitoring and inkjet printing and spin-coating [103, 104]. However, there are various reports which confirm toxicity of QDS in aquatic organism [105, 106]. Guo et al. [107] reported toxicity of Graphene quantum dots in zebrafish embryo. The results indicate as exposure concentration of Graphene quantum dots increased, the hatching and heart rate decreased, which lead an increase in mortality rate. High concentration exposure of Graphene quantum dots (200 μg/mL) resulted in various embryonic deformities comprising pericardial edema, vitelline cyst, bent spine, and bent tail. In rainbow trout, the total metallothionein and lipid peroxidation increase on exposure to 2 μg/L of QDS after 48 h [38]. Lewinski et al. [106] exposed QDS to *A. franciscana* and *D. magna* for 24 h with concentration of 0.6 mg/L and the organisms were fed to adult and juvenile zebrafish for 21 days. The results revealed no mortality was recorded after exposure, but the accumulation efficiency was 8% and 4% for adult and juvenile zebrafish, respectively. Kim et al. [108] use different light conditions on QDS LC_{50} on *D. magna*, they indicated that after 48 h exposure, toxicity increased by changing the light condition from dark to white fluorescence to UV-B light, and to natural sunlight. In another study toxicity of quantum dots and cadmium were assessed in early ontogenesis of rainbow trout *(Oncorhynchus mykiss)*. In this research Cibulskaite et al. [109] reported short-term exposure (upto 96-h) of sublethal concentrations of QDS and cadmium increased mortality of embryos and larvae, distressed function of the gill ventilation frequency, heart rate and affected behavioural resOponse in fish larvae. The results concluded that the toxic effects of QDS and cadmium on rainbow trout larvae depended on the type of chemical substance used, stage of development and exposure time. Comparative studies of the effects of QDS and cadmium on rainbow trout in embryological development stages revealed that larvae were more sensitive to cadmium and QDS as compared to embryos. Gagne et al. [39] studied toxicity of cadmium telluride quantum dots in vitro of rainbow trout hepatocytes. QDS were found to be lethal to fish hepatocytes at a concentration of 0.1 mg/l and QDS increased the levels of metallothioneins, labile zinc, DNA strand breaks and heat shock proteins. In another comparative study on two bivalves (*Mytilus edulis and Elliptio complanata*) and a fish (*Oncorhynchus mykiss*) with capped cadmium sulphide/cadmium telluride quantum dots concentration and size dependent toxicity was determined. Large cadmium sulphide/cadmium telluride quantum dot aggregates (25 nm < size <100 nm) reduced phagocytosis more than smaller nanoparticles (<25 nm).

The mechanism of toxicity QDS in aquatic organisms can be correlated with the higher animals that are explained by two ways, (i) existence of Cd^{+2} in QDS cause toxicity by interfaring with DNA repair mechanism and acts as a substitution for physiologic zinc in the body and (ii) QDS containing CdSe and CdTe upon photoactivation leads to their oxidation and generates an excited electron that is transferred to molecular oxygen and its reaction with biological molecule produce free radicles [110, 111].

All the above studied related with toxicity of QDS on aquatic organisms can be conclude with statement higher concentration of QDS cause oxidative stress, stress-related gene expression and bioaccumulation in different aquatic species and further more research is needed to study effect of QDS on aquatic organism.

16.9 Other Metallic Nanoparticles (Iron, Copper and Zinc Oxides) and Aquatic Organism

Iron nanoparticles are highly reactive due to their large surface area. In the existence of oxygen and water, they quickly oxidize to form free iron ions. They are widely used in biomedical applications, such as drug delivery, cellular labelling, tissue repair, in vitro bio separation, hyperthermia and for remediation of industrial sites contaminated with water pollution [32, 112] Various studies have been carried out on ecological effects of iron oxide nanoparticles (FeNPS) on aquatic organisms. Zhu et al. [33] reported greater than 10 mg/L of FeNPS cause developmental toxicity in zebrafish embryo causing mortality, hatching delay and malformation. In medaka (*Oryzias latipes*) after exposed to lethal and sub-lethal concentration FeNPS caused significant mortality and result further revealed coating NPs with carboxymethyl cellulose had less toxicity in comparison to uncoated NPs [32]. In another study, iron oxide nanoparticles were prepared using biosynthesis method and its haematological effects on *Oreochromis mossambicus* was carried out. The results indicated that after 96 h exposure of *Oreochromis mossambicus*, haematological (RBC, WBC, Hb, HCT) and biochemical parameters (SGOT, SGPT) significantly changed in treated groups than control [34]. Remya et al. [113] examined the chronic toxicity effects FeNPS of (500 mg l^{-1}) on certain hematological, ionic regulatory and gill Na^+/K^+ ATPase activity of an Indian major carps (*Labeo rohita*) for a period of 25 days. A significant upsurge in hemoglobin (Hb) content, red blood cell (RBC) count and hematocrit (Ht) value was observed throughout the period of study when compared to control groups. On the other hand, mean cellular volume, mean cellular hemoglobin (except on 5th day), mean cellular hemoglobin concentration levels and white blood cell (WBC) counts were found to be declined. FeNPS also caused alterations in iono regulation resulting in hyponatremia (Na^+), hypochloremia (Cl^-) (except on 5th day) and hypokalemia (K^+) (except up to 15th day). Overall results validate that high FeNPS concentrations in the aquatic environment may have adverse physiological effects on fish. However more research is needed to determine safe concentration and exposure method for determining toxicity of FeNPS in aquatic organism.

Like many other forms of NPs, copper nanoparticles (CuoP) can be prepared through natural or chemical synthesis. CuoP are of interest due to their historical application as colouring agents and their bactericide and excellent thermal conductivity [114]. There are fewer studies on CuoP toxicity on aquatic organism. Song et al. [115] studied acute toxicity of spherical 50 nm CuoP in juvenile rainbow trout, fathead minnow (*Pimephales promelas*) and zebrafish. After exposure of 96 h LC_{50} of the CuoP were 0.68 ± 0.15, 0.28 ± 0.04 and 0.22 ± 0.08 mg Cu/L for rainbow trout, fathead minnow and zebrafish, respectively. Furthermore, their results revealed soluble copper was one of main component for the acute toxicity of the copper nanoparticles suspensions. Both CuoP suspension and copper nitrate harmed the gill filaments and gill pavement cells, with alterations in sensitivity for these effects between the fish species. In another report, juvenile of orange-spotted grouper (*Epinephelus coioides*) were exposed to 20 or 100 μg Cu L^{-1} as either copper sulphate ($CuSO_4$) or CuoP for

25 days and growth performance reduced with increasing $CuSO_4$ or CuoP dose, more so in the $CuSO_4$ than CuoP treatment. Further, both forms of copper exposure suppressed activities of digestive enzymes (protease, amylase, and lipase) found in liver, stomach, and intestine in fish. Overall results of this study unveiled CuoP had a parallel type of toxic effects as $CuSO_4$, but soluble copper was more toxic than CuoP. Zhao et al. [116] assess the effect of lethal and sub-lethal concentration of CuoP in common carp (*Cyprinus carpio*). They indicated that after 4 days exposure, no toxic effects were detected but after 30 days disclosure to sub-lethal concentration, growth reduced and copper accumulation followed an order: intestine > gill > muscle > skin > liver > brain.

Zinc oxide nanoparticles (ZNPs) have been used in a board range of commercial processes and in industrial products in recent years such as, optoelectronics, cosmetics, catalysts, ceramics, glass, cement, rubber, paints pigments [117] and in aquaculture [118]. Like most of the nanoparticles ZNPs toxicological studies showed that ZNPs had adverse effects on aquatic environmental species. Abdel-Khalek et al. [119] evaluate and compare between the possible ecotoxicological effects of zinc bulk particles and ZNPs on Nile tilapia (*O. niloticus*). Fish exposed to zinc bulk particles and ZNPs aroused a major decline in total lipids, total protein and globulin contents, concurring with an augmentation in serum glucose, albumin, creatinine and uric acid concentrations, as well as activities of liver enzymes. Additionally, ZNPs significantly casued an upsurge in liver and gills glutathioneperoxidase (GPx), catalase (CAT), superoxide dismutase (SOD) activities and malondialdehyde (MDA) levels. At the end of this study it was concluded ZNPs have more toxic impacts than the zinc bulk particles. Hao et al. [29] demonstrated in a study on juvenile of common carp (*C. carpio*) exposed to ZNPs and results revealed that after 30 days exposure, severe histopathological alteration was observed and intracellular oxidative stress induced in response to ZNPs.

All the above studies unveiled that metallic nanoparticles cause effect on embryo development, oxidative stress and bioaccumulation which may depend on exposure time and different concentration or dose. Hence more studies are required to determine safe concentration or dose and exposure method or time before using for aquaculture activity.

16.10 Polymeric Nanoparticles (Chitosan, Poly (Lactic-co-glycolic Acid) Alginate)

Several different polymers both natural and synthetic have been utilized in formulating biodegradable nanoparticles [120]. Of these polymers, chitosan, poly (lactide-coglycolide) (PLGA) and alginate and have been the most extensively explored for drug delivery [121, 122].

Chitosan is a natural cationic biopolymer derived from hydrolysis of chitin. It can be obtained from natural sources, namely crab and shrimp shell wastes. Chitosan has been widely used in pharmaceutical and medical areas, due to its favourable biological properties such as biodegradability, biocompatibility, low toxicity, hemostatic, bacteriostatic, fungistatic, anticancerogenic properties [123–125]. Nanoparticles of chitosan has been extensively examined for its potential in the development of the controlled release drug delivery system like peptides, protein antigens, oligonucleotides and genes in aquatic animals [121, 126, 127]. However there are some reports that show chitosan nanoparticles possess potential risks to aquatic animal and environment Hu et al. [17] revealed chitosan nanoparticles caused 100% mortality in zebrafish embryos when exposed with 40 mg/L of 200 nm size chitosan nanoparticles. However, there is little information available regarding toxicity of chitosan nanoparticles in aquatic animals.

Poly lactic-co-glycolic acid (PLGA) has been among the most attractive polymer for drug delivery and tissue engineering in the last few decades [128]. PLGA is produced by co-polymerization of two diverse monomers, the cyclic dimers (1, 4-dioxane-2,5-diones) of glycolic acid and lactic acid. PLGA is mostly used polymer because it is biodegradable and biocompatibe and can be easily hydrolyed to monomers, lactic acid, and glycolic acid. In aquafarming PLGA have been mainly used for drug and vaccine delivery. It is easily removed from the body through the Kreb cycle and due to this reason, there is minimal toxicity related with the use of PLGA as a nanodelivery system in animals [129].

Alginate is a natural polysaccharide product derived from brown seaweed *like Macrocystis pyrifera* and *Laminaria*. Alginate is widely used in food industry and pharmaceutical for human and animal as a thickener, emulsifier and stabilizer agent [130, 131]. Calcium and sodium alginate have been used for oral delivery of protein, vaccines as well probiotics in aquaculture [132].

Information about toxicity of polymeric nanoparticles in aquatic organism is scanty. Typical inorganic nanoparticles as confirmed by various previous studies are more toxic than the polymeric nanoparticles. Therefore, more studies are required to determine toxicity of polymeric nanoparticles before using different aquaculture practise.

16.11 Genotoxicity of Aquatic Animals with References to Different Nanoparticles

Genotoxicity or Genetic toxicology is the study of interaction of DNA-damaging agents with the cells genetic material in relation to subsequent effects on the health of the organism [133]. The chemical which can induce mutations or so called indicator effects (for example, induction of DNA modifications, DNA repair, or recombination) is called as genotoxins or genotoxic chemical [134]. Ecogenotoxicology (genetic ecotoxicology) is an approach that applies the principles and techniques of genetic toxicology to evaluate the possible effects of environmental pollution in the form of genotoxic agents on the health of the ecosystem [135]. Genotoxicity assays address different interactions between agents and the DNA. A wide variety of prokaryotic and eukaryotic genotoxicity assays has been developed so far. Wurgler and Kramers [136] explained that genotoxic chemicals can damage the genetic material of humans and other organisms living in the environment. Genotoxicity on organisms may result gamete loss, decrease in reproductive performance, abnormal development and embryo mortality, lethal mutations as well as increased or decreased genetic diversity [137, 138]. It also induces alterations in the germ line which in turn alters the genetic make-up of populations. Changes in the genetic constitution may cause (i) the induction of (pesticide) resistance, (ii) the enhanced virulence of pathogens, (iii) changes of host ranges of pathogenic forms or the appearance of new virus types and (iv) subtle changes in parasite—host or predator—prey relationships. Structural changes to the integrity of DNA caused by genotoxic agents are useful endpoints for assessing exposure to hazardous environmental pollutants on human health and biota [139, 140].

To assess the toxicity in aquatic environment, invertebrate models are used [141, 142]. Dixon et al. [143] gave a review about the genotoxicty assessment in marine higher vertebrates. Current awareness of the potential hazards of pollutants in the aquatic environment has caused much attention in the use of fish as indicators for monitoring carcinogens, teratogens, clastogens, and mutagens [144]. Different genotoxicity tests and their applications to environmental monitoring and assessment have been reported in fish by various authors [145–147]. Shugart and Theodorakis [140] examined DNA for structural modifications indicating damage caused by a genotoxic agent (adduct, strand breakage, and photoproduct) in whales, turtle and sunfish. Exposure to DNA damaging agents may result in the formation of carcinogen-DNA adducts, which are possible indicators for carcinogens and have been detected in mussels [148] and fish [149, 150]. Different fishes and shellfishes have been studied for monitoring pollution in water in developing countries [151, 152]. De Flora et al. [153] studied the effect of genotoxins in the river water in the early stage of rainbow trout (*Oncorhynchus mykiss*). Two fresh water fish species (*Astyanax bimaculatus and Hoplias malabaricus*) have been used as bioindicators to study the effect of mutagenic pollutants from river water [152].

Many studies were carried out to assess the genotoxicity of nanoparticle by comet assay due to increase in uses of nanoparticles. Comet assay has been used for evaluation of the genotoxicity of some nanoparticles included in drug delivery systems such as single-walled carbon nanotubes (SWCNT), C60 fullerenes, titanium dioxide (TiO_2), carbon black particles, etc. [154, 155]. Reeves et al. [156] found cytotoxic and genotoxic effects of TiO_2 nanoparticles on goldfish skin cells (GFSk-S1) at 10 and 100 μg/ml. DNA damage in freshwater crustacean *Daphnia magna* was reported when treated with silver nanoparticles at 1 μg/l and silver nitrate at 1.5 μg/l [157]. Flower et al., [158] recorded DNA damage in human peripheral blood cells due to silver nanoparticles of 50 and 100 μg/mL. Silva et al. [159] studied the genotoxic effect of wild rodents (*Ctenomys torquatus*, "tuco-tuco") from a region close to a strip coal mine, where they have found the wild rodent having more DNA damage than the other rodents. When *Mytilus edulis* which is an important pollution indicator organism, were exposed to Hydrogen peroxide and N nitrosodimethylamine (NDMA), it induced DNA damage in gill cells [160]. In several other molluscs, such as, *Patunopecten yessoensis* and *Tapes japonica* [161] Oyster (*Crassostrea virginica*) [162] DNA damage have been studied in gill cells and haemocytes when exposed chemicals. In fishes, such as, Flounder (*Pleuronectes americanus*) [162] when exposed to hydrogen peroxide at concentration range 5–500 μM found DNA damage in red blood cells. Nwani et al. [163] examined DNA damage in freshwater fish *Channa punctatus* when treated with carbosulfan in erythrocytes and gill cells in vivo. Ferraro et al. [164] examined significant increase of tailed nucleoids in the erythrocytes of fish (*Hoplias malabaricus*) when treated with inorganic lead (PbII) and tributyltin (TBT). Matsumoto et al. [165] found DNA damage in erythrocytes of Tilapia (*Oreochromis niloticus*) due to chromium. In our previous report genotoxic effects of the AgNPS in the fish liver tissue using single-cell gel electrophoresis revealed DNA damage on exposure to concentrations of 400 and 800 μg/l as given in Fig. 16.3.

16.12 Molecular Response of Stress-Related Gene with Respect to Different Nanoparticles

Along with the cytotoxic and genotoxic assessment methods, gene expression analysis has also been increasingly used in aquatic ecotoxicology, as it offers high sensitivity and mechanistic values in the diagnosis of environmental contamination [166]. Several studies have been carried out to understand the responses to chemical stressors at the molecular level in aquatic invertebrates [157]. Liu et al. [167] studied the stress gene expression in mice when treated with inorganic arsenicals. Arsenic-induced stress proteins such as HO-1, HSP70, HSP90, metallothionein, the metal responsive transcription factor MTF-1, nuclear factor kappa B and c-Jun/AP-1. Fujita et al. [168] studied gene expression profiling of the rat lung after whole-body inhalation exposure to C60 fullerene (0.12 mg/m^3; 4.1 × 104 particles/cm^3, 96 nm diameter). It revealed that few genes involved in the inflammatory response, oxidative stress, apoptosis and metalloendopeptidase activity were up-regulated at both 3 days and 1 month post-exposure. Cerium oxide nanoparticles with different sizes when exposed to human lung epithelial cells (BEAS-2B) led to cell death, ROS increase, GSH decrease and the inductions of oxidative stress-related genes such as hemeoxygenase-1, catalase, glutathione S-transferase, and thioredoxin reductase [169]. Nair et al. [170] studied the toxicity effect of silver nanoparticles in aquatic midge, *Chironomus riparius* where alteration in gene expression was reported, where they observed that AgNPs could modulate the ecdysone nuclear receptor and have significant implications in different developmental stages in *C. riparius*. Oxide nanoparticles of around 100 nm size, namely, mesoporous silica

(MCM-41), iron oxide (Fe_2O_3-NPs), and zinc oxide (ZnO-NPs), was evaluated in the human embryonic kidney cell line HEK293, which showed cytoxic effect and identified a number of up- and down-regulated genes that were found to be associated with inflammation, stress, and the cell death and defense response [171].

Various stress related genes such as Heat shock protein (hsp 70), cytochrome p450 1A(cyp1a), cyp3a45, Glutathione peroxidase (gpx), the metallothionein (MT), glutathione S transferase (GST), p53, transferrin (TF) genes, etc. are involved on exposure to a toxic particle. Metallothionein (MT), Transferrin and Glutathione peroxidise has found to expressed in gill and liver tissue of zebrafish, rainbow trout, medaka, etc. leading to oxidative stress and apoptisis due to nanoparticles exposure [23, 77].

Metallothionein is a biomarker often used as an indicator of metal exposure since it is highly specific and sensitive to metals and is induced in response to elevated metal concentrations in tissues or living cells [172]. Many researchers described the up-regulation of the MT gene when exposed to AgNPs [23, 38]. They recommended that MT being a metal-binding protein plays a major role in silver metabolism and detoxification of AgNPs by binding to Ag+ ions. Choi et al. [81] studied that the metal-sensitive metallothionein 2 (MT2) mRNA induced in the liver tissues of AgNP-treated zebrafish.

HSP70s, a class of molecular chaperones, is often associated with cellular response to harmful stress or to adverse life conditions. Due to the high conservation, these proteins have been used extensively to monitor the impact of environmental factors on various species [39]. Chae et al. [173] observed down regulation of HSP 70 genes in liver tissue of medaka fish when exposed to silver nanoparticles. Similar results were obtained by Scown et al. [77] who reported the inhibition of HSP70 genes when rainbow trout are exposed to silver nanoparticles of different sizes. They suggested that release of Ag+ ion leads to toxicity in fishes.

Transferrin (TRF) is a blood plasma protein which plays an essential function in the transport of iron through the blood to the liver, spleen and bone marrow. TRF is regarded as an immune system-related gene as iron metabolism is vital for cell proliferation and is associated with the innate immune system. The binding of iron to the regulatory regions of TF inhibits the gene expression [174]. Chae et al. [173] examined that the mRNA level of the TRF gene in medaka fish was dramatically lower in the liver when exposed to Ag-NPs. These findings suggest that TRF expression is controlled by the amount of iron required for the cell to maintain its metabolism. TRF protein is involved in detoxification of Ag+ ions, whereas silver nanoparticles with their high surface areas and surface free energies, might affect the protein synthesis, causing it to malfunction to protect the cells.

Recently Sharma et al. [23] reported down-regulation of stress-related genes (MT, HSP and TRF) due to the production of free radicals and reactive oxygen species when fishes were exposed with different concentration of AgNPs. In another similar study, on zebrafish, Choi et al. [175] reported ZNPs affect genes (*aicda*, *cyb5d1*, *edar*, *intl2*, *ogfrl2* and *tnfsf13b*) linked to inflammation and the immune system, causing in yolk-sac edema and pericardia edema in embryonic/larval developmental stages of zebrafish. The toxic effect of some nanoparticles on aquatic organisms is presented in Table 16.2.

16.13 Conclusion and Future Direction

Despite the wide spread applications of nanoparticles in various fields such as, medical science, space science, semiconductor technology, information technology, cellular and molecular biology, agriculture and animal science etc., numerous reports about side effects of these materials on biological systems are available which need to be addressed. In addition to physicochemical properties, the toxicity of nanomaterials on aquatic organisms depends mainly on size of nanoparticles and concentration or doses of exposure time. Disclosure to these nanoparticles has been found to cause genotoxic and cytotoxic effects in various aquatic organisms. Since these nanoparticles are used nowadays in commercial products, there is a chance of reaching these nanoparticles in aquatic ecosystem and cause nanotoxicity to the aquatic organisms which may enter to human body through food chain.

Therefore, before using nanoparticles in any commercial product, all nanoproducts should be evaluated through toxicological studies. These toxicological studies will provide baseline for protection of both human and aquatic environment. Even though, many studies have evaluated the impact of nanomaterials on aquatic organism, however there no authorizing conclusion on their related risks. Nevertheless, method for synthesis of nanomaterials, doses and concentration of nanoparticles, route of exposure, coating material and bioaccumulation of nanomaterials in human and aquatic organisms need to be carefully considered and well studies before using nanomaterials for any commercial application. Both in vivo and in vitro assays should be used for toxicity assessment of nanomaterials. Hence more research is needed to solve the problems and overcome the challenges associated with nanonmaterial, their possible impact on aquatic organism and aquatic ecosystem.

References

1. Navalakhe RM, Nandedkar TD (2007) Application of nanotechnology in biomedicine. Ind J ExpBiol 45:160–165
2. Bhushan B (2010) Introduction of nanotechnology (ed) Handbook of nanotechnology, XLVIII. Springer, p 1964
3. NANOFORUM (2006) Nanotechnology in consumer products. Nanoforum Report. Eur Nanotechnol Gateway
4. Silva GA (2006) Neuroscience nanotechnology: progress, opportunities and challenges. Nat Rev Neurosci 7:65–74
5. Feynman R (1960) There's plenty of room at the bottom: an invitation to enter a new field of physics. Eng Sci 23(5):22–36
6. Taniguchi N (1974) On the basic concept of nanotechnology. In: Proceedings of the international congress on Prod Eng. JSPE, Tokyo, Japan, pp 457–462
7. Peterson CI (2004) Nanotechnology: from feynman to the grand challenge of molecular manufacturing. IEEE technology and society magazine, winter
8. Boxall ABA, Chaudhry Q, Sinclair C et al (2007) Current and future predicted environmental exposure to manufactured nanoparticles. Report by the Central Science Laboratory (CSL) York for the Department of the Environment and Rural Affairs (DEFRA), UK
9. Stone V, Nowack B, Baun A et al (2010) Nanomaterials for environmental studies: classification, reference material issues, and strategies for physico-chemical characterisation. Sci Total Environ 408:1745–1754
10. Sun H, Choy TS, Zhu DR et al (2009) Nano-silver-modified PQC/DNA biosensor for detecting *E. coli* in environmental water. Biosens Bioelectron 24:1405–1410
11. Kirkpatrick CJ, Bonfield W (2010) NanoBioInterface: a multidisciplinary challenge. J R Soc Interface 7:S1–S4
12. Yang Z, Liu ZW, Allaker RP et al (2010) A review of nanoparticle functionality and toxicity on the central nervous system. J R Soc Interface 7:S411–S422
13. Rather M, Sharma R, Aklakur M et al (2011) Nanotechnology: a novel tool for aquaculture and fisheries development. A prospective mini-review. Fish Aquacult J 16:1–5
14. Ferosekhan S, Gupta S, Singh RA et al (2014) RNA-loaded chitosan nanoparticles for enhanced growth, immunostimulation and disease resistance in fish. Curr Nanosci 10(3):453–464
15. Aklakur M, Rather MA, Kumar N (2015) Nanodelivery: an emerging avenue for nutraceuticals and drug delivery. Crit Rev Food Sci Nutr 56(14):2352–2361
16. Bullock G, Blazer V, Tsukuda S et al (2000) Toxicity of acidified chitosan for cultured rainbow trout (Oncorhynchus mykiss). Aquaculture 185:273–280
17. Hu YL, Qi W, Han F et al (2011) Toxicity evaluation of biodegradable chitosan nanoparticles using a zebrafish embryo model. Int J Nanomedicine 6:3351–3359
18. Rafiee A, Alimohammadian MH, Gazori T et al (2014) Comparison of chitosan, alginate and chitosan/alginate nanoparticles with respect to their size, stability, toxicity and transfection. Asian Pac J Trop Dis 4(5):372–377
19. Krishnaraj C, Harper SL, Yun SI (2015) In vivo toxicological assessment of biologically synthesized silver nanoparticles in adult zebrafish (*Danio rerio*). J Hazard Mater 301:480–491
20. Osborne OJ, Lin S, Chang CH et al (2015) Organ-specific and size-dependent Ag nanoparticle toxicity in gills and intestines of adult zebrafish. ACS Nano 9(10):9573–9584
21. Bilberg K, Malte H, Wang T et al (2010) Silver nanoparticles and silver nitrate cause respiratory stress in Eurasian perch (*Percafluviatilis*). Aquat Toxicol 96:159–165
22. Rajkumar KS, Kanipandian N, Thirumurugan R (2016) Toxicity assessment onhaemotology, biochemical and histopathological alterations of silver nanoparticles exposed freshwater fish Labeorohita. Appl Nanosci 6(1):19–29
23. Sharma N, Rather MA, Ajima MNO et al (2016) Assessment of DNA damage and molecular responses in *Labeorohita* (Hamilton, 1822) follow-

ing short-term exposure to silver nanoparticles. Food Chem Toxicol 96:122–132
24. Jang MH, Kim WK, Lee SK et al (2014) Uptake, tissue distribution, and depuration of total silver in common carp *(Cyprinus carpio)* after aqueous exposure to silver nanoparticles
25. Powers CM, Levin ED, Seidler FJ et al (2010) Silver exposure in developing zebrafish produces persistent synaptic and behavioral changes. Neurotoxicol Teratol 2:329–332
26. Laban G, Nies EL, Turco RF et al (2010) The effects of silver nanoparticles on fathead minnow (*Pimephalespromelas*) embryos. Ecotoxicology 19:185–195
27. Dedeh A, Ciutat A, Treguer-D M et al (2015) Impact of gold nanoparticles on zebrafish exposed to a spiked sediment. Nanotoxicology 9:71–80
28. Asharani PV, Wu YL, Gong Z (2008) Toxicity of silver nanoparticles in zebrafish models. Nanotechnology 19(25):255102
29. Hao L, Chen L, Hao J et al (2013) Bioaccumulation and sub-acute toxicity of zinc oxide nanoparticles in juvenile carp (*Cyprinus carpio*): a comparative study with its bulk counterparts. Ecotoxicol Environ Saf 91:52–60
30. Johnston BD et al (2010) Bioavailability of nanoscale metal oxides TiO2, CeO2, and ZnO to fish. Environ Sci Technol 44:1144–1151
31. Bai W, Zhang Z, Tian W et al (2010) Toxicity of zinc oxide nanoparticles to zebrafish embryo: a physicochemical study of toxicity mechanism. J Nanopart Res 12:1645–1654
32. Chen PJ, Su CH, Tseng CY et al (2011) Toxicity assessments of nanoscalezerovalent iron and its oxidation products in medaka (*Oryziaslatipes*) fish. Mar Pollut Bull 63:339–346
33. Zhu X, Tian S, Cai Z (2012) Toxicity assessment of iron oxide nanoparticles in zebrafish (Danio rerio) early life stages. PLoS One 7(9):e46286
34. Karthikeyeni S, Vijayakumar S, Vasanth S et al (2013) Biosynthesis of iron oxide nanoparticles and its haematological effects on fresh water fish *Oreochromismossambicus*. J Acad Indus Res 10:645–649
35. Zhang W, Qiao X, Chen J (2007) Synthesis of nanosilver colloidal particles in water/oil microemulsion. Colloids Surf A Physicochem Eng Asp 299:22–28
36. Clemente Z, Castro V, Moura M, Jonsson C, Fraceto L (2014) Toxicity assessment of TiO2 nanoparticles in zebrafish embryos under different exposure conditions. Aquat Toxicol 147:129–139
37. Smith CJ, Shaw BJ, Handy RD (2007) Toxicity of single walled carbon nanotubes to rainbow trout (Oncorhynchus mykiss): respiratory toxicity, organ pathologies, and other physiological effects. Aquat Toxicol 82:94–109
38. Louis S, Gagné F, Auclair J et al (2010) The characterisation of the behaviour and gill toxicity of CdS/CdTe quantum dots in rainbow trout (*Oncorhynchusmykiss*). Int J Biomed Nanosci Nanotechnol 1(1):52–69
39. Gagne F, Auclair J, Turcotte P et al (2008) Ecotoxicity of CdTe quantum dots to freshwater mussels: impacts on immune system, oxidative stress and genotoxicity. Aquat Toxicol 86(3):333–340
40. Luna-Velasco A, Field JA, Cobo-Curiel A et al (2011) Inorganic nanoparticles enhance the production of reactive oxygen species (ros) during the autoxidation of L-3,4-dihydroxyphenylalanine (L-dopa). Chemosphere 85(1):19–25
41. Cho WS, Duffin R, Thielbeer F et al (2012) Zeta potential and solubility to toxic ions as mechanisms of lung inflammation caused by metal/metal oxide nanoparticles. Toxicol Sci 126(2):469–477
42. Fard JK, Jafari S, Eghbal MA (2015) A review of molecular mechanisms involved in toxicity of nanoparticles. Adv Pharm Bull 5:447–454
43. Hutter E, Fendler JH (2004) Exploitation of localized surface plasmon resonance. Adv Mater 16:1685–1706
44. Sudrik S, Chaki NK, Chavan VB et al (2006) Silver nanocluster redox-couple-promoted nonclassical electron transfer: an efficient electrochemical Wolff rearrangement of alpha-diazoketones. Chem Eur J12:859–864
45. Choi Y, Ho N, Tung C (2007) Sensing phosphatase activity by using gold nanoparticles. Angew Chem Int Ed 46:707–709
46. Fabrega J, Luoma SN, Tyler CR et al (2011) Silver nanoparticles: behaviour and effects in the aquatic environment. Environ Int 37:517–531
47. Kholoud MM, El-Nour A, Eftaiha A et al (2010) Synthesis and applications of silver nanoparticles. Arab J Chem 3:135–140
48. Capek I (2004) Preparation of metal nanoparticles in water-in-oil (w/o) microemulsions. Adv Colloid Interf Sci 110:49–74
49. Frattini A, Pellegri N, Nicastro D et al (2005) Effect of amine groups in the synthesis of Ag nanoparticles using aminosilanes. Mater Chem Phys 94:148–152
50. Arora S, Jain J, Rajwade JM et al (2009) Interactions of silver nanoparticles with primary mouse fibroblasts and liver cells. Toxicol Appl Pharm 236:310–318
51. Cohen MS, Stern JM, Vanni AJ et al (2007) In vitro analysis of a nanocrystalline silver-coated surgical mesh. Surg Infect 8:397–403
52. Lee HY, Park HK, Lee YM et al (2007) A practical procedure for producing silver nanocoated fabric and its antibacterial evaluation for biomedical applications. Chem Commun (Camb) 28:2959–2961
53. Cheng D, Yang J, Zhao Y (2004) Antibacterial materials of silver nanoparticles application in medical appliances and appliances for daily use. Chin Med Equip J4:26–32
54. Jain P, Pradeep T (2005) Potential of silver nanoparticle coated polyurethane foam as an antibacterial water filter. Inter science. Biotechnol Bioeng 90:59–63

55. Zhang Y, Sun J (2007) A Study on the bio-safety for nano-silver as anti-bacterial materials. Chin J Med Instrumen 31:35–38
56. Purcell TW, Peters JJ (1998) Sources of silver in the environment. Environ Toxicol Chem 17:539–546
57. Sanudo-Willhelmy S, Flegal R (1992) Anthropogenic silver in the Southern California bight: a new tracer of sewage in coastal waters. Environ Sci Technol 26:2147–2151
58. Benn TM, Westerhoff P (2008) Nanoparticle silver released into water from commercially available sock fabrics. Environ Sci Technol 42:4133–4139
59. Geranio L, Heuberger M, Nowack B (2009) The behavior of silver nanotextiles during washing. Environ SciTechnol 43:8113–8118
60. Gottschalk F, Sonderer T, Scholz RW et al (2009) Modeled environmental concentrations of engineered nanomaterials (TiO_2, ZnO, Ag, CNT, fullerenes) for different regions. Environ Sci Technol 43:9216–9222
61. Ratte HT (1999) Bioaccumulation and toxicity of silver compounds: a review. Environ Toxicol Chem 18:89–108
62. Luoma SN, Ho YB, Bryan GW (1995) Fate, bioavailability and toxicity of silver in estuarine environments. Mar Pollut Bull 31:44–54
63. Adams NWH, Kramer JR (1998) Reactivity of Ag+ ion with thiol ligands in the presence of iron sulfide. Environ Toxicol Chem 17:625–629
64. Erickson RJ, Brooke LT, Kahl MD et al (1998) Effects of laboratory test conditions on the toxicity of silver to aquatic organisms. Environ Toxicol Chem 17:572–578
65. Levard C, Hotze EM, Lowry GV et al (2012) Environmental transformations of silver nanoparticles: impact on stability and toxicity. Environ Sci Technol 46:6900–6914
66. Dale AL, Lowry GV, Casman EA (2013) Modeling nanosilver transformations in freshwater sediments. Environ Sci Technol 47:122920–122928
67. Quik JTK, Stuart MC, Wouterse M et al (2012) Natural colloids are the dominant factor in the sedimentation of nanoparticles. Environ Toxicol Chem 2012(31):1019–1022
68. Kohler AR, Som C, Helland A et al (2008) Studying the potential release of carbon nanotubes throughout the application life cycle. J Clean Prod 16:927–937
69. Luoma SN, Rainbow PS (2008) Metal contamination in aquatic environments: science and lateral management. Cambridge University Press, Cambridge
70. Luoma SN, Rainbow PS (2005) Why is metal bioaccumulation so variable? Biodynamics as a unifying concept. Environ Sci Technol 39:1921–1931
71. Wood CM, Hogstrand C, Galvez F et al (1996) The physiology of waterborne silver toxicity in freshwater rainbow trout (*Oncorhynchusmykiss*). The effects of ionic Ag^+. Aquat Toxicol 35:93–109
72. Zhou B, Nichols J, Playle RC et al (2005) An in vitro biotic ligand model (BLM) for silver binding to cultured gill epithelia of freshwater rainbow trout (Oncorhynchusmykiss). Toxicol Appl Pharmacol 202:25–37
73. Birge W, Zuiderveen J (1995) The comparative toxicity of silver to aquatic biota. In: Proceedings, 3rd argentum international conference on the transport, fate, and effects of silver in the environment, Washington DC, pp 99–108
74. Hogstrand C, Wood CM (1998) Toward a better understanding of the bioavailability, physiology and toxicity of silver in fish: implications for water quality criteria. Environ Toxicol Chem 17:547–561
75. Grosell M, De Boeck G, Johannsson O et al (1999) The effects of silver on intestinal ion and acid–base regulation in the marine teleost fish, Papophrysvetulus. Comp Biochem Physiol C Toxicol Pharmacol 124:259–270
76. Yeo MK, Kang M (2008) Effects of nanometer sized silver materials on biological toxicity during zebrafish embryogenesis. Bull Korean Chem Soc 29:1179–1184
77. Scown TM, Santos ME, Johnston BD (2010) Effects of aqueous exposure to silver nanoparticles of different sizes in rainbow trout. Toxicol Sci 115:521–534
78. Ringwood AH, McCarthy M, Bates TC et al (2010) The effects of silver nanoparticles on oyster embryos. Mar Environ Res 69:549–551
79. Wu Y, Zhou Q, Li H et al (2010) Effects of silver nanoparticles on the development and histopathology biomarkers of Japanese medaka (Oryziaslatipes) using the partial-life test. Aquat Toxicol 100:160–167
80. Farkas J, Christian P, Urrea JAG et al (2010) Effects of silver and gold nanoparticles on rainbow trout (Oncorhynchusmykiss) hepatocytes. Aquat Toxicol 96(1):44–52
81. Choi JE, Kim S, Ahn JH et al (2010) Induction of oxidative stress and apoptosis by silver nanoparticles in the liver of adult zebrafish. Aquat Toxicol 100:151–159
82. Bilberg K, Hovgaard MB, Besenbacher F et al (2012) In vivo toxicity of silver nanoparticles and silver ions in Zebra fish (*Daniorerio*). J Toxicol 29:1–9
83. Rayner TA, Mackevica A, Wejdling A et al (2010) Uptake and toxicity of silver ions, nanoparticals and microparticles in *NereisVirens* in a sediment environment project spring
84. Zhang R, Piao JM, Kim CK et al (2012) Endoplasmic reticulum stress signalling is involved in silver nanoparticles-induced apoptosis. Int J Biochem Cell Biol 44:224–234
85. Sanford RV (2010) State of the science literature review: everything nanosilver and more. In: Varner K (ed) Scientific, technical, research, engineering, and modeling support final report. US Environmental Protection Agency, Office of Research and Development, Washington DC, pp 1–197
86. Reidy B, Haase A, Luch A et al (2013) Mechanisms of silver nanoparticle release, transformation and toxicity: a critical review of current knowledge and recommendations for future studies and applications. Materials 6:2295–2350

87. Roh JY, Sim SJ, Yi J et al (2009) Ecotoxicity of silver nanoparticles on the soil nematode Caenorhabditis elegans using functional ecotoxicogenomics. Environ Sci Technol 43:3933–3940
88. Gopinath P, Gogoi SK, Sanpuic P et al (2010) Signaling gene cascade in silver nanoparticle induced apoptosis. Colloids Surf B Biointerfaces 77:240–245
89. Van Aerle R, Lange A, Moorhouse A et al (2013) Molecular mechanisms of toxicity of silver nanoparticles in zebrafish embryos. Environ Sci Technol 47:8005–8014
90. Hughes GA (2005) Nanostructure-mediated drug delivery. Nanomedicine 1(1):22–30
91. Joshi H, Shirude PS, Bansal V et al (2004) Isothermal titration calorimetry studies on the binding of amino acids to gold nanoparticles. J Phys Chem B 108(31):11535–11540
92. Shukla R, Bansal V, Chaudhary M et al (2005) Biocompatibility of gold nanoparticles and their endocytotic fate inside the cellular compartment: a microscopic overview. Langmuir 21(23):10644–10654
93. Unfried K, Albrecht C, Klotz LO et al (2007) Cellular responses to nanoparticles: target structures and mechanisms. Nanotoxicology 1(1):52–71
94. Aillon KL, Xie Y, El-Gendy N et al (2009) Effects of nanomaterial physicochemical properties on in vivo toxicity. Adv Drug Deliv Rev 61(6):457–466
95. Pan Y, Leifert A, Ruau D et al (2009) Gold nanoparticles of diameter 1.4 nm trigger necrosis by oxidative stress and mitochondrial damage. Small 5(18):2067–2076
96. Truong L, Zaikova T, Richman EK et al (2012) Media ionic strength impacts embryonic responses to engineered nanoparticle exposure. Nanotoxicology 6(7):691–699
97. Tedesco S, Doyle H, Redmond G et al (2008) Gold nanoparticles and oxidative stress in *Mytilus edulis*. Mar Environ Res 66:131–133
98. García-Negrete CA, Blasco J, Volland M et al (2013) Behaviour of Au-citrate nanoparticles in seawater and accumulation in bivalves at environmentally relevant concentrations. Environ Pollut 174:134–141
99. Bar-Ilan O, Albrecht RM, Fako VE et al (2009) Toxicity assessments of multisized gold and silver nanoparticles in zebrafish embryos. Small 5(16):1897–1910
100. Joubert Y, Pan JF, Buffet PE et al (2013) Subcellular localization of gold nanoparticles in the estuarine bivalve *Scrobiculariaplana* after exposure through the water. Gold Bull 46(1):47–56
101. Ashoori RC (1996) Electrons in artificial atoms. Nature 379(6564):413–419
102. Collier CP, Vossmeyer T, Heath JR (1998) Nanocrystal super lacttices. Annu Rev Phys Chem 49:371
103. Carrillo C, Moliner CM, Simonet Y et al (2011) Capillary electrophoresis method for the characterization and separation of CdSe quantum dots. Anal Chem 83(7):2807–2813
104. Coe-Sullivan S, Steckel JS, Woo WK et al (2005) Large-area ordered quantum-dot monolayers via phase separation during spin-casting. Adv Funct Mater 15(7):1117–1124
105. Feswick A, Griffitt RJ, Siebein K (2013) Uptake, retention and internalization of quantum dots in Daphnia is influenced by particle surface functionalization. Aquat Toxicol 130:210–218
106. Lewinski NA, Zhu H, Ouyang CR et al (2011) Trophic transfer of amphiphilic polymer coated CdSe/ZnS quantum dots to *Daniorerio*. Nanoscale 3:3080–3083
107. Guo H, Hong Z, Yi R (2015) Core-shell collagen peptide chelated calcium/calcium alginate nanoparticles from fish scales for calcium supplementation. J Food Sci 80:41–50
108. Kim J, Park Y, Yoon TH et al (2010) Phototoxicity of CdSe/ZnSe quantum dots with surface coatings of 3-mercaptopropionic acid or tri-n-octylphosphine oxide/gum arabic in *Daphniamagna* under environmentally relevant UV-B light. Aquat Toxicol 97:116–124
109. Cibulskaite Z, Kazlauskienė N, Rotomskis R et al (2015) Toxicity of quantum dots and cadmium to rainbow trout (Oncorhynchusmykiss) in early ontogenesis. In: Front Mar Sci conference abstract: XV European Congress of Ichthyology. https://doi.org/10.3389/conf.FMARS.2015.03.0019
110. Chen N, He Y, Su Y et al (2012) The cytotoxicity of cadmium-based quantum dots. Biomaterials 33:1238–1244
111. Singh BR, Singh BN, Khan W et al (2012) ROS-mediated apoptotic cell death in prostate cancer LNCaP cells induced by biosurfactant stabilized CdS quantum dots. Biomaterials 33:5753–5767
112. Zhang WX (2003) Nanoscale iron particles for environmental remediation: an overview. J Nanopart Res 5(3/4):323–332
113. Remya AS, Ramesh M, Saravanan M et al (2015) Iron oxide nanoparticles to an Indian major carp, Labeorohita: impacts on hematology, iono regulation and gill Na+/K+ ATPase activity. J King Saud Uni-Sci 27(2):151–160
114. Buffet PE, Tankoua OF, Pan JF et al (2013) Behavioural and biochemical responses of two marine invertebrates *Scrobiculariaplana*and *Hediste*diversicolor to copper oxide nanoparticles. Chemosphere 84:166–174
115. Song L, Vijver MG, Peijnenburg WJ et al (2015) A comparative analysis on the in vivo toxicity of copper nanoparticles in three species of freshwater fish. Chemosphere 139:181–189
116. Zhao J, Wang Z, Liu X et al (2011) Distribution of CuO nanoparticles in juvenile carp (*Cyprinuscarpio*) and their potential toxicity. J Hazard Mater 197:304–310
117. Chen TH, Lin CC, Meng PJ (2014) Zinc oxide nanoparticles alter hatching and larval locomotor

activity in zebrafish (*DanioRerio*). J Hazard Mater 277(30):134–134
118. Faiz H, Zuberi A, Nazir S et al (2015) Zinc oxide, zinc sulfate and zinc oxide nanoparticles as source of dietary zinc: comparative effects on growth and hematological indices of Juvenile Grass Carp (*Ctenopharyngodonidella*). Int J Agric Biol 17:568–574
119. Abdel-Khalek A, Kadry M, Hamed A et al (2015) Ecotoxicological impacts of zinc metal in comparison to its nanoparticles in Nile tilapia; Oreochromisniloticus. J Basic Appl Zool 72:113–125
120. Moghimi SM, Hunter AC, Murray JC (2001) Long-circulating and target-specific nanoparticles: theory to practice. Pharmacol Rev 53:283–318
121. Rather MA, Bhat IA, Gireesh-Babu P et al (2016) Molecular characterization of kisspeptin gene and effect of nano–encapsulted kisspeptin-10 on reproductive maturation in *Catlacatla*. Domes Anim Endocrinol 56:36–47
122. Jain RA (2000) The manufacturing techniques of various drug loaded biodegradable poly (lactide-*co*-glycolide) (PLGA) devices. Biomaterials 21(23):2475–2490
123. Hejazi R, Amiji M (2003) Chitosan-based gastrointestinal delivery systems. J Control Release 89:151–165
124. Rather MA, Sharma R, Gupta S et al (2013) Chitosan-nanoconjugated hormone nanoparticles for sustained surge of gonadotropins and enhanced reproductive output in female fish. PLoS One 8:e57094
125. Bhat IA, Rather MA, Saha R et al (2016) Expression analysis of Sox9 genes during annual reproductive cycles in gonads and after nanodelivery of LHRH in *Clariasbatrachus*. Res Vet Sci 106:100–106
126. Rajeshkumar S, Venkatesan C, Sarathi M et al (2009) Oral delivery of DNA construct using chitosan nanoparticles to protect the shrimp from white spot syndrome virus (WSSV). Fish Shellfish Immunol 26:429–437
127. Ramya VL, Sharma R, Gireesh-Babu P et al (2014) Development of chitosan conjugated DNA vaccine against nodavirus in *Macrobrachium rosenbergii* (De Man, 1879). J Fish Dis 37:815–824
128. Gentile P, Chiono V, Carmagnola I et al (2014) An overview of poly (lactic-co-glycolic) acid (PLGA)-based biomaterials for bone tissue engineering. Int J Mol Sci 15(3):3640–3659
129. Panyam J, Labhasetwar V (2003) Biodegradable nanoparticles for drug and gene delivery to cells and tissue. Adv Drug Deliv Rev 55:329–347
130. George M, Abraham TE (2006) Polyionic hydrocolloids for the intestinal delivery of protein drugs: alginate and chitosan-a review. J Control Release 114:1–14
131. Klinkesorn U, McClements DJ (2009) Influence of chitosan on stability and lipase digestibility of lecithin-stabilized tuna oil-in-water emulsions. Food Chem 114:1308–1315
132. Polk AE, Amsden B, Scarratt DJ et al (1994) Oral delivery in aquaculture: controlled release of proteins from chitosan-alginate microcapsules. Aquacult Eng 13(4):311–323
133. RosasLedesma P, Le6nRubio JM, Alarc6n FJ et al (2012) Calcium alginate capsules for oral administration of fish probiotic bacteria: assessment of optimal conditions for encapsulation. Aquac Res 43(1):106–116
134. Shugart LR, Theodorakis CW (1998) New trends in biological monitoring: application of biomarkers to genetic ecotoxicology. Biotherapy 11:119–127
135. Belfiore NM, Anderson SL (1998) Genetic patterns as a tool for monitoring and assessment of environmental aspects: the example of genetic toxicology. Environ Monit Assess 51:465–479
136. Wurgler FE, Kramers PGN (1992) Environment effects of genotoxins (ecogenotoxicology). Mutagenesis 7:321–341
137. Anderson SL, Sadinski WJ, Shugart LBP et al (1994) Genetic and molecular ecogenotoxicology: a research framework. Environ Health Perspect 102:9–12
138. Kohn HW (1983) The significance of DNA-damage assays in toxicity andcarcinogenecity assessment. Ann N Y Acad Sci 407:106–118
139. Shugart LR (1990) DNA damage as an indicator of pollutant-induced genotoxicity. In: Landis WG, van der Schalie WH (eds) 13th symposium on aquatic toxicology and risk assessment: sublethal indicators of toxic stress. ASTM, Philadelphia, pp 348–355
140. Shugart LR, Theodorakis CW (1994) Environmental genotoxicity: probing the underlying mechanisms. Environ Health Perspect 102:13–17
141. Chandler GT, Coull BC, Schizas NV et al (1997) A culturebased assessment of the effects of Chlorpyrifos on multiple meiobenthic copepods using microcosms of intact estuarine sediments. Environ Toxicol Chem 16:2339–2346
142. Kovatch CE, Chandler GT, Coull BC (1999) Utility of a full lifecycle copepod bioassay approach for assessment of sediment- associated contaminant mixtures. Mar Pollut Bull 38:692–701
143. Dixon DR, Pruski AM, Dixon LRJ et al (2002) Marine invertebrate eco-genotoxicology: a methodological overview. Mutagen 17:495–507
144. Obiakor MO, Okonkwo JC, Nnabude PC et al (2012) Ecogenotoxicology: micronucleus assay in fish erythrocytes as in situ aquatic pollution biomarker: a review. J Anim Sci Adv 2:123–133
145. Hartmann A, Golet EM, Gartiser S et al (1999) Primary DNA damage but not mutagenicity correlates with ciprofloxacin concentrations in German hospital waste waters. Arch Environ Contam Toxicol 36:115–119
146. Vargas VM, Migliavacca SB, de Melo AO et al (2001) Genotoxicity assessments in aquaticenvironments under the influence of heavy metals and organic contaminants. Mutat Res 490:141–158

147. Mitchelmore CL, Chipman JK (1998) DNA strand breakage in aquatic organisms and the potential value of the comet assay in environmental monitoring. Mutat Res 399:135–147
148. Harvey JS, Lyons BP, Waldock M et al (1997) The application ofthe 32P-post labeling assay to aquatic monitoring. Mutat Res 378:77–88
149. El Adlouni C, Tremblay J, Walsh P et al (1995) Comparative study of DNA adducts levels in white sucker fish (*Catostomuscommersoni*) from the basin of the St. Lawrence River (Canada). Mol Cell Biochem 148:133–138
150. Erickson G, Larsson A (2000) DNA adducts in perch (*Percafluviatilis*) living in coastal water polluted with bleached pulp mill effluents. Ecotoxicol Environ Saf 46:167–173
151. Saotome K, Hayashi M (2003) Application of a sea urchin micronucleus assay to monitoring aquatic polllution: influence of sample osmolality. Mutagen 18:73–76
152. Pantaleao SM, Alcantara AV, Alves JP et al (2006) The piscine micronucleus test to assess the impact of pollution on the Japaratuba River in Brazil. Environ Mutagen 47:219–224
153. Flora S, Vigario LD, Agostini F et al (1993) Multiple biomarkers in fish exposed in situ to polluted river water. Mutat Res 319:167–177
154. Jacobsen NR, Saber AT, White P et al (2007) Increased mutant frequency by carbon black, but not quartz, in the lacZ and cII transgenes of muta mouse lung epithelial cells. Environ Mol Mutagen 48:451–461
155. Vandghanooni S, Eskandani M (2011) Comet assay: a method to evaluate genotoxicity of nano-drug delivery system. Bioimpacts 1:87–97
156. Reeves JF, Davies SJ, Dodd NJF et al (2008) Hydroxyl radicals (•OH) are associated with titanium dioxide (TiO2) nanoparticle-induced cytotoxicity and oxidative DNA damage in fish cells. Mutat Res 640:113–122
157. Park SY, Choi J (2010) Geno- and ecotoxicity evaluation of silver nanoparticles in freshwater crustacean *Daphnia magna*. Environ Eng Res 15:23–27
158. Flower NAL, Brabu B, Revathy M et al (2012) Characterization of synthesized silver nanoparticles and assessment of its genotoxicity potentials using the alkaline comet assay. Mutat Res 742:61–65
159. Silva J, de Freitas TRO, Marinho JR et al (2000) An alkaline single-cell gel electrophoresis (comet) assay for environmental biomonitoring with native rodents. Genet Mol Biol 23:241–245
160. Wilson JT, Pascoe PL, Parry JM et al (1998) Evaluation of the comet assay as a method for the detection of DNA damage in the cells of a marine invertebrate, *MytilusedulisL.* (Mollusca: Pelecypoda). Mutat Res 399:87–95
161. Sasaki YF, Izumiyama F, Nishidate E et al (1997) Detection of genotoxicity of polluted sea water using shellf fish and the alkaline single-cell gel electrophoresis (SCE) assay: a preliminary study. Mut Res 393:133–139
162. Nacci DE, Cayula S, Jackim E (1996) Detection of DNA damage in individual cells from marine organisms using the single cell gel assay. Aquat Toxicol 35:197–210
163. Nwani CD, Lakra WS, Nagpure NS et al (2010) Mutagenic and genotoxic effects of carbosulfan in freshwater fish *Channapunctatus* (Bloch) using micronucleus assay and alkaline single-cell gel electrophoresis. Food Chem Toxicol 48:202–208
164. Ferraro MVM, Fenocchio AS, Mantovani MS et al (2004) Mutagenic effects of tributyltin and inorganic lead (Pb II) on the fish H.malabaricus as evaluated using the comet assay and the piscine micronucleus and chromosome aberration tests. Genet Mol Biol 27:103–107
165. Matsumoto ST, Mantovani MS, Malaguttii MIA et al (2006) Genotoxicity and mutagenicity of water contaminated with tannery effluents, as evaluated by the micronucleus test and comet assay using the fish *Oreochromisniloticus*and chromosome aberrations in onion root-tips. Genet Mol Biol 29:148–158
166. Ankley GT, Daston GP, Degitz SJ et al (2006) Toxicogenomics in regulatory ecotoxicology. Environ Sci Technol 40:4055–4065
167. Liu J, Maria B, Kadiiska MB et al (2001) Stress-related gene expression in mice treated with inorganic arsenicals. Toxicol Sci 61:314–320
168. Fujita K, Morimoto Y, Ogami A et al (2009) Gene expression profiles in rat lung after inhalation exposure to C60 fullerene particles. Toxicology 258:47–55
169. Park EJ, Choi J, Park YK et al (2008) Oxidative stress induced by cerium oxide nanoparticles in cultured BEAS-2B cells. Toxicology 245:90–100
170. Nair PMG, Park SY, Lee SW et al (2011) Differential expression of ribosomal protein gene, gonadotrophin releasing hormone gene and Balbiani ring protein gene in silver nanoparticles exposed *Chironomusriparius*. Aquat Toxicol 10:31–37
171. Dua P, Chaudhari KN, Lee CH et al (2011) Evaluation of toxicity and gene expression changes triggered by oxide nanoparticles. Bull Kor Chem Soc 32:2051
172. Olafson RW, McCubbin WD, Kay CM (1988) Primary- and secondarystructural analysis of a unique prokaryotic metallothionein from a *Synechococcus* sp. cyanobacterium. Biochem J 251:691–699
173. Chae YJ, Pham CH, Lee J et al (2009) Evaluation of the toxic impact of silver nanoparticles on *Japanese* medaka (*Oryziaslatipes*). Aquat Toxicol 94:320–327
174. Xu L, Takemura T, Xu M et al (2011) Toxicity of silver nanoparticles as assessed by global gene expression analysis. Mater Express 1:74–79
175. Choi JS, Kim RO, Yoon S et al (2016) Developmental toxicity of zinc oxide nanoparticles to zebrafish (*Daniorerio*): a transcriptomic analysis. PloS ONE 11(8):e0160763

17 Cytotoxicity and Physiological Effects of Silver Nanoparticles on Marine Invertebrates

Adriano Magesky and Émilien Pelletier

Abstract

Silver nanoparticles (AgNPs) incorporation in commercial products is increasing due to their remarkable physical and chemical properties and their low cost on the market. Silver has been known for a long time to be highly toxic to bacterial communities, aquatic organisms, and particularly to marine biota. Strong chloro-complexes dominate Ag speciation in seawater and facilitate its persistence in dissolved form. It has a great impact on marine organisms because low concentration of silver can lead to strong bioaccumulation, partly because the neutral silver chloro complex ($AgCl_0$) is highly bioavailable. Owing to the fact that estuaries and coastal areas are considered as the ultimate fate for AgNPs, the study of their toxic effects on marine invertebrates can reveal some environmental risks related to nanosilver exposure. In an attempt to reach this goal, many invertebrate taxa including mollusks, crustaceans, echinoderms and polychaetes have been used as biological models. The main findings related to AgNP toxicity and marine invertebrates are summarized hereafter. Some cellular mechanisms involving nano-internalization (cellular uptake, distribution and elimination), DNA damaging, antioxidant cellular defenses and protein expression are discussed. Physiological effects on early stage development, silver metabolic speciation, immune response, tissue damaging, anti-oxidant effects and nano-depuration are also described. Finally, we paid attention to some recent interesting findings using sea urchin developmental stages and their cells as models for nanotoxicity investigation. Cellular and physiological processes characterizing sea urchin development revealed new and multiple toxicity mechanisms of both soluble and nano forms of silver.

A. Magesky (✉) · É. Pelletier
Institut de sciences de la mer de Rimouski (ISMER), Université du Québec à Rimouski, 310 Allée des Ursulines, Rimouski, QC, G5L 2Z9, Canada
e-mail: adrianomagesky@gmail.com

Q. Saquib et al. (eds.), *Cellular and Molecular Toxicology of Nanoparticles*, Advances in Experimental Medicine and Biology 1048, https://doi.org/10.1007/978-3-319-72041-8_17

Keywords
Silver nanoparticles · Marine invertebrates · Sea urchins · Development · Dissolved silver

17.1 Scientific Context

Silver nanoparticles (AgNPs) have been produced for more than a decade to fulfill a wide range of applications in the modern life. Hence, they have been incorporated to many consumer products such as medical devices, cosmetics, food-packaging materials, electronics, and household appliances due to their antimicrobial, thermal, optical and electrical outstanding features [1]. Many questions about their real environmental risks have been also raised and especially for marine fauna. Several authors observed that a substantial part of AgNPs incorporated to commercial products are eventually released into the sewer system and the wastewater treatment stations [2–6]. Silver (Ag) is a trace metal that can be found in many geological structures in a native state or combined with other elements forming oxides, sulfites and salts, and is naturally present in freshwater and marine environments at very low concentration.

17.1.1 Fate of Silver in Estuarine Waters

Dissolved silver has a non-conservative chemical behavior in estuaries due to the high reactivity of silver ions with counter ions present in seawater and also with negatively charged organic ligands (as thiol groups) either present on soluble organic matter or living cell surface. In freshwater system, organic matter and sulfide might dominate Ag speciation and reduce its bioavaibility [7], but at low salinity, neutral chemical complexes such as AgCl (aq) and AgHS (aq) are rather formed and slow AgNP dissolution is taking place. With an increasing salinity towards the ocean, Ag^+ availability is changing while silver chloride complexes ($AgCl^0$, $AgCl_2^-$, $AgCl_3^{2-}$ and $AgCl_4^{3-}$) and possibly silver-thiol complexes become dominant. It turns out that in saltwater, some soluble complexes and Ag^+ associated with refractory organic substances may remain bioavailable, but eventually tend to slowly settle down in the benthic environment [8].

Estuaries are often close to highly populated urban areas, with surface water contamination being ubiquitous. So, it has been suggested that the growing use of silver nanoparticles might increase the concentration of silver in natural waters. This appears to be particularly true in those areas where sewage treatment is missing or when treatment plants are less effective to remove Ag from wastewater before discharging in receiving environment. Even though environmental impacts of AgNPs remain unknown under field conditions, previous knowledge about environmental fate and toxic effects of dissolved silver provides a baseline for risk assessment to aquatic organisms [7].

The stratified estuarine water column is clearly divided into a warmer upper mixed layer and a cooler and saltier lower layer. Between these layers, there is a thin interface layer, a region of rapid changes of temperature and density. It has been observed that aggregated AgNPs get mostly trapped at the mixing layer due to the density gradient and less than 5% can reach the seawater layer [9]. As a result, planktonic organisms living in or close by the halocline can be exposed to AgNPs. Sea urchin larvae that eventually migrate through halocline layer to find rich feeding ground is a good example [10]. Eventually, AgNPs associated with organic matter will form large aggregates (a few μm) and precipitate onto surface sediment exposing thus the benthic fauna [11]. This scenario is representative of stratified estuaries and coastal waters where a significant part of marine fauna is often located.

17.1.2 Occurrence of AgNPs in Natural Waters

Several studies have demonstrated that AgNPs from many commercial products will eventually be released in sewage treatment effluents and wastewater [2–6]. In fact, the quantification and characterization of Ag released from nanoAg products can provide some insights of the effects of AgNPs in the environment. As an example, fabrics with different methods of silver incorporation into or onto fibers may release distinct amounts of dissolved silver or AgNPs [2]. Particulate Ag in influent wastewater can be detected at a level of 6.9 ± 5.0 mg.kg^{-1} whereas dissolved Ag could reach up to 0.27 ± 0.11 μg.L^{-1} [12]. Up to 90% of the total Ag in wastewater can be found bound to particles and a large portion (more than 80%) can be efficiently taken out by efficient treatment systems, although 20% can be directly released into natural waters [12]. In fact, the average Ag concentrations currently measured in natural waters where wastewater treatment plants normally operate can be around 1 ng. L^{-1} for soluble Ag and 1.0 mg.kg^{-1} for particulate Ag. So, silver concentrations in wastewater effluents are higher than those from natural waters with a real potential for contamination of rivers and estuaries.

Soluble silver in marine waters is an effective indicator of anthropogenic inputs considering that its chemistry favours a relatively stable dissolved form that is easily dispersed [13]. The concentration of Ag in open sea has been estimated to be as low as 0.03–0.1 ng.L^{-1}, but an increase of 50% has been recently reported in some coastal regions (0.14–1.29 ng.L^{-1}) [14, 15]. Living in the sea surface microlayer (1–1000 μm thick), bacteria, phytoplankton and marine invertebrates in early stages of development normally present enrichment factors from 10^2 to 10^4, 1 to 10^2 and 1 to 10, respectively [16]. Many organic and inorganic contaminants as Ag complexes can be found at this microlayer sometimes at μg.L^{-1} level [17, 18], which could allow a constant chemical exposure of both food and feeding developing invertebrates. It should also be considered that dissolved and/or particulate organic matter resulting from planktonic activity is a major component of the ocean surface microlayer [19] and could contribute to stabilize AgNP dispersion and optimize its bioavailability.

17.2 Toxicity Effects of AgNPs on Marine Invertebrates After Short and Chronic Exposures

17.2.1 Basic Mechanisms

Toxicity mechanisms of nanomaterials are related to their physicochemical properties in the biological medium [20]. Chemical and physical mechanisms are expected to take place at the nano-biointerface (as the surface of animal tissues) [21]. Therefore, biochemical processes driven by the production of reactive oxygen species (ROS), dissolution and release of toxic ions, disturbance of the electron/ion cell membrane transport activity, oxidative damage through catalysis, lipid peroxidation or nano-surfactant properties correspond to a significant part of nanotoxicity. Physical mechanisms directly involving the size and surface properties of nanoparticles can lead to structural disruption of cell membrane (by formation of nano-holes, e.g.), impairment of membrane activity, failure of transport processes and modification of protein structure and folding [20, 21].

Once internalized by living cells, nanosilver can act as a main source of ionic silver. This so-called Trojan-horse mechanism seems to operate in the way that allows H_2O_2 to react with AgNPs to form more free moving Ag inside the cells [22]. Thus, the generation of oxidative stress by over production of ROS has been suggested as the major event responsible for DNA damages, activation of antioxidant enzymes and depletion of antioxidant molecules, and binding and disabling of proteins. Nonetheless, AgNPs also undergo complex and dynamic processes inside cells which include internalization and accumulation by subcellular compartments, exclusion and degradation inside endo-lysosomal vesicles with release of Ag^{+} ions, and chemical reactions of both silver forms with Cl^{-} ions and thiol-containing molecules such as cysteine and glutathione [23]. Therefore, to determine main mechanisms driving nanosilver toxicity, it is necessary to assess Ag chemical transformation in biological tissues and identify which cellular processes are directly involved with the physicochemical transformation of nanoparticles over time.

Although short-term exposures provide a fast response from organisms, long-term exposures definitely correspond to a more realistic scenario where marine invertebrates would be regularly intoxicated by low AgNP concentrations entering in the marine environment by intermittent spills. It can also clearly reveal which defensive mechanisms would be employed to face a sustained stress and, what could be their metabolic costs. Long-term in vitro studies have shown that a chronic exposure caused by 50 nm-AgNPs (400 $pg.mL^{-1}$) did not induce cytotoxic response in human cells, but instead generated an augmented stress response, substantial modification to gene regulation and an increased EGF (epidermal growing factor) signaling efficiency [24]. EGF is a growth factor that stimulates cell growth, proliferation and differentiation. Different growth factors have also been found in marine invertebrates. The detection of growth factor of molecules in various tissues means that they can be important modulators of wound repair, regeneration and immune functions. Yet, there is no study about the chronic effects of AgNPs on the expression of growth factors in marine invertebrates. Thus, if an extended inflammatory response caused by nanoAg would be harmful for their immune system in long-term exposures remains an unresolved question.

17.2.2 Overview of Toxicity Studies

Silver has been shown to be highly harmful to marine fauna. Physiological and cellular toxicity effects caused by nanoAg and soluble Ag have been described for mollusks, echinoderms, crustaceans, polychaetes and cnidarians species (Table 17.1). Most of the studies have been carried out with mollusks (36%) as biological models and species of marine invertebrates at adult stage (54%). A part of these studies (22%) had exclusively echinoderms as models whereas 17%, 8% and about 3% of the work have been using polychaetes, crustaceans and cnidarians, respectively. Some other studies have more than one zoological group in analysis (14%). Only 35% of the research efforts have been focusing on toxicity mechanisms caused by small AgNPs (<15 μm) although they are known to be much more toxic for animal cells than larger ones. So far, short-term effects corresponded to slightly more than one half of the studies. Different aspects of nanoAg toxic kinetics such as accumulation, biodistribution, and AgNP speciation have been addressed as well as several metabolic effects related to nanoAg exposure as development impairment, cellular metabolism disturbance, immune response failure, tissue damaging with abnormal protein expression, DNA damage, apoptosis, anti-stress heat shock proteins expression, short anti-oxidant responses and depuration processes [25–43]. Nonetheless, we found a limited number of studies with cold water marine fauna, since most acute exposures have been conducted at ambient temperature (±20 °C) with species mostly living in temperate seawater.

17.2.3 Bivalves

Following different short time exposures to nanosilver, mollusks seem to show strong Ag accumulation and protein levels in their digestive organ, phagocytosis inhibition in hemolymph directly exposed to AgNPs (as for *Crassostrea virginica* within 140 min, e.g.), some stimulation of phagocytosis in hemocytes of exposed mussels (as for *Mytilus galloprovincialis* within 24 h time exposure, e.g.), efficient depuration processes, sensitivity of gill cells with impairment of Na-K-ATPase activity, ROS production, anti-oxidant catalase activity, dis-

Table 17.1 Physiological and cellular toxicity effects caused by nanoAg and soluble Ag in marine organisms

Zoological groups	Biological models (species)	Number of organisms exposed and exposure media	AgNP size in nanopure water	AgNP nominal concentration	Observed cellular and physiological effects	References
Mollusks	Oyster (*Crassostrea virginica*)	4 L with 50 embryos mL^{-1}	15 ± 6 nm	0.0016–1.6 μg.L^{-1} (embryos), 0.0016–16 μg.L^{-1} (adults)	Embryonic development impaired; high metallothionein mRNA levels in both embryos and adult oysters; lysosomal integrity affected in hepatopancreas	Ringwood et al. [25]
	Mussel (*Mytilus edulis*)	5 individuals in 500 mL	20–35 nm	0.7 μg.L^{-1}	Accumulation of >60% of Ag in soft tissues in nanoAg and Ag^{+} treatments with maximum concentrations located in the digestive organ; low bulk activity of Ag in the extrapallial fluid	Zuykov et al. [26]
	Estuarine snail (*Peringia ulvae*)	48 individuals per concentration	16.5 ± 4.5 nm	1.25–200 μg L^{-1}	Fast elimination of Ag from AgNPs in the first days of depuration compared to dissolved silver	Khan et al. [53]
	Oyster (*Crassostrea virginica*)	10 μL of hemolymph (per replicate) extracted from 10 oysters	26 ± 1.2 nm	1; 5 and 400 μg. L^{-1}	Reduction of phagocytosis in exposed hemolymph	Abbott Chalew et al. [28]
	Mussel (*Mytilus galloprovincialis*)	50 mussels in 20 mL	<100 nm	10 μg.L^{-1}	DNA damage in hemolymph cells	Gomes et al. [44]
					Alteration of proteins associated to stress response, cytoskeleton and cell structure (actin and α-tubulin)	Gomes et al. [54]
	Deposit-feeding clam (*Macoma balthica*)	30 clams in total (5 replicates)	20–80 nm and 2–3.5 μm	200 μg.Ag.g dw sed	Less than 10% of mortality, high Ag body burden without toxicity	Dai et al. [43]
	Oyster (*Crassostrea virginica*)	5–6 oysters, volume not given	20–30 nm	0.02; 0.2; 2.0 and 20 μg.L^{-1}	Total protein levels increased in hepatopancreas pointing to tissue damaging and metabolic impairment	McCarthy et al. [31]
	Scallop (*Chlamys islandica*)	10 individuals in 30 L	10–20 nm and 70–80 nm	110 and 151 ng. Ag.L^{-1}	Strong accumulation of AgNPs in short time followed by efficient depuration process	Al-Sid-Cheikh et al. [34]
	Bivalve (*Scrobicularia plana*)	12 individuals in 2 L (waterborne exposure), 10 individuals in 900 mL (dietary exposure)	40–45 nm	10 μg. Ag.L^{-1}	Anti-oxidant enzymes activated (catalase, glutathione S-transferase, superoxide dismutase); feeding behavior impaired	Buffet et al. [55]

(continued)

Table 17.1 (continued)

Zoological groups	Biological models (species)	Number of organisms exposed and exposure media	AgNP size in nanopure water	AgNP nominal concentration	Observed cellular and physiological effects	References
	Gastropod (*Littorina littorea*)	1 individual in 1.5 L	45 ± 33 nm	10 and 20 μg.L^{-1}	Presence of silver in the head, gills and foot; little measurable Ag accumulation in the kidney or stomach when introduced by Ag nano-form	Li et al. [30]
	Mussel (*Mytilus galloprovincialis*)	20 L of seawater (2.5 mussels/L^{-1})	41.68 ± 9.62 nm	10 μg.L^{-1}	Oxidative stress; antioxidant enzymes activation; lipid peroxidation higher in gills; metallothionein seem to be the main detoxification mechanism of nanoAg in the gills	Gomes et al. [45]
	Mussel (*Mytilus galloprovincialis*)	Hemolymph of 50 specimens (2 × 10^5 cells.mL exposed)	<10 nm; 20 nm, 80 nm and 2 μm (bulk)	0.15, 0.31, 0.62, 1.25 and 2.5 mg.L	Higher sensitivity of gill cells compared to hemocytes, anti-oxidant defenses (ROS production, catalase activity), DNA damage in both cells; activation of lysosomal activity, disruption of actin cytoskeleton, stimulation of phagocytosis in hemocytes; increase of MXR transport activity and inhibition of Na-K-ATPase in gill cells	Katsumiti et al. [45]
	Mussel (*Mytilus galloprovincialis*)	20 L of seawater (2.5 mussels.L^{-1})	42 ± 10 nm	10 μg.L^{-1}	Up-regulation of cytochrome CYP4y1 and cathepsin; down-regulation of catalase, glutathione transferase, caspase, Hsp70 and elongation factor	Bebianno et al. [46]
	Bivalve (*Scrobicularia plana*)	12 bivalves in 3.0 L	2–10 nm (±5 nm in average)	10 μg.L^{-1}	Behavior impairments; significant bioaccumulation of Ag in cytosolic fraction of the soft tissues of the digestive gland; reduced anti-oxidative capacities and impairment of metabolic activity	Bertrand et al. [61]
Cnidarians, crustaceans, polychaetes,	Coral (*Acropora japonica*)	200 fertilized eggs in 200 mL, 20 larvae into 2 mL; 80 metamorphosing larvae into 4 mL; 25 settled polyps in 10 mL	57.2 ± 3.6 nm	0.5, 5, 50, and 500 μg.L^{-1}	Larval deformation and metamorphosis failure under 50 μg·L^{-1} exposures; high mortality of gametes, larvae and polyps with 500 μg·L^{-1},growth impairment of polyps	Suwa et al. [56]

	Brine shrimp (*Artemia salina*)	20 cysts in 2 mL; 10 nauplii in 2 mL	30–40 nm	0.2–1.2 μg.L^{-1}	With increasing concentrations of nanoAg, several lethal and sublethal effects in nauplii emerged (mortality, nano-aggregation in gut region, apoptosis and DNA damage); *Artemia* cysts hatching decreased	Arulvasu et al. [36]
	Copepod (*Tisbe battagliai*)	5 individuals in 2.5 mL	56.9 ± 7.8 nm	1; 3.2; 5.6; 10; 32; 56 and 100 μg.L^{-1}	Mortality increased in a concentration-dependent manner for AgNPs	Macken et al. [82]
	Brine shrimp (*Artemia salina*), barnacle (*Amphibalanus amphitrite*)	Not given	1–10 nm	0.1–100 mg.L^{-1}	Mortality (100 mg·L^{-1}), abnormal development (0.1 mg·L^{-1}), high mortality of barnacle nauplii for all tested solutions; larvae of brine shrimp less sensitive to nanoAg	Falugi et al. [27]
	Amphipod (*Ampelisca abdita*), mysid (*Americamysis bahia*) and polychaete (*Nereis virens*)	10 amphipods, 10 mysids, 1 worm; 20 g of contaminated sediment	30 nm	7.5 mg.kg dw of AgNPs into 200 mL (for worms) and 20 g of amended sediment (0.75; 7.5 and 75 mg.kg dw) into 60 mL (for crustacea)	Worms exposed to cit-AgNPs had a similar distribution among Ag metal, AgCl and Ag_2S; whereas those exposed to PVP-AgNPs contained mostly Ag metal and AgCl, no toxicity for amphipods and mysids at concentrations up to 75 mg·Kg^{-1} dw;	Wang et al. [11]
	Barnacle (*Balanus amphitrite*), gasteropod (*Crepidula onyx*) and polychaete (*Hydroides elegans*)	20 larvae in 20 mL	30–50 nm	1, 10 and 100 μg.L^{-1}	Retardation in growth and development; reduction of larval settlement rate; transfer of Ag by phytoplankton cells, internalization of nanoAg by vacuoles of epithelial cell in the digestive tract of *C. onyx*	Chan and Chiu [59]
	Crustaceans (*Amphibalanus amphitrite* and *Artemia salina*), jellyfish (*Aurelia aurita*)	10–15 crustacea larvae, 1 ephyra	990 nm	100 μg.L^{-1}	Mortality, cnidaria and crustacean behavior alteration and sea urchin sperm motility change	Gambardella et al. [38]

(continued)

Table 17.1 (continued)

Zoological groups	Biological models (species)	Number of organisms exposed and exposure media	AgNP size in nanopure water	AgNP nominal concentration	Observed cellular and physiological effects	References
	Polychaeta (*Hediste diversicolor*) and bivalve (*Scrobicularia plana*)	20 individuals per species in each core (1000 individuals. m^{-2})	40–45 nm	10 $\mu g.L^{-1}$	DNA damages in the digestive gland of *S. plana*, and higher presence of phenoloxidase and lysozyme activities in both species	Buffet et al. [57]
	Gastropod (*Potamopyrgus antipodarum*) and polychaeta (*Capitella teleta*, *Capitella* sp. S)	10 worms and 15 snails in 50 mL each with 5 g dw of sediment	10–15 nm	100 µg.Ag/g dw sediment	Low Ag accumulation; no apparent toxicity to snails; survival and growth affected (worms)	Ramskov et al. [59]
Polychaeta	Polychaeta (*Nereis diversicolor*)	Individual worms exposed to 5 g of wet sediment in 60 mL of seawater	30±5 nm	250 $ng.Ag/g^{-1}$ with 5 g for each worm exposed	Internalization of AgNPs by gut epithelium; association of nanoAg with inorganic granules, organelles and heat denatured proteins	García-Alonso et al. [50]
	Polychaeta (*Nereis diversicolor*)	Individual worms exposed to 320 g of wet sediment in 600 mL of seawater	<100 nm	1, 5, 10, 25, 50 µg.Ag/g dw sediment (dry weight)	Highest genotoxicity effects compared to dissolved silver and micron-silver (2–3.5 µm)	Cong et al. [52]
	Polychaeta (*Laeonereis acuta*)	5 worms in 5 mL	33.5 nm and 205.6 nm	0.01; 0.1; 1.0 mg.L	Symbiotic bacteria reduced; posterior region of the worms with lower antioxidant capacity	Marques et al. [53]
	Polychaeta (*Platynereis dumerilii*)	300 eggs, 60 individuals for each larval stage, 60 juveniles, 20 adults	16.5 nm (cit-AgNPs), 13.1 nm (HA-AgNPs)	1, 20, 40, 60, 80, 120 and 500 $\mu g.L^{-1}$	Mortality, abnormal development and high uptake rate particularly for HA-AgNPs	García-Alonso et al. [34]
	Polychaete (*Nereis (Hediste) diversicolor*)	1 worm in 600 mL and 320 g wet weight of sediment	20–80 nm	5, 10, 25, 50 and 100 µg.Ag/g dw sediment (dry weight)	Lysosomal membrane permeability and DNA damage of *Nereis* coelomocytes increased in a concentration-dependent manner	Cong et al. [49]

	Polychaeta (*Nereis diversicolor*)	25 worms into 1 kg of sediment covered by 450 mL of non-aerated artificial estuarine water	63 ± 27 nm; 202 ± 56 mm (bulk)	2.5, 5, 10 mg. Ag/g^{-1} dw sediment	Significant increases in total glutathione concentrations, superoxide dismutase depletion; inhibition of catalase	Cozzari et al. [51]
Echinoderms	Sea urchin (*Paracentrotus lividus*)	Not given	5–35 nm	0.3 mg.L^{-1}	Agglomerated AgNPs found in larvae tissue; Ag oxidized species found with S and O/N ligands in larvae; loss of calcite, signs of Ag metabolization	Piticharoenphun et al. [29]
	Sea urchin (*Paracentrotus lividus*)	10 µL of dry sperm from 3 different specimens into 1 mL of AgNPs	1–10 nm	0.1–1000 µg.L^{-1}	Developmental anomalies, larval enzymatic activity affected (acetylcholinesterase and propionylcholinesterase); potential neurotoxic damage	Gambardella et al. [32]
	Sea urchin (*Paracentrotus lividus*)	100 embryos/larvae into 6 well multiwell-plates (volume of seawater not given)	5–35 nm	~0.3 mg.L^{-1}	Development arrest before gastrulation is completed; strong morphological changes on skeletal parts	Šiller et al. [33]
	Sea urchins (*Arbacia lixula, Paracentrotus lividus* and *Sphaerechinus granularis*)	Initial 50 µL aliquot of oocytes in 1500 mL^{-1} and 10 µL of diluted sperm (number of exposed eggs not given)	60 ± 3 nm	1–100 µg.L^{-1}	Larval deformities and developmental arrest	Burić et al. [39]
	Sea urchin (*Paracentrotus lividus*)	10 µL of dry sperm from 3 different specimens into 1 mL of AgNPs	1–10 nm	0.1–1000 µg.L^{-1}	Altered fibronectin suggesting a role in alteration of skeletogenic cells migration; different degree of anomalies related to Golgi apparatus modification; AgNPs interfered with biomineralization	Gambardella et al. [40]

(continued)

Table 17.1 (continued)

Zoological groups	Biological models (species)	Number of organisms exposed and exposure media	AgNP size in nanopure water	AgNP nominal concentration	Observed cellular and physiological effects	References
	Sea urchin (*Strongylocentrotus droebachiensis*)	30 individuals in 5.5 mL	15 ± 5 nm	50 and 100 $\mu g.L^{-1}$	Developmental deformities in different embryonic stages, vacuolization of blastocoelar cells, lack of skeleton, abnormal clusters of primary mesenchyma cells, swollen process	Magesky and Pelletier [42]
	Sea urchin (*Strongylocentrotus droebachiensis*)	30 individuals in 5.5 mL	15 ± 5 nm	100 $\mu g.L^{-1}$	Failure of metamorphosis completion; lethargy, oedema, necrosis and immobility in juveniles; cellular immune reaction in both larvae and juveniles	Magesky et al. [60]
	Sea urchin (*Strongylocentrotus droebachiensis*)	5 juveniles in 5.5 mL	14 ± 6 nm	100 $\mu g.L^{-1}$	Increasing levels of Hsp60 and Hsp70 expression, coelomic contamination and cell immune response (spherulocytes and amoebocytes)	Magesky et al. [43]

ruption of actin cytoskeleton, DNA damage, increase of P-glycoprotein (P-gp) mediated multixenobiotic resistance (MXR) transport activity, and developmental abnormalities [25, 26, 28, 31, 34, 41]. Feeding habits of mollusks might be a factor that influences AgNP toxicity as well. When exposed to AgNPs over 15 days at ±17 °C, *Mytilus galloprovincialis* can exhibit the typical effects related to nanosilver toxicity such as oxidative stress, DNA damage, lipid peroxidation in gills, up- and down-regulation of several antioxidant enzymes, down-regulation of stress-induced heat shock protein 70 (Hsp70) and metallothionein activation [44–46]. Differently from suspension feeders as *Mytilus galloprovincialis*, the deposit feeder clam *Macoma balthica* exposed over 35 days did not show any evidence of toxicity, nor DNA damage at 15 °C [47]. It may be indicative of the efficiency of protective mechanisms of *M. balthica* to cope with toxic effects or even a lower uptake rate of nanoparticles amended with sediments. It has been also demonstrated that an active threshold avoidance response by burrowed *M. balthica* took place in sediments containing the highest metal levels as Cu, Pb, Zn, Cr and Ag [48].

17.2.4 Annelids

Polychaetes form another zoological group of interest for nanosilver toxicity studies. In this case, short term toxicity mechanisms for AgNPs are still unknown. So far, it was reported that early life stages of development of *Platynereis dumerilli* exposed to humic acid-coated AgNPs (HA-AgNPs) endured the highest toxic effects if compared to the soluble Ag counterpart within 48 h [37]. The Ag speciation in biological tissues was demonstrated in *Nereis virens* and can be presented as Ag metal, AgCl and Ag_2S once exposed to citrate-AgNPs (cit-AgNPs) over 28 days at 22 °C [11]. At a similar exposure time, coelomocytes of *Nereis* (*Hediste*) *diversicolor* experienced lysosomal membrane permeability and DNA damage in a concentration-dependent manner of polyvinylpyrrolidone-coated AgNPs (PVP-AgNPs) at 15 °C [49]. Experiments with *Nereis diversicolor* demonstrated that for some deposit feeders the ingestion of nano-contaminated sediment is crucial for uptake and cellular internalization of AgNPs [50]. From the intestinal lumen, AgNPs were directly internalized by gut epithelial cells and then became associated with the apical plasma membrane (in endocytic pits and in endosomes). Interestingly, the subcellular fractionation of both soluble Ag and nanosilver in *Nereis diversicolor* is different: AgNPs were associated with inorganic granules (~40.6 $\mu g.g^{-1}$.Ag), organelles (~85.6 $\mu g.g^{-1}$.Ag) and heat denatured proteins (~36.7 $\mu g.g^{-1}$.Ag) whereas soluble Ag was mainly taken by the metallothionein-like proteins fraction (~78.7 $\mu g.g^{-1}$.Ag). The antioxidant system that arises in *Nereis diversicolor* as a consequence of toxic effects of bulk AgNPs particles (~63 nm and ~202 mm) throughout 11 days of exposure (2.5–10 mg.Ag g^{-1} dw) at 10 °C seems to rely on glutathione, and be related Se-independent and Se-dependent glutathione peroxidases (GPx and SeGPx) increasing levels at day 7 [51]. However, it seems that DNA damage in coelomocytes of *Nereis diversicolor* can significantly arise after 10 days once exposed to lower concentrations of nanoAg in sediments (<100 nm, 25 µg.Ag g dw) at 15 °C [52].

17.2.5 Benthic and Pelagic Invertebrates

Some interesting findings have been reported for benthic and planktonic invertebrates after chronic exposures to AgNPs (Table 17.1), but more mechanistic studies about the innate immune response to AgNPs and their interactions with cellular compartments are still highly needed. Mechanisms of cellular internalization of AgNPs and elimination, growth rate disturbance, metamorphosis incompletion, behavior impairment, silver metabolic speciation, genotoxicity, alteration of proteins expression, immune cell reaction and antioxidant defenses have been described as specific biological effects related to nanoparticulate silver and/or its leaching ions [11, 43–46, 49–61].

Recent work using 3-month old sea urchins (*Strongylocentrotus droebachiensis*) showed an inflammatory response involving red spherulocytes and amoebocytes emerged within 12 h of dissolved silver exposure [43]. With 96-h exposure, poly(allylamine)-coated AgNPs (14 ± 6 nm, 100 μg.L^{-1}) already formed a trail of foreign bodies grabbed on the mesenteries indicating a facilitated pathway to contaminate sea urchin coelomic sinuses. After 12-day exposure to poly(allylamine)-coated AgNPs, a strong phagocytic activity emerged in these sinuses (Fig. 17.1).

Though it is not certain how the particle coating and the size range of AgNPs modulated the phagocytic response in such a case. In comparison, titanium oxide nanoparticles (TiO_2NPs, 10–65 nm; 1–5 μg.L^{-1}) also stimulated phagocytic response of immune cells of sea urchin *Paracentrotus lividus* and increased a Toll-like receptor (Pl-TLR) mRNA and protein levels after 1 day of exposure [62]. The same study also reported an anti-inflammatory effect with a significant inhibition in the p38 MAPK (mitogen-activated protein kinases) levels. Similar effects can also arise in crustacean species. It has been reported that an inhibition of hemocyte phagocytosis by AgNPs (26 ± 1.2 nm, 400 μg.L^{-1}) happened in adults of *Crassostrea virginica* after very short time exposures (15–120 min) [28]. By the same token, hemocytes phagocytic activity in *Mytilus galloprovincialis* was significantly increased at 41% only by 20 nm-maltose-stabilized AgNPs (Mal-AgNPs, 1.25 mg L^{-1}) compared to <10 nm, 80 nm and 2 μm- Mal-AgNPs at 18 °C [41]. The release of Ag^+ ions from Mal-AgNPs after 1 h in seawater was fast and continuous. This study demonstrated a high toxicity potential of organic-coated AgNPs. In fact, Mal-AgNPs (20 nm) were much more toxic to gill cells and hemocytes than the Ag bulk form. Also, DNA damages occurred in hemocytes (1.25 and 2.5 mg L^{-1}/Ag) and in gill cells (2.5 mg L^{-1}/Ag) [41]. In green sea urchins, circulating coelomocytes from internal haemal vessel fully contaminated by nanoAg (~8 nm) underwent an apoptotic-like process with evidence of cell blebbing and nanoparticles located inside nucleous [43].

It is observed from these numerous studies that abiotic factors interfere strongly with the physicochemical state of nanomaterials, and any variation in exposure conditions may complicate the interpretation of the experimental results. Some factors such as dissolution, agglomeration

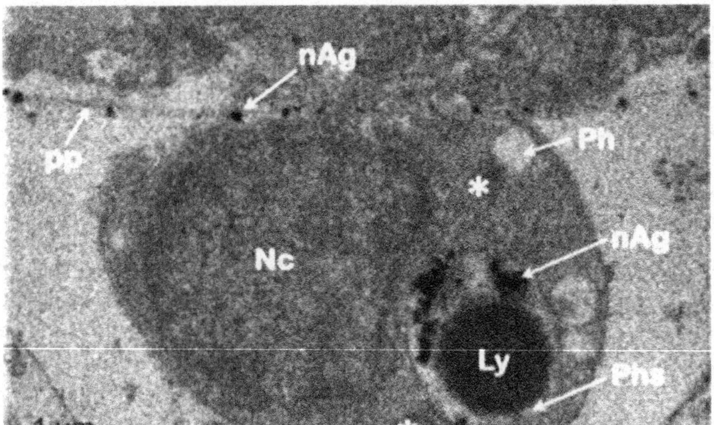

Fig. 17.1 Coelomocyte from the coelomic sinus of *S. droebachiensis* juvenile showing a typical phagocytic pathway in presence of nanosilver (nAg) in a transmission electron photomicrograph (non-contrasted sample) (Magesky et al. [43]). White asterisk indicate early lysosomes close to a phagosome and the larger phagolysosome. *Ly* lysosome, *nAg* nanosilver, *Nc* cell nucleous, *Ph* phagosome, *Phs* phagolysosome, *pp* parietal peritoneum

and sedimentation or the adsorption of ions and macromolecules should be identified and quantified in order to understand bioavailability and toxicity of silver [63]. Additionally, some standard experimental conditions such as the number of exposed organisms, temperature range, exposure time and volume of exposure media must be detailed to make comparable results from different short or long term exposures. For example, the maximum density of embryos and larvae in test solutions should be between 15 and 30 individuals/mL. Concentrations up to 50 organisms/mL may show less toxic effects than normally expected due to the reduced toxic charge and body burden retained by each organism.

17.3 Physicochemical Interactions of AgNPs with Membranes and Biomolecules

17.3.1 Physical and Chemical Processes

Once in the environment, nanomaterials undergo several transformation processes that would dictate their biotic interactions and toxicity. These physical and chemical processes are highly complex as illustrated in Fig. 17.2.

The affinity of nanoparticles (NPs) to biological surface can change as a function of the conditions in the surrounding medium, such as pH, ionic strength or presence of colloids. Forming a

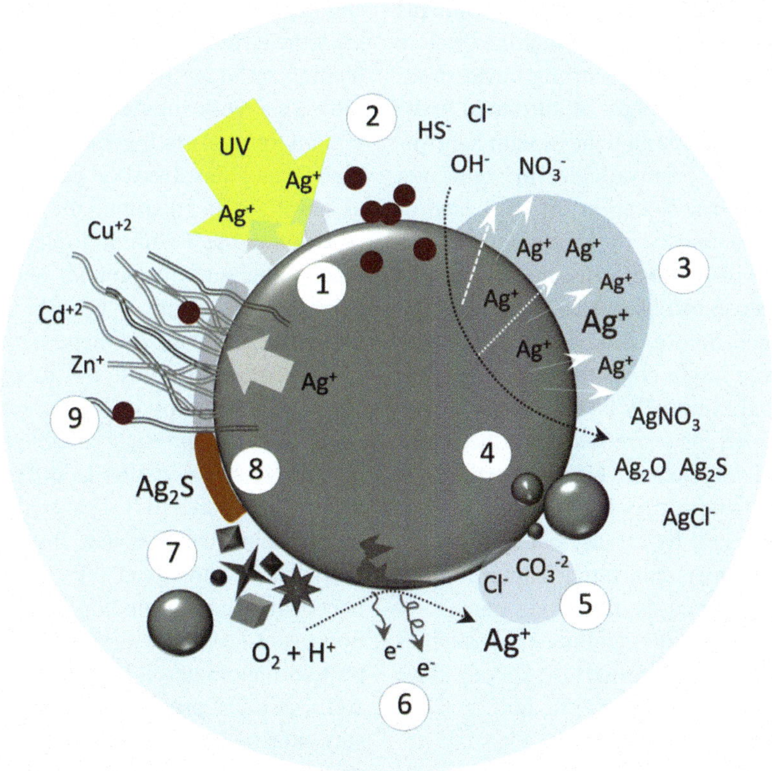

Fig. 17.2 A schematic model for AgNP and multiple features and mechanisms that may influence nanosilver toxicity in aquatic media (Modified from Lapresta-Fernandez et al. [85]). *1* Photo-oxidation. *2* Interaction with organic matter. *3* Dissolution of metal core by anionic attack and releasing of Ag^+ ions by oxidative processes. *4* Aggregation with other AgNPs. *5* Ionic adsorption on AgNP surface. *6* Oxidation and leaching of Ag^+ ions from the irregular surface of nanoparticle. *7* Different shapes that may influence internalization of AgNPs by cells. *8* Sulfidation of AgNPs (Ag_2S-NPs) can slow down or prevent oxidation and ionic leaching. *9* Organic coating providing stability for nanoparticles, but also acting as ligands for other toxic metals or organic molecules

corona-like structure on NP surface, natural colloids may strongly interfere with their final fate in natural environment. In fact, the effects of NOM (natural organic matter) can both enhance and reduce aggregation, as it depends on other ambient environmental conditions. When NOM is present in the medium, it leads to the formation of more negatively charged particles which enhances particle stability via electrosteric stabilization as well as steric hindrance [64]. The presence of an organic matter corona in some NPs is expected to mitigate their toxicity by hindering their direct contact with cells in an exposure medium [65]. Notwithstanding, organisms ingesting AgNPs enrobed by NOM will unequivocally endure nano-intoxication as well. As an example, mussels seem to be particularly susceptible to NPs due to their sessile nature and filter feeding habits [44, 53]. Macrobenthic organisms as blue mussels exposed to estuarine waters charged with AgNPs showed strong bioaccumulation of silver [9]. This might be attributed to the uptake of setting particles amended with nanosilver and relatively bioavailable in seawater. Actually, a fast translocation of Ag^{+} and/or AgNPs from the tissues of bivalves (gills) in direct contact with the exposure media towards hepatopancreas can be observed as well [26, 34].

As mentioned before, filtering organisms as sea urchin larvae might cross the estuarine halocline to feed and eventually be exposed to silver. A threshold of particle size could drive the intake of NOM-AgNPs by larvae. Not all particles in solution are suitable for food or are of a size or shape to be ingested by larvae, but later stages before metamorphosis can rather capture larger particles [66]. Some cases of indirect transfer of AgNPs by means of biological material (such as phytoplankton cells) ingested by larvae have been described in the literature [58, 60].

17.3.2 Interactions with Cells and Tissues

Phytoplankton contaminated with poly(allylamine)-coated AgNPs can efficiently transfer silver to echinoplutei gut (Fig. 17.3). From digestive tissues, 15 nm-poly(allylamine)-coated AgNPs and some aggregates (~230 nm) reach the coelomic epithelium, the proctodaeum epithelium and the mesothelium as aggregates. After contamination took place, the large perivisceral cavity of larvae (blastocoel) became burden of nanoAg (130–400 nm). In such conditions, many larval cell-types as coelomic epithelial cells, phagocytic amoeboids, spherulocytes, petaloid-like cells, peritoneocytes, etc. internalized nano-aggregates of different sizes. Intoxication with nanoAg prevented metamorphosis to occur and caused a twofold higher mortality in clean conditions compared to soluble Ag. Larvae experiencing nanosilver contamination under good feeding conditions will first struggle to complete metamorphosis at 8 °C and then suffer lethal side effects of silver in the meantime. In a like manner, 30–50 nm oleic acid and polyvinylpyrrolidone-coated AgNPs reduced settlement rate of larvae of polychaete *H. elegans* feeding on algae in a 5-day experiment (25 °C) [58]. This work suggested that Ag uptake in phytoplankton cells is higher than in larvae so both AgNPs and dissolved Ag^{+} can be easily delivered to larval gut and disrupt their metabolism.

Upon contact with biological fluids, NPs primarily interact with proteins and other biomolecules giving them a new biological identity different from their synthetic nature. The affinity of a protein to NPs controls their behavior (adsorption/desorption, complexation/dissociation) when biophysical interactions take place during NP translocation to different physiological compartments [67]. These interactions rely on the protein structure and the physicochemical properties of involved NPs. It has been established that the net binding energy of an adsorption event (ΔG_{ads}) will guide the stability of the protein-nanomaterial interactions [68]. When ΔG_{ads} is high, proteins adsorbed on NPs surface stay attached. On the contrary, with low ΔG_{ads} proteins likely lose bound and return to solution. Hence, proteins adsorbing NPs with high affinity form the hard corona and do not get easily desorbed. Those with low affinity form the soft corona, which consists of loosely bound proteins. How these layers communicate in a dynamic way with each other and with the nanomaterial surface is still subject to debate.

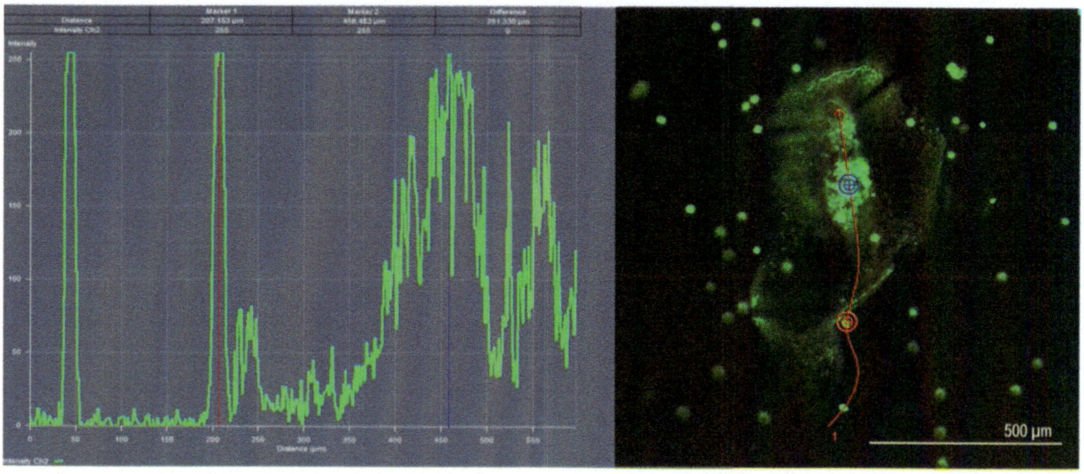

Fig. 17.3 Profile view of levels of fluorescence signal emitted by poly(allylamine)-coated AgNPs-FITC fluorescent marker in metamorphic larva of sea urchin *S. droebachiensis* feeding on phytoplankton. Image was taken at 24 h of exposure in vivo (8 °C) in confocal microscopy (10×, scale bar: 500 µm). Intensity of the signal is given on y axis, distance (in µm) is on x axis. The *red line* (on the right) runs through larva and phytoplankton showing levels of fluorescence emitted by both (shown on the left). The distance of red target marker (phytoplankton cell) from blue target marker (larval stomach) was ~251 µm (Magesky unpublished results)

17.3.3 Effects of Interactions with Biomolecules

Transcytosis of AgNPs through body cavities provides different physiological environments that will likely influence aggregation-disaggregation processes, corona formation and so toxicity [69]. So far, there are no available studies reporting those effects on nanoAg toxicity in marine invertebrates. An important AgNP effect on cells corresponds to nanoscale holes formation in living cell membranes [70]. This effect has been studied with polymeric nanomaterials from 1.2 to 22 nm in mammal cells, but not described in marine species. Moreover, it can be suggested that the translocation of nano-aggregates through biological tissues might contribute to some physiological stress possibly regulated over time. It is also difficult to known how different cell-types contaminated with AgNPs would respond to silver toxicity in marine invertebrates. In sea urchins, peritoneal cells seem to have a role in AgNP dispersion inside body cavities after internalization; circulating coelomocytes loaded with nanoAg experienced from physical disturbance of cell membrane to cell disaggregation [43]. A second effect is the Ag core dissolution itself. The Trojan-horse mechanism in which Ag^+ ions are released over time may be especially toxic in the intracellular milieu, but this effect appear to be less clear in body cavities. For instance, sea urchin larva seems to react to ingested 5–35 nm-AgNPs. The X-ray absorption near edge structure (XANES) spectra of Ag in contaminated echinoplutei of *Paracentrotus lividus* showed multiple aggregates of AgNPs oxidized and complexed with S and O/N ligands as well as a loss of calcite [29]. In fact, sulphur-containing compounds are found in excess in nanoAg exposed-echinoplutei thus indicating a biological response to reduce the concentration of soluble Ag.

Noteworthy, three major findings about protein coronas as a modulator of AgNP sulphidation in macrophages have been established [71]: (1) protein coronas modulate nano-Ag_2S formation at AgNPs; (2) ion release from AgNPs is necessary for nano-Ag_2S formation, and (3) protein corona-mediated sulphidation impacts cell toxicity. As above mentioned, NPs get covered with biomolecules when lying in biological media. As soon as polymer-coating AgNPs enter serum medium (1%), the polymer coating is replaced by hard and soft coronas in about 1 h. The mechanism proposed implies that leaching ions released from the nano-surface get trapped first in the long-lived protein corona (hard corona) where Ag_2S may be formed if there is enough reduced sulphur and Ag^+ available [71]. On the contrary, sulphide formation does not happen on the soft corona where the turnover of biomolecules is faster. Instead, the loosely aggregated proteins transport Ag^+ away from the particle, which decreases local ion concentration. Hence, nano-Ag_2S formation at AgNPs decreases with the presence of soft corona and increases in free circulating proteins. Authors suggested that sulphidation involving dynamic corona formation in living organisms can reduce AgNP toxicity at lethal and sub-lethal doses by preventing cell death, and pro-inflammatory cell cytokine production. Such mechanisms were not described for sea urchins, but sulphur-containing compounds found in nanoAg-exposed echinoplutei might be a good indication of corona mitigation of silver toxicity in this taxon.

17.4 Cellular Toxicity Mechanisms of Polymer-Coated AgNPs During Sea Urchin Developmental Stages

For decades, sea urchins have been used as classical biological model in development research, cell biology and toxicological studies. With sea urchin genome so closely related to humans and other deuterostomian animals, their cells are considered as a powerful model for nanotoxicity investigations without ethical normative issues [62]. Many cellular processes of developmental stages of green sea urchin (*S. droebachiensis*) can be severely disrupted by silver contamination.

17.4.1 Embryotoxicity of AgNPs

Major events of embryogenesis from blastula to midgastrula stage were studied in presence of poly(allylamine)-coated AgNPs (100 $\mu g.L^{-1}$) at 5 °C [42]. Early mesenchyme blastula was the most sensitive stage exposed to nanoAg. Throughout 96-h assay, deformed larvae showed abnormal swimming behavior, strong inhibition of biomineralization, low concentration of secondary mesenchyme cells in blastocoel, abnormal clusters of primary mesenchyme cells (PMCs), low migration of pigment cells in the ectodermal layer from animal pole, no differentiation of pigment cells at the vegetal pole, dysfunctional digestive tract, and impairment of blastocoelar cells. Early on, all developmental stages were severely impacted by soluble Ag, whose effects were faster and much more severe when compared to nanosilver. Under nanoAg conditions, Ag^+ free levels steadily increased during exposure time. Poly(allylamine)-coated AgNPs (~8 to 17 nm) structure is basically a silver core embedded with polyallylamine (PAAm) firmly integrated within the nanoparticles giving them an aspect of a hairy ball. The negatively charged polyallylamine chains may hold around 2.5% of Ag^+ ions ready to be released in seawater [42]. The chemical behavior of poly(allylamine)-coated AgNPs is characterized by a slow dissolution of nano-core particle over 16 days and a rapid release of Ag^+ entrapped by polymer chains within 24 h. This entrapped silver ready to be released can be considered as a major source of toxicity of PAAm-AgNPs in short-time exposures. Due to a fast solubilization in brackish waters, Ag^+ quickly interfered with several sensitive processes that are crucial for embryogenesis and larval formation.

While embryos undergo gastrulation process, they remained closed to circulating seawater until the stomodeum (mouth opening) appeared in the prisma larvae. We believe that soluble forms of Ag could particularly interfere with the calcium uptake before feeding larvae formation since skeleton formation depends on it. We further hypothesized that ectoderm would act as a barrier to AgNPs at some point when compared to Ag^+. So taking invaginating blastulae stage as a model to explain the interaction of silver with PMCs (Fig. 17.4), it can be established that the main disruptor of biomineralization process came from Ag^+ ions. We suggested that Ag^+ passing through ectoderm layer would provoke early on some agonistic competition with Ca^{2+} ions that are normally imported into blastocoel to form calcite granules inside PMCs. Abnormal syncytia of

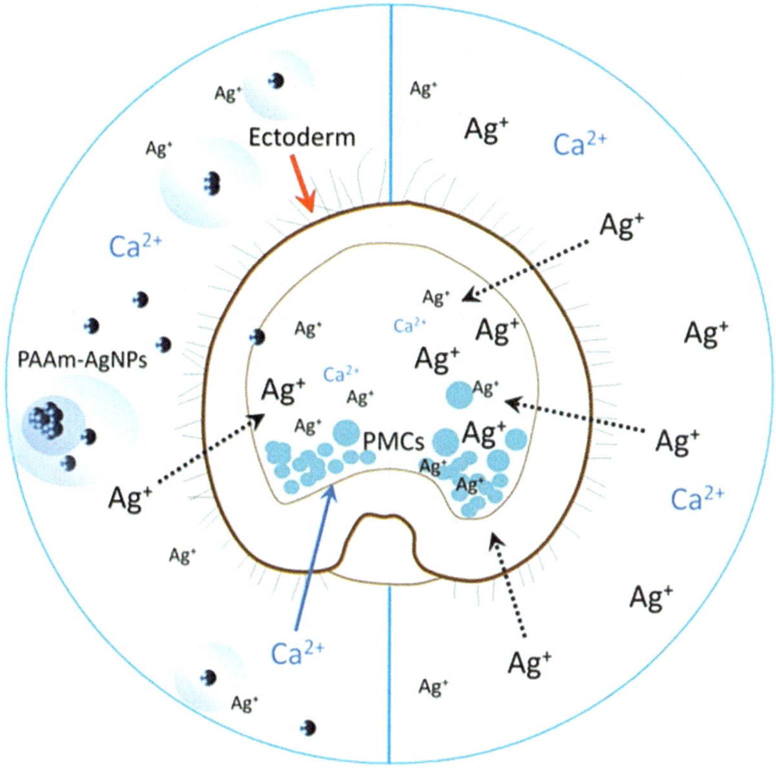

Fig. 17.4 Mechanistic model proposed for Ag^+ ions interactions with ectoderm of invaginating blastulae and their primary mesenchyme cells (PMCs) (Magesky and Pelletier [42]). NanoAg treatment is seen on the left, $AgNO_3$ one is on the right. Ectoderm layer is indicated by red arrow; dissolved silver pathway is showed by black dashed arrows while Ca^{2+} one is in blue arrow. Polymer-coated AgNPs are represented by blue balls and dissolution process is shown by rounded bluish surrounding areas. The major sphere around blastula indicates the interaction layer in seawater. Dissolved silver targets PMCs by disrupting cell membranes of forming syncytium and thus spicule mineralization

mesenchyme cells held growing clusters unable to properly mineralize spicules under silver stress [42].

Currently unknown is how Ag can disturb skeletogenic cells (Fig. 17.5). Confocal images revealed that some Ca^{2+} ions are stored by calcifying PMCs, but abnormal clusters are unable to deliver Ca^{2+}-loaded vesicles to build the spicules [42]. A first hypothesis chiefly relies on Ca^{2+} trafficking towards skeletogenic syncytium cells and the competition with Ag^+ for transmembrane ion pumps on cell membrane. Alternatively, it could be due to an impairment of intracellular compartments involved with Ca^{2+} exportation for skeleton growth. Cell contact and filopodial formation are clearly affected by silver. As both are crucial steps for skeletogenic cells organization and spicule calcification in blastocoel, so membrane disturbances could be tentatively responsible for growing agglomerations of abnormal calcifying cells (Fig. 17.6).

Nucleous of some Ag-chemically affected cell clusters are less marked by DNA fluorescent dyes, which may indicate failure of cell machinery to keep up with their regular activity in sea urchin embryos. Either way, Ag permanently disrupted skeleton mineralization. Comparatively, Mn^+ contamination (1.12 mM) prevented embryos of sea urchin *Paracentrotus lividus* to form spicules and interfered with the expression of PMCs-specific genes [72]. Notwithstanding, impaired skeleton of Mn^+-exposed embryos can be rescued from chemical insult when Mn is diluted and washed out. After 16-h exposure, all

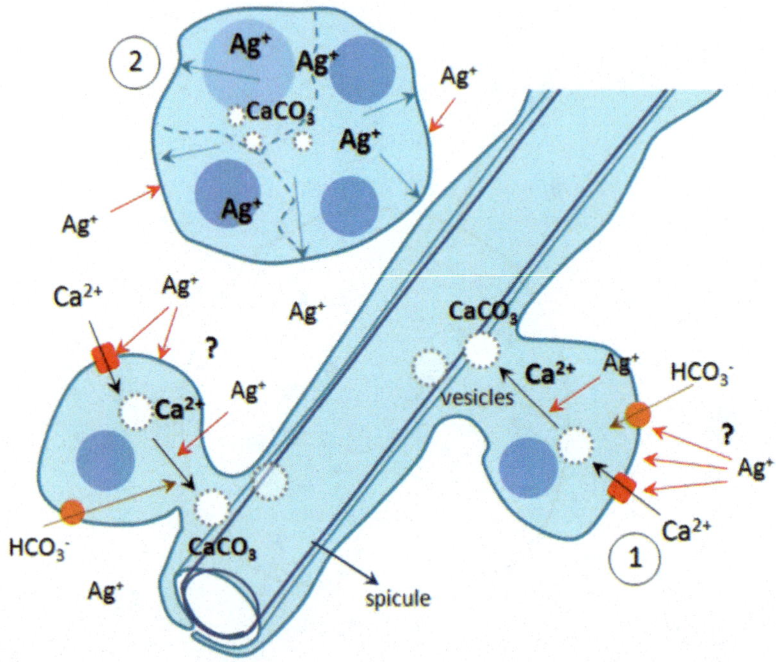

Fig. 17.5 Mechanistic model for Ag^+ ions interactions with calcifying PMCs inside blastocoel (calcification process modified from Stumpp et al. [84]). *1* Soluble Ag might have disruptive effects on membrane transporters, Ca^{2+} trafficking into cells, uptake of HCO_3^- and final precipitation of $CaCO_3$ for spicule construction. *2* Aberrant clusters failed to mineralize skeleton, albeit small granules of $CaCO_3$ precipitated inside swollen cells. Blue arrows indicate swollen process of abnormal clusters, the red ones represent the possible targets of Ag^+, orange arrows show HCO_3^- uptake for amorphous calcium carbonate precipitation inside the cells (vesicles in the figure), the black ones indicate Ca^{+2} influx

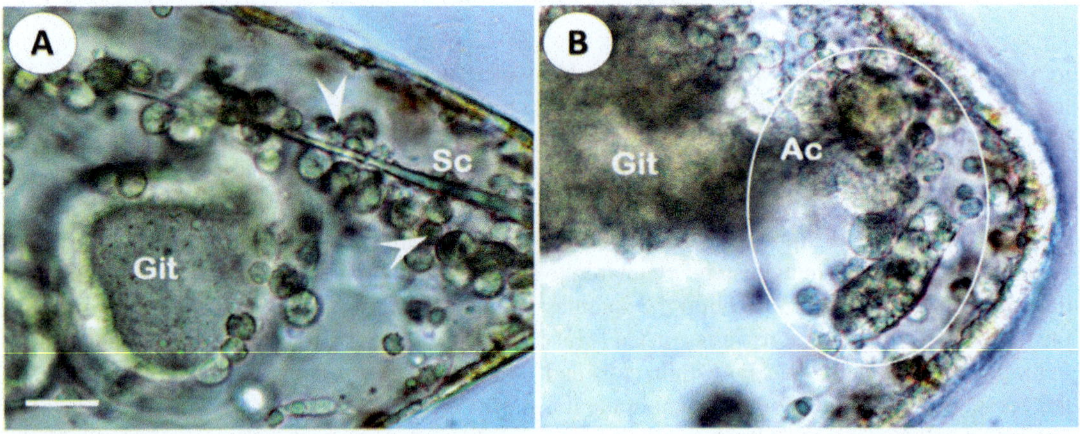

Fig. 17.6 Morphology of skeletogenic cells from posterior region of a healthy larva and a 100 μg.L^{-1}/nanoAg-treated larva after 96 h (Magesky unpublished results). (**a**) Uncontaminated echinopluteus showing calcifying syncytium and spicule. (**b**) Contaminated larva lacking skeleton; disturbed congregation of spiculogenic cells are seen inside white spiral. Abbreviations- *Ac* abnormal clusters, *Git* gastrointestinal tract, *Sc* spicule (Scale bar: 200 μm, 20×)

embryos were able to develop normally. In general, PMCs, pathways related to skeleton mineralization and enzymatic activity seemed to be strongly impaired by AgNPs in embryos [27, 32, 33, 39]. In addition, cell-types involved in the organogenesis of early larvae undergo a permanent metabolic impairment after AgNPs contamination [42].

During gastrulation process, PMCs ingress into blastocoel following fibronectin pathways and form the matrix on which calcium carbonate crystallizes to build up larval skeleton [73]. At least for early echinopluteus larvae, the presence of fibronectin-like molecules in the skeletogenic syncytium of abnormal embryos shows the toxic effects of AgNPs (1–10 nm) in the deposition of extracellular matrices (such as fibronectin) [40]. Even well-established calcifying clusters of older echinoplutei are impaired by soluble Ag and nanoAg. Unfed 6-arm echinoplutei became intoxicated within a few hours (12 h) after being contaminated by Ag. Aberrant clusters are not formed as observed in prisma larvae and embryos, but mineralization is permanently blocked and skeletogenic cells got individually dispersed [42].

With stomodeum formation, a new route for nano-contaminated water emerges. During larval development, cell specification processes of hindgut continue throughout late gastrula and prisma stages; foregut and midgut undergo to morphological changes readily from late gastrula on, in a period when gut regions are molecularly defined [74]. The formation of larval gut depends on continuous movements of endodermal cells requiring constant regulation of adhesive properties of archenteron cells. Hence, mouth formation would likely provide a passageway for a facilitated contaminant uptake allowing silver to directly interfere with gastrointestinal tract regionalization processes. Mid-to-late prisma larvae exposed to soluble Ag had poorly developed stomach whereas poly(allylamine)-coated AgNPs-treated larvae showed only a little delay in stomach development. Silver can also have its toxicity worsened when combined with functionalized single-walled carbon nanotubes (f-SWCNTs). In this case, mid-to-late prisma were shapeless and their gastrointestinal tract became a deformed hollow tube with no cardiac sphincter throughout 96 h exposure. Notably, even when nanoAg are mixed with f-SWCNTs, free Ag^+ levels are higher if compared to single exposures of nanoAg. Dynamic interactions of Ag^+ with f-SWCNTs increased silver toxicity likely providing an ion capture on nanotubes walls and an optimized contaminant transfer to the organisms [42].

17.4.2 Toxicity of AgNPs After Metamorphosis

Along early life of green sea urchins, it has been observed that some developmental stages might be more directly affected by nanotoxicity than others. After metamorphosis, newly settled urchins could face major threats to their survival. While transitioning from a planktonic to a benthonic lifestyle, early postmetamorphic juveniles need a fast immune system organization to cope with natural diseases and ubiquitous pollution. Juveniles *S. droebachiensis* have cryptic habits and are particularly sedentary [75], thus they may become an easy target for AgNPs discharged into seawater when wastewater treatment plants do not operate properly.

Three-month old sea urchin juveniles have contrasting immune cell strategies and stress induced heat shock proteins (Hsp) expression to deal with soluble Ag and poly(allylamine)-coated AgNPs [43]. Given the stronger expression of Hsp70 and Hsp60 in poly(allylamine)-coated AgNPs-treated urchins (100 $\mu g.L^{-1}$) at 48 h exposure, the hypothesis of harmful effects mainly driven by nanoAg form itself was confirmed. In turn, dissolved silver toxicity rapidly arises if compared to AgNPs ones. Among several immune cell-types recognized in adults of echinoderm species, red spherulocytes and amoebocytes are known as the main cell effectors during clotting reactions caused by chemical injuries or infestation of biological agents. Both cells rapidly migrated (within 12 h) towards tissue wounds provoked by soluble Ag then large cell plugs arose. On the other hand, circulating coelomocytes internalized AgNPs greatly dispersed in coelomic sinuses and haemal ducts of juveniles. Of a particular significance were the extruding vesicles caused by ~8 nm-AgNPs embedding on cell membrane of coelomocytes [43]. As observed, nanosilver was first concentrated in small aggregates then eliminated from cell membrane before internalization. Similarly, some human immune cells such as neutrophils, monocytes and macrophages seemed to have extracellular traps acting as physical barriers that prevent positively charged 15 nm-polymeric-coated gold (Au) nanospheres to undergo endocytosis [76].

17.4.3 Comparative Toxicity of Dissolved and Particulate Silver

A schematic model has been constructed in an attempt to summarize and compare both Ag^+ and polymer-coated AgNPs toxicity effects in coelomocytes of sea urchin (Fig. 17.7). Although this scheme already appears as quite complex, it probably represents only a part of all mechanisms involved with AgNPs toxicity in sea urchin coelomocytes. Because AgNPs translocate through primary and secondary cavities in sea urchin body, their physico-chemical properties may be

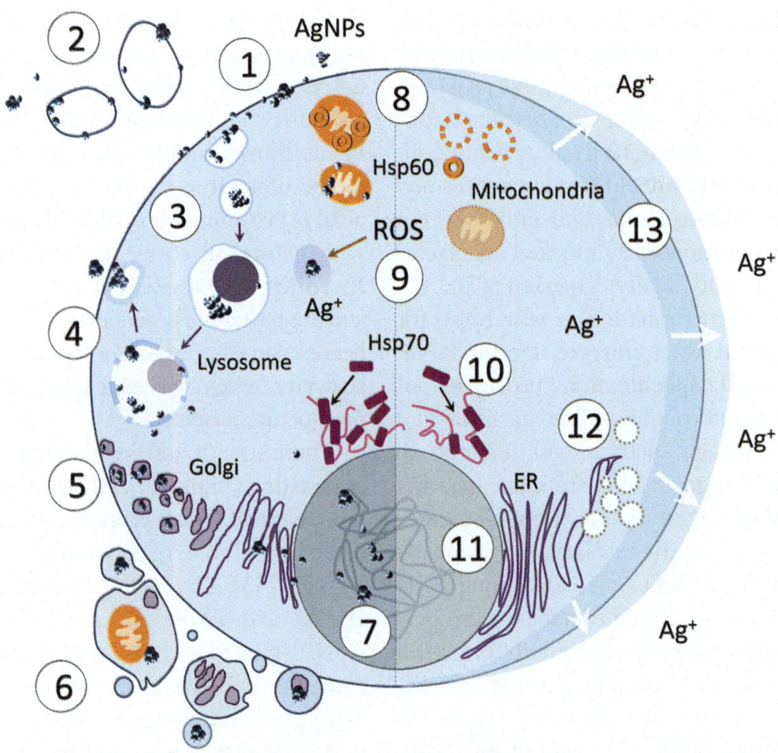

Fig. 17.7 Schematic representation summarizing nanosilver (on the left) and dissolved silver (on the right) main cytotoxic effects in Ag-treated organisms (mostly juveniles of green sea urchin). *1* Internalization of small aggregates by non-phagocytic pathway after small organic coated AgNPs embedding on cellular surface; *2* Extruding vesicles from coelomocyte membrane; *3* Phagocytic pathway: loading vesicles with nanosilver are seen in larger phagolysosome; *4* Dispersion of AgNPs in cytoplasm or elimination to extracellular milieu; *5* Secreting vesicles with loaded nanosilver; *6* Apoptotic-like process; *7* Dispersion of AgNPs in nucleus (chromatin and perinuclear region); *8* Hsp60 up-regulation in nano-exposed organisms and down-regulation in dissolved silver exposed ones; *9* ROS production and mitochondrial dysfunction; *10* Hsp70 up-regulation in both nanoAg and soluble Ag-exposed organisms (nano-treated larger urchins had higher levels of Hsp70). *11* Nuclear and DNA disaggregation; *12* Ca^{2+}-vesicles held by calcifying cells unable to build spicules (mechanism described in developing larvae); *13* Cellular swelling (oncosis)

affected by the surrounding coelomic fluid and circulating biomolecules. Consequently, processes like aggregation and disaggregation, dissolution, biocorona envelopment, metabolization/detoxification of Ag^+ released from Ag nanoforms can individually dominate and/or be concomitant for different groups of cells. The first effects of AgNPs and its leaching Ag^+ ions on cell membrane, cytosol and nucleous likely happen at different times at the beginning of toxic insult ($t = 0$ h), but it must steadily evolve to a highly toxic scenario depending on the time of exposure, the kinetics of biochemical reactions at low or high temperatures, and contaminant concentration. Coelomocytes flowing nearby facilitated routes for contaminant uptake (as mouth) would likely be more severely affected. Toxicity arises as a function of nanomaterial size (small single particles and/or small and large aggregates), surface area, redox potential, surface functionalization and composition (coating and/or biocorona formed) [77].

It has been suggested that NPs could interact with different proteins within different cells, so some changes in their intracellular localization would naturally occur as a function of the cell-type involved [78]. Peritoneocytes are secretory cells distributed along mesothelium of body cavities in echinoplutei and sea urchin juveniles [60]. When nanosilver flowed through larval perivisceral cavity and juvenile inner secondary cavities, small aggregations of AgNPs (~30 to 80 nm) can be caught up by the cellular secretory pathway. Accordingly, extracellularly secreted vesicles seemed to act as nano-transporters. In this way, NPs would likely reach new tissues and contribute to phagocytosis in larger spaces of coelom [60]. With regard to circulating coelomocytes, simultaneous processes involving phagocytic and non-phagocytic pathways would likely take place in highly contaminated cavities. Endocytosis of small aggregates probably emerged as a consequence of a threshold of tension forces on cell membrane basically created by aggregations and/or single AgNPs close to each other [43, 79]. Furthermore, assessing polymer-coated AgNP resilience-time captured by coelomocytes as aggregates of different sizes may reveal the cellular costs involved with digestion, retention and disposal of nanoAg. Inside intracellular milieu, free AgNPs can interact with several cellular components and eventually reach the nucleus. Even though the elimination rate of poly(allylamine)-coated AgNPs engulfed by coelomocytes has not been examined yet, removal of large NPs may demand more energy and cell receptors than smaller ones [78].

17.4.4 Starvation Effects on AgNP Toxicity

Starvation can represent another factor to enhance Ag toxicity especially for smaller urchins by (i) reducing the intake of anti-oxidant food sources and then (ii) misbalancing ROS (reactive oxygen species) production and decomposition reactions [43]. Hence, ROS produced beyond cellular reparative mechanisms might become a driving force to not only interfere with protein molecular conformation, but with dissolution of AgNPs as well [22, 80]. In fact, protein and peptide-nanoparticle interactions can cause deformities on the protein structure and modify their biological functions [81]. Nevertheless, the degree of the conformational changes depends on the structure of the protein and the chemistry of NPs [68]. In comparison to soluble Ag, cellular effects of AgNPs may be considered a much more severe toxicant when considering their potential actions such as (a) acting as single particles and/or aggregates, (b) dissolving in the intracellular milieu and/or body cavities, (c) physically interacting with structure of biomolecules at several cellular compartments, (d) producing systemic contamination in the body by dynamic translocation, and (e) targeting multiple groups of cells.

17.5 Remaining Questions and Conclusion

Henceforward, a complete understanding of chemical behavior of AgNPs inside cellular and acellular environment of marine invertebrates is still an objective to pursue. With regard to polymer-coated AgNPs, it is uncertain whether free single nanoparticles flowing through intracellular milieu would be easily attacked by reactive oxygen species and what could be the protective role of the organic coating. Despite this, it has been shown that reactive intracellular H_2O_2 promotes oxidation of internalized AgNPs with final Ag^+ release [22]. Secondly, if the Ag core slowly dissolves through time, more studies are necessary to understand how and at which time Ag^+ ions could be trapped and released from polymer coating. With protein corona enveloping nanoparticles, at which point the organic shell would allow a fast release of Ag ions and/or a reduction of toxicity in marine invertebrate models? While clear answers to these questions cannot be given yet, more studies focusing on the toxicity mechanisms of organic-coated AgNPs in developing marine invertebrates could reveal insights about the real environmental threats could be involved with nanosilver contamination.

With increasing salinity towards the ocean, small invertebrates living in brackish waters might be particularly targeted by AgNPs [82]. If released Ag^+ ions react quickly with other seawater elements as Cl^- near the particle surface, a faster dissolution might happen [64]. Marine invertebrates exhibited lethal and multiple sublethal effects driven by silver in both nano and ionic forms at relatively high concentrations compared to low levels presently encountered in field conditions. Long-term assays of AgNPs at ng/L^{-1} need to be carried out to elucidate: (i) how metabolism can possibly shift without apparent morphological impairment [83] (ii) how inflammatory processes may compromise immune system effectiveness. Some cellular processes (as apoptosis, ROS over-production, phagocytosis, membrane disruption, NP internalization and disposal etc.) mostly observed in cells of mussels and sea urchins contaminated by AgNPs show strong similarity with those already described for human cells. However, marine invertebrates have several cell-types that react differently to AgNPs. So additional knowledge on small (<15 nm) AgNPs effects on different cell-types of marine invertebrate cells is mandatory. The fate of nanosilver in cellular compartments of coelomocytes (polychaetes and echinoderms) and hemocytes (mollusks and crustaceans) might reveal interesting strategies used by marine invertebrates to mitigate the effects of nanotoxicity.

References

1. Wijnhoven SWP, Peijnenburg W, Herberts CA et al (2009) Nano-silver – a review of available data and knowledge gaps in human and environmental risk assessment. Nanotoxicology 3(2):109–138
2. Geranio L, Heuberger M, Nowack B et al (2009) The behavior of silver nanotextiles during washing. Environ Sci Technol 43:8113–8118
3. Kaegi R, Voegelin A, Sinnet B et al (2011) Behavior of metallic silver nanoparticles in a pilot wastewater treatment plant. Environ Sci Technol 49(9):3902–3908
4. Quadros ME, Marr LC (2011) Silver nanoparticles and total aerosols emitted by nanotechnology-related consumer spray products. Environ Sci Technol 45(24):10713–10719
5. Angel BM, Batley GE, Jarolimek CV et al (2013) The impact of size on the fate and toxicity of nanoparticulate silver in aquatic systems. Chemosphere 93(2):359–365
6. Kaegi R, Voegelin A, Ort C et al (2013) Fate and transformation of silver nanoparticles in urban wastewater systems. Water Res 47(12):3866–3877
7. Fabrega J, Luoma SN, Tyler CR et al (2011) Silver nanoparticles: behaviour and effects in the aquatic environment. Environ Int 37(2):517–531
8. Luoma SN, Ho YB, Bryan GW et al (1995) Fate, bioavailability and toxicity of silver in estuarine environments. Mar Pollut Bull 31(1–3):44–54
9. Pelletier É et al (2015) The fate of nanomaterials at the river/ocean frontier using large mesocosms: the case story of silver nanoparticles. In: SETAC Europa 25th annual meeting, Barcelona
10. Metaxas A, Young CM (1998) Behaviour of echinoid larvae around sharp haloclines: effects of the salinity gradient and dietary conditioning. Mar Biol 131:443–459
11. Wang H, Ho KT, Scheckel KG et al (2014) Toxicity, bioaccumulation, and biotransformation of silver nanoparticles in marine organisms. Environ Sci Technol 48:13711–13717
12. Deycard VN, Schäfer J, Petit JC et al (2017) Imputs, dynamics and potential impacts of silver (Ag) from urban wastewater to a highly turbid estuary (SW France). Chemosphere 167:501–511
13. Sanudo-Wllhelmy SA, Flegal AR (1992) Anthropogenic silver in the Southern California bight: a new tracer of sewage in coastal waters. Environ Sci Technol 26:2147–2151

14. Ranville MA, Flegal AR (2005) Silver in the North Pacific Ocean. Geochem Geophys 6:Q03M01
15. Gallon C, Flegal AR (2015) Sources, fluxes and biogeochemical cycling of silver in the oceans. In: Whitacre DM (ed) Reviews of environmental contamination and toxicology, v. 235. Springer, Cham
16. Cunliffe M, Engel A, Sanja F et al (2012) Review: sea surface microlayers – a unified physicochemical and biological perspective of the air-ocean interface. Prog Oceanogr 109:104–116
17. Wurl O, Obbard JP (2004) A review of pollutants in the sea-surface microlayer (SML): a unique habitat for marine organisms. Mar Pollut Bull 48(11–12):1016–1030
18. Cunliffe M et al (2013) Sea surface microlayers: a unified physicochemical and biological perspective of the air-ocean interface. Prog Oceanogr 109:104–116
19. Soloviev A, Lukas R (2014) The near-surface layer of ocean: structure, dynamics and applications. Springer, Dordrecht. 552 pages
20. McShan et al (2014) Molecular toxicity mechanism of nanosilver. J Food Drug Anal 22:116–127
21. Elsaesser A, Howard CV (2012) Toxicology of nanoparticles. Adv Drug Deliv Rev 64(2):129–137
22. Hsiao IL et al (2015) Trojan-horse mechanism in the cellular uptake of silver nanoparticles verified by direct intra- and extracellular silver speciation analysis. Environ Sci Technol 49:3813–3821
23. Wang Y et al (2015) An overview on nanotoxicity and nanomedicine research: principles, progress and implications on cancer therapy. J Mater Chem B 3:7153–7172
24. Comfort KK et al (2014) Less is more: long-term in vitro exposure to low levels of silver nanoparticles provides new insights for nanomaterial evaluation. ACS Nano 4:3260–3271
25. Ringwood AH et al (2010) The effects of silver nanoparticles on oysters embryos. Mar Environ Res 69:S49–S51
26. Zuykov M et al (2011) Colloidal complexed silver and silver nanoparticles in extrapallial fluid of *Mytilus edulis*. Mar Environ Res 71:17–21
27. Falugi C et al (2012) Toxicity of metal oxide nanoparticles in immune cells of the sea urchin. Mar Environ Res 76:114–121
28. Abbott Chalew TE et al (2012) Pilot study on effects of nanoparticle exposure on *Crassostrea virginica* hemocyte phagocytosis. Mar Pollut Bull 64:2251–2253
29. Piticharoenphun S et al (2012) Agglomeration of silver nanoparticles in sea urchin. Int J Environ Pollut Remed 1:44–50
30. Li H et al (2013) Accumulation of aqueous and nanoparticulate silver by the marine gastropod *Littorina littorea*. Water Air Soil Pollut 224:1354
31. McCarthy M et al (2013) Tissue specific responses of oysters, *Crassostrea virginica*, to silver nanoparticles. Aquat Toxicol 138–139:123–128
32. Gambardella C et al (2013) Developmental abnormalities and changes in cholinesterase activity in sea urchin embryos and larvae from sperm exposed to engineered nanoparticles. Aquat Toxicol 130–131:77–85
33. Šiller L et al (2013) Silver nanoparticle toxicity in sea urchin *Paracentrotus lividus*. Environ Pollut 178:498–502
34. Al-Sid-Cheikh M et al (2013) Tissue distribution and kinetics of dissolved and nanoparticulate silver in Iceland scallop (*Chlamys islandica*). Mar Environ Res 86:21–28
35. Marques BF et al (2013) Toxicological effects induced by the nanomaterials fullerene and nanosilver in the *Polychaeta laeonereis acuta* (Nereididae) and in the bacteria communities living at their surface. Mar Environ Res 89:53–62
36. Arulvasu C et al (2014) Toxicity effect of silver nanoparticles in brine shrimp Artemia. Sci World J 2014:10. Article ID 256919
37. García-Alonso J et al (2014) Toxicity and accumulation of silver nanoparticles during development of the marine polychaete *Platynereis dumerilii*. Sci Total Environ 476–477:688–695
38. Gambardella C et al (2015) Effect of silver nanoparticles on marine organisms belonging to different trophic levels. Mar Environ Res 111:41–49
39. Burić P et al (2015) Effect of silver nanoparticles on Mediterranean sea urchin embryonal development is species specific and depends on moment of first exposure. Mar Environ Res 111:50–59
40. Gambardella C et al (2015) Exposure of *Paracentrotus lividus* male gametes to engineered nanoparticles affects skeletal bio-mineralization processes and larval plasticity. Aquat Toxicol 158:181–191
41. Katsumiti A et al (2015) Mechanisms of toxicity of Ag nanoparticles in comparison to bulk and ionic Ag on mussel hemocytes and gill cells. Plos One 10:1–30
42. Magesky A, Pelletier É (2015) Toxicity mechanisms of ionic silver and polymer-coated silver nanoparticles with interactions of functionalized carbon nanotubes on early development stages of sea urchin. Aquat Toxicol 167:106–123
43. Magesky A et al (2017) Silver nanoparticles and dissolved silver activate contrasting immune responses and stress-induced heat shock protein expression in sea urchin. Environ Toxicol Chem 36(7):1872–1886
44. Gomes T et al (2013) Genotoxicity of copper oxide and silver nanoparticles in the mussel *Mytilus galloprovincialis*. Mar Environ Res 84:51–59
45. Gomes T et al (2014) Effects of silver nanoparticles exposure in the mussel *Mytilus galloprovincialis*. Mar Environ Res 101:208–214
46. Bebianno M et al (2015) Is gene transcription in mussel gills altered after exposure to Ag nanoparticles? Environ Sci Pollut Res Int 22:17425–17433
47. Dai L et al (2013) Effects, uptake, and depuration kinetics of silver oxide and cooper oxide nanoparticles in a marine deposit feeder, *Macoma balthica*. Environ Sci Technol 1:760–767

48. McGreer ER (1979) Sublethal effects of heavy metal contaminated sediments on the bivalve *Macoma balthica* (L.) Mar Pollut Bull 10:259–262
49. Cong Y et al (2014) Toxicity and bioaccumulation of sediment-associated silver nanoparticles in the estuarine polychaete, *Nereis (Hediste) diversicolor*. Aquat Toxicol 156:106–115
50. García-Alonso J et al (2011) Cellular internalization of silver nanoparticles in gut epithelia of the estuarine polychaete *Nereis diversicolor*. Environ Sci Technol 45:4630–4636
51. Cozzari M et al (2015) Bioaccumulation and oxidative stress responses measured in the estuarine ragworm (*Nereis diversicolor*) exposed to dissolved, nano- and bulk-sized silver. Environ Pollut 198:32–40
52. Cong Y et al (2011) Toxic effects and bioaccumulation of nano-, micron- and ionic-Ag in the polychaete, *Nereis diversicolor*. Aquat Toxicol 105:403–411
53. Khan FR et al (2012) Bioaccumulation dynamics and modeling in an estuarine invertebrate following aqueous exposure to nanosized and dissolved silver. Environ Sci Technol 46:7621–7628
54. Gomes T et al (2013) Differential protein expression in mussels *Mytilus galloprovincialis* exposed to nano and ionic Ag. Aquat Toxicol 136–137:79–90
55. Buffet P-E et al (2013) Biochemical and behavioural responses of the endobenthic bivalve *Scrobicularia plana* to silver nanoparticles in seawater and microalgal food. Ecotoxicol Environ Saf 89:117–124
56. Suwa R et al (2014) Effects of silver nanocolloids on early life stages of the scleractinian coral *Acropora japonica*. Mar Environ Res 99:198–203
57. Buffet P-E et al (2014) A marine mesocosm study on the environmental fate of silver nanoparticles and toxicity effects on two endobenthic species: the ragworm *Hediste diversicolor* and the bivalve mollusc *Scrobicularia plana*. Sci Total Environ 470–471:1151–1159
58. Chan CYS, Chiu JMY (2015) Chronic effects of coated silver nanoparticles on marine invertebrate larvae: a proof of concept study. PLoS One 10:0132457
59. Ramskov T et al (2015) Accumulation and effects of sediment-associated silver nanoparticles to sediment-dwelling invertebrates. Aquat Toxicol 166:96–105
60. Magesky A et al (2016) Physiological effects and cellular responses of metamorphic larvae and juveniles of sea urchin exposed to ionic and nanoparticulate silver. Aquat Toxicol 174:208–227
61. Bertrand C et al (2016) The influence of salinity on the fate and behavior of silver standardized nanomaterial and toxicity effects in the estuarine bivalve *Scrobicularia plana*. Environ Toxicol Chem 35:2550–2561
62. Pinsino A et al (2015) Titanium dioxide nanoparticles stimulate sea urchin immune cell phagocytic activity involving TLR/p38 MAPK-mediated signaling pathway. Sci Rep 5:14492. https://doi.org/10.1038/srep14492
63. Schultz AG et al (2014) Aquatic toxicity of manufactured nanomaterials: challenges and recommendations for future testing. Environ Chem 11:207–226
64. Garner KL, Keller AA (2014) Emerging patterns for engineered nanomaterials in the environment: a review of fate and toxicity studies. J Nanopart Res 16:2503
65. Docter D et al (2015) The nanoparticle biomolecule corona: lessons learned – challenge accepted? Chem Soc Rev 44:6094–6121
66. Strathmann R (1971) The feeding behavior of planktotrophic echinoderm larvae: mechanisms, regulation, and rates of suspension feeding. J Exp Mar Biol Ecol 6:109–160
67. Walkey CD, Chan CW (2012) Understanding and controlling the interaction of nanomaterials with proteins in a physiological environment. Chem Soc Rev 41:2780–2799
68. Norde W, Giacomelli CE (2000) BSA structural changes during homomolecular exchange between the adsorbed and the dissolved states. J Biotechnol 79:259–268
69. Lazarovits J et al (2015) Nanoparticle-blood interactions: the implications on solid tumor targeting. Chem Commun 51:2756–2767
70. Leroueil PR et al (2007) Nanoparticle interaction with biological membranes: does nanotechnology present a Janus face? Acc Chem Res 40:335–342
71. Miclăus T et al (2016) Dynamic protein coronas revealed as a modulator of silver nanoparticle sulphidation in vitro. Nat Commun 7:11770. https://doi.org/10.1038/ncomms11770
72. Pinsino A et al (2011) Manganese interferes with calcium, perturbs ERK signaling, and produces embryos with no skeleton. Toxicol Lett 123(1):217–230
73. Wilt FH (1999) Matrix and mineral in the sea urchin larval skeleton. J Struct Biol 126:216–226
74. Annunziata R et al (2014) Pattern and process during sea urchin gut morphogenesis: the regulatory landscape. Genesis 52:251–268
75. Scheibling RE, Hatcher BG (2013) Chapter 26: Strongylocentrotus droebachiensis. In: Sea urchins: biology and ecology, v.38. John Lawrence, Elsevier, San Diego
76. Bartneck M et al (2010) Phagocytosis independent extracellular nanoparticle clearance by human immune cells. Nano Lett 10(1):59–64
77. Ahmed KBR et al (2017) Silver nanoparticles: significance of physicochemical properties and assay interference on the interpretation of in vivo cytotoxicity studies. Toxicol In Vitro 38:179–192
78. Sakhtianchi R et al (2013) Exocytosis of nanoparticles from cells: role in cellular retention and toxicity. Adv Colloid Interf Sci 201–202:18–29
79. Yue T, Xianren Z (2012) Cooperative effect in receptor-mediated endocytosis of multiple nanoparticles. ACS Nano 4:3196–3205

80. Kalmar B, Greensmith L (2009) Induction of heat shock proteins for protection against oxidative stress. Adv Drug Deliv Rev 61:310–318
81. Shemetov AA et al (2012) Molecular interaction of proteins and peptides with nanoparticles. ACS Nano 6:4585–4602
82. Macken A et al (2012) Effects of salinity on the toxicity of ionic silver and Ag-PVP nanoparticles to *Tisbe battagliai* and *Ceramim tenuicorne*. Ecotoxicol Environ Saf 86:101–110
83. Zhang B et al (2017) Metabolic responses of the growing *Daphnia similis* to chronic AgNPs exposure as revealed by GC-Q-TOF/MF and LC-Q-TOF/MF. Water Res. https://doi.org/10.1016/j.watres.2017.02.046
84. Stumpp M et al (2012) Acidified seawater impacts sea urchin larvae pH regulatory systems relevant for calcification. PNAS 109:18192–18197
85. Lapresta-Fernández A et al (2012) Nanoecotoxicity effects of engineered silver and gold nanoparticles in aquatic organisms. Trends Anal Chem 32(797):40–59

A *Drosophila* Model to Decipher the Toxicity of Nanoparticles Taken Through Oral Routes

18

S. Aurosman Pappus and Monalisa Mishra

Abstract

In recent era, nanoparticles (NPs) are widely used in food, medicine and body implants. Besides it's wide use being a foreign particle it may have some noxious effect on the body. To understand the mechanistic role of NPs toxicity, *Drosophila* appeared to be a superior model organism. Toxicity of several nanoparticles were accessed using *Drosophila*. The NPs, after oral route of exposure enter into the gut, crosses the barrier of peritrophic membrane and induces apoptosis. The toxicity of NPs within gut resulted in developmental delay, with decrease in pupa count, fly hatching along with weight loss. The adult fly hatched after nanoparticle treatment shows increasing phenotypic defect in various sensory organs as well as in different body parts. Besides phenotypic defect some of the nanoparticle results altered behavioural phenotypes like larva crawling or adult climbing. Alteration of both phenotypic as well as behavioural assay clearly hints that signalling pathway like Notch, Wnt, EGFR etc. get affected due to exposure of nanoparticle. Results from various labs prove that nanoparticle can mediate developmental defect by altering signalling pathways. Since many of the signalling pathways are conserved the effect seen in model organisms cannot be overlooked. All the nanoparticles used in food and medicine should be modified to nullify the toxic effect before used in food and medicine.

S. Aurosman Pappus
Department of Biological Sciences, IISER Kolkata,
Nadia, West Bengal, India

M. Mishra (✉)
Neural Developmental Biology Lab, Department of
Life Science, National Institute of Technology,
Rourkela, Odisha, India
e-mail: mishramo@nitrkl.ac.in

Q. Saquib et al. (eds.), *Cellular and Molecular Toxicology of Nanoparticles*, Advances in Experimental Medicine and Biology 1048, https://doi.org/10.1007/978-3-319-72041-8_18

Keywords
Drosophila · Nanoparticles · Wnt · Drosophila genotoxicity · Notch · EGFR · Nanotoxicity

18.1 Introduction

NPs are widely used in various foods, medicine and in some non-edible products associated with mouth [1, 2]. In food industry, it is either used for preservation or to enhance the quality of food [3]. In medicine, it is used for efficient drug delivery [4]. Its antibacterial property allows us to use it in various mouth implants [5]. The direct and indirect use of NPs in oral cavity finds its way to enter into the food chain. Being a foreign particle, NPs may interfere with the cellular as well as biochemical machinery of the body and thus ultimately will change its functionality. Thus, it is essential to check the toxic effect of these NPs before widely introduced into our food chain. Various model organisms are used to access the toxicity of the nanoparticles. The current article provokes *Drosophila* model to study nanoparticles toxicity. A range of genes and transcription factor required for the development is conserved between vertebrate and *Drosophila* [6, 7]. Furthermore, *Drosophila* gut share homology with the vertebrate gut in several ways. Besides structural homology, innate immunological pathways are also conserved in both *Drosophila* and vertebrate [8]. All together suggests that *Drosophila* can be used as an ideal model organism to learn the toxicity of nanoparticles taken through oral route.

18.2 Nanoparticles Cross Through Oral Route: Food, Medicine, Mouth Implants

NPs are widely used in medicine to increase the efficiency. Especially NPs used in oral formulation of drugs are considered to be advantageous over other routes. There are two main reasons for this (1) convenience and compliance for patient and (2) cheap to manufacture since sterility is not a criteria for synthesis [9]. Besides medicine, NPs are used to enhance flavor, color or as health supplement [10–12]. Various non-edible substances like nano-coating on food, drink containers, and toothbrushes, baby bottles and pacifiers also enhances the chances of entering of nanoparticle through oral cavity [13]. Consumption of animal products containing nanoparticle within it also allow indirect entry of nanoparticle through the oral cavity [14, 15]. There are many direct/indirect ways via which NPs entered into our gastro intestinal (GI) tract. To discuss the entire nanoparticle entered through the oral cavity is beyond the scope of the chapter. Hence, we are focusing the ones which are widely used in food, medicine or in implants.

In food industry AgNPs are used to keep the food bacteria-free due to its antibacterial property [16]. ZnO NPs have anti-UV and IR reflective property. It has further sterilizing and higher temperature tolerance characteristic, which allows it to be used in skin cream and ointment [17, 18]. Silica NPs are used to detect *Salmonella* in food [19]. TiO_2 NPs have O_2 scavenging property, so used to prevent the food from oxidation [20]. Metal chalcogenide nanoporous ceramic pallets are used in oil to prevent the clumping. Nanoporous graphene is used to avert desalination of water [21]. Sodium alginate NPs are used as insecticide on food crops. NPs like nanocharcoals enhance the creaminess of food due to uniform sized emulsion. Engineered nanoparticles are used to enhance food uptake in a cellular level [16, 22–24]. Various applications of NPs in food industry provide dramatic increase of NPs in food preparation, preservation and processing.

NPs are widely used in cancer treatment for efficient drug delivery purpose. In such cases NPs not only increases the effectiveness of treatment, but also minimizes the damage to other normal cells [25–28]. Polymeric-nanoparticles, viral-nanoparticles, carbon-nanoparticles, liposomes and magnetic nanoparticles are usually used as nanocarriers to deliver drugs to specific cells so that they can stay longer time in the circulating blood. They act as a carrier for active drugs [29–34]. Carrier NPs can detect the cancer microenvironment from the normal environment and thus deliver the drugs more efficiently. Quantum dots are used to treat antibiotic resistant infection; whereas polymer coated iron oxide NPs and silver NPs are helpful in the treatment of chronic bacterial infection [35–37]. Nanoparticles like hydroxyapatite are also used by dentists as mouth implants as they have a Ca/P proportion of 1.67, which is matching with bone apatite and enamel of teeth [34, 38–40]. NPs like Chlorhexidine hexa-metaphosphate are used as a novel coating to prevent bacterial growth in dental implants [5]. ZrNPs are used for dental filling and implants [41]. Application of NPs in various fields suggests that a fair number of nanoparticles are being ingested into our body on a regular basis. However, due to their extremely small size, NPs have the potential to easily penetrate into a cell and cause damage. Thus, a toxicity assessment of the nanoparticles is of extreme importance using a model organism.

18.3 Advantage of *Drosophila* over Other Models

To check the toxicity of NPs various in vivo and in vitro models are used. Animals like mouse, *Drosophila*, *Caenorhabditis elegans*, zebra fish and hydra were used to decipher the toxicity of nanoparticles. Assessment of toxicity using a mammalian cell line would be the most effective, in terms of accuracy, maintenance but it can answer the uptake only at a cellular level. To observe the effects at tissue level, a mammalian model organism like mouse is generally used. After oral intake of the nanoparticles, the tissues are harvested and then observed for abnormalities. But with mouse, given that the body organization is complex and presence of a large genome, it is difficult to study any specific gene or cellular pathway. Invertebrate model like *C. elegans* (roundworm) was used to check the toxicity of NPs like gold, ZnO, silica, titania, silver and cerium. *Danio rerio* (zebrafish) are also used as model organisms to study NP toxicity. NPs like sodium, gold and silver are successfully checked for their toxicity using Zebrafish model [42, 43]. Besides its advantages, there are many disadvantages for this model organism. Among invertebrate lower metazoans like hydra was used to check the toxicity of NPs. Hydra possess only two cell layers (ectoderm and endoderm). Endoderm faces towards the gastric cavity, and thus, it is easy to track NPs after exposure. The uptake of Hydra cells for macromolecules is greatly enhanced by the positive net charge. Thus Hydra is used by various labs to study the developmental and genotoxicity of various chemicals [44–46]. Zinc doped TiO_2 NPs were checked using Hydra as a model organism [44, 47]. Besides its advantages, it is microscopic and morphogenesis of complex organ cannot be answered using this model.

Among all the models *Drosophila* proves to be the most suitable one to study NP toxicity, since it has a simpler body organization and only four chromosomes. The advantages includes (1) Low maintenance cost and easy to handle, (2) short life span, which enables scientists to observe any trend over generations, (3) small number of chromosome with fully sequenced genome, (4) Genetic manipulation is easy, (5) *Drosophila* genome and gene causes disease in human being share 75% functional homology, (6) various proteins involved in gene regulation, expression and metabolism are found to be present in both *Drosophila* and humans [48], (7) developmental stages are extensively studied, (8) ethical issue is not associated while giving high dose to the animal.

18.3.1 Mode of Transportation of Nanoparticle Through Human and *Drosophila* Gut

Nanoparticle entered through the oral cavity has to pass through different parts of digestive system including the absorption site. *Drosophila* guts share homology with vertebrate gut in various ways although differences were reported from microbiota and immune response. Major homology are (1) foregut, midgut and hindgut are analogous to vertebrate esophagus, small intestine and large intestine [49], (2) internal layering of the vertebrate gut lumen is lined by mucus layer. *Drosophila* gut is lined by peritrophic membrane which is composed of chitin and glycoprotein [50, 51] (3) Like vertebrate the digestion within the GI tract is regulated by pH as well as various enzymes, although the enzymes present in *Drosophila* and vertebrate are different [52]. Like vertebrates, the pH of the GI tract varies in different segments, which is crucial in the absorption of nanoparticle. Various parts of gut and it's pH is show in Fig. 18.1 [53]. The nanoparticle and gut interaction is explained on the basis of charge it possesses. Positively charged NPs can pass the neutral pH of midgut easily and reach the acidic middle midgut. In highly acidic environment the proteins and extracellular matrix present in peritrophic membrane become highly positively charged. Thus, the matrix protein repels the nanoparticle resulting faster passage within the middle to posterior midgut. Anterior part of the posterior midgut has neutral pH where the NPs stay a while before enter into alkaline posterior segment of pH 10. The NPs stays in that region for more than 10 h. The peritrophic membrane of this region is highly negatively charged thus the nanoparticle stay there for longer time period due to electrostatic interaction. Nutrient absorbing cells are present in this region and helps in the absorption of nanoparticle [54] (Fig. 18.2). The positively charged NP become neutral at this stage and passes the hindgut [53].

The fate of nanoparticle within the gut can be checked by feeding fluorescent tagged NPs to the

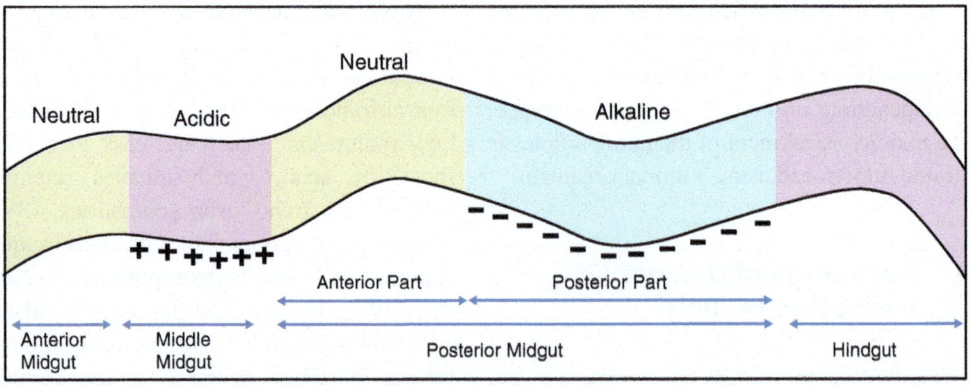

Fig. 18.1 Variation of pH in the gut of *Drosophila*

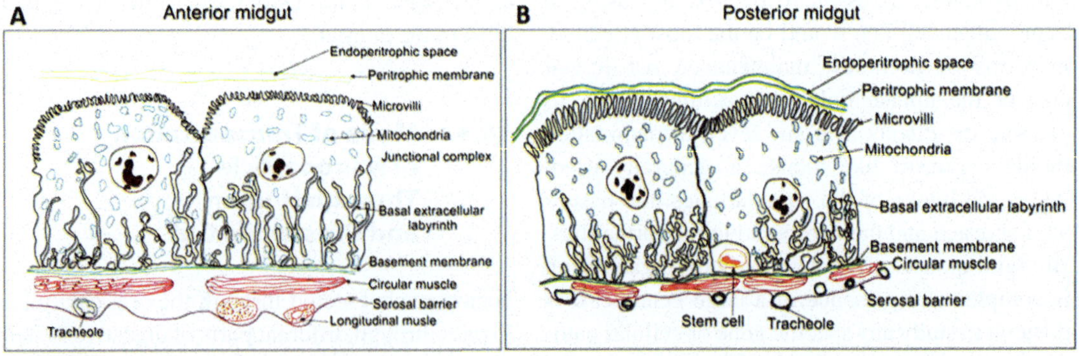

Fig. 18.2 T.S of midgut of *Drosophila*. (**a**) *Anterior midgut.* (**b**) *Posterior midgut. Note the difference in the length of microvilli between anterior and posterior midgut*

larvae and image the digestive system using time lapse imaging. Larval midgut are dissected and imaged with confocal microscopy for the absorption of NP within the gut [53]. Recently fate of chitosan NPs within the gut was investigated by feeding fluorescent tagged NPs to *Drosophila* larvae. Also the damage caused by NPs to the gut cells can be observed by staining the cells with trypan blue, where the dead cells will retain the stain [55].

18.4 Assessment of NP Toxicity Using Biochemical Methods

Nanoparticles, due to their extremely small size, can easily pass through the plasma membrane. NPs can induce reactive oxygen species (ROS) inside the cell and cause cellular damage by generating $OH^{\bullet}$ and $O_2^{\bullet}$ radicals. The amount of ROS generated within a cell is dependent on the concentration and type of NPs. There are several mechanisms by which NPs can trigger ROS formation. Some of them includes (a) prooxidant functional groups on the reactive surface of NP, (b) transition metal based NPs can produce ROS by active redox cycling on the nanoparticle surface and NPs can cause impaired Electron Transport System in the mitochondria, resulting in ROS production [56]. ROS within the cell can damage membrane, DNA and proteins by caspases activation and trigger apoptosis. AgNPs can induce upregulation of HSP70 along with p53 and p38 proteins [57]. All these proteins are potential markers for cellular stress and DNA damage. Likewise, gold NPs also perturb metabolic processes in a cell along with HSP70 upregulation and ROS formation [58]. NPs like carbon, ZnO and Titania also work in a similar mechanism resulting oxidative stress and ROS production in a concentration and NP-dependent manner [57, 59, 60]. Several dyes are used to detect the amount of ROS within the cell. 2,7-dichlorofluorescein diacetate (DCFDA) dye is one among them and used to measure the amount of intracellular ROS in NP treated *Drosophila* [46, 61]. High amount of ROS results enhanced activity of antioxidant enzymes like superoxide dismutase (SOD) and catalase. So alternatively, the intracellular ROS can be determined by checking the extent of increase in SOD and catalase [62]. The antioxidant nature of vitamin C or vitamin C palmitate has been shown to suppress the ROS production due to NP treatment [63–65]. Flies treated with AgNP and vitamin C showed less SOD formation than the flies treated with NP alone [66].

Nanoparticles can directly interact with the genetic material and cause genotoxicity in fly by interacting with the nucleus and causing mutations and damage to the cell [65, 67]. A similar situation is reported when sodium citrate-capped AuNPs were exposed to flies through the oral route [68, 69]. Higher DNA damage was observed in the gastrointestinal tissues which can be checked by single cell gel electrophoresis assay or TUNEL assay. AuNPs treatment to *Drosophila*, shows greater degree of DNA damage with smaller sized NP treatment [58, 68]. Various intestinal cell culture study suggests pretreatment of NPs with digestive enzymes to reduce the toxicity of NPs [70]. In contrast, ZnO NPs when mixed with fatty acids increases the cytotoxicity [71]. Treatment NPs (TiO_2, Ag, ZnO, and SiO_2) increases the cytokine secretion in intestinal cell lines [72, 73]. Growth patterns of intestinal cell, size of the cell, and rate of cell proliferation of intestinal cell further determine the NP-induced cytotoxicity [74]. Colon cells are found to be more resistant towards Ag [75].

Oral intake of NPs at early developmental time tends to affect the neural system. Neuronal defect at early time point can be checked by larvae crawling behavioral assay. Wild type third instar larvae of *Drosophila* tend to crawl in a straight line. However, NP-treated (Titania, Ag, SiO_2 and Alumina) larvae shows irregular behavior, indicating malformed mechanosensory neurons [76]. Nanoparticles can adhere to macromolecular surface like that of the lipid membrane, alter the function of several ion channels resulting faulty neural transmission [77]. Electrophysiology recordings of Alumina NP treated *Drosophila* shows altered the oscillations in the local interneurons of the antennal lobe [78] which is the largest chordotonal organ. Silver

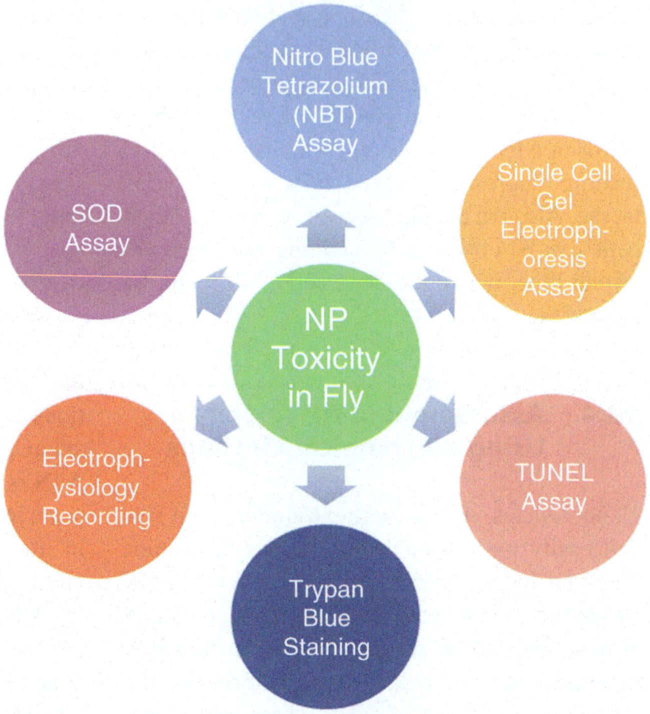

Fig. 18.3 Biochemical methods to measure NP toxicity in *Drosophila*

NPs alters neural stem cell differentiation and signaling, although the exact mechanism is yet to be known [79]. Iron NPs affect the expression of ID genes (Inhibitor of DNA binding protein), which play a vital role in regulation of nervous system, hence causing neural disorders [80]. NP interacts with various biochemical pathways and result certain defect in the developmental and maintenance of neurons. The most common assays to detect and measure toxicity in *Drosophila* are represented in Fig. 18.3.

18.5 Toxicity at Phenotypic Level

In the previous section, we have discussed how nanoparticles can cause genotoxicity, by damaging or mutating the genetic material. The defective genotype resulted in a faulty phenotype, which is observed in NP-treated *Drosophila*. Phenotypic defects are mainly observed in wings, bristles, thorax, abdomen and eyes (Fig. 18.4). Genotoxicity can usually be measured by the wing spot analysis. This test is also referred as somatic mutation and recombination test (SMART) [81, 82]. By this assay, the somatic recombination and genetic aberrations like deletion, point mutations and translocations can be identified [83]. Oral consumption of AgNP induces genotoxic effect, resulting increase in number of spots in the wing. Such phenotype is formed by the significant increased expression of multiple wing hairs and flare-3, two recessive proteins in the wings [84]. The aberrant phenotypic defect caused by AuNPs is inherited through generations. AuNPs treatment causes deformed wing and eyes phenotype. The surface analysis of eyes reveals the presence of different structures in the center of the eye which are different from the ommatidia, the functional units of the eye [85]. Cobalt NPs (CONP) can also induce higher number of spots in the wing, which can be detected by wing spot assay. CoNP further can alter the spatial arrangement of wing hairs along with their distribution on the wing surface [67]. Titania and silver NPs can cause abnormality in the bristles, decreasing the sensory efficiency in *Drosophila*. Bristles are either found to be broken or missing in such mutants [76]. Besides bristle antenna is also assumed to be affected as a conse-

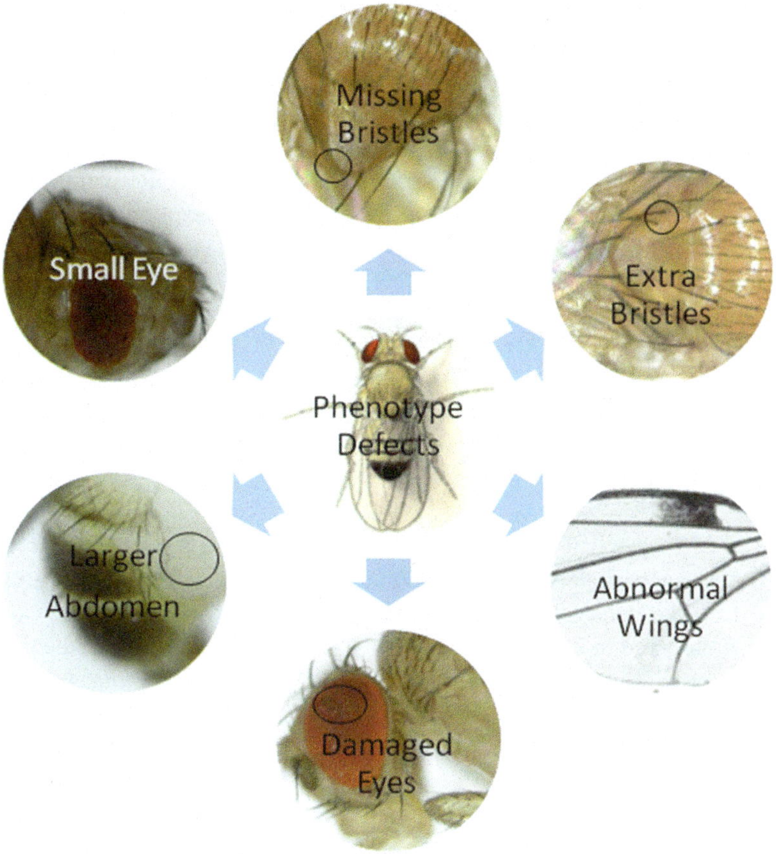

Fig. 18.4 Phenotypic defects seen in adult fly due to oral intake of nanoparticle

quence of NP treatment as evidenced by the negative geotropic assay. Flies are geotrophic in nature but with a malformed antenna a fly become positively geotropic. Flies treated with AuNPs, CoNPs, AgNPs, Carbon NPs, Titania NPs shows positive geotaxis behavior which hints for a defective antenna [86–88]. The NPs causing defect in antennae also causes defect in the formation of abnormal bristles. As bristles are essential for sensory purposes, with treatment of NPs sensing ability must be compromised in flies. AuNPs, Hydroxyapatite and Titania along with AgNPs cause the bristles either to be incomplete (broken) or fully absent [76, 85, 88]. Hydroxyapatite NPs causes the formation of thick wing hairs [89]. Abnormalities in wings also include missing or incomplete venations, abnormal wing size and formation of wing spots (due to irregular spatial distribution of wing hair). The wing hairs are composed of actin; a defective wing hair indicates the interaction of NPs in actin formation and its proper distribution [90]. Defects in cuticle development and melanization are also reported from AgNP treatment as a result of metabolic defects. Flies treated with AgNPs show comparatively softer cuticles with improper pigmentations [88, 91]. This phenotype occurs due to less amount of copper since AgNPs affect copper intake by inhibiting the copper transporter. Melanin synthesis, requires a tyrosinase which is dependent on copper, so unavailability of copper resulted pigmentation defect or less amount of pigment in the body [91]. A recent study on ZnO NP causes defect in bristles and wing structure in F1 generation [92].The phenotypes become severe in successive generation. A recent study by Mishra et al. [93] reported that oral intake of Zirconia NPs can cause phenotypic defect in the eye, wing and bristle of the fly. Cell death is also observed in the gut after oral treatment with zir-

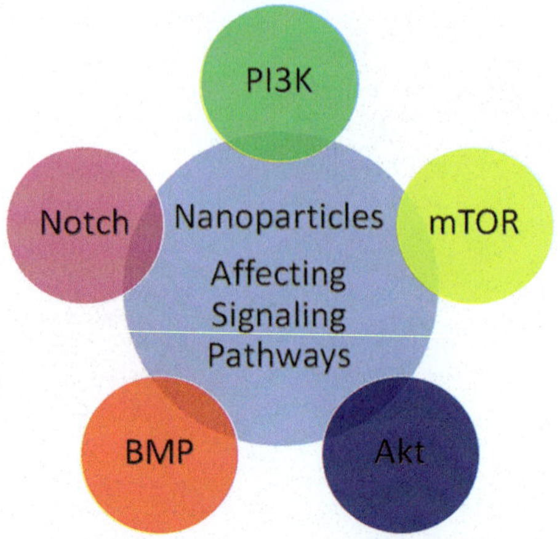

Fig. 18.5 Signaling pathways affected due to various NPs in *Drosophila*

conia nanoparticle. Besides structure behavioral defect like defect in climbing against gravity is also observed in Zirconia treated flies.

18.6 Signaling Pathways Affected Due to NPs

Cells in a developing embryo are constantly communicating with each other, and send molecules or receive signals which are crucial for normal development. These signals are universally known as signaling pathways. Many intracellular signaling pathways are known, and some of which are activated with response to secreted growth factors. Often the concentration of secreted factors appears as a gradient and specifies the cell fate. Thus, any alteration in signaling pathway resulted in a defective adult fly. Recent reports suggest NP can alter the signaling pathways of *Drosophila* and resulted phenotypic defect in fly. All the pathways affected due to NP intake is summarized in Fig. 18.5. Signaling pathways like PI3K/Akt/mTOR are the key players in regulating the energy metabolism of *Drosophila* [94]. AuNP has been reported to interfere with this signaling pathway and enhances the lipid and fatty acid production [95]. EGFR and Notch signaling pathways are essential in bristle formations. NPs like silver, gold, titania and cerium oxide causes bristle abnormalities, by interfering with these two signaling pathways [76, 88, 96]. NPs cause abnormal wing pattern by mutating a gene called *PCV* (posterior cross vein). Along with *PCV*, other two signaling pathways BMP (bone morphogenic protein) and Notch regulate the spacing, wing hair and pattern formations in the wing [80, 97]. mTOR pathway is associated with immune response, cell differentiation, and carcinogenesis in mammalian system. mTOR signaling get affected by action of iron oxide, zinc oxide, silica, copper oxide and gold NPs [98]. In *Drosophila* gold NP can affects the mTOR signaling pathway [99]. Since many of signaling pathways are conserved a similar effect may be anticipated in higher organism as well.

18.7 NPs Affecting Survivorship and Fecundity of *Drosophila*

NPs affect the survivability and fecundity of *Drosophila*. NPs distress various developmental stages and alter the survivorship. Oral administration of various NPs known to decreases the survivorship of *Drosophila*. AgNPs significantly affects the survival rate in a concentration dependent manner. However, the similar result was not observed when the flies were treated with similar doses of $AgNO_3$ [100]. This suggests that, only the nano-sized materials can decline survivorship. AgNP treatment affects larvae survivability and delayed the pupae formation. This suggests that AgNP can affect different developmental stages of *Drosophila*. CdSe-ZnS quantum dots and AuNPs known to decline the lifespan of the flies treated with NPs during early developmental stages [101]. This advocates probably early developmental stages are vulnerable towards NP stress. Similar type of results was obtained from silver, gold and TiO_2 treatment [58, 66, 76, 101]. Sliver NPs significantly decreases eggs number resulting less pupae and adults. Oral intake of TiO_2 NPs causes progeny loss in *Drosophila* [66]. Files exposed to GO during early development shows significantly altered behaviors in comparison to the normal flies. GONPs treatment of 10, and 100 μg/ml does not affect larva-pupa transition stage. However, NP of 500 and 1000 μg/ml GO, affects the larvae-pupae transition stage significantly. This result suggests that

GO affects the survivability only at higher concentration [85].

Fecundity stands for the reproduction rate in a given population. *Drosophila*, proved to be an ideal model for fecundity tests due to NP treatment because of its quicker reproductive cycle and high number of offspring. AgNPs and AuNPs were found to cause long term effects in subsequent generations. To prove this statement, AgNP was treated to the parents and the F1 generations were transferred to a food media without NPs. Till F4 generations, there was very low reproductive rate, showing the generation spanning long term toxicity of AgNPs. However, from the F5 generation the fecundity rate increased and by F8, it became normal [66]. A recent study reported that ZnO NP can cause more severe phenotypic defect in F4 generation [102].

18.8 Conclusion

Besides of wide application of NP the toxicity was overlooked since years. Recent report confirms the toxicity in various in vivo and in vitro models. In current era *Drosophila*, has been used as a powerful model to address the toxicity of various nanoparticles. Effect of NP on various factors like survivorship, ROS formation, fecundity and genotoxicity has been demonstrated by various authors using *Drosophila* model. The toxicity observed in *Drosophila* studies provide information regarding the fate of NPs once entered into the living system. Since it is cost effective many NPs can be accessed within a shorter time period. More importantly, *Drosophila* share functional similarity of 75% diseased genes with human being. So, the toxicity observed in *Drosophila* studies can't be neglected. The results obtained from toxic study suggest further modification of NPs to decrease the toxicity and increase the effectiveness in the field of medical and engineering. The assessment of NP toxicity using *Drosophila* will open up more than a few opportunities for a better, healthier society.

Acknowledgements The authors are thankful to Mr. Abhinandan Patnaik for providing the *Drosophila* defective eye images and Mr. Unnikanan P. for providing the defective abdomen images. S. A. Pappus is thankful to DST INSPIRE which enabled him to study the effect of oral intake of NP on *Drosophila*.

References

1. Sanguansri P, Augustin MA (2006) Nanoscale materials development–a food industry perspective. Trends Food Sci Technol 17(10):547–556
2. Panyam J, Labhasetwar V (2003) Biodegradable nanoparticles for drug and gene delivery to cells and tissue. Adv Drug Deliv Rev 55(3):329–347
3. Krivorotova T, Cirkovas A, Maciulyte S et al (2016) Nisin-loaded pectin nanoparticles for food preservation. Food Hydrocoll 54:49–56
4. Douglas S, Davis S, Illum L (1986) Nanoparticles in drug delivery. Crit Rev Ther Drug Carrier Syst 3(3):233–261
5. Wood NJ, Jenkinson HF, Davis SA et al (2015) Chlorhexidine hexametaphosphate nanoparticles as a novel antimicrobial coating for dental implants. J Mater Sci Mater Med 26(6):1–10
6. Campos-Ortega JA, Hartenstein V (2013) The embryonic development of Drosophila melanogaster. Springer, Berlin
7. Pandey UB, Nichols CD (2011) Human disease models in Drosophila melanogaster and the role of the fly in therapeutic drug discovery. Pharmacol Rev 63(2):411–436
8. Medzhitov R, Preston-Hurlburt P, Janeway CA (1997) A human homologue of the Drosophila Toll protein signals activation of adaptive immunity. Nature 388(6640):394–397
9. Yun Y, Cho YW, Park K (2013) Nanoparticles for oral delivery: targeted nanoparticles with peptidic ligands for oral protein delivery. Adv Drug Deliv Rev 65(6):822–832
10. García M, Forbe T, Gonzalez E (2010) Potential applications of nanotechnology in the agro-food sector. Food Sci Technol (Campinas) 30(3):573–581
11. Ranjan S, Dasgupta N, Chakraborty AR et al (2014) Nanoscience and nanotechnologies in food industries: opportunities and research trends. J Nanopart Res 16(6):2464
12. Fröhlich E, Roblegg E (2012) Models for oral uptake of nanoparticles in consumer products. Toxicology 291(1):10–17
13. Benn T, Cavanagh B, Hristovski K et al (2010) The release of nanosilver from consumer products used in the home. J Environ Qual 39(6):1875–1882
14. Gaiser BK, Fernandes TF, Jepson M et al (2009) Assessing exposure, uptake and toxicity of silver and cerium dioxide nanoparticles from contaminated environments. Environ Health 8(1):S2

15. Fondevila M, Herrer R, Casallas M et al (2009) Silver nanoparticles as a potential antimicrobial additive for weaned pigs. Anim Feed Sci Technol 150(3):259–269
16. Bouwmeester H, Dekkers S, Noordam M et al (2007) Health impact of nanotechnologies in food production. RIKILT, Wageningen
17. Bussiere P-O, Peyroux J, Chadeyron G et al (2013) Influence of functional nanoparticles on the photostability of polymer materials: recent progress and further applications. Polym Degrad Stab 98(12):2411–2418
18. Shi L, Shan J, Ju Y et al (2012) Nanoparticles as delivery vehicles for sunscreen agents. Colloids Surf Physicochem Eng Aspects 396:122–129
19. Wang Z, Xu H, Wu J et al (2011) Sensitive detection of Salmonella with fluorescent bioconjugated nanoparticles probe. Food Chem 125(2):779–784
20. Xiao-e L, Green AN, Haque SA et al (2004) Light-driven oxygen scavenging by titania/polymer nanocomposite films. J Photochem Photobiol A Chem 162(2):253–259
21. Cohen-Tanugi D, Grossman JC (2012) Water desalination across nanoporous graphene. Nano Lett 12(7):3602–3608
22. Qin J-J, Oo MH, Kekre KA (2007) Nanofiltration for recovering wastewater from a specific dyeing facility. Sep Purif Technol 56(2):199–203
23. Neethirajan S, Jayas DS (2011) Nanotechnology for the food and bioprocessing industries. Food Bioprocess Technol 4(1):39–47
24. Farhang B (2007) Nanotechnology and lipids. Lipid Technol 19(6):132–135
25. Yezhelyev MV, Gao X, Xing Y et al (2006) Emerging use of nanoparticles in diagnosis and treatment of breast cancer. Lancet Oncol 7(8):657–667
26. Davis ME, Shin DM (2008) Nanoparticle therapeutics: an emerging treatment modality for cancer. Nat Rev Drug Discov 7(9):771–782
27. Rowe MD, Thamm DH, Kraft SL et al (2009) Polymer-modified gadolinium metal-organic framework nanoparticles used as multifunctional nanomedicines for the targeted imaging and treatment of cancer. Biomacromolecules 10(4):983–993
28. Kennedy LC, Bickford LR, Lewinski NA et al (2011) A new era for cancer treatment: gold-nanoparticle-mediated thermal therapies. Small 7(2):169–183
29. Kumari A, Yadav SK, Yadav SC (2010) Biodegradable polymeric nanoparticles based drug delivery systems. Colloids Surf B: Biointerfaces 75(1):1–18
30. Malam Y, Loizidou M, Seifalian AM (2009) Liposomes and nanoparticles: nanosized vehicles for drug delivery in cancer. Trends Pharmacol Sci 30(11):592–599
31. Manchester M, Singh P (2006) Virus-based nanoparticles (VNPs): platform technologies for diagnostic imaging. Adv Drug Deliv Rev 58(14):1505–1522
32. Elhissi A, Ahmed W, Hassan IU et al (2011) Carbon nanotubes in cancer therapy and drug delivery. J Drug Deliv 2012:837327
33. Dobson J (2006) Magnetic nanoparticles for drug delivery. Drug Dev Res 67(1):55–60
34. Alexiou C, Schmid RJ, Jurgons R et al (2006) Targeting cancer cells: magnetic nanoparticles as drug carriers. Eur Biophys J 35(5):446–450
35. Gupta AK, Naregalkar RR, Vaidya VD et al (2007) Recent advances on surface engineering of magnetic iron oxide nanoparticles and their biomedical applications. Nanomedicine 2:23–39
36. Morones JR, Elechiguerra JL, Camacho A et al (2005) The bactericidal effect of silver nanoparticles. Nanotechnology 16(10):2346
37. Huh AJ, Kwon YJ (2011) "Nanoantibiotics": a new paradigm for treating infectious diseases using nanomaterials in the antibiotics resistant era. J Control Release 156(2):128–145
38. Sadat-Shojai M, Atai M, Nodehi A et al (2010) Hydroxyapatite nanorods as novel fillers for improving the properties of dental adhesives: synthesis and application. Dent Mater 26(5):471–482
39. Domingo C, Arcıs R, Osorio E et al (2003) Hydrolytic stability of experimental hydroxyapatite-filled dental composite materials. Dent Mater 19(6):478–486
40. Asgary S, Eghbal MJ, Parirokh M (2008) Sealing ability of a novel endodontic cement as a root-end filling material. J Biomed Mater Res A 87(3):706–709
41. Piconi C, Maccauro G (1999) Zirconia as a ceramic biomaterial. Biomaterials 20(1):1–25
42. Asharani P, Wu YL, Gong Z et al (2008) Toxicity of silver nanoparticles in zebrafish models. Nanotechnology 19(25):255102
43. Bar-Ilan O, Albrecht RM, Fako VE et al (2009) Toxicity assessments of multisized gold and silver nanoparticles in zebrafish embryos. Small 5(16):1897–1910
44. Pachura-Bouchet S, Blaise C, Vasseur P (2006) Toxicity of nonylphenol on the cnidarian Hydra attenuata and environmental risk assessment. Environ Toxicol 21(4):388–394
45. Khan FR, Paul KB, Dybowska AD et al (2015) Accumulation dynamics and acute toxicity of silver nanoparticles to Daphnia magna and Lumbriculus variegatus: implications for metal modeling approaches. Environ Sci Technol 49(7):4389–4397
46. Zeeshan M, Murugadas A, Ghaskadbi S et al (2016) ROS dependent copper toxicity in Hydra-biochemical and molecular study. Comp Biochem Physiol Part C Toxicol Pharmacol 185:1–12
47. Yeo M-K, Kang M (2010) The effect of nano-scale Zn-doped TiO2 and pure TiO2 particles on Hydra magnipapillata. Mol Cell Toxicol 6(1):9–17
48. Stuart JM, Segal E, Koller D et al (2003) A gene-coexpression network for global discovery of conserved genetic modules. Science 302(5643):249–255

49. Apidianakis Y, Rahme LG (2011) Drosophila melanogaster as a model for human intestinal infection and pathology. Dis Model Mech 4(1):21–30
50. Lehane M (1997) Peritrophic matrix structure and function. Annu Rev Entomol 42(1):525–550
51. Kuraishi T, Binggeli O, Opota O et al (2011) Genetic evidence for a protective role of the peritrophic matrix against intestinal bacterial infection in Drosophila melanogaster. Proc Natl Acad Sci 108(38):15966–15971
52. Lemaitre B, Miguel-Aliaga I (2013) The digestive tract of Drosophila melanogaster. Annu Rev Genet 47:377–404
53. Jiang S, Teng CP, Puah WC et al (2015) Oral administration and selective uptake of polymeric nanoparticles in Drosophila larvae as an in vivo model. ACS Biomater Sci Eng 1(11):1077–1084
54. Park J-H, Kwon JY (2011) Heterogeneous expression of Drosophila gustatory receptors in enteroendocrine cells. PLoS One 6(12):e29022
55. Nazir A, Mukhopadhyay I, Saxena D et al (2003) Evaluation of toxic potential of captan: Induction of hsp70 and tissue damage in transgenic drosophila melanogaster (hsp70-lacZ) Bg9. J Biochem Mol Toxicol 17(2):98–107
56. Manke A, Wang L, Rojanasakul Y (2013) Mechanisms of nanoparticle-induced oxidative stress and toxicity. Biomed Res Int 2013:942916
57. Ahamed M, Posgai R, Gorey TJ et al (2010) Silver nanoparticles induced heat shock protein 70, oxidative stress and apoptosis in Drosophila melanogaster. Toxicol Appl Pharmacol 242(3):263–269
58. Pompa PP, Vecchio G, Galeone A et al (2011) In vivo toxicity assessment of gold nanoparticles in Drosophila melanogaster. Nano Res 4(4):405–413
59. Panacek A, Prucek R, Safarova D et al (2011) Acute and chronic toxicity effects of silver nanoparticles (NPs) on Drosophila melanogaster. Environ Sci Technol 45(11):4974–4979
60. Philbrook NA, Winn LM, Afrooz AN et al (2011) The effect of TiO 2 and Ag nanoparticles on reproduction and development of Drosophila melanogaster and CD-1 mice. Toxicol Appl Pharmacol 257(3):429–436
61. Labuschagne CF, Brenkman AB (2013) Current methods in quantifying ROS and oxidative damage in Caenorhabditis elegans and other model organism of aging. Ageing Res Rev 12(4):918–930
62. Pandey A, Chandra S, Chauhan LKS et al (2013) Cellular internalization and stress response of ingested amorphous silica nanoparticles in the midgut of Drosophila melanogaster. Biochim Biophys Acta (BBA) Gen Subj 1830(1):2256–2266
63. Hirst SM, Karakoti A, Singh S et al (2013) Biodistribution and in vivo antioxidant effects of cerium oxide nanoparticles in mice. Environ Toxicol 28(2):107–118
64. Ahmad J, Ahamed M, Akhtar MJ et al (2012) Apoptosis induction by silica nanoparticles mediated through reactive oxygen species in human liver cell line HepG2. Toxicol Appl Pharmacol 259(2):160–168
65. Sul OJ, Kim JC, Kyung TW et al (2010) Gold nanoparticles inhibited the receptor activator of nuclear factor-κb ligand (RANKL)-induced osteoclast formation by acting as an antioxidant. Biosci Biotechnol Biochem 74(11):2209–2213
66. Posgai R, Cipolla-McCulloch CB, Murphy KR et al (2011) Differential toxicity of silver and titanium dioxide nanoparticles on Drosophila melanogaster development, reproductive effort, and viability: size, coatings and antioxidants matter. Chemosphere 85(1):34–42
67. Vales G, Demir E, Kaya B et al (2013) Genotoxicity of cobalt nanoparticles and ions in Drosophila. Nanotoxicology 7(4):462–468
68. Vecchio G, Galeone A, Brunetti V et al (2012) Concentration-dependent, size-independent toxicity of citrate capped AuNPs in Drosophila melanogaster. PLoS One 7(1):e29980
69. Sabella S, Brunetti V, Vecchio G et al (2011) Toxicity of citrate-capped AuNPs: an in vitro and in vivo assessment. J Nanopart Res 13(12):6821–6835
70. Böhmert L, Girod M, Hansen U et al (2014) Analytically monitored digestion of silver nanoparticles and their toxicity on human intestinal cells. Nanotoxicology 8(6):631–642
71. Cao Y, Roursgaard M, Kermanizadeh A et al (2015) Synergistic effects of zinc oxide nanoparticles and fatty acids on toxicity to Caco-2 cells. Int J Toxicol 34(1):67–76
72. Kaiser JP, Roesslein M, Diener L et al (2013) Human health risk of ingested nanoparticles that are added as multifunctional agents to paints: an in vitro study. PLoS One 8(12):e83215
73. De Angelis I, Barone F, Zijno A et al (2013) Comparative study of ZnO and TiO2 nanoparticles: physicochemical characterisation and toxicological effects on human colon carcinoma cells. Nanotoxicology 7(8):1361–1372
74. Fröhlich E, Meindl C, Roblegg E et al (2012) Cytotoxity of nanoparticles is influenced by size, proliferation and embryonic origin of the cells used for testing. Nanotoxicology 6(4):424–439
75. Gopinath P, Gogoi SK, Chattopadhyay A et al (2008) Implications of silver nanoparticle induced cell apoptosis for in vitro gene therapy. Nanotechnology 19(7):075104
76. Sabat D, Patnaik A, Ekka B et al (2016) Investigation of titania nanoparticles on behaviour and mechanosensory organ of Drosophila melanogaster. Physiol Behav 167:76–85
77. Haney MJ, Zhao Y, Li S et al (2011) Cell-mediated transfer of catalase nanoparticles from macrophages to brain endothelial, glial and neuronal cells. Nanomedicine 6(7):1215–1230
78. Huang N, Yan Y, Xu Y et al (2013) Alumina nanoparticles alter rhythmic activities of local interneurons

in the antennal lobe of Drosophila. Nanotoxicology 7(2):212–220
79. Cooper RJ (2016) Adult neural stem cell differentiation and signaling is disrupted by low-level silver nanoparticle exposure in vitro
80. Zou J, Wang X, Zhang L et al (2015) Iron nanoparticles significantly affect the in vitro and in vivo expression of Id genes. Chem Res Toxicol 28(3):373–383
81. Rodrigues de Andrade HH, Reguly ML, Lehmann M (2004) Wing somatic mutation and recombination test. Drosophila Cytogenet Protoc 247:389–412
82. Graf U, Moraga AA, Castro R et al (1994) Genotoxicity testing of different types of beverages in the Drosophila wing somatic mutation and recombination test. Food Chem Toxicol 32(5):423–430
83. Graf U, Würgler F, Katz A et al (1984) Somatic mutation and recombination test in Drosophila melanogaster. Environ Mol Mutagen 6(2):153–188
84. Demir E, Vales G, Kaya B et al (2011) Genotoxic analysis of silver nanoparticles in Drosophila. Nanotoxicology 5(3):417–424
85. Vecchio G, Galeone A, Brunetti V et al (2012) Mutagenic effects of gold nanoparticles induce aberrant phenotypes in Drosophila melanogaster. Nanomed Nanotechnol Biol Med 8(1):1–7
86. Liu X, Vinson D, Abt D et al (2009) Differential toxicity of carbon nanomaterials in Drosophila: larval dietary uptake is benign, but adult exposure causes locomotor impairment and mortality. Environ Sci Technol 43(16):6357–6363
87. Raj A, Shah P, Agrawal N (2016) Ingestion of gold nanoparticles (AuNPs) affects survival in Drosophila in a dosedependent manner. Int J Sci Res 5(6)
88. Key SCS, Reaves D, Turner F et al (2011) Impacts of silver nanoparticle ingestion on pigmentation and developmental progression in Drosophila. Atlas J Biol 1(3):52–61
89. Pappus SA, Ekka B, Sahu S et al (2017) A toxicity assessment of hydroxyapatite nanoparticles on development and behaviour of Drosophila melanogaster. J Nanopart Res 19(4):136
90. Ren N, He B, Stone D et al (2006) The shavenoid gene of Drosophila encodes a novel actin cytoskeleton interacting protein that promotes wing hair morphogenesis. Genetics 172(3):1643–1653
91. Armstrong N, Ramamoorthy M, Lyon D et al (2013) Mechanism of silver nanoparticles action on insect pigmentation reveals intervention of copper homeostasis. PLoS One 8(1):e53186
92. Chandra M, Anand KS (2017) Assessment of nicotine dependence in subjects with vascular dementia. Int J Res Med Sci 3(3):711–714
93. Monalisa M, Sabat D, Ekka B et al (2017) Oral intake of zirconia nanoparticle alters neuronal development and behaviour of Drosophila melanogaster. J Nanopart Res 19(8):282
94. Zhang H, Cicchetti G, Onda H et al (2003) Loss of Tsc1/Tsc2 activates mTOR and disrupts PI3K-Akt signaling through downregulation of PDGFR. J Clin Invest 112(8):1223–1233
95. Li JJ, Hartono D, Ong C-N et al (2010) Autophagy and oxidative stress associated with gold nanoparticles. Biomaterials 31(23):5996–6003
96. Alaraby M, Hernández A, Annangi B et al (2015) Antioxidant and antigenotoxic properties of CeO2 NPs and cerium sulphate: studies with Drosophila melanogaster as a promising in vivo model. Nanotoxicology 9(6):749–759
97. de Celis JF (2003) Pattern formation in the Drosophila wing: the development of the veins. BioEssays 25(5):443–451
98. Hulea L, Markovic Z, Topisirovic I et al (2016) Biomedical potential of mTOR modulation by nanoparticles. Trends Biotechnol 34(5):349–353
99. Wang B, Chen N, Wei Y et al (2012) Akt signaling-associated metabolic effects of dietary gold nanoparticles in Drosophila. Sci Rep 2:563
100. Kaya B, Marcos R, Yanikoğlu A et al (2004) Evaluation of the genotoxicity of four herbicides in the wing spot test of Drosophila melanogaster using two different strains. Mutat Res Genet Toxicol Environ Mutagen 557(1):53–62
101. Gorth DJ, Rand DM, Webster TJ (2011) Silver nanoparticle toxicity in Drosophila: size does matter. Int J Nanomedicine 6:343–350
102. Anand AS, Prasad DN, Singh SB et al (2017) Chronic exposure of zinc oxide nanoparticles causes deviant phenotype in Drosophila melanogaster. J Hazard Mater 327:180–186

19 Using of Quantum Dots in Biology and Medicine

Svetlana Pleskova, Elza Mikheeva, and Ekaterina Gornostaeva

Abstract

Quantum dots are nanoparticles, which due to their unique physical and chemical (first of all optical) properties, are promising in biology and medicine. There are many ways for quantum dots synthesis, both in the form of nanoislands self-forming on the surfaces, which can be used as single-photon emitters in electronics for storing information, and in the form of colloidal quantum dots for diagnostic and therapeutic purposes in living systems. The paper describes the main methods of quantum dots synthesis and summarizes medical and biological ways of their use. The main emphasis is laid on the ways of quantum dots surface modification. Influence of the size and form of nanoparticles, charge on the surfaces of quantum dots, and cover type on the efficiency of internalization by cells and cell compartments is shown. The main mechanisms of penetration are considered.

Keywords

Quantum dots · Nanoparticles · Synthesis · Core · Shell

19.1 Introduction

One of the most intensively developing directions of nanotechnology is synthesis and practical use of the quantum dots (QDs). QDs are the fluorescent semiconductor nanocrystals consisting of atoms elements II–IV or III–V groups of Mendeleev periodic table and having the size less than the radius of Bohr's exciton for this material [1, 2]. They have the properties of controllable photoluminescence due to the effect of dimensional quantization [3]. Particular interest in QDs is explained by considerable difference of their exciton discrete spectrum from a bulk crystal spectrum of the same chemical. This difference results in the change of QDs optical properties in comparison with the bulk material. In particular, depending on chemical composition and QDs size the emission band may be on any site of a

S. Pleskova (✉) · E. Mikheeva · E. Gornostaeva
Lobachevsky Nizhny Novgorod State University, Nizhniy Novgorod, Russia
e-mail: pleskova@mail.ru

Q. Saquib et al. (eds.), *Cellular and Molecular Toxicology of Nanoparticles*, Advances in Experimental Medicine and Biology 1048, https://doi.org/10.1007/978-3-319-72041-8_19

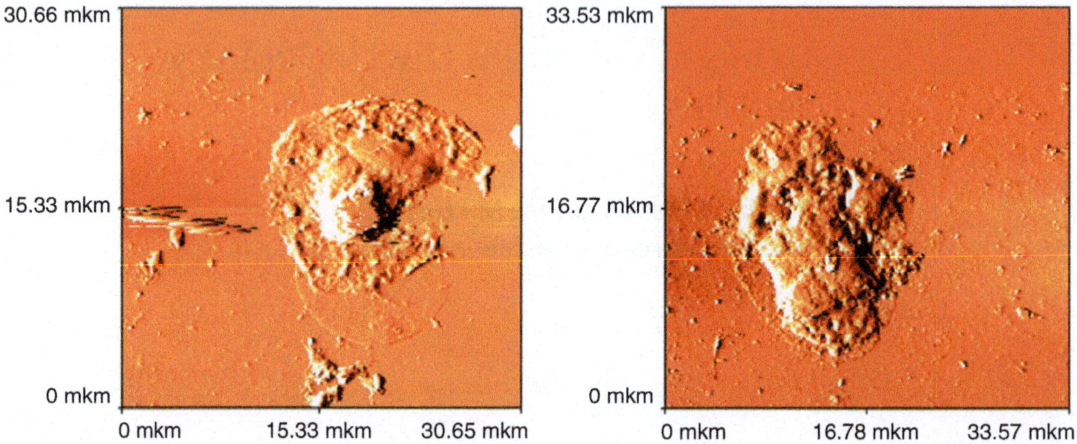

Fig. 19.1 The necrotic death of neutrophil granulocytes after incubation of cells with quantum dots CdSe/ZnS-MPA 620 (cells incubated with QDs in concentration 0.1 mg/ml for 30 min, after that fixed with glutaraldehyde 2.5%). Cells were scanned by atomic-force microscopy (NT-MDT)

spectrum from ultra-violet to infrared. This feature allows receiving marks of various colors for optical coding, to use QDs in optoelectronics, and apply them to study the structure of biological cells, by marking the different structure [4].

The progress in science and technology of QDs was observed after 1984 when Luís Brus received dependence between the size and width of the energy gap band for semiconductor nanoparticles [5]. Nearly a decade was necessary for further advance in the study of QDs. The results of these studies was the synthesis of colloidal QDs of CdX (X = S, Se, Te) with the reconstructed absorption band and emission. Up to now CdX are the most studied QDs having excellent optical and electrochemical properties. However, their use in biology and medicine was impossible because of toxicity of cadmium ion as a part of QDs core (Fig. 19.1).

To improve biocompatibility, and quantum yield of fluorescence as well as stability of these nanocrystals, they were encapsulated with the formation of nanocrystals of core-shell type. Efficiency of luminescence was considerably improved due to passivation on QDs surface of the semiconductor with a large of an energy gap band due to which leaching of metal ions from a core was blocked by this structure [5]. At first QDs of CdSe/ZnS and CdSe/CdS were most intensively studied. Later, many other QDs of "core-shell" type, such as CdSe/ZnSe, CdTe/CdS, CdTe/ZnS, and even CdTe/CdS/ZnS "core/shell/shell" were developed. In traditional QDs, cadmium is the main element of the composition. Nevertheless, it is well known that leakage of cadmium ions is the main cause of QDs cytotoxicity on the basis of cadmium which complicates their further use in vivo or in vitro. With increase in demand for more biocompatible QDs, there was a shift of priority towards synthesis of cadmium-free QDs, allowing their use in biology. QDs containing silicon (Si-QDs), containing carbon (C-QDs), graphene quantum dots (GQDs), Ag_2Se, Ag_2S, InP, $CuInS_2/ZnS$ allowing their use as luminescent probes for biosensors and bioimaging were developed.

19.2 Synthesis of QDs

QDs can be formed on a surface by a method of molecular-beam epitaxy growing with the subsequent etching. Such structures are used in semiconductor electronics as optical converters in light-emitted diode sources and photovoltaic cells [6]. However, obviously only colloidal QDs are used for biomedical studies and practical used. In solutions QDs are stabilized due to the ligands covering them, and depending on ligand structure they form either organic or

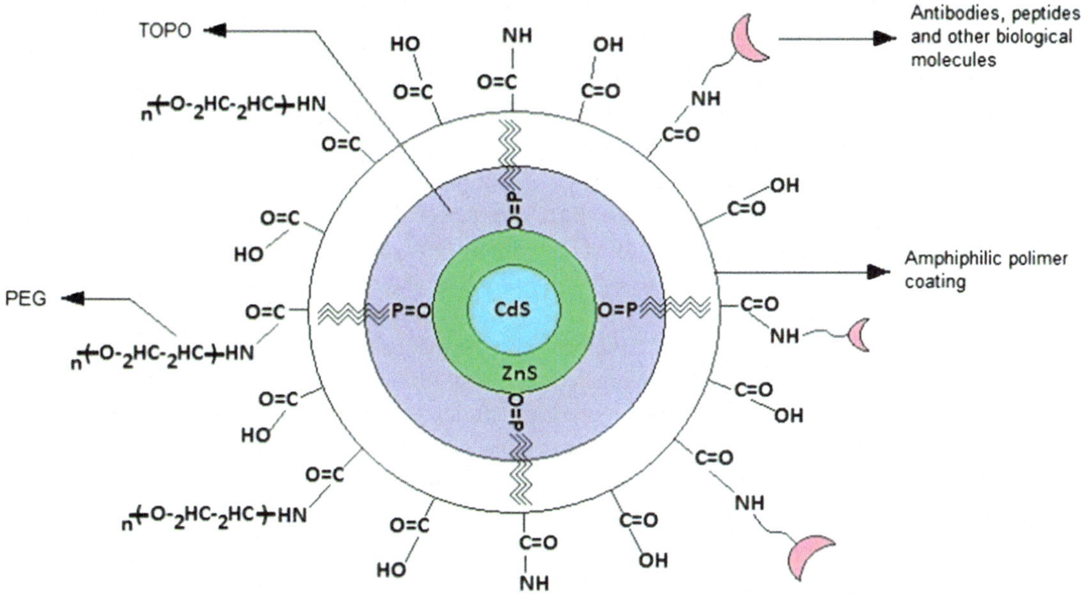

Fig. 19.2 Schematic representation of a quantum dot: a core in the center (blue), over it a shell (green), above a quantum dot is functionalized by various molecules

aqueous colloidal solutions [7]. In live systems QDs aqueous colloidal solutions are required, since, first, all reactions (including the marked QDs) proceed in the hydrophilic environment, secondly, it is necessary to get rid of toxic organic chemicals. The best of QDs proved to be QDs consisting of base material (core), usually either cadmium telluride (CdTe) or cadmium selenide (CdSe), covered with shell, for example, zinc sulfide (ZnS) or cadmium sulfide (CdS), as well as QDs of a complex composition a core-shell-shell, for example, CdTe/CdS/ZnS which have high fluorescence yield (Fig. 19.2).

Increase in quantum yield at synthesis of QDs "core-shell" type is due to:

(i) Passivation of uncompensated chemical bonds on a nanocrystal surface (trapping states in the energy gap band).
(ii) Protection of a QD's core from oxidation in external environment by covering.
(iii) Blocking of nonradiative recombination process (due to the development by the cover of a potential barrier for an exciton in QDs core) [6].

Trioctylphosphine oxide (TOPO) is usually used in QDs synthesis, but it giving them hydrophobic character. For transition to aqueous colloidal solution by means of various exchange reactions QDs surface is covered with a hydrophilic covering, for example mercaptoacids, polyethyleneglycol (PEG), bovine serum albumin (BSA) and others [8]. Ligand replacement is performed covering QDs with amphiphilic polymers, or making micellar encapsulation [3]. Conversion in an aqueous phase is often results in a considerable decrease in luminescence brightness [9]. Increase in QDs hydrodynamic diameter is another consequence of initial ligand replacement for a polymer coating [3]. QD direct synthesis in an aqueous phase is also possible, though in this case, quantum yield and stability are lower than in the QDs originally obtained in an organic phase. Besides, it is difficult to obtain QDs of different diameter in aqueous media, while it is possible to do it in organic media [7]. Anyhow, the used precursors, solvents, reaction temperature, injection parameters (when using injection methods of QDs colloidal synthesis) directly influence morphology, chemical and optical properties of QDs, their average size, its standard deviation, place and spectral bandwidth, photoluminescence and its quantum yield [6]. Nucleic acids, DNA and RNA for example, can serve a matrix for semiconductor nanocrystals synthesis. The DNA specific groups can control synthesis and influence the emission properties of the received product. Thus, it was shown that in case of QDs PbS and CdS synthesis only GTF use as a mononucleotide ligands leads to formation of fluorescent nanocrystals [3].

Due to the active target teranostics development, functionalized QDs, i.e. QDs conjugated with biomolecules (Fig. 19.2) are widely used now. Most often: nucleic acids, peptides, proteins act as biomolecules [3]. Conjugation can be carried out due to hydrophobic or electrostatic interaction, however the most stable effect can be achieved only due to covalent binding, that is why it is most often used in the synthesis of functionalized QDs [7]. Both functionalized, and not functionalized QDs have should be stable and not subject to aggregation, their hydrodynamic diameter has to be less than 5.5 nm to provide a normal excretion by kidneys [3]. For stabilization of QDs they are quite often covered by PEG that has as positive effect (QDs are stable, do not aggregate, are not uptake by reticuloendothelial system), and negative (hydrodynamic diameter increases, an exit in a tumor is at a loss). Besides, it is necessary to remember that in a blood stream the serum proteins, such as albumine, immunoglobulins, proteins of complement system, fibrinogen, apolipoprotein, transferrin, hemoglobin envelop nanoparticles, causing their aggregation and forming a so-called "protein crown" [10]. It not only significantly increases the hydrodynamic diameter of QDs, but also carries out a peculiar role of "Trojan Horse", causing capture of nanoparticles by the mechanism of classical endocytosis [11]. It occurs first of all because "the protein crown" includes such classical opsonin as immunoglobulins and proteins of complement system in the structure. All features of QD interaction with living systems should be considered at synthesis of nanoparticles for medical and biological use.

19.3 The Prospects of QDs Modification for Biomedical Use

To maintain the smallest possible QD hydrodynamic diameter and at the same time not to lose the possibility of samples multiplex marking (now the wavelength of QD emission is regulated generally by the QD size (Fig. 19.3), it is sug-

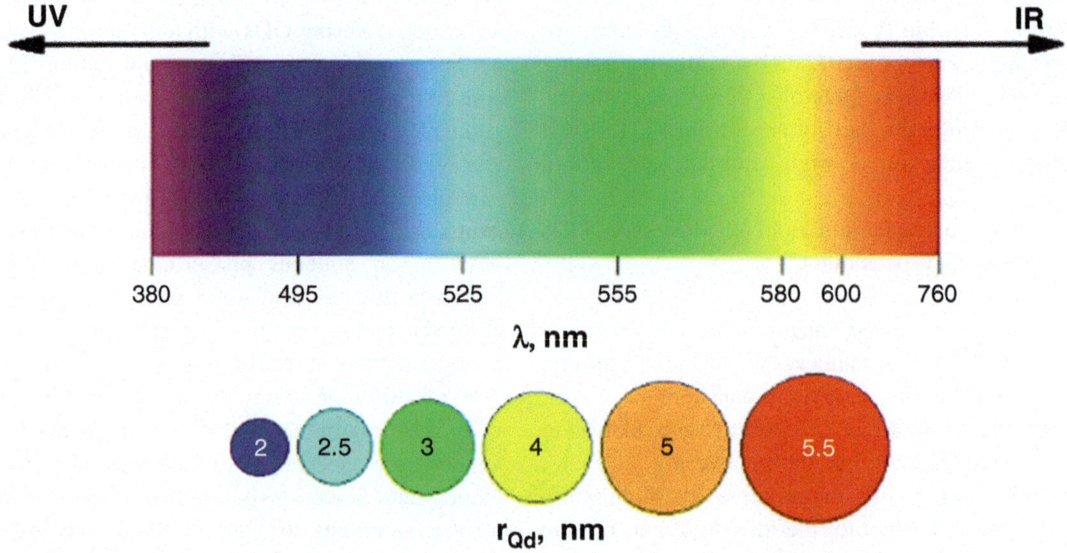

Fig. 19.3 Dependence of emission (in the visible light) from the size of quantum dots

gested to use $CdSe_xS_{1-x}$, $CdTe_xSe_{1-x}$, $Hg_xCd_{1-x}Te$ and $Hg_xCd_{1-x}Se$ cores instead of a classical CdSe core. It will allow to maintain constant and small diameter of QDs core, and to carry out fluorescence control by means of variations of core chemical composition.

Traditional cores are used as germinal structures for cultivation of QDs analogs of various form: rods, hexagonal, tetrapodal and star-like structures.

Use of polydentate ligands, which unlike classical ones (PEG and others) do not move from QDs surface outside but "envelop", promote QDs small size maintenance. Polydentate ligands use has the following advantages:

(i) The hydrophobic barrier layer is excluded;
(ii) QDs small hydrodynamic diameter remains;
(iii) High colloidal stability, resistance to photobleaching is provided;
(iv) High quantum yield is maintained;
(v) Due to QDs "enveloping" its nonspecific binding with organic molecules is excluded;
(vi) Preservation of small hydrodynamic diameter provides rapid renal clearance.

One of the main directions is modification of QDs chemistry for the decrease toxic effects, while preserving QDs unique optical properties [12].

19.4 Uses of QDs in Biology and Medicine

The point of the greatest interest for researchers is the possibility to study various biochemical, physical, kinetic processes in cells or the whole body by means of fluorescent marks. The existing organic fluorophors are of limited use owing to such restrictions as fast photobleaching, wide range of emission and necessity for constant selection of a suitable source of excitement. QDs do not have these disadvantages. The main advantages and disadvantages of QDs use in biology and medicine are summarized in Table 19.1.

QDs are used in classical biology for studying transport mechanisms in a cell [13], including endocytosis [14], functional heterogeneity of cells [15], diffusion movements of membrane transport proteins [16], intracellular organelle marking [17]. In medicine QDs are used for contrasting of blood and lymph vessels (including microvessels) [18], but first of all for multiplex molecular diagnostics and visualization *in vivo* [19–23].

One of earliest studies devoted to a possibility of QDs use for molecular diagnostics showed that conjugation of peptides with QDs leads to their selective accumulation in vessels of tumors and other tissues [23]. Later the studies showing a possibility of "QDs-peptide" conjugates use for specific visualization of tissues *in vivo* were performed. Thus, in the work by Cai et al. [24] QDs conjugated with tripeptide *arg-gly-asp* were used [24]. The last is an antagonist of the integrin $a_v\beta_3$ which, in turn, is selectively expressed on the surface of tumor cells and vessels. As a result of intravenous administration of conjugates of QDs with the specified tripeptide it was possible to get an ideal fluorescent picture of hypodermic glioblastoma in mice *in vivo*. QDs were also conjugated with monoclonal antibodies to membrane prostate-specific antigen for detection of a prostate cancer in mice *in vivo* [25]. In other study, QDs conjugates with antibodies against an α-fetoprotein were used to diagnose hepatoma *in vivo* [26]. The conjugates have to specifically interact with the target and do so in a stable manner, whilst possessing a low level of nonspecific binding [27]. Tada et al. [28], using a method of high-speed confocal microscopy of a skin fold, studied the movement of a single QD conjugated with an antibody to HER-2 in mice with breast cancer: QD circulation in blood vessel lumen, QD extravazation, its binding with a membrane antigen and the movement from tumor cell membrane to perinuclear zone were observed [28]. QDs enable not only to localize a tumor in the organism, but also to estimate the level of expression of various proteins, as well as the activity of individual cells and the processes that have an impact on tumor behavior and its response to the action of therapeutic agents [27]. The receptor part of signal proteins that are overexpressed on tumor cell membranes is used most often as a specific target. The level of expression of these cellular molecular oncomarkers, determined

Table 19.1 Advantages and shortcomings of quantum dots of biomedical practice

Advantages of quantum dots	Disadvantages of quantum dots
High photostability, resistance to photobleaching is 100–1000 times higher, than at organic fluorophores	Multiexponential decline of fluorescence and blinking of separate QDs
Narrow and symmetric peak of emission, Stokes' shift more than 200 nm (ease to detection)	High background level of deduction and accumulation of QDs in reticuloendothelial system
High quantum yield, long lifetime	Instability and increase hydrodynamic diameter after interaction with the serum proteins
Possibility of emission control by changing QDs size and structure (Fig. 19.3)	High toxicity of QDs when they using in *in vivo* systems
Excitation and emission in the visible light range (ease to detection), a possibility of emission in the field of "an optical window"	Incomplete elimination of QDs after injection into an organism
The wide absorption spectrum (operability) can be pumped up QDs of the different size and structure, and in a biological sample to investigate structures, marked by QDs of different color	Possibility of nonspecific binding with any organic molecules
Resistance to chemical and biological degradation	Instability of a colloid system of QDs in the wide range pH and ionic surrounding of the solutions that usually the main characteristic of biological systems
An opportunity to functionalized of QDs by biomolecules for creating of target-delivery system	"Binding" of optical and biomedical properties to hydrodynamic diameter

directly in the tumor tissue, characterizes the molecular profile of each individual tumor and is used to determine the immune status of the tumor and the individualization of therapeutic treatment [29]. But QDs can be using not only for specific diagnostic in target system, but also for nonspecific estimation of cancer cells motility and migration which are associated with metastases and the formation of secondary tumors. The cells nonspecifically incorporate QDs as they crawl over them, leaving behind QDs free zones representing the pattern of phagokinetic uptake of QDs [30]. Cancer cells uptake QDs more actively than normal cells. In particular, MDA-MB-231 tumor cells uptake more QDs than nontumorigenic MCF 10A cells [31].

Due to high quantum yield, QDs can be used not only for diagnostics, but also for photodynamic therapy of malignancies. Thus, QDs are bifunctional agents (therapeutic + diagnostic = theranostic). As antibodies, which carry out target delivery of QDs to tumor specific antigens (and sometimes in are therapeutic agents themselves) are not always type-specific; they are, as a rule, highly immunogenic. Therefore, it is well to use low-immunogenic specific molecules for target delivery. They can be covalent-bound with QDs, or connected via adapters. In case of large molecules (immunoglobulins), the probability of conformational changes of a molecule due to covalent binding with QDs is unlikely, while in case of small molecules use of adapter is essential for preservation of specificity and affinity of interaction. Biotin-streptavidin or barnase-barstar can be used as adapters ("molecular zipping") [27]. Now modular systems of delivery are being developed. For example, in work Wang et al. [32] reported on the synthesis of the multifunctional module uniting QDs (detection), magnetic nanoparticles (targeting in magnetic field) and paxitacel (the therapeutic agent) [32]. Biotin, which specifically binds with biotin receptor hyperexpressing on the surface of a malignantly transformed cell, may be used as a target molecule in such module. Bagalkot et al. [33] presented a novel and simple proof of concept QD-aptamer conjugate that can image and deliver anticancer drugs to prostate cancer cells and sense the delivery of drugs to the targeted tumor cells based on the mechanism of fluorescence resonance energy transfer (FRET) [33]. Their conjugate is comprised of three components:

(i) QDs, which function as fluorescent imaging vehicles;
(ii) RNA aptamers covalently attached to the surface of QD, which serve a dual function as targeting molecules and as drug carrying vehicles;
(iii) Doxorubicin (Dox), which is a widely used anthracycline drug with known fluorescent properties that intercalates within the double-stranded CG sequences of RNA and DNA as a therapeutic agent.

The assembly of this system results in the formation of a Bi-FRET complex: a donor-acceptor model FRET between QD and Dox, where the fluorescence of QD is quenched as a result of Dox absorbance, and a donor-quencher model FRET between Dox and aptamer, where Dox is quenched by double-stranded RNA aptamer. Therefore, both QD and Dox of the conjugate are in the fluorescence "OFF" state when the QD-Apt is loaded with Dox [QD-Apt(Dox)]. After the particle is taken up by targeted cancer cells, Dox is gradually released from the conjugate, which induces the activation of QD and Dox fluorescence to the "ON" state [33]. QDs can be used to sensitize either a photodynamic therapy agent via a FRET mechanism or molecular oxygen through a triplet energy transfer mechanism. Photodynamic therapy is a cancer treatment that takes advantage of the interaction between light and a photosensitizing agent to initiate apoptosis of cancer cells. The photosensitizing agent becomes activated by light but does not react directly with cells and tissues. Instead, it transfers its triplet state energy to nearby oxygen molecules to form reactive singlet oxygen ($^1O_2$) species, which cause cytotoxic reactions in the cells [34, 35]. Also QDs may increase therapeutic efficiency of radiotherapy by selectively scattering and/or absorbing X-rays and gamma rays causing localized damage to DNA and other targeted organelles of cancer cells and thus decreasing total radiation dose to minimize side effects of ionizing radiation on cancer patients [36]. Toxically potential of QDs also can be used for cancer treatment, because this kind of nanoparticles causes epigenetic changes. Choi et al. [37] showed that exposure of MCF-7 cells to hardly detectable intracellular QDs can cause epigenetic changes and trigger p53 posttranslational modifications and its translocation to mitochondria [37]. The activation of p53 results in upregulation of several p53-regulated proapoptotic genes: Puma, Noxa, and Bax. Thus, there are many different approaches for using QDs in the diagnosis and therapy of cancer. Depending on the delivering system, QDs can penetrate into the cell or localize only on their membrane. Total use of QDs for the diagnosis and treatment of cancer in dependence of the localization of the nanoparticles is shown in Fig. 19.4.

Unfortunately, tissue vital fluorescent diagnostics by means of QDs, which has been successfully demonstrated in several works for small laboratory animals, cannot be directly extrapolated to clinical practice because of the limited depth of optical signal penetration. In medicine, this technique can be used for identification of superficial tissue formations (skin and subcutaneous tumors), intraoperative diagnostics and visualization of the zones available at endoscopy [38]. QDs blinking caused by sporadic changes of radiating and nonradiating QDs state is another its drawback [39]. Blinking results in a rupture of QDs fluorescence which is especially important in the field of detection of one molecule, one QD. In some cases, for example, in immunocytologic studies, rupture in fluorescence helps to distinguish the signal coming from QDs from the signal received as an artifact. Hohng and Ha [40] assumed that QDs blinking can be suppressed as a result of passivation of QDs surface by thiol groups [40].

The interesting way of QDs use in ophthalmology is described in works of Chashchin et al. [41, 42]. QDs is suggested to be used for visualization of a vitreous body and epiretinal membranes at vitreoretinal interventions, for delivery medicines to eye tissues, including to the retina and cornea. The authors synthesized QDs on the basis of indium phosphide and intend to use them for retina photoreceptors stimulation in treatment of such pathologies as a pigmented retinitis, retina detachment, diabetic retinopathy and a macular degeneration.

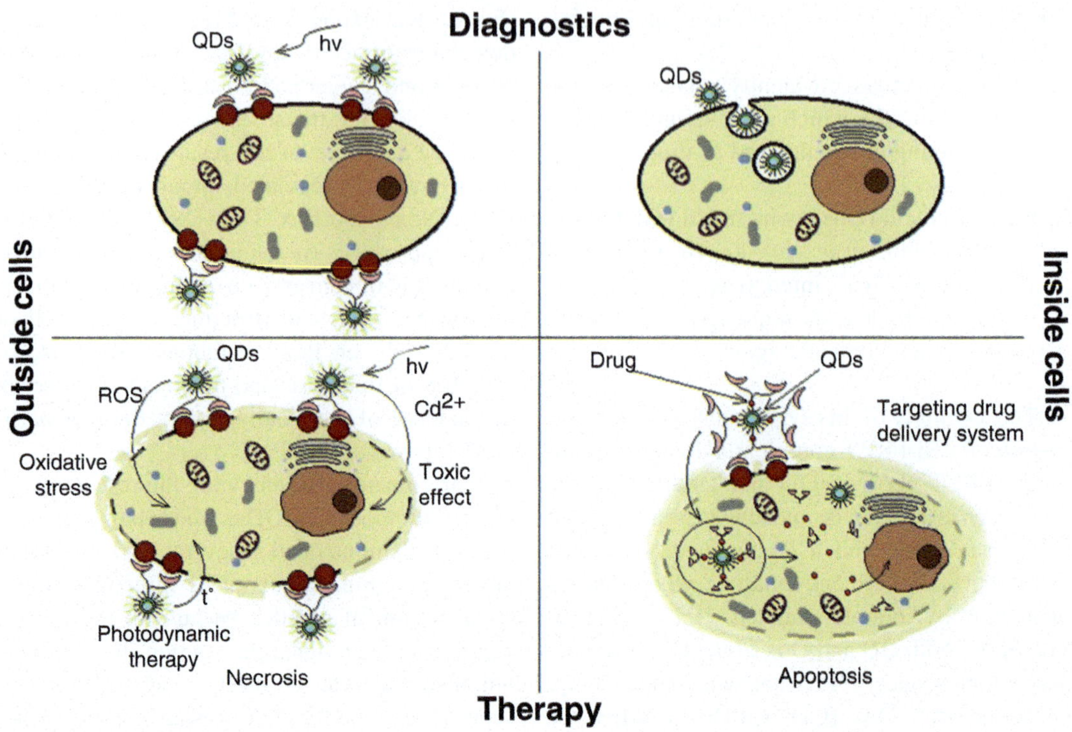

Fig. 19.4 Using of QDs for diagnostics (upper line) and therapy (lower line) of cancer: the upper left square-QDs due to specific delivery system bind to oncomarkers on the membranes of tumor cells, and after excitation, narrow-band diagnostic emission is observed, which allows localizing the tumor. This system is also used to assess the expression density of tumor markers; the upper right square–QDs can penetrate non specifically, directly into the malignant cell, depending on the degree of aggregation of nanoparticles, it can be raft-dependent, caveolae- dependent, clathrin-dependent, or independent penetration; the lower left square – the total possible therapeutic effect of QDs specifically interaction through antibodies with malignant cell oncomarkers (toxic effects, hyperthermic effects, sensitizing effects for photodynamic and radiotherapy); the lower right square – realization FRET and drug-delivery strategy into malignant cell by QDs

One of the main QDs disadvantages is their high toxicity described in many works [43, 44]. It is the main limiting factor for QDs use in medicine, however it is possible to use all QDs advantages in developing diagnostic systems *in vitro*. In particular, QD are used for the development of diagnostic test systems, and with their help, the following substances have already been detected: chloramphenicol in milk and IgE in blood serum [45], sulfamethazine in chicken meat [46], chlorpyriphos in drinking water [47], *Listeria monocytogenes* surface antigens [48], clenbuterol in pig urine [49], progesterone in cow milk [50]. However, despite great advantage of QDs conjugate use with organic molecules, in the immunoenzymatic and immunocytochemistry immunoassay, the work by Korzhevsky et al. [51] reveals a number of drawbacks of QDs conjugation with streptavidin: its less stability at long-term storage in comparison with organic fluorochrome conjugates, poor aliquot reagents preservation, impossibility of long preservation of fluorescence of stained preparations at storage, incompatibility with a number of the commercial media intended for keeping of drugs.

QDs use in protein and DNA-biochips allow not only to improve significantly the sensitivity of diagnostic test systems (for example, protein biochips are more sensitive two orders of magnitude than a traditional solid-phase enzyme multiplied immunoassay), but also to perform breakthrough functional studies in the field of genomics and proteomics. Thus, using of protein biochips it is possible to find and identify proteins and peptides in a proteome, carry out functional analysis of proteins and determine profiles of their expression and the level of their phosphorylation. Biochips are used for determining biochemical activity of a proteome, including interactions of protein-protein, protein-nucleic acids, protein-phospholipids, protein-drug substances. Additional intensity of fluorescence is achieved by the use of adapters [52].

Relatively large total area of the surface, combined with the universal controlled chemistry makes QDs the best decision in the field of development of selectively acting nanodrugs. QDs physical and chemical properties, including size, form, surface charge and type of covering, play an important role in definition and forecasting variants of cellular internalization, pharmacokinetics and biodistribution of certain QDs kinds. Knowing main mechanisms of internalization and tracing of nanoparticles, it is possible to evaluate their biological compatibility and safety [53, 54].

Nabiev et al. [55] revealed dependence of QDs endocytosis and intracellular tracing on their size. QDs CdSe were quickly absorbed by macrophages and depending on their size accumulated in different cellular compartments. The smallest green QDs were localized mainly in nuclei and nucleolus, and the process of absorption was multistage, including endocytosis, active cytoplasmic transport and entering the nucleus through nuclear pores. Red QDs concentrated in cell cytoplasm. In the study of CdTe QDs during their incubation with N9 microglial cells, Lovrić et al. [56] revealed a similar dependence of CdTe QDs distribution on their size: small-sized QDs with green fluorescence (2.2 ± 0.1 nm in diameter) concentrated in the nucleus and near it after 1 h of incubation, and large-sized QDs with red fluorescence (5.2 ± 0.1 nm in diameter) were found in the cytosol and did not enter the nucleus. QDs of the small size can easier enter into the nuclei and other compartments of cells, than a nanoparticle of the larger size. To test this hypothesis, QDs with green fluorescence were covered with BSA. As a result, QDs size increased, and they no longer entered into the nuclei. In other studies it was shown that, the size of nanoparticles significantly influence the binding and activation of membrane receptors, and the subsequent protein expression [57].

The possible explanation of the fact that nanoparticle internalization kinetics depends on the size of nanoparticles is that multivalent cation-active QD interactions with cells are more easily available to particles with the larger surface of contact, than to the particles with a smaller one. Champion and Mitragotri [54] showed by an example of alveolar macrophages that phagocytosis of polystyrene particles is influenced by a form, but not by the size. However, it is known that particle size primarily influences on phagocytosis completeness by cells.

Equally important characteristic of QDs is the charge on their surface, because at the initial stage of their contact with a negatively charged cell membrane electrostatic forces arise, which can prevent internalization of QDs by the cells. Shan et al. [58] showed that with QD surface covered with carboxyl groups, no endocytosis was observed. This allowed suggesting that positively charged QDs are uptake due to electrostatic interactions with a negatively charged HeLa cell membrane. Using the method of atomic force microscopy they showed that for absorption of one QD by a cell 0.4 s are enough [58]. However, in work by Hoshino et al. [59] it was shown that QDs carboxylated on a surface uptake the mouse EL-4 cells by endocytosis. Besides, Jaiswal et al. [60] revealed that negatively charged QDs CdSe/ZnS-dihydrolipoic acid are easily absorbed by mammal cells. Nabiev et al. [55] not only discovered endocytosis of the negatively charged CdTe-thioglycolic acid QDs by cells, but also demonstrated that after absorption intracellular QD transport does not stop in lysosomes. This QDs type enters into cytoplasm and accumulates in perinuclear area. Cationic particles can bind with negatively charged groups on cell surfaces (for example, by sialic acid) and to travel through the plasma membrane, unlike low level of interaction and internalization by the cells of neutral and negatively charged particles. Harush-Frenkel et al. [61] studied endocytosis mechanisms of the charged particles. Their results showed that negatively charged nanoparticles are less effectively uptake by the cells whereas positively charged particles are uptake quickly. Besides, at inhibition of clathrin-dependent way of absorption in cells, internalization of positively charged particles proceeds on compensation ways with high speed.

Size, form, and charge of nanomaterials contribute significantly to their interaction with the cells. However, functional groups on QDs surface determine many important properties of nanomaterials, such as solubility and ability to interact with cell surface. As a rule, nanomaterial incubation in the cells media leads to plasma and/or proteins adsorption on nanoparticles surface [62]. As a result, QDs endocytosis can proceed by a receptor-mediated way. However, protein adsorption on the surface can lead to agglomeration of nanoparticles and their removal into reticuloendothelial system. Nonspecific interactions can also lead to nanoparticles binding with a cellular membrane that will make marking and detection inefficient. To avoid such problems, QDs can be covered with neutral ligands (for example, PEG) which do not interact with proteins. When comparing QDs covered and not covered with PEG it was found that in the latter endocytosis was higher [63].

Thus, development of new approaches to QDs synthesis and covering will promote not only their use as fluorescent markers in diagnostic test systems and experimental biology, but also will allow use them in therapeutic systems *in vivo*.

Acknowledgements This work was supported by the Russian Science Foundation, project № 16-14-10179.

References

1. Yashin KD, Osipovich VS, Pitsuk SE (2007) Structure of the nanocrystals of cadmium selenit received by the method of colloid chemistry for the use in medical diagnostics. Rep BGUIR 3:74–79
2. Guo Z, Tan L (2009) Fundamentals and applications of nanomaterials. Artech House, London, pp 1–249
3. Musihin SF, Alexandrova OA, Luchinin VV et al (2012) Semiconductor colloidal nanoparticles in biology and medicine. Fund Biomed Eng 5–6:47–55
4. Rempel SV, Razvodov AA, Nebogatikov MC et al (2013) Sizes and fluorescence of quantum dots of cadmium sulfide. Solid State Phys 55(3):567–571
5. Zhu JJ, Li JJ, Huang HP et al (2013) Quantum dots for DNA biosensing. Springer Briefs Mol Sci. https://doi.org/10.1007/978-3-642-44910-9-2
6. Samohvalov PS, Artemyev MV, Nabiev IR (2013) Modern methods of luminescent semiconductor nanocrystals synthesis for biomedical usage. Russ Nanotechnol 8(5–6):119–129
7. Speranskaya ES, GoryachevaYu I (2013) Fluorescent quantum dots: synthesis, modification and use in the immunoanalysis. Russ Nanotechnol 8(11–12):7–17
8. Oleynikov VA, Sukhanova AV, Nabiyev IR (2007) Fluorescent semiconductor nanocrystals in biology and medicine. Russ Nanotechnol 2(1–2):160–173
9. Gao X, Chan WCW, Nie S (2002) Quantum-dot nanocrystals for ultrasensitive biological labeling and multicolor optical encoding. J Biomed Opt 7:532–537
10. Duran N, Silveira CP, Duran M et al (2015) Silver nanoparticle protein corona and toxicity: mini-review. J Nanobiotechnol 13:1–17
11. Casals E, Pfaller T, Duschl A et al (2011) Hardening of the nanoparticle-protein corona in metal (Au, Ag) and oxide (Fe_3O_4, CoO, and CeO_2) nanoparticles. Small 7:3479–3486
12. Kairdolf BA, Smith AM, Stokes TH et al (2013) Semiconductor quantum dots for bioimaging and biodiagnostic applications. Annu Rev Anal Chem 6(1):143–162
13. Belyaeva TN, Salova AV, Leonteva EA et al (2009) Inappropriate quantum points in vital confocal microscopic examinations of cells. Tsitologiia 51:838–848
14. Lidke DS, Nagy P, Heintzmann R et al (2004) Quantum dot ligands provide new insights into erbB/HER receptor-mediated signal transduction. Nat Biotechnol 22(2):198–203
15. Pleskova SN, Balalayev IV, Gushchina YY et al (2009) Distinctions in functional activity of neutrophilic granulocytes in reactions with semiconductor quantum dots. Morphology 3:47–49
16. Crane JM, Haggie PM, Verkman AS (2009) Quantum dot single molecule tracking reveals a wide range of diffusive motions of membrane transport proteins. Proc SPIE 7489. https://doi.org/10.1117/12.816900
17. Hanaki K, Momo A, Oku T et al (2003) Semiconductor quantum dot/albumin complex is a long-life and highly photostable endosome marker. Biochem Biophys Res Commun 302:496–501
18. Lim YT, Kim S, Nakayama A et al (2003) Selection of quantum dot wavelengths for biomedical assays and imaging. Mol Imaging 2(1):50–64
19. Wu X, Liu H, Liu J et al (2002) Immunofluorescent labeling of cancer marker Her2 and other cellular targets with semiconductor quantum dots. Nat Biotechnol 21:41–46
20. Rhyner MN, Smith AM, Gao XH et al (2006) Quantum dots and multifunctional nanoparticles: new contrast agents for tumor imaging. Nanomedicine 1:209–217

21. Xing Y, Chaudry Q, Shen C et al (2007) Bioconjugated quantum dots for multiplexed and quantitative immunohistochemistry. Nat Protoc 2:1152–1165
22. Kim S, Lim YT, Soltesz EG et al (2004) Near-infrared fluorescent type II quantum dots for sentinel lymph node mapping. Nat Biotechnol 22:93–97
23. Gao X, Cui Y, Levenson RM et al (2004) In vivo cancer targeting and imaging with semiconductor quantum dots. Nat Biotechnol 22:969–976
24. Cai W, Shin DW, Chen K et al (2006) Peptide-labeled near-infrared quantum dots for imaging tumor vasculature in living subjects. Nano Lett 6:669–676
25. Akerman ME, Chan WC, Laakkonen P et al (2002) Nanocrystal targeting in vivo. Proc Natl Acad Sci U S A 99:12617–12621
26. Yu X, Chen L, Li K et al (2007) Immunofluorescence detection with quantum dot bioconjugates for hepatoma in vivo. J Biomed Opt 12(1):014008
27. Zdobnova TA, Lebedenko EN, Deev SM (2011) Quantum dots for molecular diagnostics of tumors. Acta Nat 3(8):30–50
28. Tada H, Higuchi H, Wanatabe TM et al (2007) In vivo real-time tracking of single quantum dots conjugated with monoclonal anti-HER2 antibody in tumors of mice. Cancer Res 67:1138–1144
29. Smith AM, Dave S, Nie S et al (2006) Multicolor quantum dots for molecular diagnostics of cancer. Expert Rev Mol Diagn 6(2):231–244
30. Parak WJ, Boudreau R, Le Gros M et al (2002) Cell motility and metastatic potential studies based on quantum dot imaging of phagokinetic tracks. Adv Mater 14:882–885
31. Alivisatos AP, Gu W, Larabell C (2005) Quantum dots as cellular probes. Annu Rev Biomed Eng 7:55–76
32. Wang J, Wang F, Li F et al (2016) Multifunctional poly(curcumin) nanomedicine for dual-modal targeted delivery, intracellular responsive release, dual-drug treatment and imaging of multidrug resistant cancer cells. J Mater Chem 4:2954–2962
33. Bagalkot V, Zhang L, Levy-Nissenbaum E et al (2007) Quantum dot-aptamer conjugates for synchronous cancer imaging, therapy, and sensing of drug delivery based on bi-fluorescence resonance energy transfer. Nano Lett 7(10):3065–3070
34. Samia ACS, Chen X, Burda C (2003) Semiconductor quantum dots for photodynamic therapy. J Am Chem Soc 125(51):15736–15737
35. Bakalova R, Ohba H, Zhelev Z et al (2004) Quantum dots as photosensitizers? Nat Biotechnol 22(11):1360–1361
36. Juzenas P, Chen W, Sun YP et al (2008) Quantum dots and nanoparticles for photodynamic and radiation therapies of cancer. Adv Drug Deliv Rev 60:1600–1614
37. Choi AO, Brown SE, Szyf M et al (2008) Quantum dot-induced epigenetic and genotoxic changes in human breast cancer cells. J Mol Med 86:291–302
38. Shlyakhto EV (ed) (2009) Nanotechnologies in biology and medicine. Publishing House, St. Petersburg
39. Kaji N, Tokeshi M, Baba Y (2007) Quantum dots for single bio-molecule imaging. Anal Sci 23:21–24
40. Hohng S, Ha T (2004) Near-complete suppression of quantum dot blinking in ambient conditions. J Am Chem Soc 126:1324–1325
41. Chashchin GV, Ponomarev VO, Nosov SV et al (2011) Prospect of use of artificial fluorophor (quantum dots) in ophthalmology. Bull Reg Pub Inst 14:394–396
42. Chashchin GV, Ponomarev VO, Nosov SV (2012) New inorganic fluorophor and the photosensitive retinal-containing proteins as a basis of element base for quantum nanosurgery of the eye retina. Bull Reg Pub Inst 12:225–228
43. Nel A, Xia T, Madler L (2006) Toxic potential of materials at the nanolevel. Science 311:622–627
44. Medina C, Santos-Martinez MJ, Radomski A et al (2007) Nanoparticles: pharmacological and toxicological significance. Br J Pharmacol 150:552–558
45. Taranova NA, Berlina AN, Zherdeev AV et al (2012) Quantum dota as a marker in immunochromatographic diagnostic test systems. Nanotechnol Health Protect 4:44–47
46. Ding SG, Chen JX, Jiang HY et al (2006) Application of quantum dot-antibody conjugates for detection of sulfamethazine residue in chicken muscle tissue. J Agric Food Chem 54:6139–6142
47. Chen YP, Ning BA, Liu N et al (2010) A rapid and sensitive fluoroimmunoassay based on quantum dot for the detection of chlorpyrifos residue in drinking water. J Environ Sci Health B 45:508–515
48. Tully E, Hearty S, Leonard P et al (2006) The development of rapid fluorescence-based immunoassays, using quantum dot-labeled antibodies for the detection of *Listeria monocytogenes* cell surface proteins. Int J Biol Macromol 39:127–134
49. Wang XL, Tao GH, Meng YH (2009) A novel CdSe/CdS quantum dot-based competitive fluoroimmunoassay for the detection of clenbuterol residue in pig urine using magnetic core/shell Fe_3O_4/Au nanoparticles as a solid carrier. Anal Sci 25:1409–1413
50. Trapiella-Alfonso L, Costa-Fernandez JM, Pereiro R et al (2011) Development of a quantum dot-based fluorescent immunoassay for progesterone determination in bovine milk. Biosens Bioelectron 26:4753–4759
51. Korzhevsky E, Kirik OV, Gilyarov AV et al (2010) Use of semiconductor nanocrystals (quantum dots) in immunocytochemistry studies. Morphology 3:71–74
52. Oleynikov VA (2011) Semiconductor fluorescent nanocrystals (quantum dots) in protein biochips. Bioorg Chem 37(2):171–189
53. Rejman J, Oberle V, Zuhorn IS et al (2004) Size-dependent internalization of particles via pathways of clatrin-and caveolae-mediated endocytosis. Biochem J 377:159–169
54. Champion JA, Mitragotri S (2006) Role of target geometry in phagocytosis. Proc Natl Acad Sci U S A 103:4930–4934
55. Nabiev I, Mitchell S, Davies A et al (2007) Non functionalized nanocrystals can exploit a cell's active

transport machinery delivering them to specific nuclear and cytoplasmic compartments. Nano Lett 7:3452–3461
56. Lovrić J, Bazzi HS, Cuie Y et al (2005) Differences in subcellular distribution and toxicity of green and red emitting CdTe quantum dots. J Mol Med 83:377–385
57. Jiang W, Kim BYS, Rutka JT et al (2008) Nanoparticle-mediated cellular response is size-dependent. Nat Biotechnol 3:145–150
58. Shan Y, Hao X, Shang X et al (2011) Recording force events of single quantum-dot endocytosis. Chem Commun 47:3377–3379
59. Hoshino A, Fujioka K, Oku T et al (2004) Quantum dots targeted to the assigned organelle in living cells. Microbiol Immunol 48:985–994
60. Jaiswal JK, Mattoussi H, Mauro JM et al (2003) Long-term multiple color imaging of live cells using quantum dot bioconjugates. Nat Biotechnol 21:47–51
61. Harush-Frenkel O, Debotton N, Benita S et al (2007) Targeting of nanoparticles to the clathrin-mediated endocytic pathway. Biochem Biophys Res Commun 353:26–32
62. Khan JA, Pillai B, Das TK et al (2007) Molecular effects of uptake of gold nanoparticles in Hela cells. Chem Biol Chem 8:1237–1240
63. Xie J, Xu C, Kohler N et al (2007) Controlled PEGylation of monodisperse Fe_3O_4 nanoparticles for reduced non-specific uptake by macrophage cells. Adv Mater 19:3163–3166

Index

A
Abbott Chalew, T.E., 289
Abdel-Khalek, A., 275
Abdelmonem, A.M., 184
Åberg, C., 188
Abid, A.D., 33
Activator protein-1 (AP-1), 47–48, 92
Ahmad, B., 143–158
Ahmad, F., 175–195
Ahmad, J., 163–172
Ahsan, S.M., 175–195
Alatar, A. A., 163–172
Alginate, 276
Al-Khedhairy, A. A., 163–172
Al-Sid-Cheikh, M., 289
Amani, A., 5
Amin, N., 32
Anderson, D.S., 33
Annelids, 295
Ansari, S.M., 143–158, 163–172
Aoyama, M., 190
AP-1, *see* Activator protein-1
Apoptosis, 46, 220, 258
Aquatic organism, nanomaterials effects, 267–268
  copper nanoparticles, 274
  genotoxicity, 276–277
  gold nanoparticles, 271–273
  iron nanoparticle, 274
  nanomaterial classification, 265
  nanotechnology, 265
  polymeric nanoparticles, 275–276
  quantum dots (QDS), 273–274
  silver nanoparticles
    antioxidant enzymes, 271
    bactericidal properties, 269
    bioaccumulation, 270
    bioavailability and toxicity, 270
    chemical transformations, 270
    Crassostrea virginica, 271
    Eurasian perch (*Perca fluviatilis*), 270
    fathead minnows (*Pimephales promelas*) embryos, 270
    genotoxic and cytotoxic effects, 266
    mechanism of, 271
    physico-chemical properties, 269
    rain-bow trout (*Onchorhyncus mykiss*), 271
    rohu (*Labeo rohita*), 271
    worm (*Nereis virens*), 271
    in zebrafish, 270
  stress gene expression, 277–278
  toxicity mechanisms, 266
  toxicological profiling, 265–266
  zinc oxide nanoparticles, 275
Aquatic toxicity
  antibacterial water filters, 253
  bactericidal effect of, 252
  biocide mechanisms, 252
  ceramics, water treatment, 253
  in vitro and in vivo studies, 254
  phytoplankton biomass and biodiversity, 253
Arora, H., 11
Arulvasu, C., 291
Asare, N., 258
Asati, A., 221
Asharani, P.V., 267, 272
Ashitate, Y., 31, 32
Ashkarran, A.A., 231
Atherosclerosis, 61
Atudorei, V., 32
Auffan, M., 219
AuNPs, *see* Gold nanoparticles
Aurosman Pappus, S., 311–319
Autophagy, 220, 243–244, 259
  classification, 72–73
  dysfunction, human endothelial cells, 65–66
  in nanoparticles-induced toxicity
    A549 human lung cells, 73
    BEAS-2B human bronchial epithelial cells, 73
    cationic polystyrene nanoparticles, 74
    cellular mechanisms, 76–78
    copper oxide nanoparticles, 73–74
    dual role, 73
    HeLa cells, 74
    hepatoma HepG2 cells, 73
    molecular mechanisms, 78–81
    physicochemical mechanisms, 74–76
    zinc oxide nanoparticles, 74
Azam, A., 143–158

Q. Saquib et al. (eds.), *Cellular and Molecular Toxicology of Nanoparticles*, Advances in Experimental Medicine and Biology 1048, https://doi.org/10.1007/978-3-319-72041-8

B

Babadi, V.Y., 109
Bachler, G., 30, 33
Baeza-Squiban, A., 21–34
Bagalkot, V., 328
Baharifar, H., 5
Bai, W., 268
Baratli, Y., 203
Bar-Ilan, O., 273
Bayat, N., 62
Bebianno, M., 290
Becker, G., 190
Beer, C., 232
Behzadi, S., 184
Benameur, L., 219
Benthic and pelagic invertebrates
  abiotic factors, 296
  bioavailability and silver toxicity, 297
  immune response, 295
  sea urchins (*Strongylocentrotus droebachiensis*), 296
Bermejo-Nogales, A., 105
Bertrand, C., 290
Bhat, I.A., 263–279
Bilberg, K., 267
Binding affinities, 176
Biochips, 331
Biocorona, 28
Biological identity, 28
Biomolecular corona, 176
Bivalves, 287, 295
Blackburn, K.L., 32
Blood-brain barrier, AgNPs
  BMECs, 234, 235
  dysfunctional BBB, 236
  morphological changes, 236
  physical and biochemical barriers, 234
  tight junction, 234–235
Boggetto, N., 26
Boland, S., 21–34
Brain microvascular endothelial cells (BMECs), 234
Brandão, F., 199–209
Buffet, P.-E., 289, 292
Bullock, G., 267
Buriç, P., 293
Buyukhatipoglu, K., 207

C

Cai, W., 327
Campbell, A., 186, 188
Cao, Y., 59–67
Carbon based nanoparticles, 2
Carbon nanotubes (CNTs), 124
Casaulta, M., 33
Caspases, 242
Caveolae-mediated endocytosis, 25
Cell death
  apoptosis, 242–243
  autophagy, 243–244 (*see also* (Autophagy)
Cell membrane impairment, 204
Cellular mechanisms, autophagy
  endoplasmic reticulum stress, 77–78
  lysosome impairment, 76–77
  mitochondria dysfunction and mitophagy, 77
Cell viability, 208
Central nervous system, 233–234
Cerium dioxide nanoparticles (nanoceria)
  AL-PEG-600 coating, 112
  antioxidant properties, 216
  application, 111
  applications, 216
  biodistribution of, 217
  biomolecules interaction, 216
  cancer cell toxicity, 112
  DNA damage response
    apoptosis, 220
    autophagy, 220
    cyto-genotoxic effects, 219
    DNA fragmentation, 220
    enzyme mimetic properties, 221
    genotoxic effect, 219, 220
    indirect mechanism, 219
    oxidative stress and inflammatory processes, 220
    radioprotective activity, 220
    ROS generation, 219
    ROS production, 221
  dynamics, 216
  exposure, 216
  genotoxicity, 221–223
  in vivo particokinetics, 218
  kinetics of, 216
  low dose exposure, 112
  microenvironments, 217
  neuroprotective role, 113
  pH, 216, 217
  physical-chemical characteristics, 218–219
  thiol-containing biomolecules, 217
  toxic effects, 216
  UV radiation, 216
Cervicovaginal mucus (CVM), 41
Chae, Y.J., 278
Champion, J.A., 331
Chan, C.Y.S., 291
Chaperone-mediated autophagy, 73
Chashchin, G.V., 329
Chen, J., 6
Chen, P.J., 268
Chen, Y., 79
Chitosan, 49, 275
Chiu, J.M.Y., 291
Choi, A.O., 329
Choi, H.S., 31, 32
Choi, J.E., 46, 47, 271, 278
Cibulskaite, Z., 273
Cicha, I., 206
Clark, E.D., 32
Clathrins, 25
Clemente, Z., 268

Clyne, A.M., 207
Coccini, T., 204
Coenegrachts, L., 30
Cohen, J.M., 30
Comet assay, 12, 62
Cong, Y., 292
Contremoulins, V., 26
Copper nanoparticles (CuoP), 274
Copper oxide nanoparticles (CuO NPs), 110–111
Corbalan, J.J., 65
Costa, C., 199–209
Couto, D., 206
Cowie, H., 63
Cozzari, M., 293
Cronholm, P., 231
Cytotoxicity, 61–62
  in vitro, 5
  MTT assay, 11
  silica, 6
  SPIONs, 6

D

Dąbrowska-Bouta, B., 234
D'Agata, A., 106
Dai, L., 289
Dalai, S., 105
Dale, A.L., 270
Daniel, K.M., 7
Danielsen, P.H., 61
Dapivirine-loaded NPs, 51
Das, G.K., 33
da Silva, V.S., 251–259
das Neves, J., 51
Dedeh, A., 267
De Flora, S., 276
De Matteis, V., 1–14
Derk, R., 30
de Souza, T.A.J., 215–224, 251–259
Dhoble, S., 37–54
Di Bona, K.R., 109
2′,7′-Dichlorofluorescein diacetate (DCF-DA) assay, 12
Dixon, D.R., 275
DNA damage, 86
  genotoxicity, 47, 62–63, 66
  mitochondrial, 47
  ROS, 89, 92
  TiO_2NPs, 7
  ZnO-NPs, 49
DNA fragmentation, 49
Docter, D., 183
Doxorubicin (Dox), 329
Drosophila model, nanoparticles toxicity
  biochemical methods, assessment of, 315–316
  EGFR and Notch signaling pathways, 318
  genotoxicity, 316
  mammalian cell line, 313
  mode of transportation, 313–315
  phenotypic defects, 316–318
  signaling pathways, 318
  survivorship and fecundity, 318–319
Duan, J., 62, 64
Duran, N., 233
Dwivedi, S., 143–158

E

ECs, *see* Endothelial cells
Elder, A., 31, 32
Elia, G., 181
Endocytosis, 91
Endoplasmic reticulum (ER)
  AgNPs, 258, 271
  autophagy, 77–78, 243, 258
  nanoceria, 217
  stress, 77–78, 81, 115
Endothelial cells (ECs)
  eNOS, 60, 61
  human ECs (*see* Human endothelial cells, NPs toxicity)
  monocytes, 60
Endothelial NOS (eNOS) uncoupling, 60, 61, 64
Engineered nanoparticles (ENPs), 60
  antioxidant and anti-inflammatory properties, 220
  applications, 2
  biodistribution, 8–10
  biological applications, 220
  bottom-up approach, 2
  cellular uptake
    AuNPs, 10, 11
    endocytosis process, 10
    phagocytosis, 9–10
    pinocytosis, 10
    plasma membrane, 9
    SNPs, 10
    TiO_2-ARS, 11
    TiO_2NPs, 11
  chronic inflammatory response, 222
  classification, 2
  cytotoxicity, 219
  definition, 1
  epigenetic alterations, 222
  microenvironments and pH, 217
  morphological shapes, 2, 3
  nano risk assessment, 12–13
  physico-chemical properties, 2
    charge, 6
    crystalline structure, 7
    media effects, 7
    shape, 5–6
    size, 5
    surface modification, 6–7
  plasma membrane disruption, 217
  route of entry
    ingestion, 8
    inhalation, 7–8
    skin penetration, 8, 9

Engineered nanoparticles (ENPs) (*cont.*)
safety, risk and toxicology studies, 216
therapeutic purposes, 224
top-down approach, 2
toxicity
comet assay, 12
DCF-DA assay, 12
in vitro and in vivo studies, 4
LDH leakage assay, 11
MTT assay, 11
nanotoxicology, 4
in silico toxicology, 12–13
toxicological effects in human, 2–3
ENPs, *see* Engineered nanoparticles
Eom, H.J., 46, 47
Epidermal growing factor (EGF), 288
Epidermal growth factor receptor (EGFR), 48
ER, *see* Endoplasmic reticulum
Erbe, F., 32
Espinosa-Cristobal, L.F., 231

F
Fabrega, J., 270
Faisal, M., 163–172
Falk, R., 32
Falugi, C., 291
Federici, G., 105
Fernández-Bertólez, N., 199–209
Ferraro, M.V.M., 277
First apoptosis signal (FAS), 46
Fischer, D., 182
Fisichella, M., 155
Fleischer, C.C., 43
Flower, N.A.L., 277
Fluorescence quenching molecules, 25
Fluorescence resonance energy transfer (FRET), 328
Fourches, D., 13
Franchi, L.P., 215–224, 251–259
Fujita, K., 277

G
Gagne, F., 268, 273
Gaharwar, U.S., 109
Gambardella, C., 291, 293
Garaud, M., 222
Garbuzenko, O.B., 33
García-Alonso, J., 292
Garcia-Negrete, C.A., 272
Ge, C., 62
Geiser, M., 33
Gelein, R., 31, 32
Genomic alterations, 205–206
Genotoxicity, 46–47, 62–63
Genter, M.B., 49
George, I., 30
Geys, J., 30
Girigoswami, K., 99–116
Glial cells, 238
Gliga, A.R., 231, 258
Gold nanoparticles (AuNPs)
aquatic organism, nanomaterials effects, 271–273
protein corona, 178–179
Gomes, T., 289, 290
Gonnissen, D., 207
Gornostaeva, E., 323–332
Grzelak, A., 85–93
Guo, H., 61, 62, 65, 273
Guo, L., 49

H
Haase, A., 238
Hackenberg, S., 49
Haider, M.S., 253
Han, L., 64
Hao, L., 268, 275
Harper, B., 6
Harush-Frenkel, O., 331
Ha, T., 329
Hata, K., 190
Heat shock proteins (HSP), 172
Heme oxygenase-1 (HO-1) protein, 65
Higashisaka, K., 190
High molecular weight kininogen (HMWK), 178
Hirn, S., 33
Hoet, P.H., 31, 32
Hohng, S., 329
Hoshino, A., 331
Hoylaerts, M.F., 32
Huang, C.L., 242
Human cardiac microvascular endothelial cells (HCMECs), 61
Human dermal microvascular endothelial cells (HDMVCs), 62
Human embryonic lung fibroblasts (IMR-90), 153
Human endothelial cells, NPs toxicity
cytotoxicity, 61–62
endothelial activation, 63–64
genotoxicity, 62–63
immortalized EC cell lines, 66
impaired NO signaling, 64
mechanisms
autophagy dysfunction, 65–66
inflammation, 65
oxidative stress, 64–65
NOS dysfunction, 64
three-tier model, 66, 67
Human umbilical vein endothelial cells (HUVECs)
cytotoxicity, 61–62
endothelial activation, 63–64
Fe_3O_4 NPs, 66
genotoxicity, 63
HO-1 protein, 65
NOS, 64
silica NPs, 65
U937 adhesion, 65

Hussain, F., 5
Hu, Y.L., 267, 275
Hydrophobic/hydrophilic ENPs, 9
Hydrophobicity, 27
Hypoxia-inducible factor-1α, 79–80

I

Impaired NO signaling, 64
Inorganic nanoparticles, 2
In silico toxicology, 12–13
Internalisation, NM
  endocytic pathways, 25–26
  mechanism, 24–25
  physico-chemical characteristics, 26–28
  translocation and systemic distribution, 24
Intracellular endocytosis, 42–43
Iron oxide nanoparticles (IONPs), 60
  aquatic animals, 108
  aquatic organism, nanomaterials effects, 274
  biomedical applications, 200
  biomedical field, applications, 108
  breast cancer cell lines, 107
  cellular and molecular effects
    autophagy, 206
    cell cycle alterations, 206
    cell cycle arrest, 206
    cell death, 206
    cell membrane disruptions, 204–205
    cellular stress, 202
    direct and indirect contact, 202
    exposure, 202
    genomic alterations, 205–206
    JNK and p53-mediated pathways, 207
    mitochondrial alterations, 203–204
    morphological alterations, 206
    NADPH oxidase, 207
    oxidative damage, 203
    viability, 202–203
  cell viability, 108
  characteristics, 200
  heavy metal, removal of, 108
  HepG2 cells, 107
  internalization of, 115
  rat and mouse models, 109–110
  structure and composition, 201
  toxicity mechanism, 208
  VEGF, 108

J

Jaiswal, J.K., 331
Jang, M.H., 267
Jiang, H.S., 253
Johnson, L., 74
Johnson-Lyles, D.N., 77
Johnston, B.D., 268
Ju, D., 71–82

K

Karthikeyeni, S., 268
Katsumiti, A., 290
Kedziorek, D.A., 151
Keller, J., 33
Kettler, K., 5
Khan, F.R., 289
Kiliç, G., 204, 205
Kim, J., 273
Kim, T.H., 258
Kissel, T., 182
Klemetsrud, T., 50
Kobayashi, T., 33
Konduru, N., 33
Korzhevsky, E., 330
Kramers, P.G.N., 276
Kreyling, W.G., 1–34
Krishnaraj, C., 267
Kuharev, J., 183
Kumari, M., 109, 110

L

Laban, G., 267, 270
Lactate dehydrogenase (LDH)
  leakage assay, 11
  release, 61–62
Laffon, B., 199–209
Lai, T.H., 65
Lai, X., 206
Lapresta-Fernandez, A., 297
Lee, J.H., 31, 32
Levard, C., 270
Lewinski, N.A., 273
Li, H., 290
Liang, S., 61, 65
Li, L., 205
Lin, J., 80
Lipka, J., 33
Liu, F., 65
Liu, J., 277
Liu, X., 258
Liu, Y., 73, 78, 153
Li, Y., 71–82, 258
Li, Z., 63
Losert, S., 30, 33
Louis, S., 268
Louro, H., 123–139
Lovric, J., 331
Lundqvist, M., 181
Luoma, S.N., 270
Luther, E.M., 241
Lysosome impairment, 76–77

M

Macken, A., 291
Macroautophagy, 72–73
Macropinocytosis, 24, 43

Macropinosomes, 25
Magesky, A., 285–306
Mahmoudi, M., 184
Ma-Hock, L., 33
Majumdar, A., 37–54
Mammalian target of rapamycin (mTOR) pathway, 78–79
Mansouri, E., 105
MAPK/ERK signaling pathway, 79, 92
Marine invertebrates, AgNPs effects
  annelids, 295
  benthic and pelagic invertebrates, 295–297
    abiotic factors, 296
    bioavailability and silver toxicity, 297
    immune response, 295
    sea urchins (*Strongylocentrotus droebachiensis*), 296
  biomolecules, interaction effects, 299–300
  bivalves, 287, 295
  cells and tissue interaction, 298–299
  mechanisms, 287–288
  physical and chemical processes, 297–298
  physiological and cellular toxicity effects, 288–294
  sea urchin developmental stage
    embryotoxicity, 300–303
    metamorphosis, 303–304
    starvation effects, 305
Marques, B.F., 292
Matsumoto, S.T., 277
Matuszak, J., 62
Ma, X., 76
McCarthy, M., 289
McCoy, M.T., 12
Menard, N., 62
Metal oxide nanoparticle toxicity
  cerium oxide nanoparticles, 111–113
  CuO and Cu_2O toxicity, 110–111
  iron oxide nanoparticles, 106–110
  mechanism, 113–115
  titanium dioxide (TiO_2), 103–106
  zinc oxide (ZnO), 101–103
Microarray analysis
  BV2 microglia, 154
  cluster analysis, of genes, 153
  cytokine gene, 151
  downregulated genes, 149
  gene expression, 153
  GO analysis, 153
  HaCaT cells, 154
  human embryonic lung fibroblasts, 153
  human skin fibroblasts (HSF42), 153
  human THP-1 derived macrophages, 149
  KEGG pathway enrichment analysis, 154, 157
  SWCNT, 151
  transcriptomic profiling, 148–149
  upregulated genes, 149, 153
Microautophagy, 72
Mikheeva, E., 323–332
Mills, N.L., 32
Mini organ cultures (MOCs), 49
Mishra, M., 311–319
Mitochondria dysfunction, 77
Mitochondrial alterations, 203–204
Mitochondrial permeability transition (MPT), 242
Mitogen-activated protein kinases (MAPKs), 48, 79
Mitophagy, 77
Mitragotri, S., 331
Mittal, S., 219
Mizuhara, T., 11
Molecular mechanisms, autophagy
  hypoxia-inducible factor-1α, 79–80
  MAPK/ERK signaling pathway, 79
  oxidative stress, 80–81
  PI3K/Akt/mTOR signaling pathway, 78–79
  toll-like receptor signaling pathways, 79
Möller, W., 33
Monopoli, M.P., 186, 188
Montiel-Davalos, A., 63
Moyano, D.F., 11
MTT assay, 11
Mucoadhesive drug delivery systems, 40
Mucoadhesive polymers, 49
Mucosal system
  glycoprotein mucin, 39
  micro-and nano-delivery technologies, 40
  mucoadhesive drug delivery systems, 40
  mucous and pharmacology, 39
  mucous membranes, 39
  transmucosal routes
    nasal mucosa, 40
    NPs fate via (*see* (Transmucosal nanoparticles))
    ocular mucosa, 40
    oral mucosa, 40–41
    pulmonary mucosa, 41
    rectal and vaginal mucosa, 41
Multiple-Path Particle Dosimetry Model (MPPD), 24
Multi-walled carbon nanotubes (MWCNTs), 62, 63, 124, 134–135
Musarrat, J., 143–158, 163–172
Myelin, 237–238
Myelin basic protein (MBP), 237

N

Nabiev, I., 331
N-acetylcysteine (NAC), 65
NADPH oxidase (NOX), 46
Nair, P.M.G., 277
Nanocarriers (NC), 256–257
Nanogenotoxicity, 46–47, 221–223
Nanomaterials (NMs)
  cell environment, in biological effects
    bronchial epithelial cell line (Beas-2B), 131–133
    DLS analysis, 132
    lung adenocarcinoma epithelial cell line (A549), 131–133
    MWCNTs, 134–135
    toxicological assays, 134

CNTs, 124
compositions, 21
cytotoxicity, 131
definition, 1
diversity, 22
exposure and health effects, 22
genotoxic effects
biological assays, 126
correlation analysis, 127
cytokinesis blocked micronucleus assay, 126
nano-bio interactions, 131
physicochemical characteristics, 128, 131
sedimentation studies, 128–130
TiO_2, 127, 130
genotoxicity assays, 124
genotoxic properties, 124
internalisation
endocytic pathways, 25–26
mechanism, 24–25
physico-chemical characteristics, 26–28
translocation and systemic distribution, 24
in vivo genotoxicity assays
cellular uptake, 136
histopathological and cytological analyses, 136
LacZ-transgenic mouse line, 135–136
long-term toxicological consequences, 137
3D-cell cultures, 137
nano-bio interactions, 125
nano-cellular interactions, 138–139
nanotoxicology, 22
physicochemical characterization, 125–126
physicochemical properties, 124
public health, 124
respiratory tract
clearance mechanisms, 21
particle deposition, 23–24
structure, 22–23
translocation
biodistribution, 28
epithelial barriers, 28
in vitro studies, 28–31
in vivo studies, 31–34
Nanoparticle-induced redox imbalance
redox balance, signal transmission
amino acids oxidation, 87
exemplary transcription factors, 87
methionine, 87
protein phosphorylation process, 87
redox homeostasis, 86, 87
superoxide dismutase and catalase, 86, 87
26S proteasome, 88
ubiquitin-activating enzymes, 88
redox status, 86
ROS (*see* Reactive oxygen species)
Nanoparticles (NPs)
aerosolization, 144
Ag NPs, 60
animal products, 312
blood brain barrier, 144
cancer treatment, 312
carrier NPs, 313
definition, 1, 59, 144
engineered NPs, 60
extracellular and intracellular environmental changes, 144
health risks, 144
health supplements, 312
human ECs (*see* Human endothelial cells, NPs toxicity)
industrialization, 264
inorganic and organic forms, 264
IONPs, 60
metal-based NPs, 60
oral formulation of drugs, 312
pathophysiological changes, 144
physiochemical properties, 164
protein corona, formation of (*see* Protein corona)
quantum dots, 313
quantum mechanism, 264
reactive oxygen species (ROS), 144
scientific context
in estuarine waters, 286
in natural waters, 287
silica NPs, 312
sodium alginate NPs, 312
surface area/mass ratio, 264
toxicological potential of, 155, 157
ZrNPs, 313
Nanoparticles-induced toxicity, autophagy
A549 human lung cells, 73
BEAS-2B human bronchial epithelial cells, 73
cationic polystyrene nanoparticles, 74
cell death style, 72
cellular mechanisms, 72, 76–78
copper oxide nanoparticles, 73–74
dual role, 73
HeLa cells, 74
hepatoma HepG2 cells, 73
molecular mechanisms, 72, 78–81
physicochemical mechanisms, 72, 74–76
zinc oxide nanoparticles, 74
Nanoparticle toxicity, 91
Nanoporous graphene, 312
Nanotechnology
in aquaculture and fisheries, 265
definition, 264
Nanotoxicology, 4, 22, 72
Naoghare, P.K., 8
Nasal mucoadhesive drug delivery system, 40, 49
Nasal mucosa, 40, 49–50
Natural organic matter (NOM) interaction, 254–255
Naudin, G., 30
Necrosis, 258
Nemmar, A., 31, 32
Neurons and synapses, 236–237

Nickel oxide nanoparticles (NiO-NPs) toxicity, in HEPG2 cells
apoptotic and oxidative stress genes, 171
CASP8 expression, 171
cell culture and exposure, 165–166
characterization, 165, 167
CO_2 capture technologies and mineralization processes, 164
cytotoxicity and apoptosis responses, 165
direct aerial emission, 164
DNA damage, 167, 170
EGR1, *MDM2*, *GADD45A* and *DDIT3* gene upregulation, 171
environment and human health, 164
in vitro DNA damage analysis, by Comet assay, 166
long term intratracheally instillation, 164
NFKB activation, 171
oxidative stress and toxicity pathway genes, 170
pulmonary pathology, 164
qPCR array of, 168–169
quantitative and qualitative analysis, of intracellular ROS, 167–168
ROS measurements, 166
RT^2 profiler PCR array analysis, 166–167
transcriptomic alterations and DNA damage, 172
ultrafine-size particle, 165
xenobiotic metabolism, 165
Nicotinamide adenine dinucleotide phosphate (NADPH) oxidase, 207
Ning, L-H., 47
N-methyl-D-aspartate receptors (NMDARs), 232
NMs, *see* Nanomaterials
NOS dysfunction, 64
NPs, *see* Nanoparticles
Nrf-2 pathway, 93
Nuclear factor kappa-light-chain-enhancer of activated B cells (NF-κB), 47, 92–93
Nwani, C.D., 277

O
Oberdorster, G., 32
Ocular mucosa, 40, 48–49
Oral buccal mucosa, 40–41, 50
Orecna, M., 62
Organic nanoparticles, 2
Osborne, O.J., 267
Osima, Y., 33
Osmond-McLeod, M.J., 154
Ostling, J.K., 12
Oxidative damage, 203
Oxidative stress, 86, 87, 220, 240–241
NP-induced oxidative stress, 88–90
production (*see* Reactive oxygen species)

P
Pal, S., 8
Pandey, A.K., 219
Pang, L., 257
Panyam, J., 43
Paolini, A., 206
Paracellular endocytosis, 43
Paracelsus, 252
Parivar, K., 109
Park, C.H., 256
Park, E.J., 73
Park, H-G, 106
Pásaro, E., 199–209
Patil, S., 219
Patravale, V., 37–54
Paulraj, R., 109
Paunesku, T., 11
Payne, C.K., 43
Pelletier, E., 285–306
Petters, C., 203
Phagocytosis, 24–25, 43
Phagosomes, 24
Phillips, D.M., 51
Phosphatidylinositol 3-kinase (PI3K), 78–79
Photodynamic therapy, 329
Photosensitizer, 229
Photosensitizing agent, 329
Physicochemical mechanisms, autophagy
charge, 75
dispersity, 74
size, 74–75
surface chemistry, 75–76
Physiologically-based toxicokinetic (PBTK) model, 231
Pierscionek, B.K., 220
PI3K/Akt/mTOR signaling pathway, 78–79
Pinocytosis, 42–43
Piticharoenphun, S., 293
Pleskova, S., 323–332
Poly lactic-co-glycolic acid (PLGA), 275–276
Polymeric nanoparticles, 275–276
Pongrac, I.M., 206
Post-synaptic (PSD-95) proteins, 237
Potter, T.E., 30
Pouya, S., 10
Powers, C.M., 267
Prasad, R.Y., 7, 242
Professional phagocytes, 24
Protein corona
binding affinities, 176
biological environments, 176
complement factors and coagulation proteins, 178
composition of
physiological function, 179
SiNPs, bioinformatic classification, 180–181
consequences
biocompatibility and toxicity, 192–193
cellular targeting, 188–190
cellular uptake, 190–192
drug release effect, 192
nanoparticle size effects, 188
zeta-potential effects, 188
fabrication, of corona free nanoparticles, 193–195
factors
biological environment, 185–186

exposure time, 182–184
nanoparticle size, 181
plasma concentration, 186–188
surface properties, 181–182
temperature, 184–185
gold nanoparticles, 178–179
hard corona, 176
HMWK, 178
Lyz adsorption and denaturation, 179
protein conformation, 178
silver nanoparticles, 232–233
soft corona, 176
structure and composition, 176
thermodynamic feasibility, 178
Vroman effect, 176
Protein tyrosine phosphatases (PTP), 48
Prow, T.W., 49
Pu, D., 13
Puisney, C., 21–34
Pulmonary mucosa, 41, 50–53

Q
Quantum dots (QDs), 273–274, 313
biology and medicine, 329
advantages, 327, 328
biotin-streptavidin/barnase-barstar, 328
cancer cells uptake QDs, 328
cancer treatment, 329
CdSe, 331
components, 328–329
disadvantages, 327, 330
α-fetoprotein, 327
FRET, 328
high quantum yield, 328
monoclonal antibodies, 327
ophthalmology, 329
photodynamic therapy, 329
photosensitizing agent, 329
protein and DNA-biochips, 330–331
radiotherapy, therapeutic efficiency of, 329
tissue vital fluorescent diagnostics, 329
tripeptide *arg-gly-asp*, 327
biomedical uses, 326–327
CdSe/ZnS and CdSe/CdS, 324
chemical composition, 323
optical and electrochemical properties, 324
in optoelectronics, 324
synthesis
CdTe/CdS/ZnS, 325
conjugation, 326
hydrodynamic diameter, 325
hydrophilic environment, 325
nucleic acids, DNA and RNA, 326
"protein crown," 326
semiconductor electronics, 324
stabilization, 326
trioctylphosphine oxide, 325
Quik, J.T.K., 270

R
Radiosensitizers, 229
Rafiee, A., 267
Rainbow, P.S., 270
Rajiv, S., 205
Rajkumar, K.S., 267, 271
Ramsden, C.S., 105
Ramskov, T., 292
Rangasamy, S., 8
Rao, C.M., 175–195
Rather, M.A., 263–279
RAW264.7 cells, 76, 78
Rayner, T.A., 271
Reactive nitrogen species, 86
Reactive oxygen species (ROS), 46, 144
autophagy, molecular mechanisms, 80–81
generation
antimicrobial activity, 89
detection methods, 88
induction mechanism, 89
iron nanoparticles, 89
metalloid nanoparticles, 89
particle size, 89–90
signal transduction pathways, 91–93
uncoated metal nanoparticles, 89
intracellular signal transduction, 86
primary free radicals, 86
redox sensitive signal cascades activation, 86
Rectal and vaginal mucosa, 41, 51, 54
Reddy, U.A., 109
Reeves, J.F., 277
Remya, A.S., 274
Respiratory tract
clearance mechanisms, 21
particle deposition, 23–24
structure, 22–23
Rinaldi, R., 1–14
Ringwood, A.H., 289
Rinna, A., 79
Rivet, C.J., 202
Robinson, S.D., 32
Rocha, T.L., 215–224
ROS, *see* Reactive oxygen species
Roser, M., 182
Ruenraroengsak, P., 30

S
Saha, K., 11
Saikia, J., 10
Salvati, A., 188
Sanchez-Crespo, A., 32
Saquib, Q., 143–158, 163–172
Sarlo, K., 32
Schöttler, S., 190
Schütz, C.A., 202
Schutz, I., 77
Scown, T.M., 278
Seiffert, J., 5
Seitz, J., 32

Semmler-Behnke, M., 32, 33
Shannahan, J.H., 230, 232
Shan, Y., 331
Sharma, N., 263–279
Sharma, R., 263–279
Sharp, Z., 32
Shi, J., 63, 65
Shugart, L.R., 276
Siddiqui, M.A., 163–172
Silica nanoparticles (SiNPs), bioinformatic classification, 180–181
Šiller, L., 293
Silva, J., 277
Silva, V., 31, 32
Silver nanoparticles (AgNPs), 49, 312
antibacterial potential, 252
apoptosis, 258
aquatic organism, nanomaterials effects
antioxidant enzymes, 271
bactericidal properties, 269
bioaccumulation, 270
bioavailability and toxicity, 270
chemical transformations, 270
Crassostrea virginica, 271
Eurasian perch (*Perca fluviatilis*), 270
fathead minnows (*Pimephales promelas*) embryos, 270
genotoxic and cytotoxic effects, 266
mechanism of, 271
physico-chemical properties, 269
rain-bow trout (*Onchorhyncus mykiss*), 271
rohu (*Labeo rohita*), 271
worm (*Nereis virens*), 271
in zebrafish, 270
aquatic toxicity
antibacterial water filters, 253
bactericidal effect of, 252
biocide mechanisms, 252
ceramics, water treatment, 253
in vitro and in vivo studies, 254
phytoplankton biomass and biodiversity, 253
autophagy, 259
blood-brain barrier, 234–236
glial cells, 238
myelin, 237–238
neurons and synapses, 236–237
cellular targets, in CNS, 233–234
commercial and medical applications
AIDS and meningitis, 230
Alzheimer's disease detection, 229
Anticoat™ Flex 3 Dressing, 228
antimicrobial properties, 228
biodiagnostic and biosensing applications, 229
bio-diagnostics, imaging and therapy, 228
conventional antimicrobial drugs, 228
conventional therapies, 230
neurotransmitters and glucose detection, 229
optical properties, 229
photosensitizer, 229
radiosensitizers, 229
wound healing, 228
cytotoxic effect, 252, 258
industrial applications, 252
liberated silver ions, 231–232
marine invertebrates, toxic effects on
annelids, 295
benthic and pelagic invertebrates, 295–297
biomolecules, interaction effects, 299–300
bivalves, 287, 295
cells and tissue interaction, 298–299
cellular toxicity mechanisms, 300–305
embryotoxicity, 300–303
mechanisms, 287–288
metamorphosis, 303–304
physical and chemical processes, 297–298
physiological and cellular toxicity effects, 288–294
starvation effects, 305
molecular mechanisms, neurotoxic effects
cell death, 242–244
inflammation, 242
mitochondrial damage, 240
oxidative stress, 240–241
receptors and channel interaction, 239–240
nanocarriers (NC), 256–257
natural organic matter (NOM) interaction, 254–255
necrosis, 258
oxidants and antioxidants, 258
protein corona, 232–233
ROS, 258
size and shape, 230–231
surface coating, 231
Trojan Horse mechanism, 255–256
Singh, N.P., 12
Single cell gel electrophoresis assay, *see* Comet assay
Single-walled carbon nanotubes (SWCNTs), 62, 124
Skalska, J., 227–244
Smistad, G., 50
Smith, C.J., 268
Snyder-Talkington, B.N., 63
Soenen, S.J.H., 202, 206
Song, L., 274
Son, J., 6
Souza, L.R.R., 251–259
Sperling, R.A., 33
SPIONs, *see* Superparamagnetic iron oxide nanoparticles
Src family kinase, 48
Stigler, J., 181
Stoeger, T., 33
Stress gene expression, 277–278
Stružyñska, L., 227–244
Stumpp, M., 302
Sun, C., 232
Sundarraj, K., 109
Sun, J., 61, 65, 207
Superparamagnetic iron oxide nanoparticles (SPIONs), 6–7
biomedical applications, 107
breast cancer cell lines, 107
cellular toxicity, 108

cell viability, 107
DNA damage, 107
flow cytometry analysis and gene expression, 107
imaging applications, 107
magnetic moments, 107
serum free and normal serum containing medium, 107
uptake and metabolism, 107
Superparamagnetism, 200
Suwa, R., 290
Su, Y., 62, 64
Szalay, B., 110

T
Tada, H., 327
Tak, Y.K., 8
Talkar, S., 37–54
Teixeira, J.P., 199–209
Tenzer, S., 183
Teubl, B.J., 50
Theodorakis, C.W., 276
Thorley, A.J., 30
Thurn, K.T., 11
Tice, R.R., 12
Titanium dioxide (TiO_2) nanoparticle toxicity, 312
aquatic models, 104–106
cosmetic products, 104
dentistry, uses in, 104
DNA damage, 104
fibrinogen (FG), 104
HepG2 cells, 103
immune responses, 104
ionizing radiation, 104
NRF2 and BRCA1 gene expression studies, 103
oxidative stress, 104
rutile, brokite and anatase, 103
Toll-like receptor signaling pathways, 79
Tomassone, M.S., 33
Toxicogenomics
hybridization technologies, 145
microarray analysis
BV2 microglia, 154
cluster analysis, of genes, 153
cytokine gene, 151
downregulated genes, 149
gene expression, 153
GO analysis, 153
HaCaT cells, 154
human embryonic lung fibroblasts (IMR-90), 153
human skin fibroblasts (HSF42), 153
human THP-1 derived macrophages, 149
KEGG pathway enrichment analysis, 154, 157
SWCNT, 151
transcriptomic profiling, 148–149
upregulated genes, 149, 153
in nanoparticles research, 144–145
RNA-seq analysis
eukaryotic green alga *Chlamydomonas reinhardtii*, 146–147
gene ontology (GO) analysis, 147
KEGG, 147
MCF-7 cells, 147
nano-hybrid, 147
P12 exposure, 147, 149
splice junctions and novel exons, 146
TiO_2-NPs exposure, 147
Transcellular endocytosis, 43
Transferrin (TRF), 278
Translocation, NM
biodistribution, 28
epithelial barriers, 28
in vitro studies
alveolar epithelial cells, 29–30
Calu-3 cell line, 30
characteristics, 30–31
paracellular route, 28, 29
protein ZO-1 immunolabeling, 30
transcytosis, 28, 29
in vivo studies, 31–34
Transmission electron microscopy (TEM), 233
Transmucosal nanoparticles
biomarkers, 53
cellular uptake, 42
drug absorption, 41
molecular and cellular interactions
degradation, 43–44
internalisation pathways, 42–43
protein binding, 43
mucosal barriers, 42
nanocarriers, 54
nanotoxicology, 54
toxicity mechanism
apoptosis, 46
cell surface interaction with proteins, 47
genotoxicity, 46–47
interaction with cellular signalling pathway, 47–48
mitochondrial interaction, 47
physiochemical properties, 44–45
ROS, 46
toxicological aspects
nasal mucosa, 49–50
ocular mucosa, 48–49
oral buccal mucosa, 50
pulmonary mucosa, 50–53
rectal/vaginal mucosa, 51, 54
Trioctylphosphine oxide (TOPO), 325
Trojan Horse mechanism, 255–256, 287, 299
Tropsha, A., 13
Truong, L., 272
Tumour necrosis factor (TNF) signal, 46

U
Umehara, Y., 30, 33

V

Valdiglesias, V., 199–209
Vanbilloen, H., 32
van de Meent, D., 5
Vanquickenborne, B., 31, 32
Vascular endothelial growth factor (VEGF)-induced angiogenesis, 229
Veltman, K., 5
Vercammen, J., 30
Vermeij, E.A., 110
Vranic, S., 26, 30
Vroman effect, 176

W

Wahab, R., 163–172
Walczyk, D., 186, 188
Walker, V .G., 62
Wallace, K., 7
Wang, H., 291
Wang, J.Y., 6
Wang, K., 153
Wang, L., 30
Wang, Q., 76, 79
Wang, X., 105
Wang, Y., 328
Wang, Z., 256, 257
Wei, F., 78
Wiebert, P., 32
Wiegman, C., 5
Winkler, J., 33
Winzen, S., 190
Wood, C.M., 270
World Health Organization (WHO), 269
Worthington, K.L., 51
Wu, H., 77
Wu, J., 207
Wurgler, F.E., 276

X

Xenobiotic metabolism, 165
X-ray absorption near edge structure (XANES), 299
Xu, H., 63

Y

Yang, J., 6
Yang, L., 62
Yan, J., 106
Yan, M., 61
Yazdimamaghani, M., 10
Yin, N., 240
Yu, Y., 74, 80

Z

Zacharopoulos, V.R., 51
Zhang, W., 268
Zhang, Y.L., 77
Zhao, J., 275
Zhou, W., 74
Zhou, Z., 6
Zhu, L., 106
Zhu, M.T., 63, 64, 207
Zhu, X., 106, 268, 274
Zinc oxide nanoparticles (ZnO-NPs), 49, 275, 312
  aquatic organism, nanomaterials effects, 275
  photocatalysis, 102
  toxicity, 102–103
  UV radiation and absorption, 101
  wurtzite and zinc blend, 101
Zuberek, M, 85–93
Zuykov, M., 289

The manufacturer's authorised representative in the EU is Springer
Nature Customer Service Centre GmbH, Europaplatz 3, 69115 Heidelberg,
Germany. If you have any concerns regarding our products, please
contact ProductSafety@springernature.com

Printed and bound by CPI Group (UK) Ltd, Croydon, CR0 4YY
30/09/2026
02225435-0019